Notes On

ACOUSTICS

Notes On

ACOUSTICS

By

Uno Ingard

INFINITY SCIENCE PRESS LLC

Hingham, Massachusetts
New Delhi, India

INFINITY SCIENCE PRESS LLC
11 Leavitt Street
Hingham, MA 02043
Tel. 877-266-5796 (toll free)
Fax 781-740-1677
info@infinitysciencepress.com
www.infinitysciencepress.com

This book is printed on acid-free paper.

Uno Ingard. *Notes on Acoustics*.
ISBN: 978-1-934015-08-7

The publisher recognizes and respects all marks used by companies, manufacturers, and developers as a means to distinguish their products. All brand names and product names mentioned in this book are trademarks or service marks of their respective companies. Any omission or misuse (of any kind) of service marks or trademarks, etc. is not an attempt to infringe on the property of others.

Library of Congress Cataloging-in-Publication Data
Ingard, K. Uno
   Acoustics / Uno Ingard.
      p. cm.
   Includes bibliographical references and index.
   ISBN-13: 978-1-934015-08-7 (hardcover : alk. paper)
   1. Sound.   I. Title.
   QC225.15.I56 2008
   534–dc22   2007019297

8 9 10 4 3 2 1

Our titles are available for adoption, license or bulk purchase by institutions, corporations, etc. For additional information, please contact the Customer Service Dept. at 877-266-5796 (toll free).

Preface

Having been involved periodically for many years in both teaching and research in acoustics has resulted in numerous sets of informal notes. The initial impetus for this book was a suggestion that these notes be put together into a book. However, new personal commitments of mine caused the project to be put on hold for several years and it was only after my retirement in 1991 that it was taken up seriously again for a couple of years.

In order for the book to be useful as a general text, rather than a collection of research reports, new material had to be added including examples and problems, etc. The result is the present book, which, with appropriate choice of the material, can be used as a text in general acoustics. Taken as such, it is on the senior undergraduate or first year graduate level in a typical science or engineering curriculum. There should be enough material in the book to cover a two semester course.

Much of the book includes notes and numerical results resulting to a large extent from my involvement in specific projects in areas which became of particular importance at the early part of the jet aircraft era. In subsequent years, in the 1950's and 1960's, much of our work was sponsored by NACA and later by NASA.

After several chapters dealing with basic concepts and phenomena follow discussions of specific topics such as flow-induced sound and instabilities, flow effects and nonlinear acoustics, room and duct acoustics, sound propagation in the atmosphere, and sound generation by fans. These chapters contain hitherto unpublished material.

The introductory material in Chapter 2 on the oscillator is fundamental, but may appear too long as it contains summaries of well known results from spectrum analysis which is used throughout the book. As examples in this chapter can be mentioned an analysis of an oscillator, subject to both 'dynamic' and 'dry' friction, and an analysis of the frequency response of a model of the eardrum.

In hindsight, I believe that parts of the book, particularly the chapters on sound generation by fans probably will be regarded by many as too detailed for an introductory course and it should be apparent that in teaching a course based on this book, appropriate filtering of the material by the instructor is called for.

As some liberties have been taken in regard to choice of material, organization, notation, and references (or lack thereof) it is perhaps a fair assessment to say that the 'Notes' in the title should be taken to imply that the book in some respects is less formal than many texts.

In any event, the aim of the book is to provide a thorough understanding of the fundamentals of acoustics and a foundation for problem solving on a level compatible with the mathematics (including the use of complex variables) that is required in a typical science-engineering undergraduate curriculum. Each chapter contains examples and problems and the entire chapter 11 is devoted to examples with solutions and discussions.

Although great emphasis is placed on a descriptive presentation in hope of providing 'physical insight' it is not at the expense of mathematical analysis. Admittedly, inclusion of all algebraic steps in many derivations can easily interrupt the train of thought, and in the chapter of sound radiation by fans, much of this algebra has been omitted, hopefully without affecting the presentation of the basic ideas involved.

Appendix A contains supplementary notes and Appendix B a brief review of the algebra of complex numbers.

Acknowledgment.

I wish to thank colleagues and former students at M.I.T. as well as engineers and scientists in industry who have provided much of the stimulation and motivation for the preparation of this book. Special thanks go to several individuals who participated in some of the experiments described in the book, in particular to Stanley Oleson, David Pridmore Brown, George Maling, Daniel Galehouse, Lee W. Dean, J. A. Ross, Michael Mintz, Charles McMillan, and Vijay Singhal. At the time, they were all students at M.I.T.

A grant from the Du Pont Company to the Massachusetts Institute of Technology for studies in acoustics is gratefully acknowledged.

Uno Ingard, Professor Emeritus, M.I.T.
Kittery Point, May, 2008

Contents

List of Figures

Chapter 1

Introduction

1.1 Sound and Acoustics Defined

In everyday conversational language, 'acoustics' is a term that refers to the quality of enclosed spaces such as lecture and concert halls in regard to their effect on the perception of speech and music. It is supposed to be used with a verb in its plural form. The term applies also to outdoor theaters and 'bowls.'

From the standpoint of the physical sciences and engineering, acoustics has a much broader meaning and it is usually defined as the science of waves and vibrations in matter. On the microscopic level, sound is an intermolecular collision process, and, unlike an electromagnetic wave, a material medium is required to carry a sound wave.[1] On the macroscopic level, acoustics deals with time dependent variations in pressure or stress, often cyclic, with the number of *cycles per second, cps or Hz,* being the *frequency.*

The frequency range extends from zero to an upper limit which, in a gas, is of the order of the intermolecular collision frequency; in normal air it is $\approx 10^9$ Hz and the upper vibration frequency in a solid is $\approx 10^{13}$ Hz. Thus, acoustics deals with problems ranging from earthquakes (and the vibrations induced by them) at the low-frequency end to thermal vibrations in matter on the high.

A small portion of the acoustic spectrum, ≈ 20 to $\approx 20,000$ Hz, falls in the audible range and 'sound' is often used to designate waves and vibrations in this range. In this book, 'sound' and 'acoustic vibrations and waves' are synonymous and signify mechanical vibrations in matter regardless of whether they are audible or not.

In the audible range, the term 'noise' is used to designate 'undesireable' and disturbing sound. This, of course, is a highly subjective matter. The control of noise has become an important engineering field, as indicated in Section 1.2.6. The term noise is used also in signal analysis to designate a random function, as discussed in Ch. 2.

Below and above that audible range, sound is usually referred to as *infrasound* and *ultrasound*, respectively.

---

[1] Since molecular interactions are electrical in nature, also the acoustic wave can be considered electromagnetic in origin.

There is an analogous terminology for electromagnetic waves, where the visible portion of the electromagnetic spectrum is referred to as 'light' and the prefixes 'infra' and 'ultra' are used also here to signify spectral regions below and above the visible range.

To return to the microscopic level, a naive one-dimensional model of sound transmission depicts the molecules as identical billiard balls arranged along a straight line. We assume that these balls are at rest when undisturbed. If the ball at one end of the line is given an impulse in the direction of the line, the first ball will collide with the second, the second with the third, and so on, so that a wave disturbance will travel along the line. The speed of propagation of the wave will increase with the strength of the impulse. This, however, is not in agreement with the normal behavior of sound for which the speed of propagation is essentially the same, independent of the strength. Thus, our model is not very good in this respect.

Another flaw of the model is that if the ball at the end of the line is given an impulse in the opposite direction, there will be no collisions and no wave motion. A gas, however, can support both compression and rarefaction waves.

Thus, the model has to be modified to be consistent with these experimental facts. The modification involved is the introduction of the thermal random motion of the molecules in the gas. Through this motion, the molecules collide with each other even when the gas is undisturbed (thermal equilibrium). If the thermal speed of the molecules is much greater than the additional speed acquired through an external impulse, the time between collisions and hence the time of communication between them will be almost independent of the impulse strength under normal conditions. Through collision with its neighbor to the left and then with the neighbor to the right, a molecule can probe the state of motion to the left and then 'report' it to the right, thus producing a wave that travels to the right. The speed of propagation of this wave, a sound wave, for all practical purposes will be the thermal molecular speed since the perturbation in molecular velocity typically is only one millionth of the thermal speed. Only for unusually large amplitudes, sometimes encountered in explosive events, will there be a significant amplitude dependence of the wave speed.

The curious reader may wish to check to see if our definitions of sound and acoustics are consistent with the dictionary versions. The American Heritage Dictionary tells us that
(a): "Sound is a vibratory disturbance in the pressure and density of a fluid or in the elastic strain in a solid, with frequencies in the approximate range between 20 and 20,000 cycles per second, and capable of being detected by the organs of hearing," and
(b): "Loosely, such a disturbance at any frequency."

In the same dictionary, *Acoustics* is defined as
1. "The scientific study of sound, specially of its generation, propagation, perception, and interaction with materials and with other forms of radiation. Used with a singular verb."
2. "The total effect of sound, especially as produced in an enclosed space. Used with a plural verb."

1.1.1 Frequency Intervals. Musical Scale

The lowest frequency on a normal piano keyboard is 27.5 Hz and the highest, 4186 Hz. Doubling the frequency represents an interval of *one octave*. Starting with the lowest C (32.7 Hz), the keyboard covers 7 octaves. The frequency of the A-note in the fourth octave has been chosen to be 440 Hz (international standard). On the *equally tempered chromatic scale*, an octave has 12 notes which are equally spaced on a logarithmic frequency scale.

A frequency interval $f_2 - f_1$ represents $\log_2(f_2/f_1)$ octaves (logarithm, base 2) and the number of *decades* is $\log_{10}(f_2/f_1)$. A frequency interval covering one nth of an octave is such that $\log_2(f_2/f_1) = 1/n$, i.e., $f_2/f_1 = 2^{1/n}$. The center frequency of an interval on the logarithmic scale is the (geometrical) mean value, $f_m = \sqrt{f_1 f_2}$.

Thus, the ratio of the frequencies of two adjacent notes on the equally tempered chromatic scale (separation of 1/12th of an octave) is $2^{1/12} \approx 1.059$ which defines a *semitone* interval, half a *tone*. The intervals in the *major scale* with the notes C, D, E, F, G, A, and B, are 1 tone, 1 tone, 1/2 tone, 1 tone, 1 tone, 1 tone, and 1/2 tone.

Other measures of frequency intervals are *cent* and *savart*. One cent is 0.01 semitones and one savart is 0.001 decades.

1.1.2 Problems

1. **Frequencies of the normal piano keyboard**
   The frequency of the A note in the fourth octave on the piano is 440 Hz. List the frequencies of all the other notes on the piano keyboard.

2. **Pitch discrimination of the human ear**
   Pitch is the subjective quantity that is used in ordering sounds of different frequencies. To make a variation Δf in frequency perceived as a variation in pitch, $\Delta f/f$ must exceed a minimum value, *the difference limen for pitch*, that depends on f. However, in the approximate range from 400 to 4000 Hz this ratio is found to be constant, ≈ 0.003 for sound pressures in the normal range of speech. In this range, what is the smallest detectable frequency variation $\Delta f/f$ in (a) octaves, (b) cents, (c) savarts?

3. **Tone intervals**
   The 'perfect fifth,' 'perfect fourth,' and 'major third' refer to tone intervals for which the frequency ratios are 3/2, 4/3, and 5/4. Give examples of pairs of notes on the piano keyboard for which the ratios are close to these values.

4. **Engineering acoustics and frequency bands**
   (a) Octave band spectra in noise control engineering have the standardized center frequencies 31.5, 63, 125, 250, 500, 1000, 2000, 4000, and 8000 Hz. What is the bandwidth in Hz of the octave band centered at 1000 Hz.
   (b) One-third octave bands are also frequently used. What is the relative bandwidth $\Delta f/f_m$ of a 1/3 octave where $\Delta f = f_2 - f_1$ and f_m is the center frequency?

1.2 An Overview of Some Specialties in Acoustics

An undergraduate degree in acoustics is generally not awarded in colleges in the U.S.A., although general acoustics courses are offered and may be part of a departmental requirement for a degree. On the graduate level, advanced and more specialized courses are normally available, and students who wish to pursue a career in acoustics usually do research in the field for an advanced degree in whatever department they belong to. Actually, the borderlines between the various disciplines in science and engineering are no longer very well defined and students often take courses in departments different from their own. Even a thesis advisor can be from a different department although the supervisor usually is from the home department. This flexibility is rather typical for acoustics since it tends to be interdisciplinary to a greater extent than many other fields. Actually, to be proficient in many areas of acoustics, it is almost necessary to have a working knowledge in other fields such as dynamics of fluids and structures and in signal processing.

In this section we present some observations about acoustics to give an idea of some of the areas and applications that a student or a professional in acoustics might get involved with. There is no particular logical order or organization in our list of examples, and the lengths of their description are not representative of their relative significance.

A detailed classification of acoustic disciplines can be found in most journals of acoustics. For example, the *Journal of the Acoustical Society of America* contains about 20 main categories ranging from Speech production to Quantum acoustics, each with several subsections. There are numerous other journals such as *Sound and Vibration* and *Applied Acoustics* in the U.K. and *Acustica* in Germany.

1.2.1 Mathematical Acoustics

We start with the topic which is necessary for a quantitative understanding of acoustics, the physics and mathematics of waves and oscillations. It is not surprising that many acousticians have entered the field from a background of waves acquired in electromagnetic theory or quantum mechanics. The transition to linear acoustics is then not much of a problem; one has to get used to new concepts and solve a number of problems to get a physical feel for the subject. To become well-rounded in aeroacoustics and modern problems in acoustics, a good knowledge of aerodynamics and structures has to be acquired.

Many workers in the field often spend several years and often a professional career working on various mathematical wave problems, propagation, diffraction, radiation, interaction of sound with structures, etc., sometimes utilizing numerical techniques. These problems frequently arise in mathematical modeling of practical problems and their solution can yield valuable information, insights, and guidelines for design.

1.2.2 Architectural Acoustics

Returning to the two definitions of acoustics above, one definition refers to the perception of speech and music in rooms and concert halls. In that case, as mentioned, the plural form of the associated verb is used.

Around the beginning of the 20th century, the interest in the acoustical characteristics of rooms and concert halls played an important role in the development of the field of acoustics as a discipline of applied physics and engineering. To a large extent, this was due to the contributions by Dr. Wallace C. Sabine, then a physics professor at Harvard University, with X-rays as his specialty. His acoustic diversions were motivated initially by his desire to try to improve the speech intelligibility in an incredibly bad lecture hall at Harvard. He used organ pipes as sound sources, his own hearing for sound detection, and a large number of seat cushions (borrowed from a nearby theater) as sound absorbers. To eliminate his own absorption, he placed himself in a wooden box with only his head exposed. With these simple means, he established the relation between reverberation time and absorption in a room, a relation which now bears his name. The interest was further stimulated by his involvement with the acoustics of the Boston Symphony Hall. These efforts grew into extensive systematic studies of the acoustics of rooms, which formed the foundation for further developments by other investigators for many years to come.

It is a far cry from Sabine's simple experiments to modern research in room acoustics with sophisticated computers and software, but the necessary conditions for 'good' acoustics established by Sabine are still used. They are not sufficient, however. The difficulty in predicting the response of a room to music and establishing subjective measures of evaluation are considerable, and it appears that even today, concert hall designers are relying heavily on empiricism and their knowledge of existing 'good' halls as guides. With the aid of modern signal analysis and data processing, considerable research is still being done to develop a deeper understanding of this complex subject.

Architectural acoustics deals not only with room acoustics, i.e., the acoustic response of an enclosed space, but also with factors that influence the background noise level in a room such as sound transmission through walls and conduits from external sources and air handling systems.

1.2.3 Sound Propagation in the Atmosphere

Many other areas of acoustics have emerged from specific practical problems. A typical example is atmospheric acoustics. For the past 100 years the activity in this field has been inspired by a variety of societal needs. Actually, interest in the field goes back more than 100 years. The penetrating crack of a bolt of lightning and the rolling of thunder always have aroused both fear and curiosity. It is not until rather recently that a quantitative understanding of these effects is emerging.

The early systematic studies of atmospheric acoustics, about a century ago, were not motivated by thunder, however, but rather by the need to improve fog horn signaling to reduce the hazards and the number of ship wrecks that were caused by fog in coastal areas. Many prominent scientists were involved such as Tyndall and Lord Rayleigh in England and Henry in the United States. Through their efforts, many important results were obtained and interesting questions were raised which stimulated further studies in this field.

Later, a surge of interest in sound propagation in the atmosphere was generated by the use of sound ranging for locating sound sources such as enemy weapons. In this country and abroad, several projects on sound propagation in the atmosphere were undertaken and many theoretical physicists were used in these studies. In Russia, one of their most prominent quantum theorists, Blokhintzev, produced a unique document on sound propagation in moving, inhomogeneous media, which later was translated by NACA (now NASA).

The basic problem of atmospheric acoustics concerns sound propagation over a sound absorptive ground in an inhomogeneous turbulent atmosphere with temperature and wind gradients. The presence of wind makes the atmosphere acoustically anisotropic and the combination of these gradients and the effect of the ground gives rise to the formation of shadow zones. The theoretical analysis of sound propagation under these conditions is complicated and it is usually supplemented by experimental studies.

The advent of the commercial jet aircraft created community noise problems and again sound propagation in the atmosphere became an important topic. Numerous extensive studies, both theoretical and experimental, were undertaken.

The aircraft community noise problem in the US led to federal legislation (in 1969) for the noise certification of aircraft, and this created a need for measurement of the acoustic power output of aircraft engines. It was soon realized that atmospheric and ground conditions significantly affected the results and again detailed studies of sound propagation were undertaken.

The use of sound as a diagnostic tool (SODAR, SOund Detection And Ranging) for exploration of the conditions of the lower atmosphere also should be mentioned as having motivated propagation studies. An interesting application concerns the possibility of using sound scattering for monitoring the vortices created by a large aircraft at airports. These vortices can remain in the atmosphere after the landing of the aircraft, and they have been found to be hazardous for small airplanes coming in for landing in the wake of a large plane.

1.2.4 Underwater Sound, Geo-acoustics, and Seismology

The discussion of atmospheric acoustics above illustrates how a particular research activity often is stimulated and supported from time to time by many different societal needs and interests.

Atmospheric acoustics has its counterparts in the sea and in the ground, sometimes referred to as ocean and geo-acoustics, respectively. From studies of the sound transmission characteristics, it is possible to get information about the sound speed profiles which in turn contain information about the structure and composition of the medium. Geo-acoustics and seismology deal with this problem for exploring the structure of the Earth, where, for example, oil deposits are the target of obvious commercial interests.

Sound scattering from objects in the ocean, be it fish, submarines, or sunken ships, can be used for the detection and imaging of these objects in much the same way as in medical acoustics in which the human body is the 'medium' and organs, tumors, and fetuses might be the targets.

During World War II, an important battleground was underwater and problems of sound ranging in the ocean became vitally important. This technology developed rapidly and many acoustical laboratories were established to study this problem. It was in this context the acronym SONAR was coined.

More recently, the late Professor Edgerton at M.I.T., the inventor of the modern stroboscope, developed underwater scanners for the exploration of the ocean floor and for the detection of sunken ships and other objects. He used them extensively in collaboration with his good friend, the late Jacques Cousteau, on many oceanographic explorations.

1.2.5 Infrasound. Explosions and Shock Waves

Geo-acoustics, mentioned in the previous section, deals also with earthquakes in which most of the energy is carried by low frequencies below the audible range (i.e., in the infrasound regime). These are rather infrequent events, however, and the interest in infrasound, as far as the interaction with humans and structures is concerned, is usually focused on various industrial sources such as high power jet engines and gas turbine power plants for which the spectrum of significant energy typically goes down to about 4 Hz. The resonance frequency of walls in buildings often lie in the infrasonic range and infrasound is known to have caused unacceptable building vibration and even structural damage.

Shock waves, generated by explosions or supersonic air craft (sonic boom), for example, also contain energy in the infrasonic range and can have damaging effects on structures. For example, the spectrum often contain substantial energy in a frequency range close to the resonance frequency of windows which often break as a result of the 'push-pull' effect caused by such waves. The break can occur on the pull half-cycle, leaving the fragments of the window on the outside.

1.2.6 Noise Control

In atmospheric acoustics research, noise reduction was one of the motivating societal needs but not necessarily the dominant one. In many other areas of acoustics, however, the growing concern about noise has been instrumental in promoting research and establishing new laboratories. Historically, this concern for noise and its effect on people has not always been apparent. During the Industrial Revolution, 100 to 150 years ago, we do not find much to say about efforts to control noise. Rather, part of the reason was probably that noise, at least industrial noise, was regarded as a sign of progress and even as an indicator of culture.

Only when it came to problems that involved acoustic privacy in dwellings was the attitude somewhat different. Actually, building constructions incorporating design principles for high sound insulation in multi-family houses can be traced back as far as to the 17th century, and they have been described in the literature for more than 100 years.

As a historical aside, we note that in 1784 none less than Michael Faraday was hired by the Commissioner of Jails in England to carry out experiments on sound transmission of walls in an effort to arrive at a wall construction that would prevent communication between prisoners in adjacent cells. This was in accord with the then prevailing attitude in penology that such an isolation would be beneficial in as much as it would protect the meek from the savage and provide quiet for contemplation.

More recently, the need for sound insulation in apartment buildings became particularly acute when, some 50 to 60 years ago, the building industry more and more turned to lightweight constructions. It quickly became apparent that the building industry had to start to consider seriously the acoustical characteristics of materials and building constructions. As a result, several acoustical laboratories were established with facilities for measurement of the transmission loss of walls and floors as well as the sound absorptive characteristics of acoustical materials.

At the same time, major advances were made in acoustical instrumentation which made possible detailed experimental studies of basic mechanisms and understanding of sound transmission and absorption. Eventually, the results thus obtained were made the basis for standardized testing procedures and codes within the building industry.

Noise control in other areas developed quickly after 1940. Studies of noise from ships and submarines became of high priority during the second World War and specialized laboratories were established. Many mathematicians and physical scientists were brought into the field of acoustics.

Of more general interest, noise in transportation, both ground based and air borne, has rapidly become an important problem which has led to considerable investment on the part of manufacturers on noise reduction technology. Related to it is the shielding of traffic noise by means of barriers along highways which has become an industry all of its own. Aircraft noise has received perhaps even more attention and is an important part of the ongoing work on the control of traffic noise and its societal impact.

1.2.7 Aero-acoustics

The advent of commercial jet air craft in the 1950s started a new era in acoustics, or more specifically in aero-acoustics, with the noise generation by turbulent jets at the core. Extensive theoretical and experimental studies were undertaken to find means of reducing the noise, challenging acousticians, aerodynamicists, and mathematicians in universities, industrial, and governmental laboratories.

Soon afterwards, by-pass engines were introduced, and it became apparent that the noise from the ducted fan in these engines represented a noise problem which could be even more important than the jet noise. In many respects, it is also more difficult than the jet noise to fully understand since it involves not only the generation of sound from the fan and guide vane assemblies but also the propagation of sound in and radiation from the fan duct. Extensive research in this field is ongoing.

1.2.8 Ultrasonics

There are numerous other areas in acoustics ranging from basic physics to various industrial applications. One such area is ultrasonics which deals with high frequency sound waves beyond the audible range, as mentioned earlier. It contains many subdivisions. Medical acoustics is one example, in which ultrasonic waves are used as a means for diagnostic imaging as a supplement to X-rays. Surgery by means of focused sound waves is also possible and ultrasonic microscopy is now a reality. Ultrasonic 'drills,' which in essence are high frequency chip hammers, can produce arbitrarily shaped holes, and ultrasonic cleaning has been known and used for a long time.

Ultrasound is used also for the detection of flaws in solids (non-destructive testing) and ultrasonic transducers can be used for the detection of acoustic emission from stress-induced dislocations. This can be used for monitoring structures for failure risk.

High-intensity sound can be used for emulsification of liquids and agglomeration of particles and is known to affect many processes, particularly in the chemical industry. Ultrasonic waves in piezo-electric semi-conductors, both in bulk and on the surface, can be amplified by means of a superimposed electric field. Many of these and related industrial applications are sometimes classified under the heading Sonics.

1.2.9   Non-linear Acoustics

In linear acoustics, characterized by sound pressures much smaller than the static pressure, the time average value of the sound pressure or any other acoustic variable in a periodic signal is zero for most practical purposes. However, at sufficiently large sound pressures and corresponding fluid velocity amplitudes, the time average or mean values can be large enough to be significant. Thus, the static pressure variation in a standing sound wave in an enclosure can readily be demonstrated by trapping light objects and moving them by altering the standing wave field without any other material contact with the body than the air in the room. This is of particular importance in the gravity free environment in a laboratory of an orbiting satellite.

Combination of viscosity and large amplitudes can also produce significant acoustically induced mean flow (acoustic streaming) in a fluid and a corresponding particle transport. Similarly, the combination of heat conduction and large amplitudes can lead to a mean flow of heat and this effect has been used to achieve acoustically driven refrigeration using acoustic resonators driven at resonance to meet high amplitude requirements.

Other interesting effects in nonlinear acoustics include interaction of a sound wave with itself which makes an initially plane harmonic wave steepen as it travels and ultimately develop into a saw tooth wave. This is analogous to the steepening of surface waves on water. Interaction of two sound waves of different frequencies leads to the generation of sum and difference frequencies so that a low-frequency wave can be generated from two high-frequency waves.

1.2.10   Acoustic Instrumentation

Much of what we have been able to learn in acoustics (as in most other fields) has been due to the availability of electronic equipment both for the generation, detection, and analysis of sound. The rapid progress in the field beginning about 1930 was due to the advent of the radio tube and the equipment built around it. This first electronic 'revolution,' the electronic 'analog' era, was followed with a second with the advent of the transistor which led into the present 'digital' era. The related development of equipment for acoustic purposes, from Edison's original devices to the present, is a fascinating story in which many areas of acoustics have been involved, including the electro-mechanics of transducers, sound radiation, room acoustics, and the perception of sound.

1.2.11 Speech and Hearing

The physics, physiology, and psychology of hearing and speech occupies a substantial part of modern acoustics. The physics of speech involves modeling the vocal tract as a duct of variable area (both in time and space) driven at the vocal chords by a modulated air stream. A wave theoretical analysis of the response of the vocal tract leads to an understanding of the frequency spectrum of the vowels. In the analysis of the fricative sounds, such as s, sh, ch, and t, the generation of sound by turbulent flow has to be accounted for. On the basis of the understanding thus obtained, synthetic speech generators have been developed.

Hearing represents a more complicated problem, even on the physics level, which deals with the acoustics of the ear canal, the middle ear, and, in particular, the fluid dynamics in the inner ear. In addition, there are the neurological aspects of the problem which are even more complex. From extensive measurements, however, much of the physics of hearing has been identified and understood, at least in part, such as the frequency dependence of the sensitivity of the human ear, for example.

1.2.12 Musical Acoustics

The field of musical acoustics is intimately related to that of speech. The physics now involves an understanding of sound generation by various musical instruments rather than by the vocal tract. A thorough understanding of wind instruments requires an intimate knowledge of aero-acoustics. For string instruments, like the violin and the piano, the vibration and radiation characteristics of the sounding boards are essential, and numerous intricate experiments have been carried out in efforts to make the vibrations visible.

1.2.13 Phonons and Laser Light Spectroscopy

The thermal vibrations in matter can be decomposed into (random) acoustic waves over a range of wavelengths down to the distance between molecules. The experimental study of such high-frequency waves ('hypersonics') requires a 'probe' with the same kind of resolution and the use of (Brillouin) scattering of laser light is the approach that has been used (photon-phonon interaction). By analysis of the light scattered by the waves in a transparent solid (heterodyne spectroscopy), it is possible to determine the speed of sound and the attenuation in this high-frequency regime. The scattered light is shifted in frequency by an amount equal to the frequency of the acoustic wave and this shift is measured. Furthermore, the line shape of the scattered light provides another piece of information so that both the sound speed and attenuation can be determined.

A similar technique can be used also for the thermal fluctuations of a liquid surface which can be decomposed into random high-frequency surface waves. The upper frequency limit varies from one liquid to the next but the corresponding wavelength is of the order of the intermolecular distance. Again by using the technique of laser light heterodyne spectroscopy, both surface tension and viscosity can be determined. Actually, even for the interface between two liquids which do not mix, these quantities can be determined. The interfacial surface tension between water and oil, for example, is of considerable practical interest.

1.2.14   Flow-induced Instabilities

The interaction of a structure with fluid flow can lead to vibrations which under certain conditions can be unstable through feedback. The feedback can be a result of the interactions between fluid flow, sound, and the structure.

In some musical wind instruments, such as an organ pipe or a flute, the structure can be regarded as rigid as far as the mechanism of the instability is concerned, and it is produced as a result of the interaction of vorticity and sound. The sound produced by a vortex can react on the fluid flow to promote the growth of the vortex and hence give rise to a growing oscillation and sound that is sustained by the flow through this feedback.

A similar instability, which is very important in some industrial facilities, is the 'stimulated' Kármán vortex behind a cylinder in a duct. The periodic vortex can be stimulated through feedback by an acoustic cross mode in the duct if its resonance frequency is equal to (or close to) the vortex frequency. This is a phenomenon which can occur in heat exchangers and the amplitude can be so large that it represents an environmental problem and structural failure can also result.

A stimulation of the Kàrmàn vortex can result also if the cylinder is flexible and if the transverse resonance frequency of the cylinder is the same as the vortex frequency. Large vibrations of a chimney can occur in this manner and the structural failure of the Tacoma bridge is a classic example of the destructive effects that can result from this phenomenon.

In a reed type musical instrument, or in an industrial control valve, the reed or the valve plug represents a flexible portion of the structure. In either of these cases, this flexible portion is coupled to the acoustic resonator which, in the case of the plug, is represented by the pipe or duct involved. If the resonance frequencies of the structure and the pipe are sufficiently close, the feedback can lead to instability and very large vibration amplitudes, known to have caused structural failures of valves.

1.2.15   Aero-thermo Acoustics. Combustion Instability

This designation as a branch of acoustics is sometimes used when heat sources and heat conduction have a significant influence on the acoustics. For example, the sound generation in a combustor falls into this category as does the acoustic refrigeration mentioned earlier.

The rate of heat release Q in a combustor acts like a source of sound if Q is time dependent with the acoustic source strength being proportional to dQ/dt. If Q is also pressure dependent, the sound pressure produced in the combustion chamber can feed back to the combustor and modulate the acoustic output. This can lead to an instability with high amplitude sound (and vibration) as a consequence. The vibrations can be so violent that structural failure can result when a facility, such as a gas turbine power plant, is operating above a certain power setting. The challenge, of course, is to limit the amplitude of vibrations or, even better, to eliminate the instability. An acoustic analysis can shed valuable light on this problem and can be most helpful in identifying its solution.

1.2.16 Miscellaneous

As in most other fields of science and engineering, there are numerous activities dealing with regulations, codes, standards, and the like. They are of considerable importance in industry and in government agencies and there is great need for inputs from experts. Working in such a field, even for a short period, is apt to provide familiarity with various government agencies and international organizations and serve as an introduction to the art of politics.

Chapter 2

Oscillations

As indicated in the Preface, it is assumed that the reader is familiar with the content of a typical introductory course in mechanics that includes a discussion of the basics of the harmonic oscillator. It is an essential element in acoustics and it will be reviewed and extended in this chapter. The extension involves mainly technical aspects which are convenient for problem solving. Thus, the use of complex variables, in particular the complex amplitude, is introduced as a convenemt and powerful way of dealing with oscillations and waves.

With modern digital instrumentation, many aspects of signal processing are readily made available and to be able to fully appreciate them, it is essential to have some knowledge of the associated mathematics. Thus, Fourier series and Fourier transforms, correlation functions, spectra and spectrum analysis are discussed. As an example, the response of an oscillator to a completely random driving force is determined. This material is discussed in Section 2.6. However, it can be skipped at a first reading without a lack of continuity.

The material referred to above is all 'standard'; it is important to realize, though, that it is generally assumed that the oscillators involved and the related equations of motion are linear. This is an idealization, and is valid, at best, for small amplitudes of oscillations. But even for small amplitudes, an oscillator can be non-linear, and we end this chapter with a simple example. It involves a damped mass-spring oscillator. Normally, the friction force is tactily assumed to be proportional to the velocity in which case the equation of motion becomes linear and a solution for the displacement is readily found. However, consider the very simple case of a mass sliding on a table and subject not only to a ('dynamic') friction force proportional to the velocity but also to a ('static') friction force proportional to the static friction coefficient.

2.1 Harmonic Motions

A periodic motion is one that repeats itself after a constant time interval, the *period*, denoted T. The number of periods (cycles) per second, cps, is called the *frequency* f (i.e. $f = 1/T$ cps or Hz).[1] For example, a period of 0.5 seconds corresponds to a frequency of 2 Hz.

---

[1] The unit Hz after the German physicist Heinrich Hertz (1857–1894).

The periodic motion plays an important role in nature and everyday life; the spin of the earth and the orbital motion (assumed uniform) of the earth and of the moon are obvious examples. The ordinary pendulum is familiar to all but note that the period of oscillation increases with the amplitude of oscillation. This effect, however, is insignificant at small amplitudes.

To obtain periodicity to a very high degree of accuracy, one has to go down to the atomic level and consider the frequency of atomic 'vibrations.' Actually, this is the basis for the definition of the unit of time. A good atomic clock, a Cesium clock, loses or gains no more than one second in 300,000 years and the unit of time, *one second*, is defined as the interval for 9,191,631,770 periods of the Cesium atom.[2]

Harmonic motion is a particular periodic motion and can be described as follows. Consider a particle P which moves in a circular path of radius A with constant speed. The radius vector to the particle makes an angle with the x-axis which is proportional to time t, expressed as ωt, where ω is the *angular velocity* (for rectilinear motion, the position of the particle is $x = vt$, where v is the *linear* velocity). It is implied that the particle crosses the x-axis at $t = 0$. After one period T of this motion, the angle ωt has increased by 2π, i.e., $\omega T = 2\pi$ or

$$\omega = 2\pi/T = 2\pi f \tag{2.1}$$

where $f = 1/T$, is the frequency, introduced above. In general discussions, the term frequency, rather than angular frequency, is often used also for ω. Of course, in numerical work one has to watch out for what quantity is involved, ω or f.

The time dependence of the x-coordinate of the particle P defines the *harmonic motion*

$$\xi = A\cos(\omega t). \tag{2.2}$$

It is characteristic of harmonic motion that ω does not depend on time. But note that a motion can be periodic even if ω is time dependent. This is the case for the motion of a planet in an elliptical orbit, for example.

The velocity in the harmonic motion is

$$u = \dot{\xi} = -A\omega\sin(\omega t) \tag{2.3}$$

and the acceleration

$$a = \ddot{\xi} = -A\omega^2\cos(\omega t) = -\omega^2\xi. \tag{2.4}$$

It follows that the harmonic motion satisfies the differential equation

$$\ddot{\xi} = -\omega^2\xi. \tag{2.5}$$

Thus, if an equation of this form is encountered in the study of motion, we know that the harmonic motion is a solution. As we shall see, such is the case when a particle, displaced from its equilibrium position, is acted on by a restoring force

---

[2]The unit of length, one meter, is defined in such a way as to make the speed of light exactly 3×10^8 m/sec; thus, the unit of length, one meter, is the distance traveled by light in $(1/3)10^{-8}$ sec. This unit is very close to the unit of length based on the standard meter (a bar of platinum-iridium alloy) kept at the International Bureau of Weight and Measures at Sèvres, France.

that is proportional to the displacement. Then, when the particle is released, the subsequent motion will be harmonic. A mass at the end of a coil spring (the other end of the spring held fixed) is an example of such an oscillator. (It should be noted though that in practice the condition that the restoring force be proportional to the displacement is generally valid only for sufficiently small displacements.)

If the origin of the time scale is changed so that the displacement is zero at time $t = t_1$, we get

$$\xi(t) = A \cos[\omega(t - t_1)] = A \cos(\omega t - \phi) \qquad (2.6)$$

where $\phi = \omega t_1$ is the *phase angle* or phase lag. Quantity A is the ampliutde and the entire argument $\omega t - \phi$ is sometimes called the 'phase.' In terms of the corresponding motion along a circle, the representative point trails the point P, used earlier, by the angle ϕ.

Example

The velocity that corresponds to the displacement in Eq. 2.6. is $u = -A\omega \sin(\omega t)$. The speed is the absolute value $|u|$ of the velocity. Thus, to get the average speed we need consider only the average over the time during which u is positive, (i.e., in the time interval from 0 to T/2), and we obtain

$$\langle |u| \rangle = 2/T \int_0^{T/2} A\omega \sin(\omega t) dt = (2/\pi) u_{max}, \qquad (2.7)$$

where $u_{max} = A\omega$ is the maximum speed.

The *mean square value* of the velocity is the time average of the squared velocity and the *root mean square value, rms,* is the the square root of the mean square value,

$$\langle u^2 \rangle = (1/T) \int_0^T u^2 \, dt = u_{max}^2 / 2,$$
$$u_{rms} = u_{max} / \sqrt{2} \qquad (2.8)$$

where $u_{max} = A\omega$.

We shall take Eq. 2.6 to be the *definition of harmonic motion*. The velocity u is also a harmonic function but we have to express it in terms of a cosine function to see what the phase angle is. Thus, $u = \omega|\xi| \sin(\omega t) = \omega|\xi| \cos(\omega t - \pi/2)$ is a harmonic motion with the amplitude $\omega|\xi|$ and the phase angle (lag) $\pi/2$. Similarly, the acceleration is a harmonic function $a = -\omega^2|\xi| \cos(\omega t) = \omega^2|\xi| \cos(\omega t - \pi)$ with the amplitude $\omega^2|\xi|$ and the phase angle π.

One reason for the importance of the harmonic motion is that any periodic function, period T and *fundamental frequency* $1/T$, can be decomposed in a (Fourier) series of harmonic functions with frequencies being multiples of the fundamental frequency, as will be discussed shortly.

2.1.1 The Complex Amplitude

For a given angular velocity ω, a harmonic function $\xi(t) = |\xi| \cos(\omega t - \phi)$ is uniquely defined by the amplitude $|\xi|$ and the phase angle ϕ. Geometrically, it can be represented by a point in a plane at a distance $|\xi|$ from the origin with the radius vector making an angle $\phi = 0$ with the x-axis. This representation reminds us of a complex number $z = x + iy$ in the complex plane (see Appendix B), where x is the real part and y the imaginary part. As we shall see, complex numbers and their algebra are indeed ideally suited for representing and analyzing harmonic motions. This is due to the remarkable Euler's identity $\exp(i\alpha) = \cos\alpha + i\sin\alpha$, where i is the imaginary unit number $i = \sqrt{-1}$ (i.e., $i^2 = -1$). To prove this relation, expand $\exp(i\alpha)$ in a power series in α, making use of $i^2 = -1$, and collect the real and imaginary parts; they are indeed found to be the power series expansions of $\cos\alpha$ and $\sin\alpha$, respectively. With the proviso $i^2 = -1$ ($i^3 = -i$, etc.), the exponential $\exp(i\alpha)$ is then treated in the same way as the exponential for a real variable with all the associated algebraic rules.

It is sometimes useful to express $\cos\alpha$ and $\sin\alpha$ in terms of $\exp(i\alpha)$; $\cos\alpha = (1/2)[\exp(i\alpha) + \exp(-i\alpha)]$ and $\sin\alpha = (1/2i)[\exp(i\alpha) - \exp(-i\alpha)]$.

The complex number $\exp(i\alpha)$ is represented in the complex plane by a point with the real part $\cos\alpha$ and the imaginary part $\sin\alpha$. The radius vector to the point makes an angle α with the real axis. With $\Re$ standing for 'the real part of' and with $\alpha = \omega t - \phi$, the harmonic displacement $\xi(t) = |\xi| \cos(\omega t - \phi)$ can be expressed as

$$\boxed{\begin{array}{c} \textit{Definition of the complex amplitude} \\ \xi(t) = |\xi| \cos(\omega t - \phi) = \Re\{|\xi| e^{-i(\omega t - \phi)}\} = \Re\{|\xi| e^{i\phi} e^{-i\omega t}\} \equiv \Re\{\xi(\omega) e^{-i\omega t}\} \\ \xi(\omega) = |\xi| e^{i\phi} \end{array}}$$

(2.9)

At a given frequency, the *complex amplitude* $\xi(\omega) = |\xi| \exp(i\phi)$ uniquely defines the motion.[3] It is represented by a point in the complex plane (Fig. 2.1) a distance $|\xi|$ from the origin and with the line from the origin to the point making an angle ϕ with the real axis.

The unit imaginary number can be written $i = \exp(i\pi/2)$ ($=\cos(\pi/2) + i\sin(\pi/2)$) with the magnitude 1 and phase angle $\pi/2$; it is located at unit distance from the origin on the imaginary axis. Multiplying the complex amplitude $\tilde{\xi}(\omega) = |\xi| \exp(i\phi)$ by $i = \exp(i\pi/2)$ increases the phase lag by $\pi/2$ and multiplication by $-i$ reduces it by the same amount.

Differentiation with respect to time in Eq. 2.9 brings down a factor $(-i\omega) = \omega \exp(-i\pi/2)$ so that the complex amplitudes of the velocity $\dot{\xi}(t)$ and the acceleration $\ddot{\xi}(t)$ of the particle are $(-i\omega)\xi(\omega)$ and $(-i\omega)^2\xi(\omega) = -\omega^2\xi(\omega)$. The locations of these complex amplitudes are indicated in Fig. 2.1 (with the tilde signatures omitted, in accordance with the comment on notation given below); their phase lags are smaller than that of the displacement by $\pi/2$ and π, respectively; this means that they are running ahead of the displacement by these angles.

To visualize the time dependence of the corresponding real quantities, we can let the complex amplitudes rotate with an angular velocity ω in the counter-clockwise direction about the origin; the projections on the real axis then yield their time dependence.

---

[3]We could equally well have used $\xi(t) = \Re\{|\xi| \exp[i(\omega t - \phi)]\} = \Re\{[|\xi| \exp(-i\phi)] \exp(i\omega t)\}$ in the definition of the complex amplitude. It merely involves replacing i by $-i$. This definition is sometimes used in engineering where $-i$ is denoted by j. Our choice will be used consistently in this book. One important advantage becomes apparent in the description of a traveling wave in terms of a complex variable.

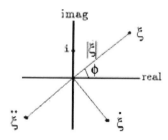

Figure 2.1: The complex plane showing the location of the complex amplitudes of displacement $\xi(\omega)$, velocity $\dot{\xi}(\omega) = -i\omega\xi(\omega) = \xi(\omega)\exp(-i\pi/2)$, and acceleration $\ddot{\xi}(\omega) = -\omega^2\xi(\omega)$.

All the terms in a differential equation for $\xi(t)$ can be expressed in a similar manner in terms of the complex amplitude $\xi(\omega)$. Thus, the differential equation is converted into an algebraic equation for $\xi(\omega)$. Having obtained $\xi(\omega)$ by solving the equation, we immediately get the amplitude $|\xi|$ and the phase angle ϕ which then define the harmonic motion $\xi(t) = |\xi|\cos(\omega t - \phi)$.

A Question of Notation

Sometimes the complex amplitude is given a 'tilde' symbol to indicate that the function $\tilde{\xi}(\omega)$ is the complex amplitude of the displacement. In other words, the complex amplitude is not obtained merely by replacing t by ω in the function ξ. However, for convenience in writing and without much risk for confusion, we adopt from now on the convention of dropping the tilde symbol, thus denoting the complex amplitude merely by $\xi(\omega)$. Actually, as we get seriously involved in problem solving using complex amplitudes, even the argument will be dropped and ξ alone will stand for the complex amplitude; the context then will decide whether $\xi(t)$ or $\xi(\omega)$ is meant.

Example

What is the complex amplitude of a displacement $\xi(t) = |\xi|\sin[\omega(t - T/6)]$, where T is the period of the motion.

The phase angle ϕ of the complex amplitude $\xi(\omega)$ is based on the displacement being written as a cosine function, i.e., $\xi(t) = |\xi|\cos(\omega t - \phi)$. Thus, we have to express the sine function in terms of a cosine function, i.e., $\sin\alpha = \cos(\alpha - \pi/2)$. Then, with $\omega T = 2\pi$, we get $\sin[\omega(t - T/6)] = \cos(\omega t - \pi/3 - \pi/2) = \cos(\omega t - 5\pi/6)$. Thus, the complex amplitude is $\xi(\omega) = |\xi|\exp(i5\pi/6)$.

2.1.2 Problems

1. **Harmonic motion, definitions**

   What is the angular frequency, frequency, period, phase angle (in radians), and amplitude of a displacement $\xi = 2\cos[100(t - 0.1)]$ cm, where t is the time in seconds?

2. **Harmonic motion. Phase angle**

The harmonic motion of two particles are $A \cos(\omega t)$ and $A \cos(\omega t - \pi/6)$.

(a) The latter motion lags behind the former in time. Determine this time lag in terms of the period T.

(b) At what times do the particles have their (positive) maxima of velocity and acceleration?

(c) If the amplitude A is 1 cm, at what frequency (in Hz) will the acceleration equal $g = 981$ cm/sec^2?

3. **Complex amplitudes and more**

Consider again Problem 1.

(a) What are the complex amplitudes of displacement, velocity, and acceleration?

(b) Indicate their location in the complex plane.

(c) What is the average speed in one period?

(d) What is the rms value of the velocity?

4. **Sand on a membrane**

A membrane is excited by an incoming sound wave at a frequency of 50 Hz. At a certain level of the sound, grains of sound on the membrane begin to bounce. What then, is the displacement amplitude of the membrane? (This method was used by Tyndall in 1874 in his experiments on sound propagation over ocean to determine the variation of the range of fog horn signals with weather and wind.)

2.1.3 Sums of Harmonic Functions. Beats

Same Frequencies

The sum (superposition) of two harmonic motions $\xi_1(t) = A_1 \cos(\omega t - \phi_1)$ and $\xi_2(t) = A_2 \cos(\omega_2 t - \phi_2)$ with the same frequencies but with different amplitudes and phase angles is a harmonic function $A \cos(\omega t - \phi)$. To prove that, use the trigonometric identity $\cos(a - b) = \cos(a) \cos(b) + \sin(a) \sin(b)$ and collect the resulting terms with $\cos(\omega t)$ and $\sin(\omega t)$ and then compare the expression thus obtained for both the sum and for $A \cos(\omega t - \phi)$. It is left as a problem to carry out this calculation (Problem 1).

If we use the complex number representation, we can express the two harmonic functions as $B_1 \exp(-i\omega t)$ and $B_2 \exp(-i\omega t)$ where B_1 and B_2 are complex, in this case $B_1 = A_1 \exp(i\phi_1)$ and $B_2 = A_2 \exp(i\phi_2)$. The sum is then $(B_1 + B_2) \exp(-i\omega t)$, with the new complex amplitude $B = B_1 + B_2$. The real and imaginary parts of B_1 are $A_1 \cos(\phi_1)$ and $A_1 \sin(\phi)$ with similar expressions for B_2 and B. By equating the real and imaginary parts in $B = B_1 + B_2$, we readily find A and ϕ.

The result applies to the sum of an arbitrary number of harmonic functions of the same frequency.

Different Frequencies

Consider the sum of two harmonic motions, $C_1 \cos(\omega t)$ and $C_2 \cos(2\omega t)$. The period of the first is T and of the second, $T/2$. The sum will be periodic with the period T since both functions repeat after this time. Furthermore, the sum will be symmetric (even) with respect to t; it is the same for positive and negative values of t since this is true for each of the components. The same holds true for the sum of any number of harmonic functions of the form $A_n \cos(n\omega t)$.

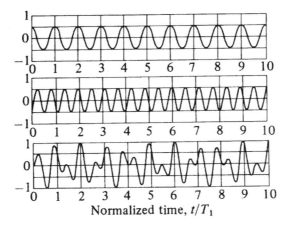

Figure 2.2: The functions $\cos(\omega_1 t)$, $\cos(\sqrt{3}\,\omega_1 t)$ and their sum (frequencies are incommensurable).

If the terms $\cos(n\omega t)$ are replaced by $\sin(n\omega t)$, the sum will still be periodic with the period T, but it will be anti-symmetric (odd) in the sense that it changes sign when t does.

If the terms are of the form

$$a_n \cos(n\omega t - \phi_n) = a_n[\cos\phi_n \cos(n\omega t) + \sin\phi_n \sin(n\omega t)],$$

where n is an integer, the sum contains a mixture of cosine and sine terms. The sum will still be periodic with the period T, but the symmetry properties mentioned above are no longer valid.

We leave it for the reader to experiment with and plot sums of this kind when the frequencies of the individual terms are integer multiples of a fundamental frequency or fractions thereof; we shall comment here on what happens when the fraction is an irrational number.

Thus, consider the sum $S(t) = 0.5\cos(\omega_1 t) - 0.5\cos(\sqrt{3}\,\omega_1 t)$. The functions and their sum are plotted in Fig. 2.2. The ratio of the two frequencies is $\sqrt{3}$, an irrational number (the two frequencies are *incommensurable*), and no matter how long we wait, the sum will not be periodic. In the present case, the sum starts out with the value 0 at $t = 0$ and then fluctuates in an irregular manner between -1 and $+1$.

On the other hand, if the ratio had been *commensurable* (i.e., a rational fraction) the sum would have been periodic; for example, a ratio 2/3 results in a period $3T_1$.

The addition of two harmonic functions with slightly different frequencies leads to the phenomenon of *beats*; it refers to a slow variation of the total amplitude of oscillation. It is strictly a kinematic effect. It will be illustrated here by the sum of two harmonic motions with the same amplitude but with different frequencies. The mean value of the two frequencies is ω, and they are expressed as $\omega_1 = \omega - \Delta\omega$ and $\omega_2 = \omega + \Delta\omega$. Using the trigonometric identity $\cos(\omega t \mp \Delta\omega t) = \cos(\omega t)\cos(\Delta\omega t) \pm$

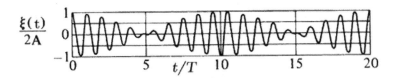

Figure 2.3: An example of beats produced by the sum of two harmonic motions with frequencies 0.9ω and 1.1ω.

$\sin(\omega t) \sin(\Delta\omega t)$, we find for the sum of the corresponding harmonic motions

$$\xi(t) = \cos(\omega t - \Delta\omega t) + \cos(\omega t + \Delta\omega t) = 2\cos(\Delta\omega t)\cos(\omega t) \qquad (2.10)$$

which can be interpreted as a harmonic motion of frequency ω with a periodically varying amplitude ("beats") of frequency $\Delta\omega$. The maximum value of the amplitude is twice the amplitude of each of the components. An example is illustrated in Fig. 2.3. In this case, with $\Delta\omega = 0.1\omega$, the period of the amplitude variation will be $\approx 10T$, consistent with the result in the figure. Beats can be useful in experimental work when it comes to an accurate comparison of the frequencies of two signals.

2.1.4 Heterodyning

The squared sum of two harmonic signals $A_1\cos(\omega_1 t)$ and $A_2\cos(\omega_2 t)$ produces signals with the sum and difference frequencies $\omega_1 + \omega_2$ and $\omega_1 - \omega_2$, which can be of considerable practical importance in signal analysis. The squared sum is

$$[A_1\cos(\omega_1 t) + A_2\cos(\omega_2 t)]^2 = A_1^2\cos^2(\omega_1 t) + A_2^2\cos^2(\omega_2 t) + 2A_1 A_2\cos(\omega_1)\cos(\omega_2 t) \qquad (2.11)$$

The time dependent part of each of the squared terms on the right-hand side is harmonic with twice the frequency since $\cos^2(\omega t) = [1 + \cos(2\omega t)]/2$. This is not of any particular interest, however. The important part is the last term which can be written

$$2A_1 A_2\cos(\omega_1 t)\cos(\omega_2 t) = A_1 A_2[\cos(\omega_1 + \omega_2)t + \cos(\omega_1 - \omega_2)t]. \qquad (2.12)$$

It contains two harmonic components, one with the sum of the two primary frequencies and one with the difference. This is what is meant by heterodyning, the creation of sum and difference frequencies of the input signals. Normally, it is the term with the difference frequency which is of interest.

There are several useful applications of heterodyning; we shall give but one example here. A photo-cell or photo-multiplier is a device such that the output signal is proportional to the square of the electric field in an incoming light wave. Thus, if the light incident on the photo-cell is the sum of two laser signals, the output will contain an electric current with the difference of the frequencies of the two signals.

Thus, consider a light beam which is split into two with one of the beams reflected or scattered from a vibrating object, such as the thermal vibrations of the surface of a liquid, where the reflected signal is shifted in frequency by an amount equal to the vibration frequency ω_0. (Actually, the reflected light contains both an up-shifted and a down-shifted frequency, $\Omega \pm \omega_0$, which can be thought of as being Doppler shifted by the vibrating surface.) Then, if both the direct and the reflected beams are incident on the photo-cell, the output signals will contain the frequency of vibration. This frequency might be of the order of $\approx 10^5$ Hz whereas the incident light frequency typically would be $\Omega \approx 10^{15}$ Hz. In this case, the shift is very small, however, only 1 part in 10^{10}, and conventional spectroscopic methods would not be able to resolve such a small shift.

With the heterodyne technique, *heterodyne spectroscopy*, this problem of resolution is solved. Since the output current contains the difference frequency ω_0, the vibration frequency, which can be detected and analyzed with a conventional electronic analyzer.

2.1.5 Problems

1. **Sum of harmonic functions**
   (a) With reference to the outline at the beginning of Section 2.1.3, show that $A_1 \cos(\omega t - \phi_1) + A_1 \cos(\omega t - \phi_2)$ can be written as a new harmonic function $A \cos(\omega t - \phi)$ and determine A and ϕ in terms of A_1, A_2, ϕ_1, and ϕ_2.
   (b) Carry out the corresponding calculation using complex amplitude description of the harmonic functions as outlined in Section 2.1.3.

2. **Heterodyning**
   In heterodyning, the sum of two signals with the frequencies ω_1 and ω_2 are processed with a square law detector producing the output sum and differences of the input signal frequencies. What frequencies would be present in the output of a cube-law detector?

2.2 The Linear Oscillator

2.2.1 Equation of Motion

So far, we have dealt only with the kinematics of harmonic motion without regard to the forces involved. The real 'physics' enters when we deal with the dynamics of the motion and it is now time to turn to it.

One reason for the unique importance of the harmonic motion is that in many cases in nature and in applications, a small displacement of a particle from its equilibrium position generally results in a restoring (reaction) force proportional to the displacement. If the particle is released from the displaced position, the only force acting on it in the absence of friction will be the restoring force and, as we shall see, the subsequent motion of the particle will be harmonic. The classical example is the mass-spring oscillator illustrated in Fig. 2.4. A particle of mass M on a table, assumed friction-less, is attached to one end of a spring which has its opposite end clamped. The displacement of the particle is denoted ξ. Instead of sliding on the table, the particle can move up and down as it hangs from the free end of a vertical spring with the upper end of the spring held fixed, as shown.

It is found experimentally that for sufficiently small displacements, the force required to change the length of the spring by an amount ξ is $K\xi$, where K is a constant. It

We start with the general expression for the harmonic displacement $\xi = A \cos(\omega_0 t - \phi)$. It contains the two constants A and ϕ which are to be determined. Thus, with $t = 0$, we obtain,

$$\xi(0) = A \cos(\phi)$$
$$u(0) = A\omega_0 \sin(\phi)$$

and

$$\tan(\phi) = u(0)/[\omega_0 \xi(0)], \quad A = \xi(0)/\cos(\phi) = \xi(0)\sqrt{1 + \tan^2(\phi)}.$$

Inserting the numerical values we find

$$\tan(\phi) = 1/200 \quad \text{and } A = 10\sqrt{1 + (1/200)^2} \approx 10[1 + (1/2)(1/200)^2].$$

The subsequent displacement is $\xi(t) = A \cos(\omega_0 t - \phi)$.

Comment. With the particular initial values chosen in this problem the phase angle is very small, and the amplitude of oscillation is almost equal to the initial displacement. In other words, the oscillator is started out very nearly from the maximum value of the displacement and the initial kinetic energy of the oscillator is much smaller than the initial potential energy. How should the oscillator be started in order for the subsequent motion to have the time dependence $\sin(\omega_0 t)$?

2.2.2 The 'Real' Spring. Compliance

The spring constant depends not only on the elastic properties of the material in the spring but also on its length and shape. In an ordinary uniform coil spring, for example, the pitch angle of the coil (helix) plays a role and another relevant factor is the thickness of the material. The deformation of the coil spring is a complicated combination of torsion and bending and the spring constant generally should be regarded as an experimentally determined quantity; the calculation of it from first principles is not simple.

The linear relation between force and deformation is valid only for sufficiently small deformations. For example, for a very large elongation, the spring ultimately takes the form of a straight wire or rod, and, conversely, a large compression will make it into a tube-like configuration corresponding to a zero pitch angle of the coil. In both these limits, the stiffness of the spring is much larger than for the relaxed spring.

It has been tacitly assumed that the spring constant is determined from a *static* deformation. Yet, this constant has been used for non-static (oscillatory) motion. Although this is a good approximation in most cases, it is not always true. Materials like rubber and plastics (and polymers in general) for which elastic constants depend on the rate of strain, the spring constant is frequency dependent. For example, there exist substances which are plastic for slow and elastic for rapid deformations (remember 'silly putty'?). This is related to the molecular structure of the material and the effect is often strongly dependent on temperature. A cold tennis ball, for example, does not bounce very well.

In a static deformation of the spring, the inertia of the spring does not enter. If the motion is time dependent, this is no longer true, and another idealization is the omission of the mass of the spring. This is justified if the mass attached to the spring is much larger than the spring mass. The effect of the spring mass will be discussed shortly.

The inverse of the spring constant K is called the *compliance*

$$C = 1/K. \tag{2.15}$$

It is proportional to the length of the spring. Later, in the study of wave motion on a spring, we shall introduce the compliance per unit length.

Frequently, several springs are combined in order to obtain a desired resulting spring constant. If the springs are in 'parallel,' the deformations will be the same for all springs and the restoring forces will add. The resulting spring constant is then the sum of the individual spring constants; the resulting spring will be 'harder.' If the springs are in 'series,' the force in each spring will be the same and the deformations add. The resulting compliance is then the sum of the individual compliances; the resulting combined spring will be 'softer' than any of the individual springs.

Effect of the Mass of the Spring

As already indicated, the assumption of a mass-less spring in the discussion of the mass-spring oscillator is of course an idealization and is not a good assumption unless the spring mass m is much smaller than the mass M of the body attached to the spring. This shows up as a defect in Eq. 2.13 for the frequency of oscillation, $\omega_0 = \sqrt{K/M}$. According to it, the frequency goes to infinity as M goes to zero. In reality, this cannot be correct since removal of M still yields a finite frequency of oscillation of the spring alone. This problem of the spring mass will be considered later in connection with wave propagation and it will be shown that for the lowest mode of oscillation, the effect of the spring mass m can be accounted for approximately, if $m/M << 1$, by adding one-third of this mass to the mass M in Eq. 2.13. Thus, the corrected expression for the frequency of oscillation (lowest mode) is

$$\omega_0 \approx \sqrt{\frac{K}{M + m/3}} = \frac{\omega_0}{\sqrt{1 + m/3M}}. \tag{2.16}$$

Air Spring

For an isothermal change of state of a gas, the relation between pressure P and volume V is simply $PV = $ constant (i.e., $dP/P = -dV/V$). For an isentropic (adiabatic) change, this relation has to be replaced by $dP/P = -\gamma(dV/V)$, where γ is the specific heat ratio C_p/C_v, which for air is ≈ 1.4.

Consider a vertical tube of length L and closed at the bottom and with a piston riding on the top of the air column in the tube. If the piston is displaced into the tube by a small among ξ, the volume of the air column is changed by $dV = -A\xi$, where A is the area of the tube. On the assumption that the compression is isentropic, the pressure change will be $dP = \gamma(PA/V)\xi$ and the corresponding force on the piston will oppose the displacement so that $F = -\gamma(PA^2/V)\xi$. This means that the spring constant of the air column is

$$K = \gamma(PA^2/V) = \gamma(PA/L), \qquad (2.17)$$

where $L = V/A$ is the length of the tube. The spring constant is inversely proportional to the length and, hence, the compliance $C = 1/K$ is proportional to the length.

After releasing the piston, it will oscillate in harmonic motion with the angular frequency $\sqrt{K/M}$. As will be shown later, the adiabatic approximation in a situation like this is valid except at very low frequencies. By knowing the dimensions of the tube and the mass M, a measurement of the frequency can be used as a means of determining the specific heat ratio γ. A modified version of this experiment, often used in introductory physics laboratory, involves a flask or bottle with a long, narrow neck in which a steel ball is used as a piston.

For a large volume change, the motion will not be harmonic since the relation between the displacement and the restoring force will not be linear. Thus, with the initial quantities denoted by a subscript 1, a general displacement ξ yields a new volume $V_2 = V_1 - A\xi$ and the new pressure is obtained from $P_2 V_2^\gamma = P_1 V_1^\gamma$. The restoring force $A(P_2 - P_1)$ no longer will be proportional to ξ and we have a non-linear rather than a linear oscillator.

2.2.3 Problems

1. **Static compression and resonance frequency**

   A weight is placed on top of a vertical spring and the static compression of the spring is found to be ξ_{st}. Show that the frequency of oscillation of the mass-spring oscillator is determined solely by the static displacement and the acceleration of gravity g.

2. **Frequency of oscillation**

   A body of mass m on a horizontal friction-less plane is attached to two springs, one on each side of the body. The spring constants are K_1 and K_2. The relaxed lengths of each spring is L. The free ends of the springs are pulled apart and fastened to two fixed walls a distance $3L$ apart.
   (a) Determine the equilibrium position of the body.
   (b) What is the frequency of oscillation of the body about the equilibrium position?
   (c) Suppose that the supports are brought close together so that the their separation will be $L/2$. What, then, will be the equilibrium position of M and the frequency of oscillation?

3. **Lateral oscillations on a spring**

   (a) In Example 2, what will be the frequency of small amplitude oscillations of M in a direction *perpendicular* to the springs?
   (b) Suppose that the distance between the end supports of the spring equals the length of the spring so that the spring is slack. What will be the restoring force for a lateral displacement ξ of M? Will the oscillation be harmonic?

4. **Initial value problem**

   The collisions in Example 8 in Ch.11 are inelastic and mechanical energy will be lost in a collision. The mechanical energy loss in the first collision is

   $$p^2/2m - p^2/2(M+m) = (p^2/2m)(M/(M+m))$$

and in the second

$$p^2/2(M + m) - p^2/2(M + 2m).$$

(a) Show that the two energy losses are the same if $m/M = 1 + \sqrt{2}$. Compare the two energy losses as a function of m/M.

(b) Suppose that n shots are fired into the block under conditions of maximum amplitude gain as explained in Example 8. What will be the amplitude of the oscillator after the n:th shot?

2.3 Free Damped Motion of a Linear Oscillator

2.3.1 Energy Considerations

The mechanical energy in the harmonic motion of a mass-spring oscillator is the sum of the kinetic energy $Mu^2/2$ of the mass M and the potential energy V of the spring. If the displacement from the equilibrium position is ξ, the force required for this displacement is $K\xi$. The work done to reach this displacement is the potential energy

$$V(\xi) = \int_0^\xi K\xi d\xi = K\xi^2/2. \tag{2.18}$$

In the harmonic motion, there is a periodic exchange between kinetic and potential energy, each going from zero to a maximum value E, where $E = Mu^2/2 + K\xi^2/2$ is the total mechanical energy. In the absence of friction, this energy is a constant of motion.

To see how this follows from the equation of motion, we write the harmonic oscillator equation (2.13) in the form

$$M\dot{u} + K\xi = 0, \tag{2.19}$$

where $u = \dot{\xi}$ is the velocity, and then multiply the equation by u. The first term in the equation becomes $Mu\dot{u} = d/dt[Mu^2/2]$. In the second term, which becomes $Ku\xi$, we use $u = \dot{\xi}$ so that it can be written $K\xi\dot{\xi} = d/dt[K\xi^2/2]$. This means that Eq. 2.19 takes the form

$$d/dt[Mu^2/2 + K\xi^2/2] = 0. \tag{2.20}$$

The first term, $Mu^2/2$, is the *kinetic energy* of the mass M, and the second term, $K\xi^2/2$, is the *potential energy* stored in the spring. Each is time dependent but the sum, the total *mechanical energy*, remains constant throughout the motion. Although no new physics is involved in this result (since it follows from Newton's law), the conservation of mechanical energy is a useful aid in problem solving.

In the harmonic motion, the velocity has a maximum when the potential energy is zero, and vice versa, and the total mechanical energy can be expressed either as the maximum kinetic energy or the maximum potential energy. The average kinetic and potential energies (over one period) are the same.

When a friction force is present, the total mechanical energy of the oscillator is no longer conserved. In fact, from the equation of motion $M\dot{u} + K\xi = -Ru$ it follows by multiplication by u (see Eq. 2.20) that

$$d/dt[Mu^2/2 + K\xi^2/2] = -Ru^2. \tag{2.21}$$

Thus, the friction drains the mechanical energy, at a rate $-Ru^2$, and converts it into heat.[4]

As a result, the amplitude of oscillation will decay with time and we can obtain an approximate expression for the decay by assuming that the average potential and kinetic energy (over one period) are the same, as is the case for the loss-free oscillator. Thus, with the left-hand side of Eq. 2.21 replaced by $d(M\langle u^2 \rangle)/dt$, and the right-hand side by $R\langle u^2 \rangle$, the time dependence of $\langle u^2 \rangle$ will be

$$\langle u(t)^2 \rangle \approx \langle u(0)^2 \rangle e^{-(R/M)t}. \tag{2.22}$$

The corresponding rms amplitude then will decay as $\exp[-(R/2M)t]$.

2.3.2 Oscillatory Decay

After having seen the effect of friction on the time dependence of the average energy, let us pursue the effect of damping on free motion in more detail and determine the actual decay of the amplitude and the possible effect of damping on the frequency of oscillation.

The idealized oscillator considered so far had no other forces acting on the mass than the spring force. In reality, there is also a friction force although in many cases it may be small. We shall assume the friction force to be proportional to the velocity of the oscillator. Such a friction force is often referred to as *viscous* or *dynamic*.

Normally, the contact friction with a table, for example, does not have such a simple velocity dependence. Often, as a simplification, one distinguishes merely between a 'static' and a 'dynamic' contact friction, the magnitude of the latter often assumed to be proportional to the magnitude of the velocity but with a direction opposite that of the velocity. The 'static' friction force is proportional to the normal component of the contact force and points in the direction opposite that of the horizontal component of the applied force.

A friction force proportional to the velocity can be obtained by means of a *dashpot damper*, as shown in Fig. 2.5. It is in parallel with the spring and is simply a 'leaky' piston which moves inside a cylinder. The piston is connected to the mass M of the oscillator and the force required to move the piston is proportional to its velocity relative to the cylinder (neglecting the mass of the piston). The cylinder is attached to the same fixed support as the spring, as indicated in Fig. 2.5. The fluid in the cylinder is then forced through a narrow channel (a 'leak') between the piston and the cylinder and it is the viscous stresses in this flow which are responsible for the friction force. Therefore, this type of damping is often referred to as *viscous*.

The friction on a body moving through air or some other fluid in free field will be proportional to the velocity only for very low speeds and approaches an approximate square law dependence at high speeds.

---

[4]When the concept of energy is extended to include other forms of energy other than mechanical, the law of conservation of energy *does* bring something new, *the first law of thermodynamics* which can be regarded as a postulate, the truth of which should be considered as an experimental fact.

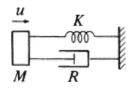

Figure 2.5: Oscillator with dash-pot damper.

With a friction force proportional to the velocity, the equation of motion for the oscillator becomes linear so that a solution can be obtained in a simple manner. For dry contact friction or any other type of friction, the equation becomes non-linear and the solution generally has to be found by numerical means, as will be demonstrated in Section 2.7.3.

With $d\xi/dt \equiv \dot{\xi}$, we shall express the friction force as $-R\dot{\xi}$ and the equation of motion for the mass element in an oscillator becomes $M\ddot{\xi} = -K\xi - R\dot{\xi}$ or, with $K/M = \omega_0^2$,

$$
\boxed{
\begin{array}{l}
\textit{Free oscillations, damped oscillator} \\
\quad \ddot{\xi} + (R/M)\dot{\xi} + \omega_0^2 \xi = 0 \\
\quad \xi(t) = A e^{-\gamma t} \cos(\omega_0' t - \phi) \\
\gamma = R/2M, \qquad \omega_0' = \sqrt{\omega_0^2 - \gamma^2}
\end{array}
} \qquad (2.23)
$$

The general procedure to solve a linear differential equation is aided considerably with the use of complex variables (Section 2.3.3). For the time being, however, we use a 'patchwork' approach to construct a solution, making use of the result obtained in the decay of the energy in Eq. 2.22 from which it is reasonable to assume that the solution $\xi(t)$ will be of the form given in Eq. 2.23, where γ, and ω_0' are to be determined. Thus, we insert this expression for $\xi(t)$ into the first equation in 2.23 and write the left-hand side as a sum of $\sin(\omega_0' t)$ and $\cos(\omega_0' t)$ functions. Requiring that each of the coefficients of these functions be zero to satisfy the equation at all times, we get the required values of γ and ω_0' in Eq. 2.23. Actually, the value of γ is the same as obtained in Eq. 2.22. The damping makes the ω_0' lower than ω_0.

When there is no friction, i.e., $\gamma = 0$, the solution reduces to the harmonic motion discussed earlier, where A is the amplitude and ϕ the phase angle. The damping produces an exponential decay of the amplitude and also causes a reduction of the frequency of oscillation. If the friction constant is large enough to that $\omega_0' = 0$, the motion is non-oscillatory and the oscillator is then said to be *critically damped*. If $\gamma > \omega_0$, the frequency ω_0' formally becomes imaginary and the solution has to be reexamined, as will be done shortly. As it turns out, the general solution then consists of a linear combination of two decaying exponential functions.

2.3.3 Use of Complex Variables. Complex Frequency

With the use of complex variables in solving the damped oscillator equation, there is no need for the kind of patchwork that was used in Section 2.3.2. We merely let the mathematics do its job and present us with the solution.

It should be familiar by now, that the complex amplitude $\xi(\omega)$ of $\xi(t)$ is defined by

$$\xi(t) = \Re\{\xi(\omega)e^{-i\omega t}\}. \tag{2.24}$$

The corresponding complex amplitudes of the velocity and the acceleration are then $-i\omega\xi(\omega)$ and $-\omega^2\xi(o)$ and if these expressions are used in Eq. 2.23 we obtain the following equation for ω

$$\omega^2 + i2\gamma - \omega_0^2 = 0 \tag{2.25}$$

in which $\gamma = R/2M$.

Formally, the solution to this equation yields complex frequencies

$$\omega = -i\gamma \pm \sqrt{\omega^2 - \gamma^2}. \tag{2.26}$$

The general solution is a linear combination of the solutions corresponding to the two solutions for ω, i.e.,

$$\xi(t) = e^{-\gamma t}\Re\{A_1 e^{i\omega_0' t} + A_2 e^{-i\omega_0' t}\}, \tag{2.27}$$

where $\omega_0' = \sqrt{\omega^2 - \gamma^2}$ and A_1 and A_2 are complex constants to be determined from initial conditions. We distinguish between the three types of solutions which correspond to $\gamma < \omega_0$, $\gamma > \omega_0$, and $\gamma = \omega_0$.

Oscillatory decay, $\gamma < \omega_0$. In this case, ω_0' is real, and the oscillator is sometimes referred to as *underdamped*; the general solution takes the form

$$\xi(t) = A\, e^{-\gamma t} \cos(\omega_0' t - \phi) \tag{2.28}$$

which is the same as in Eq. 2.23. The constants A and ϕ are determined by the initial conditions of the oscillator.

Overdamped oscillator, $\gamma > \omega_0$. The frequency ω_0' now is purely imaginary, $\omega_0' = i\sqrt{\gamma^2 - \omega_0^2}$, and the two solutions to the frequency equation (6.18) become

$$\omega_+ = -i(\gamma - \sqrt{\gamma^2 - \omega_0^2}) \equiv -i\gamma_1$$
$$\omega_- = -i(\gamma + \sqrt{\gamma^2 - \omega_0^2}0 \equiv -i\gamma_2. \tag{2.29}$$

The motion decays monotonically (without oscillations) and the corresponding general solution for the displacement is the sum of two exponential functions with the decay constants γ_1 and γ_2,

$$\xi(t) = C_1 e^{-\gamma_1 t} + C_2 e^{-\gamma_2 t}, \tag{2.30}$$

where the two (real) constants are to be determined from the initial conditions.

Critically damped oscillator, $\gamma = \omega_0$. A special mention should be made of the 'degenerate' case in which the two solutions to the frequency equation are the same, i.e., when $\gamma_1 = \gamma_2 = \omega_0$. To obtain the general solution for ξ in this case requires some thought since we are left with only one adjustable constant. The general solution must contain two constants so that the two conditions of initial displacement and velocity can be satisfied (formally, we know that the general solution to a second order differential equation has two constants of integration). To obtain the general solution we can proceed as follows.

We start from the overdamped motion $\xi = C_1 \exp(-\gamma_1 t) + C_2 \exp(-\gamma_2 t)$. Let $\gamma_2 = \gamma_1 + \Delta$ and denote temporarily $\exp(-\gamma_2 t)$ by $f(\gamma_2, t)$. Expansion of this function to the first order in Γ yields $f(\gamma_2, t) = f(\gamma_1, t) + (\partial f / \partial \Delta)_0 \Delta = \exp(-\gamma t) - t\Delta \exp(-\gamma_1 t)$. The expression for the displacement then becomes $\xi = (C_1 + C_2) \exp(-\gamma_1 t) - t(C_2 \Delta) \exp(-\gamma_1 t)$, or

$$\xi = (C + Dt)e^{-\omega_0 t}, \tag{2.31}$$

where $C = C_1 + C_2$ and $D = -C_2 \Delta$, C_2 being adjusted in such a way that D remains finite as $\Delta \to 0$. Direct insertion into the differential equation $\ddot{\xi} + 2\gamma\dot{\xi} + \omega_0^2\xi = 0$ (Eq. 2.23) shows that this indeed is a solution when $\gamma = \omega_0$.

In summary, the use of complex amplitudes in solving the frequency equation (6.18) and accepting a complex frequency as a solution, we have seen that it indeed has a physical meaning; the real part being the quantity that determines the period of oscillation (for small damping) and the imaginary part, the damping. In this manner, the solution for the displacement emerged automatically from the equation of motion.

2.3.4 Problems

1. **Oscillatory decay of damped oscillator**

   The formal solution for the displacement of a damped oscillator in free motion is given in Eq. 2.27, in which A_1 and A_2 are two independent complex constants, each with a magnitude and phase angle. Show in algebraic detail that the general solution can be expressed as in Eq. 2.28, in which A and ϕ are real constants.

2. **Critically damped oscillator. Impulse response**

   In the degenerate case of a damped oscillator when $\gamma = \omega_0$ so that $\omega_0' = 0$, the general solution for the displacement is

   $$\xi(t) = (A + Bt)e^{-\omega_0 t}, \tag{2.32}$$

   where A and B are constants to be determined by the initial conditions.
   (a) Prove this by direct insertion into the equation of motion.
   (b) The oscillator, initially at rest, is given a unit impulse at $t = 0$. Determine the subsequently motion.

3. **Paths in the complex plane**

   It is instructive to convince oneself that as γ increases, the two solutions for the complex frequency in Example 9 in Ch.11 follow along circular paths in the complex plane when the motion is oscillatory. They meet on the negative imaginary axis when the damping is critical, i.e., $\gamma = \omega_0$, and then move apart in opposite direction along the imaginary axis. Sketch in some detail the paths and label the values of γ at critical points, as you go along.

4. **Impulse response. Maximum excursion**

   The oscillator in Example 9 in Ch.11 is started from rest by an impulse of 10 Ns. For the underdamped, critically damped, and overdamped conditions in (a) and (c),

(a) determine the maximum excursion of the mass element and the corresponding time of occurrence and

(b) determine the amount of mechanical energy lost during this excursion.

5. **Overdamped harmonic oscillator**

(a) With reference to the expressions for the two decay constants in Eq. 2.29 show that if $\gamma \gg \omega_0$ we obtain $\gamma_1 \approx K/R$ and $\gamma_2 \approx R/M$.

(b) What is the motion of an oscillator, started from rest with an initial displacement $\xi(0)$, in which R is so large that the effect of inertia can be neglected?

(c) Do the same for an oscillator, started from $\xi = 0$, with an initial velocity $u(0)$, in which the effect of the spring force can be neglected in comparison with the friction force.

2.4 Forced Harmonic Motion

2.4.1 Without Complex Amplitudes

To analyze the forced harmonic motion of the damped oscillator, we add a driving force $F(t) = |F| \cos(\omega t)$ on the right-hand side of Eq. 2.23. The corresponding steady state expression for the displacement is assumed to be $\xi = |\xi| \cos(\omega t - \phi)$. Inserting this into the equation of motion, we get for the first term $-M\omega^2 |\xi| \cos(\omega t - \phi)$, for the second, $-R\omega |\xi| \sin(\omega t - \phi)$, and for the third, $K |\xi| \cos(\omega t - \phi)$. Next, we use the trigonometric identities $\cos(\omega t - \phi) = \cos(\omega t) \cos \phi + \sin(\omega t) \sin \phi$ and $\sin(\omega t - \phi) = \sin(\omega t) \cos \phi - \cos(\omega t) \sin \phi$ and express each of these three terms as a sum of $\cos(\omega t)$- and $\sin(\omega t)$-terms. Since we have only a $\cos(\omega t)$-term on the right-hand side, the sum of the sine terms on the left-hand side has to be zero in order to satisfy the equation at all times and the amplitude of the sum of the cosine terms must equal $|F|$. These conditions yield two equations from which $|\xi|$ and ϕ can be determined. It is left as a problem to fill in the missing algebraic steps and show that

$$|\xi| = \frac{|F|/\omega}{\sqrt{R^2 + (K/\omega - \omega M)^2}}$$

$$\tan \phi = \omega R/(K - \omega^2 M). \tag{2.33}$$

At very low frequencies, the displacement approaches the static value $|\xi| \approx |F|/K$ and is in phase with the driving force. At resonance, $|\xi| = |F|/(\omega R)$ which means that the velocity amplitude is $|u| = |F|/R$, with the velocity in phase with the driving force. At very high frequencies where the inertia dominates, the phase angle becomes $\approx \pi$; the displacement is then opposite to the direction of the driving force.

A good portion of the algebra has been skipped here, and what remains is a deceptively small amount. This should be kept in mind when it is compared with the complex amplitude approach used in Section 2.4.2.

The driving force $F(t) = |F| \cos(\omega t)$ and the 'steady state' motion it produces are idealizations since they have no beginning and no end. A realistic force would be one which is turned on at time $t = 0$, say, and then turned off at a later time. This introduces additional motions, so called *transients*, which have to be added to the steady state motion. An obvious indication of the shortcoming of the present analysis is that it leads to an infinite displacement at resonance if the damping is zero.

This will be clarified when we again analyze forced motion, this time with a general driving force $F(t)$ and the use of the impulse response of the oscillator (see Eq. 2.56).

2.4.2 With Complex Amplitudes

The use of complex amplitudes to solve the problem of the forced harmonic mass-spring oscillator will now be demonstrated. We choose the driving force to be $F(t) = |F| \cos(\omega t)$. The corresponding complex amplitude is $F(\omega) = |F| \exp(i\phi) = |F|$ since the phase angle ϕ is zero. The terms in the equation of motion for displacement $\xi(t)$ are replaced by the corresponding complex amplitudes and from what we have said about these amplitudes for velocity and acceleration, this complex amplitude equation of motion takes the form shown in Eq. 2.34,

$$
\boxed{
\begin{array}{c}
\textit{Forced harmonic motion} \\
(-\omega^2 M - i\omega R + K)\xi(\omega) = F \\
\xi(\omega) = F/(K - \omega^2 M - i\omega R) \\
|\xi| = |F|/\sqrt{(K - \omega^2 M)^2 + (\omega R)^2}
\end{array}
}
\qquad (2.34)
$$

Thus, as has been remarked earlier, the differential equation for the displacement $\xi(t)$ is replaced by an algebraic equation for the complex amplitude $\xi(\omega)$, as shown. Since the amplitude of a complex number is $a + ib = \sqrt{a^2 + b^2}$, the expression for the magnitude $|\xi(\omega)|$ follows, i.e., the phase angle of the numerator in $\xi(\omega)$ is zero and the phase angle of the denominator is given by $\tan(\phi_d) = -\omega R/(K - \omega^2 M)$. The phase angle of the ratio is the difference between the two which means that the phase angle of the displacement is given by

$$
\tan\phi = -\tan(\phi_d) = \omega R/(K - \omega^2 M). \qquad (2.35)
$$

As the frequency goes to zero we note that $|\xi| \to |F|/K$ and $\phi \to 0$, corresponding to the static displacement of the spring in the oscillator. As the frequency increases, however, the amplitude $|\xi|$ normally increases toward a maximum $|F|/R$ at the resonance frequency $\omega_0 = \sqrt{K/M}$ and then decreases toward zero with increasing frequency. (For large values of R, corresponding to an overdamped oscillator, the maximum turns out to be at $\omega = 0$.) The phase angle goes to π (i.e., the displacement has a direction opposite to that of the driving force).

Having obtained the complex amplitude, the time dependence of the displacement is

$$
\xi(t) = |\xi| \cos(\omega t - \phi). \qquad (2.36)
$$

It is sometimes convenient to introduce dimension-less quantities in discussing results and we rewrite Eq. 2.35 accordingly. Thus, the normalized frequency is $\Omega = \omega/\omega_0$, where $\omega_0 = \sqrt{K/M}$ is the resonance frequency. Furthermore, $D = R/\omega_0 M = R\omega_0/K$, which we call the *loss factor* or *damping factor* is a normalized measure of the resistance. The inverse of D, $Q = 1/D$, is usually called the 'Q-value' of the oscillator (Q standing for 'quality,' supposedly a term from the early days of radio to describe the selectivity of circuits). As we shall see, it is a measure of the sharpness of the response curve in the vicinity of the maximum. Furthermore, the normalized displacement amplitude is expressed as ξ/ξ', where $\xi' = |F|/K$ (i.e., the displacement obtained in a static compression of the spring with the force amplitude $|F|$). In terms of these quantities, Eq. 2.35 takes the form

$$|\xi| = \xi'/\sqrt{(1-\Omega^2)^2 + (D\Omega)^2}$$
$$\tan\phi = D\Omega/(1-\Omega^2), \tag{2.37}$$

where $\Omega = \omega/\omega_0$, $D = \omega_0 R/K = R/(\omega_0 M)$, and $\xi' = |F|/K$.

At resonance, $\Omega = 1$, we have $|x|/x' = 1/D = Q$ (i.e., the displacement $|x|$ is Q times the 'static' displacement ξ' at $\Omega = 0$).

Complex Spring Constant

The equation of motion (2.34) can be brought into the same form as for the frictionless oscillator if we introduce a complex spring constant $K_c = K - i\omega R$, in which case the complex amplitude equation of motion becomes $(-\omega^2 M + K_c)\xi = F$.

2.4.3 Impedance and Admittance

The impedance Z of the oscillator is the ratio of the complex amplitudes of the driving force and the velocity, $Z(\omega) = F(\omega)/u(\omega)$. It is a complex number $Z = |Z|\exp(i\psi)$ with the magnitude $|Z|$ and the phase angle ψ. If the phase angle of F is zero, the phase angle of the complex velocity $u(\omega)$ becomes $\phi = -\psi$.

With $u(\omega) = -i\omega\xi(\omega)$, it follows from Eq. 2.34 that

$$\boxed{\begin{array}{c} \textit{Impedance of an oscillator} \\ u(\omega) = F(\omega)/Z(\omega) \\ Z = R - i\omega M + iK/\omega = R + i(K/\omega)[1-(\omega/\omega_0)^2] \equiv R + iX \end{array}} \tag{2.38}$$

where $\omega_0 = \sqrt{K/M}$. The magnitude and phase angle of the impedance are given by

$$|Z| = \sqrt{R^2 + X^2}$$
$$\tan\psi = X/R, \tag{2.39}$$

where R is the resistance and $X = K/\omega - \omega M$, the reactance of the oscillator. At the resonance frequency, we have $X = 0$ and the impedance is purely resistive. The velocity is then in phase with the driving force. At frequencies much below the resonance frequency, the impedance is dominated by the spring and the displacement is $\xi \approx F/K$, the same as in a static deformation and $\psi = \pi/2$. The displacement is then in phase with the force and the velocity runs ahead of it by the angle $\approx\pi/2$ (the phase angle of velocity is $-\psi = -\pi/2$). At frequencies above the resonance, inertia dominates and we get $\psi \approx -\pi/2$; the phase angle of velocity is then a lag of $\pi/2$.

The magnitude of the velocity amplitude is the ratio of the magnitudes of the driving force and the impedance, and the phase angle is the difference between the phase angles of force and the impedance.

The inverse of the impedance is called the *admittance* or sometimes the *mobility*, $Y = 1/Z$. The real and imaginary parts of Y are called the *conductance* C and the *susceptance* S, $Y = C + iS$.

One useful aspect of these concepts is that the impedance or admittance of several mechanical components can be determined in terms of the impedances (admittances) of the components, which are normally known in advance. For example, if two mechanical oscillators are in 'parallel' in such a way that their displacements are the same, the total driving force will be the sum of the individual forces required to drive each oscillator separately and the impedance of the combination is the sum of the individual impedances.

2.4.4 Power Transfer

With a driving force $F(t) = |F| \cos \omega t$ and the velocity of the oscillator $u(t) = |u| \cos(\omega t - \phi)$, the time average power delivered by the force is

$$\Pi = (1/T) \int_0^T F(t)u(t)dt. \tag{2.40}$$

Using the identity $\cos A \cos B = (1/2)\cos(A + B) + (1/2)\cos(A - B)$ we have $\cos \omega t \cos(\omega t - \phi) = (1/2)\cos(2\omega t - \phi) + (1/2)\cos \phi$. The first of these terms does not contribute to the time average and

$$\Pi = (1/2) |F||u| \cos \phi. \tag{2.41}$$

With $|F|$ given, the complex amplitude analysis yields $|u|$ and ϕ so that Π can be calculated. If $|F|$ and $|u|$ are rms values, the factor 1/2 should be removed.

If we introduce $F = Zu$ and $|Z| \cos \psi = |Z| \cos \phi = R$, it follows that the power can be written $\Pi = (1/2)R|u|^2$.

Frequently, the power is expressed directly in terms of the complex amplitudes of F and u, and we include also this version, as follows. After having obtained the complex amplitude of a quantity, such as velocity $u(\omega)$, the real time function is given by $u(t) = \Re\{u(\omega)\exp(-i\omega t)\}$. It can also be expressed as

$$u(t) = (ue^{-i\omega t} + u^* e^{i\omega t})/2, \tag{2.42}$$

where u^* is the complex conjugate of u (see Appendix B). The time average of the power is

$$\Pi = \langle F(t)u(t) \rangle = (1/4)\langle (Fe^{-i\omega t} + F^* e^{i\omega t})(ue^{-i\omega t} + u^* e^{i\omega t}) \rangle$$
$$= (1/4)(Fu^* + F^* u) = (1/2)\Re\{Fu^*\}. \tag{2.43}$$

The time average of the terms containing time is zero. If rms values are used in the last expression, the factor 1/2 should be eliminated.

If we introduce $F = Zu$ into the equation, we get for the power

$$\Pi = (1/2)\Re\{Zuu^*\} = (1/2)R|u|^2 \quad (|u| \text{ rms}), \tag{2.44}$$

where we have used $uu^* = |u|^2$, $\Re\{Z\} = R$. Again, if the rms values are used, the factor (1/2) should be removed.

2.4.5 Acoustic Cavity Resonator (Helmholtz Resonator)

The derivation of the spring constant of an air spring in Eq. 2.17 for a uniform tube can easily be extended to an arbitrarily shaped air volume. In particular, consider the volume of a bottle or flask, as shown in Fig. 2.6. The volume of the flask is V, the

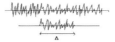

Figure 2.6: Acoustic cavity resonator (Helmholtz resonator).

area of the neck, A, and the length of the neck is ℓ. We are all familiar with how to make a bottle 'sing' by blowing across its mouth exciting the (fundamental) mode of oscillation. The resonator, often referred to as a Helmholtz resonator,[5] behaves in much the same way as a mass-spring oscillator with the air in the neck representing the mass and the air in the volume V acting like a spring.

With the density of the air denoted ρ, the mass in the neck is $M = A\rho\ell$. There is experimental evidence that the compression of the air in the volume V is adiabatic and the relation between the pressure and the volume is then $PV^\gamma = $ const, where $\gamma = C_p/C_v$ is the ratio of the specific heats at constant pressure and constant volume. For air it is ≈ 1.4. From this equation of state, it follows that a change in volume dV produces a change in pressure dP such that $dP/P = -\gamma dV/V$. For a small displacement of the air (inwards) ξ, we get $dV = -A\xi$ so that $dP = (\gamma AP/V)\xi$. The restoring force on the mass plug is $AdP = (\gamma A^2 P/V)\xi$ and the equation of motion of the air plug is $A\ell\rho\ddot{\xi} = -(\gamma A^2 P/V)\xi$, or

$$\ddot{\xi} + \omega_0^2\xi = 0, \qquad (2.45)$$

where

$$\omega_0 = \sqrt{(\gamma P/\rho)(A/V\ell)} = c\sqrt{A/(V\ell)}. \qquad (2.46)$$

The quantity c is the speed of sound in air, $c = \sqrt{\gamma P/\rho}$, a value which will be derived in the next chapter.

The frequency obtained here is the so-called *fundamental* frequency, corresponding to the lowest mode of oscillation in which the sound pressure throughout the main volume of the bottle can be considered approximately uniform. Actually, in the mass-spring oscillator we made the similar assumption that the force is the same along the spring (i.e., the force transmitted from the beginning of the spring is the same as that which appears at the end of the spring). This assumption is a consequence of the omission of the mass of the spring.

Like any other enclosure (such as a concert hall), the bottle has many other acoustic modes of oscillation with corresponding characteristic frequencies. We shall have occasion to discuss this problem in later chapters.

---

[5]Hermann Ludwig Ferdinand von Helmholtz, 1821-1894, Baron, German physician, physicist, mathematician, and philosopher.

The expression for the resonance frequency in Eq. 2.46 can be improved somewhat by including an end correction to the length ℓ which is of the order of the neck diameter. It accounts for the 'induced mass' in the oscillatory flow in the vicinity of the two ends of the neck.

2.4.6 Torsion Oscillator

A rod, clamped at one end, is acted on by a torque applied at the other end. It produces an angular displacement about the axis proportional to the torque, at least for sufficiently small displacements. The ratio of the torque and the angular displacement at the point of application of the torque is called the *torsion constant* τ. It is the analog of the spring constant and generally should be regarded as an experimentally determined quantity, although it readily can be calculated for a uniform circular rod in terms of the elastic *shear modulus* G of the rod. If the length of the rod is L and the radius a and if the angular displacement over this length is θ, the shear stress at a radial position r is $Gr\theta/L$. The corresponding torque is then

$$\tau = \int_0^a (Gr^2\theta/L)2\pi r\,dr = \beta\theta$$
$$\beta = (\pi a^4/2L)G. \tag{2.47}$$

The physical dimension of the spring constant is force divided by length and for the torsion constant β, it is torque per unit angle, i.e., force multiplied by length, since the angle is dimension-less.

The shear modulus G in N/m^2 is, for
Steel: 8.11×10^{10}
Aluminum: 2.4×10^{10}
Tungsten, drawn: 14.8×10^{10}.

To apply this result to a torsional oscillator, we consider a vertical rod which is held fixed at its upper end and supports a body, such as a circular disc or a dumbbell, at its lower end. The moment of inertia I of this body is large enough so that the moment of inertia of the rod can be neglected. The body is given an angular displacement θ and then released. The equation for θ in the ensuing motion is

$$I\ddot{\theta} = -\beta\theta$$
$$\ddot{\theta} + \omega_0^2\theta = 0, \tag{2.48}$$

where $\omega_0 = \sqrt{\beta/I}$. This is the harmonic oscillator equation, and the time dependence of the rotation angle θ is analogous to displacement in the mass-spring oscillator.

With the 'rod' being a thin wire or filament, the torsion constant can be made extremely small, and minute torques can be measured from the angular deflection. The deflection can be 'amplified' by means of a light beam reflected from a mirror attached to the oscillator. This technique has been used in sensitive galvanometers and for the measurement of light pressure. In the light pressure experiment it is advantageous to pulse the light at a frequency equal to the natural frequency of the torsion oscillator to further increase the sensitivity.

2.4.7 Electro-mechanical Analogs

In this section we shall comment briefly on the analogy between a mechanical and an electrical oscillator 'circuit.' The former is simply a mass M connected to a spring with spring constant K and acted on by a friction force Ru, proportional to the velocity u. This friction force can be provided by a 'dashpot' damper, as described in connection with Eq. 2.23.

The driving force $F(t)$ is applied to M, and with $\xi(t)$ being the displacement of M from the equilibrium position and $u = \dot{\xi}$ the velocity, the equation of motion is

$$M\dot{u} + Ru + K\xi = F \quad \text{or} \quad M\ddot{\xi} + R\dot{\xi} + K\xi = F. \tag{2.49}$$

The analogous electrical system consists of an inductance L, a resistance R, and a capacitance C, in series and a driving voltage $V(t)$. With the current through the circuit denoted $I(t)$ and the charge on the capacitor by $q(t)$, we have $I = \dot{q}$ and the voltages across the inductance, resistance, and the capacitance are $L\dot{I}$, RI, and q/C, respectively. The sum of these must equal the applied voltage $V(t)$, i.e.,

$$L\dot{I} + RI + q/C = V \quad \text{or} \quad L\ddot{q} + R\dot{q} + (1/C)q = V. \tag{2.50}$$

A comparison of these equations leads to the following correspondence between mechanical and electrical quantities: Displacement $\leftrightarrow$ electric charge, velocity $\leftrightarrow$ current, mass $\leftrightarrow$ inductance, mechanical resistance $\leftrightarrow$ electrical resistance, compliance $\leftrightarrow$ capacitance, and force $\leftrightarrow$ voltage.

Other equivalent quantities are

$\omega_0 = \sqrt{K/M} = \sqrt{1/(MC)} \Longleftrightarrow \omega_0 = \sqrt{1/(LC)}$

Kinetic energy: $Mu^2/2 \Longleftrightarrow$ Magnetic energy: $LI^2/2$

Potential energy: $K\xi^2/2 = \xi^2/(2C) \Longleftrightarrow$ Electric energy: $q^2/2C$

Power input: $Fu \Longleftrightarrow$ Power input: VI

2.4.8 Problems

1. **Vibration isolation**

   In Example 11 in Ch.11 it is stated that the driving force transmitted to the floor was smaller than the driving force if the driving frequency is larger than the resonance frequency by a factor $\sqrt{2}$. Prove this statement and plot a curve of the ratio of the transmitted force and the driving force as a function of $\Omega = \omega/\omega_0$. This is an important problem in noise and vibration control.

2. **Power transfer**

   What is the time average of the oscillatory power generated by the imbalance of the fan in Example 11 in Ch.11?

3. **Admittance and power transfer**

   Eq. 2.40 indicates that the power transfer to an oscillator can be expressed as $(1/2)R|u|^2$, where $|u|$ is the velocity amplitude. Show that the power can be written also as $(1/2)Y_r|F|^2$, where Y_r is the conductance, i.e., the real part of the admittance.

4. **Helmholtz resonator**

   A bottle with a diameter of $D = 10\,\text{cm}$ and a height $H = 20\,\text{cm}$ has a neck with a diameter $d = 1\,\text{cm}$ and a length $h = 4\,\text{cm}$. What is the resonance frequency of the air in this bottle? The speed of sound in air is $c = 340\,\text{m/sec}$.

5. **Reverberation of a Helmholtz resonator**

The resonator mode in Problem 4 is excited by a pressure impulse. It is found that the pressure amplitude decays by a factor of 10 in 10 full periods. What is the Q-value of the resonator?

6. **Forced harmonic motion**

Fill in the missing algebraic steps to prove the expression for the steady state response of an oscillator in Eq. 2.33.

2.5 Impulse Response and Applications

As already pointed out, the steady state motion in Eq. 2.33 was produced by a harmonic driving force $F(t) = |F| \cos(\omega t)$ which is an idealization since it has no beginning and no end. We now turn to the response of the oscillator to a more general and realistic driving force.

We start by considering the motion of a damped mass-spring oscillator after it is set in motion by an impulse I at time $t = t'$. We let the impulse have unit strength. Since the impulse is instantaneous, the displacement immediately after the impulse will be $\xi = 0$ and the velocity, $\dot{\xi} = 1/M$. In the subsequent motion, the oscillator is free from external forces but influenced by a spring force and a resistive force $-Ru$ proportional to the velocity u. Then, for an underdamped oscillator, the displacement will be of the form given in Eq. 2.23, i.e.,

$$\xi(t) = A \, e^{-\gamma t} \cos(\omega_0' \phi), \qquad (2.51)$$

where $\gamma = R/2M$, $\omega_0' = \sqrt{\omega_0^2 - \gamma^2}$, and $\omega_0 = \sqrt{K/M}$. As before, K is the spring constant. The amplitude A and the phase angle ϕ are determined by the displacement $\xi = 0$ and the velocity $\dot{\xi} = 1/M$ at $t = t'$.

In order to make the displacement zero at $t = t'$ we must have $\phi = \pi/2$ which means that the displacement must be of the form
$\xi = A \exp[-\gamma(t - t')] \sin[(\omega_0'(t - t')]$.
The corresponding velocity is
$\dot{\xi} = A \exp[-\gamma(t - t')][\omega_0' \cos[\omega_0'(t - t')] - \gamma \sin[\omega_0'(t - t')]$.
To make this velocity equal to $1/M$ at $t = t'$ requires that $A = 1/(M\omega_0')$. In other words, the *impulse response function*, sometimes called the *Green's function* for displacement is

$$\boxed{\begin{array}{c} \textit{Impulse response function} \\ h(t, t') = (1/\omega_0' M) \, e^{-\gamma(t-t')} \sin[\omega_0'(t - t')] \quad \text{for } t > t' \\ h(t, t') = 0 \qquad \text{for } t < t' \end{array}} \qquad (2.52)$$

$[\gamma = R/2M \cdot \omega_0' = \sqrt{\omega_0^2 - \gamma^2} \cdot \omega_0^2 = K/M]$.

The dependence on t and t' is expressed through the combination $(t - t')$ only (i.e., the time difference between the 'cause' and the 'effect'). Since we accept the *causality principle* that the effect cannot occur before the cause, we have added $h(t, t') = 0$ for $t < t'$ in the definition of the impulse response function $h(t, t')$.

2.5.1 General Forced Motion of an Oscillator

The reason for the particular importance of the impulse response function is that the response to an arbitrary driving force can be easily constructed from it.

To prove this, we consider first the displacement resulting from two unit impulses delivered at t' and t''. Although not necessary, we assume for simplicity that the displacement and the velocity of the oscillator are zero when the first impulse is delivered at $t = t'$. Then, by definition, the displacement that results at time t is the impulse response function $h(t, t')$.

At the later time t'' when the second impulse is delivered, the displacement and the velocity are both different from zero. It should be realized, however, that the *change* in the displacement produced by the second impulse does not depend on the state of motion when the impulse is delivered (because the system is linear) and the total displacement at time t will be $\xi(t) = h(t, t') + h(t, t'')$. If the impulses at t' and t'' have the values I' and I'', the displacement at t will be $I'h(t, t') + I''h(t, t'')$.

We can now proceed to the displacement produced by a general driving force $F(t)$. The effect of this force is the same as that of a succession of impulses of magnitude $F(t')\Delta t'$ over the entire time of action of the force up to the time t. The displacement, at time t, produced by one of these impulses is $h(t, t')F(t')\Delta t'$ with an analogous expression for any other impulse; this follows from the discussion of the impulse response function $h(t, t')$. The sum of the contributions from all the elementary impulses can be expressed as the integral

$$
\boxed{
\begin{array}{c}
\textit{Response to a driving force } F(t) \\
\xi(t) = \int_{-\infty}^{t} h(t, t')F(t')dt' = \int_{0}^{\infty} F(t - \tau)h(\tau)d\tau
\end{array}
}
\tag{2.53}
$$

$[h(t, t')$: See Eq. 2.52. $\tau = t - t']$.

The range of integration for t' is from $-\infty$ to t to cover all past contributions. Frequently, it is convenient to introduce a new variable, $\tau = t - t'$, in which case the range of integration is from $\tau = 0$ to infinity, as indicated in Eq. 2.53.

The validity of this result relies on the linearity of the system so that the incremental change of the displacement will be the same for a given impulse independent of the state of motion when the impulse is delivered.

2.5.2 Transition to Steady State

As an example of the use of Eq. 2.53, we consider a driving force which is turned on at $t = 0$ and defined by

$$
F(t) = F_0 \cos(\omega t) \quad t > 0
$$
$$
F(t) = 0 \quad t < 0.
\tag{2.54}
$$

Since the driving force is zero for $t < 0$, the lower bound of the range of integration in Eq. 2.53 can be set equal to 0. Using the expression for the impulse-response function in Eq. 2.52, we then get

$$
\xi(t) = \xi_0 e^{-\gamma t} \int_{0}^{t} e^{\gamma t'} \sin[\omega_0'(t - t')] \cos(\omega t')dt',
\tag{2.55}
$$

where $\xi_0 = F_0/(\omega_0'M) = A(\omega_0^2/\omega_0')$ and $A = F_0/K$. The integration is elementary and it is left as a problem to show that

$$\xi/A = \frac{1}{\sqrt{(1-\Omega^2)^2 + (D\Omega)^2}}[\cos(\omega t - \phi) - (\omega_0/\omega_0')e^{-\gamma t}\cos(\omega_0't - \beta)], \quad (2.56)$$

where $\Omega = \omega/\omega_0$, $\tan\phi = D\Omega/(1-\Omega^2)$, $\tan\beta = \gamma(1+\Omega^2)/\omega_0'(1-\Omega^2)$, $D = R/(\omega_0 M)$, and $A = F_0/K$.

The solution is the sum of two parts. The first, the *steady state solution*, has the same frequency ω as the driving force and its amplitude remains constant. The second part, often referred to as the *transient*, decays exponentially with time and can be ignored when $\gamma t >> 1$; it has a frequency ω_0', i.e., the frequency of free oscillations.

For small values of D, the maximum amplitude occurs very close to $\Omega = \omega/\omega_0 = 1$ but as the damping increases, the maximum shifts toward lower frequencies.

2.5.3 Secular Growth

There are several aspects of the solution in Eq. 2.56 that deserve special notice. One concerns the response of an undamped oscillator when the frequency of the driving force equals the resonance frequency, i.e., $\omega = \omega_0$. Our previous analysis dealt only with the steady state response from the very start and, in the absence of damping, yielded nothing but an infinite amplitude at resonance. The present approach shows how the amplitude grows with time toward the infinite value at $t = \infty$. With $\gamma = 0$, i.e., $D = 0$ and $\omega_0' = \omega_0$, the integral in Eq. 2.55 can be evaluated in a straight-forward manner and we obtain, with $F_0/(\omega_0 M) = F_0\omega_0/K = A\omega_0$,

$$\xi(t) = A\int_0^t \sin[\omega_0(t-t')]\cos(\omega_0 t')dt' = A(\omega_0 t/2)\sin(\omega_0 t). \quad (2.57)$$

The amplitude of this motion grows linearly with time, *secular growth*, toward the steady state value of infinity.[6]

2.5.4 Beats Between Steady State and Transient Motions

If the frequency ω of the driving force in Eq. 2.54 is not equal to the resonance frequency of the oscillator and if the damping is sufficiently small, the transition to steady state exhibits 'beats', i.e., variations in the amplitude. The beats result from the interference between the steady state motion with the frequency ω of the driving force and the transient motion with the frequency ω_0' of the free motion of the oscillator (see Eq. 2.56). Both of these motions are present during the transition to steady state. The curves shown refer to $\omega = 1.1\omega_0$ and the values of the damping factor $D = R/(\omega_0 M) = 0.01$ and 0.04 corresponding to the Q-values of 100 and 25. The interference between the two motions periodically goes from destructive to constructive as the phase difference $(\omega - \omega_0')t \approx (\omega - \omega_0)t$ between the two motions

[6]The general solution in Eq. 2.56 reduces to Eq. 2.57 for $\gamma = 0$ and $\omega = \omega_0$ (i.e., $\Omega = 1$). Actually, for these values the expression becomes of the form 0/0 and we have to determine the limit value as γ goes to zero, using $\exp(-\gamma t) \approx 1 - \gamma t$. Then, the steady state term is canceled out, and, since $\phi = \beta = \pi/2$ and $D = 2\gamma/\omega_0$, the remaining term indeed reduces to Eq. 2.57.

increases with time. With $\Delta\omega = \omega - \omega'_0 = 0.1\,\omega_0$, it is increased by 2π in a time interval Δt given by $\Delta\omega\Delta t = 2\pi$ which, in the present case, yields $0.1\,\omega'_0\Delta t = 2\pi$ or $\Delta t/T'_0 \approx 10$, where $T'_0 \approx T_0$ is the period of free motion. Thus, Δt is the time interval between two successive maxima or minima in the resulting displacement function, which is consistent with the result shown in the figure.

At a driving frequency below the resonance frequency, a similar result is obtained. For example, with $\omega = 0.9\,\omega_0$, the curves are much like those in the figure except that they start out in the positive rather than the negative direction.

If the driving frequency is brought sufficiently close to the resonance frequency, the time interval between beats will be so large that the amplitude of the transient will be damped so much that the beats will be less pronounced. For a more detailed discussion of this question we refer to Example 13 in Ch.11.

2.5.5 Pulse Excitation of an Acoustic Resonator

A simple and instructive demonstration of beats involves an acoustic cavity resonator exposed to repeated wave trains (pulse modulated) of sound.

A microphone in the cavity of the resonator measures the sound pressure and the corresponding signal from the microphone can be displayed on one channel of a dual beam oscilloscope. On the other channel can be shown the input voltage to the loudspeaker which produces the incident sound. The amplitude of this sound is constant during the duration of each pulse train.

The time dependence of the sound pressure in the resonator is quite different from that of the incident wave. It starts to grow toward a steady state value, but before this value has been reached, the incident pulse is terminated and the sound pressure in the cavity starts to decay. This process is repeated for each pulse.

If the 'carrier' frequency of the incident sound is equal to the resonance frequency, the growth and (exponential) decay of the sound pressure in the resonator are monotonic. During the decay, the resonator re-radiates sound which can be heard as a 'reverberation' after the incident sound has been shut off.[7]

If the frequency of the incident sound is somewhat lower than the resonance frequency, beats resulting from the interference of the free and forced oscillations occur. A similar result is obtained if the carrier frequency is somewhat higher than the resonance frequency. The sound pressure in the resonator is now much smaller than that obtained at the resonance frequency.

2.5.6 Problems

1. **Impulse response functions**

   Determine the impulse response functions of an overdamped and a critically damped oscillator.

2. **Steady state response of a harmonic damped oscillator**

---

[7]Resonators of this kind were built into the walls under the seats in some ancient Greek open air amphitheaters.

Carry out in detail the algebraic steps required to derive Eq. 2.33 for the displacement of a harmonic oscillator driven by the a harmonic force $F(t) = |F|\cos(\omega t)$. Show also that it is consistent with the steady portion of the solution in Eq. 2.56.

3. **Forced motion of oscillator. Force of finite duration**

A force driving a harmonic oscillator starts at $t = 0$ and is of the form $F(t) = |F|\cos(2\pi t/T)$ for $t < T/2$ and zero at all other times. Determine the displacement of the oscillator if (a) $T = T_0/2$, (b) $T = T_0$, and (c) $T = 2T_0$, where T_0 is the resonance period of the oscillator.

2.6 Fourier Series and Fourier Transform

We summarize in this section some well-known mathematical relations.

2.6.1 Fourier Series

As already mentioned, one of the reasons for the importance of the harmonic motion is that any periodic motion can be expressed as a sum of harmonic motions with frequencies which are multiples of the fundamental frequency. For a rigorous discussion of this important result, we refer to standard mathematics texts. We shall present here merely a brief review with examples.

Consider first a function $F(t) = C_1 \cos(\omega_1 t) + C_2 \cos(2\omega_1 t)$, the sum of two harmonic functions, the first having the (fundamental) frequency ω_1 and the second, the frequency $2\omega_1$. The function is periodic with the period $T_1 = 2\pi/\omega_1$ and it is symmetric with respect to t (i.e., it is the same for positive and negative values of t). With the cos-functions replaced by sin-functions,

$F(t)$ would be anti-symmetric in t, changing sign with t.

Varying the coefficient C_1 and C_2 will produce different shapes of the function $F(t)$. Conversely, for a given $F(t)$, the coefficients can be determined in the following manner. To obtain C_1 we multiply both sides of the equation by $\cos(\omega_1 t)$ and integrate over one period. The only contribution on the right-hand side will come from the first term which is readily seen to be $T_1 C_1/2$; the integral of the second term is zero. Similarly, multiplying by $\cos(2\omega_1 t)$ and integrating yields $T_1 C_2/2$. Thus, $C_1 = (2/T_1)\int_0^{T_1} F(t)\cos(\omega_1 t)\,dt$ and $C_2 = (2/T_1)\int_0^{T_1} F(t)\cos(2\omega_1 t)\,dt$. In this example, the mean value is zero. Had we added a constant term C_0, it would have been expressed by $C_0 = (1/T_1)\int_0^{T_1} F(t)\,dt$.

We can proceed in analogous manner for a sum of an arbitrary number of harmonic terms being mixtures of cos- and sin-functions, each term being expressed as $\cos(n\omega_1 t - \phi_n)$. Thus, with

$$F(t) = \sum_0^{\infty} a_n \cos(n\omega t - \phi_n) \qquad (2.58)$$

the coefficients in the series can be expressed as

$$a_n/2 = (1/T_1)\int_0^T F(t)\cos(n\omega_1 t - \phi_n)\,dt \quad (n > 0)$$
$$a_0 = (1/T_1)\int_0^T F(t)\,dt. \qquad (2.59)$$

The mean value of the function is a_0 which corresponds to $n = 0$.

The expansion of the periodic function $F(t)$ in Eq. 2.59 can be expressed differently by use of Euler's identity from which we have $\cos(n\omega t - \phi_n) = [\exp(i(n\omega t - \phi_n)) + \exp(-i(n\omega t - \phi_n))]/2$. By letting n be both positive and negative, we can account for both the exponential terms in this expression and put

$$F(t) = \sum_{-\infty}^{\infty} A_n\, e^{-in\omega_1 t}, \tag{2.60}$$

where A_n is complex. Comparing this expansion with that in Eq. 2.59 it follows that with m being a positive number $A_m = (a_m/2)\exp(i\phi_m)$ and $A_{-m} = (a_m/2)\exp(-i\phi_m)$.

The coefficients in the expansion (2.60) are obtained by multiplying both sides by $\exp(in\omega_1 t)$ and integrating over one period,

$$A_n = (1/T) \int F(t) e^{in\omega_1 t}\, dt. \tag{2.61}$$

This expression has the advantage over Eq. 2.59 that it is automatically valid for both $n = 0$ and $n \neq 0$. The expansion (2.60) is often referred to as a 'two-sided' Fourier expansion and (2.59) as 'one-sided.' The two expansions are compared in the following example.

The Delta Function

Consider a square wave pulse of height H and width τ such that the 'area' is $H\tau = 1$ and let $\tau \to 0$ and $H \to \infty$ in such a way that $H\tau$ remains equal to 1. Such a pulse is called a *delta function* and if it is located at $t = 0$, it is denoted by $\delta(t)$. It has the property that it is zero for all values of t except at $t = 0$ where it is infinite in such a way that $\int \delta(t)dt = 1$. The integration can be extended from minus to plus infinity.

A very useful property of the delta function is $\int \delta(t)F(t)dt = F(0)$, the only contribution to the integral coming from $t = 0$ where $\delta(t)$ is not zero; the integral then becomes $F(0)\int \delta(t)dt = F(0)$. Similarly,

$$\int \delta(t - t_1)F(t)dt = f(t_1). \tag{2.62}$$

A pulse train of delta functions similar to that in the example above can be expressed as

$$F(t) = \sum_{-\infty}^{\infty} \delta(t - nT_1) = \sum A_n e^{-in\omega_1 t}. \tag{2.63}$$

As before, the coefficients are obtained by integrating both sides over one period. We select the period from $-T_1/2$ to $T_1/2$ to obtain

$$A_n = (1/T_1) \int_{-T_1/2}^{T_1/2} \delta(t) e^{in\omega_1 t}\, dt = 1/T_1. \tag{2.64}$$

In other words, the coefficients in the expansion are all the same, $1/T_1$. Eq. 2.63 then becomes

$$F(t) = \sum_{-\infty}^{\infty} e^{-in\omega_1 t} = \frac{1}{T_1}[1 + \sum_1^{\infty} (e^{-in\omega_1 t} + e^{in\omega_1 t})] = \frac{1}{T_1}[1 + 2\sum_1^{\infty} \cos(n\omega_1 t)].$$

(2.65)

The first term corresponds to $n = 0$. The result is consistent with the solution in the pulse train example above in the limit with $H\tau = 1$ and $\tau \to 0$.

In Chapter 7 we shall have occasion to use delta functions in the analysis of sound radiation from point or line forces moving along a circle and simulating sound radiation from a fan.

2.6.2 Fourier Transform

With reference to the pulse train example above, we shall explore what happens to the Fourier series if the period T_1 goes to infinity. We start with Eq. 2.65 which, with $\omega_1 = 2\pi \nu_1$, is expressed as

$$\boxed{\begin{array}{c} \textit{Fourier series} \\ F(t) = \sum_{-\infty}^{\infty} F_n e^{-i2\pi n\nu_1 t} \\ F_n = (1/T_1) \int_{-T_1/2}^{T_1/2} F(t) e^{i2\pi n\nu_1 t} dt \end{array}}$$

(2.66)

where $\nu_1 = 1/T_1$ and n, an integer.

The separation $\nu_1 = 1/T_1$ of the frequencies of two adjacent terms can be made as small as we wish by making T_1 large and the frequency $n\nu_1 = \nu$ can be regarded as a continuous variable. The average number of terms in the series that corresponds to a small frequency interval $\Delta\nu$ is then $\Delta\nu/\nu_1 = T_1\Delta\nu$. The sum over n can then be replaced by a sum over the frequency intervals $\Delta\nu_n$ with $T_1\Delta\nu$ terms in each.

With T_1 going to infinity, the complex amplitude F_n in Eq. 2.66 goes to zero in such a way that $F_n T_1$ is finite, which was demonstrated explicitly in the pulse train example where we had $F_n T_1 = H\tau$ in the limit $T \to \infty$. We denote $F_n T_1$ by $F(\nu)$, where ν refers to the average frequency in the interval $\Delta\nu$. The sum over n in Eq. 2.66 can then be replaced by an integral over ν with $T_1\Delta\nu$ terms in the frequency interval $\Delta\nu$ as shown. Then, with $T_1 F_n$ in Eq. 2.66 replaced by $F(\nu)$, we get

$$\boxed{\begin{array}{c} \textit{Fourier transform} \\ F(t) = \int_{-\infty}^{\infty} F(\nu) e^{-i2\pi\nu t} d\nu \\ F(\nu) = \int_{-\infty}^{\infty} F(t) e^{i2\pi\nu t} dt \end{array}}$$

(2.67)

These two equations are often referred to as the *Fourier transform pair*. The quantity $F(\nu)$ is called the Fourier amplitude of $F(t)$. By using the frequency ν rather than the angular frequency in these equations, they become symmetrical without a factor of $1/2\pi$ which otherwise would be needed in Eq. 2.67.

If $F(t) = \delta(t - t')$, the Fourier transform pair becomes

$$\delta(t - t') = \int_{-\infty}^{\infty} e^{-i2\pi\nu(t-t')} d\nu$$

$$\delta(\nu) = \int_{-\infty}^{\infty} \delta(t - t') e^{i2\pi\nu t} dt = e^{i2\pi\nu t'}.$$

(2.68)

It should be kept in mind that $F(\nu)$ is a complex number; the real physical significance of it will become apparent shortly in our discussion of energy spectra. For the time being, we merely comment on the meaning of $F(\nu)$ for negative values of ν; thus, since $F(t)$ is real, it follows from Eq. 2.67 that $F(-\nu)$ must equal $F^*(\nu)$ in order for the sum of the terms for negative and positive ν to be a real number, i.e.,

$$F(-\nu) = F^*(\nu), \tag{2.69}$$

where f^* is the complex conjugate of f. We recall that the complex conjugate of $f = |f| \exp(i\phi)$ is $f^* = |f| \exp(-i\phi)$ so that $ff^* = |F(\nu)|^2 = |F(-\nu)|^2$.

Example

As an example, let $F(t)$ be a single period of the square wave function with a height H, a width τ, and with the center at $t = 0$. Since the function is zero outside this region, the Fourier integral extends from $-\tau/2$ to $\tau/2$ and the complex Fourier amplitude becomes, from Eq. 2.67,

$$F(\nu) = H \int_{-\tau/2}^{\tau/2} e^{i2\pi \nu t} \, dt = (H\tau) \frac{\sin(X)}{X}, \tag{2.70}$$

where $X = \pi \tau \nu$.

The quantity $H\tau$ is the area under the pulse. If this area is kept constant and equal to 1 as $\tau \to 0$, we get the delta function, as discussed above,

$$\delta(t) = \int_{-\infty}^{\infty} e^{-i2\pi \nu t} \, d\nu$$
$$F(\nu) = \int_{-\infty}^{\infty} \delta(t) e^{i2\pi \nu t} dt = 1. \tag{2.71}$$

It is important to notice that the infinitely narrow spike represented by the delta function occurs at the precisely determined time $t = 0$. The Fourier amplitude $F(\nu)$, on the other hand, is the same ($=1$) at all frequencies so the frequency is indeterminate. To build a delta function, all frequencies have to be included with equal 'weight.' Conversely, a precisely determined frequency corresponds to a harmonic function which extends over all times.

For a duration τ of the pulse different from zero, the Fourier amplitude decreases in an oscillatory manner toward zero as the frequency goes to infinity, as shown in Eq. 2.70. We can define a characteristic width of $F(\nu)$ by the frequency $\Delta \nu$ of the first zero of $F(\nu)$. It occurs where $\sin(X) = \pi$ (i.e., at $\pi \tau \Delta \nu = \pi$ or $\tau \Delta \nu = 1$). With τ denoted by Δt, we obtain the relation

$$\Delta t \Delta \nu = 1 \tag{2.72}$$

which is sometimes called the *uncertainty relation*.

2.6.3   Spectrum Densities; Two-sided and One-sided

As already mentioned, the Fourier amplitude $F(\nu)$ generally is a complex number. The physical meaning of it becomes clear if we calculate its magnitude, or rather its square, $|F(\nu)|^2 = F(\nu)F^*(\nu)$ in terms of $F(t)$. The calculation is straight-forward but it looks a little awkward because of all the integral signs required when we use Eq. 2.67 for $f\nu)$ and $F^*(\nu)$, the latter obtained merely by changing the sign of i in the integrals.

The squared magnitude of the Fourier amplitude can then be written

$$|F(\nu)|^2 = \int_{-\infty}^{\infty} \int_{-\infty}^{\infty} F(t)e^{i2\pi\nu t}\, dt\, F(t')e^{i2\pi\nu t'}\, dt'. \qquad (2.73)$$

Integration over ν yields

$$\boxed{\begin{array}{c} \textit{Energy relation} \\ \int_{-\infty}^{\infty} |F(\nu)|^2\, d\nu = \int_{-\infty}^{\infty}\int_{-\infty}^{\infty} F(t)F(t')\, dt\, dt' \int_{-\infty}^{\infty} e^{-i2\pi\nu(t-t')}\, d\nu \\ = \int_{-\infty}^{\infty} F(t)F(t')\delta(t-t')\, dt' = \int_{-\infty}^{\infty} F^2(t)\, dt \end{array}}, \qquad (2.74)$$

where we have used Eq. 2.68.

This illustrates the physical meaning of the Fourier amplitude. The right hand side, apart from a constant, can be thought of as the total energy transfer to the system[8] and on the left-hand side the same energy is expressed as a distribution over frequency in terms of the Fourier spectrum density $|F(\nu)|^2$.

The (Fourier) spectrum density $S_2(\nu) = |F(\nu)|^2$ in Eq. 2.74 involves the integration over frequency from $-\infty$ to ∞ and is often referred to as the two-sided spectrum density. Since $|F(\nu)|^2 = F(\nu)F^*(\nu)$ is the same for positive and negative ν, the integration can be limited to only positive value of ν, i.e.,

$$\int_{-\infty}^{\infty} |F(\nu)|^2\, d\nu = 2\int_{0}^{\infty} |F(\nu)|^2\, d\nu \equiv \int_{0}^{\infty} S_1(\nu)\, d\nu, \qquad (2.75)$$

where $S_1\nu) = 2S_2(\nu)$ is the *one-sided spectrum density*.

Example. Fourier Spectrum of Oscillatory Decay

As an example, we calculate the Fourier spectrum $|F(\nu)|^2$ of the oscillatory decay of a damped harmonic oscillator started from rest with a given displacement $F(0) = A$ from equilibrium, as discussed in Section 2.3, Eq. 2.23. In this case $F(t)$ stands for the displacement of the oscillator,

$$F(t) = Ae^{-\gamma t}\cos(\omega_0' t) \qquad (2.76)$$

for $t > 0$ and $F(t) = 0$ for $t < 0$. As before, the frequency is $\omega_0' = \sqrt{\omega_0^2 - \gamma^2}$, where $\omega_0^2 = K/M$ and the decay constant $\gamma = R/2M$.

Since $F(t) = 0$ for $t < 0$, the lower limit in the integral for the Fourier amplitude $F(\nu)$ (see Eq. 2.67) can be taken to be zero, so that the Fourier amplitude becomes

$$F(\nu) = \int_0^{\infty} e^{-\gamma t}\cos(\omega_0' t)e^{i2\pi\nu t}\, dt = (F_0/2)\int_0^{\infty}[e^{i(\omega_0')+\omega+i\gamma)t} + e^{-i(\omega_0'-\omega-i\gamma)t}]$$

$$= F_0(\gamma - i\omega)/(\omega_0^2 - \omega^2 + 2i\gamma\omega). \qquad (2.77)$$

---

[8]If $F(t)$ is an electric current flowing through a unit resistance, the integral is the total energy dissipated in the resistance.

The corresponding Fourier spectrum is then

$$|F(\nu)|^2 = (F_0/\omega_0)^2 \frac{\Omega^2 + \Gamma^2}{(1 - \Omega^2)^2 + 4\Omega^2\Gamma^2}, \qquad (2.78)$$

where $\Omega = \omega/\omega_0$ and $\Gamma = \gamma/\omega_0$. The relationship between the width of the spectrum and the 'duration' of the signal should be noticed as an example of the 'uncertainty principle,' the longer the duration of the signal, the narrower the spectrum.

2.6.4 Random Function. Energy Spectra and Correlation Function

In our analysis of the motion of the linear oscillator, we started by considering a harmonic driving force. This was followed by a study of the response to an impulse and to an arbitrary driving force $F(t)$.

In practice, however, the driving forces involved frequently vary with time in an irregular or random manner, not expressible with a regular function of time, as indicated schematically in Fig. 2.7. We encounter such a time dependence in practically every aspect of acoustics. The force on a boundary from turbulent flow and the vibration of a wheel rolling over an irregular road surface are typical examples. Often a random oscillation is superimposed on a harmonic component. The vibration and associated noise generated by a fan or compressor is an example. Musical wind instruments have noise components and the same holds true also for the the 'attack' sound by a violin. The interference of noise on transmission lines for communication and for the detection of signals in general is a common experience.

There is also an intrinsic randomness associated with the thermal motion in matter. For example, the motion of the electrons in a conductor gives rise to random fluctuations in voltage which interfere with the detection of weak signals. Sometimes, this requires experiments to be carried out at very low temperatures.

Not only the weather, but every aspect of our lives contains random components. In a random function $F(t)$, illustrated schematically in Fig. 2.7, the value at a given time cannot be predicted. Rather, the function has to be described in terms of its statistical properties. In measurements, we have at our disposal a finite sample of the function of length Δ and we can measure a statistical property of this sample, such as the mean square value, and repeat the measurement for samples of different lengths. In general, the values thus obtained depend on the sample length. In most cases of practical interest, however, we find that there exists a sample length above which the statistical properties do not change. If these are found to be independent of the time at which the sample is taken, the function is called *stationary*. In what follows, such a function will be assumed.

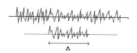

Figure 2.7: Random function of time with a sample of length Δ.

Mean and Mean Square Value

The *mean value* of $F(t)$ is

$$\langle F(t) \rangle = (1/\Delta) \int_0^\Delta F(t)\, dt. \tag{2.79}$$

Frequently, the mean value is zero.
The *mean square value* is

$$\langle F^2(t) \rangle = (1/\Delta) \int_0^\Delta F^2(t)\, dt. \tag{2.80}$$

The corresponding *root mean square value (rms)* is $F = \sqrt{\langle F^2(t) \rangle}$ which is usually the quantity displayed by an instrument.

Since we deal with a finite sample of the function, it can be regarded as being zero outside the interval Δ. This function then qualifies for a Fourier transform, and the corresponding Fourier amplitude $F(\nu)$ is given in Eq. 2.67.

The expression (2.74) involving the Fourier spectrum is still valid. If we divide both sides by the sample length Δ, we obtain the mean square value of $F(t)$

$$\langle F^2(t) \rangle = (1/\Delta) \int_{-\Delta/2}^{\Delta/2} F^2(t)\, dt = \int_{-\infty}^{\infty} |F(\nu)|^2/\Delta \equiv \int_{-\infty}^{\infty} E_2(\nu) d\nu = \int_0^{\infty} E_1(\nu)\, d\nu, \tag{2.81}$$

where $E_2(\nu) = 2|F(\nu)|^2/\Delta$ is the *two-sided* and $E_1(\nu) = 2E_2(\nu)$, the *one-sided power spectrum density* of $F(t)$. As before, we have then made use of $|F(\nu)|^2 = |F(-\nu)|^2$ (recall $|F(\nu)|^2 = F(\nu)F^*(\nu)$). If in a measurement the sample length Δ is increased, the value of $|F(\nu)|^2$ will also increase, and if the sample length is long enough (as we have assumed it to be), the increase will be proportional to Δ leaving the power spectrum densities independent of Δ so that the integration can be carried out to infinity.

In modern spectrum analyzers used in acoustics, an input signal $F(t)$ can be processed in a number of different ways, and the power spectrum density function, for example, can be determined and displayed after a short processing time of the order of Δ.

Correlation Function

Another statistical property of $F(t)$ which can readily be measured is the *correlation function*

$$\Psi(\tau) = \langle F(t)F(t+\tau) \rangle = (1/\Delta) \int_{t-\Delta/2}^{t+\Delta/2} F(t)F(t+\tau)\, dt, \tag{2.82}$$

where, as before, the angle brackets indicate time average (over a sufficiently long interval Δ). If $F(t)$ is a stationary random function, this average is independent of t and is a function only of the time displacement τ. To signify that $F(t)$ and $F(t+\tau)$ refer to the same function F, $\Psi(\tau)$ is often called the *auto-correlation function* and denoted $\Psi_{11}(\tau)$. When two functions F_1 and F_2 are involved, the quantity $\langle F_1(t)F_2(t+\tau)\rangle$ is called the *cross-correlation function* and denoted $\Psi_{12}(\tau)$.

An important property of $\Psi(\tau)$ is that its value at $\tau = 0$ is the mean square value of F, $\Psi(0) = \langle F^2(t)\rangle$. The correlation function is sometimes normalized with respect to $\Psi(0)$ so that the value at $\tau = 0$ will be unity.

Using Eq. 2.69, we can express the correlation function in terms of the Fourier amplitude $F(\nu)$ and the corresponding power spectrum density $\hat{E}(\nu)$. Thus,

$$\int_{-\infty}^{\infty} F(t)F(t+\tau)\,dt = \int_{-\infty}^{\infty} d\nu \int_{-\infty}^{\infty} d\nu'\, F(\nu) f(\nu') e^{-i2\pi \nu'\tau} \int_{-\infty}^{\infty} e^{-2\pi(\nu+\nu')t}\,dt. \tag{2.83}$$

With reference to Eq. 2.68 and by interchanging the roles of t and ν, we note that the last integral in this expression becomes $\delta(\nu + \nu')$, which is different from zero only if $\nu = -\nu'$. Consequently, the integration over ν' in Eq. 2.83 becomes $\int_{-\infty}^{\infty} f(\nu')\delta(\nu + \nu')\,d\nu' = F(-\nu) = F^*(\nu)$. Then, if we divide by Δ, the left side becomes the correlation function, and with $E_2(\nu) = F(\nu)F^*(\nu)/\Delta = |F(\nu)|^2/\Delta$, we arrive at the important result

$$\boxed{\begin{array}{c} \textit{Correlation function} \leftrightarrow \textit{spectrum density} \\ \Psi(\tau) = \int_{-\infty}^{\infty} E_2(\nu)e^{i2\pi\nu\tau}\,d\nu \\ E_2(\nu) = \int_{-\infty}^{\infty} \Psi(\tau)e^{-i2\pi\nu\tau}\,d\tau \end{array}} \tag{2.84}$$

$[\hat{E}(\nu) = |F(\nu)|^2/\Delta.\quad \Psi(\tau) = \langle F(t)F(t-\tau)\rangle$ *(correlation function)*. $\langle..\rangle$: *Time average*].

In other words, the auto-correlation function and the two-sided power spectrum density form a Fourier Transform pair. The corresponding relations for the one-sided power spectrum density are

$$\boxed{\begin{array}{c} \textit{Wiener-Kintchine relations} \\ \Psi(\tau) = \int_0^{\infty} E_1(\nu(\cos(2\pi\nu\tau)\,d\nu \\ E_1(\nu) = 4\int_0^{\infty} \Psi(\tau)\cos(2\pi\nu\tau)\,d\tau \end{array}}, \tag{2.85}$$

where $E_1(\nu) = 2E_2(\nu)$. These equations are known as the Wiener-Khintchine relations.

2.6.5 Random Excitation of the Linear Oscillator

As an example, we consider a linear oscillator driven by a random force $F(t)$ and wish to determine the correlation function for the displacement $\xi(t)$.

If the Fourier amplitude of the driving force is $F(\nu)$, where $\omega = 2\pi\nu$, the corresponding Fourier amplitude of the displacement is (see Eq. 2.34) $\xi(\omega) = F(\omega)/(K - \omega^2 M - i\omega R)$ with the two-sided power Fourier spectrum density

$$|\xi(\nu)|^2 = \frac{|F(\nu)|^2}{(K - \omega^2 M)^2 + (\omega R)^2} = \frac{|F(\nu)|^2}{M^2[(\omega_0^2 - \omega^2)^2 + 4\gamma^2\omega^2]}, \tag{2.86}$$

where $\omega = 2\pi v$, $\omega_0^2 = K/M$ and $\gamma = R/2M$.

If the power spectrum density $\hat{E}(v) = |F(v)|^2/\Delta$ of the force is constant, E_0, the correlation function, according to Eq. 2.85, is $\Psi(\tau) = (E_0)\delta(\tau)$; it is zero for all values of τ except zero.[9]

The power spectrum density is given by Eq. 2.86 and if this is used in Eq. 2.84 the correlation function for the displacement of the oscillator is found to be

$$\Psi_{osc}\tau = \frac{E_0}{4M^2\omega_0^2\gamma}\, e^{-\gamma\tau}[\cos\omega_0'\tau + \frac{\gamma}{\omega_0'}\sin(\omega_0'\tau)], \qquad (2.87)$$

where $\omega_0' = \sqrt{\omega_0^2 - \gamma^2}$ and $\gamma = R/2M$ (compare the analysis of the free damped motion of an oscillator). Again, we leave the integration involved as a problem.

The physical significance of this result is that although the driving force is completely random, the response of the oscillator is not, exhibiting substantial correlation over a range of τ of the order of $1/\gamma$. The reason for this correlation is that the oscillator in effect acts like a filter which tends to limit the spectrum density to a band centered at the frequency ω_0' and with a width proportional to γ.

The mean square value of the displacement is the value of the correlation function at $\tau = 0$,

$$\langle \xi^2 \rangle = \frac{E_0}{4M^2\omega_0^2\gamma} = |\xi(v_0)|^2\gamma, \qquad (2.88)$$

where we have used Eq. 2.86. In other words, the mean square displacement is obtained from the value of the spectrum density of the displacement at resonance multiplied by an effective bandwidth γ. (It should be borne in mind that if the spectrum density is expressed in terms of ω rather than v, we have, from $E(\omega)d\omega = E(v)dv$, $E(v) = 2\pi E(\omega)$.)

In this particular case with a completely random driving force, the correlation of the displacement depends only on the characteristics of the oscillator. If the force spectrum is itself limited to a finite band, the correlation function of the displacement will contain this characteristic as well.

2.6.6 Impulse and Frequency Response Functions; Generalization and Summary

In Section 2.5, the displacement of a harmonic oscillator caused by an impulse was calculated from elementary considerations as an initial value problem. This impulse response function was then used to determine the motion caused by a driving force of arbitrary time dependence. We now consider the response of a linear system in general, not limited to the mass-spring oscillator.

With a complex force amplitude $F(v)$ applied to a mechanical system with an input impedance $Z(v)$, the complex amplitude of the velocity is, by definition of the

---

[9] In practice, the spectrum is limited to some finite frequency range. For example, if the power spectrum density of $F(t)$ has the constant value E_0 below the frequency v_m and zero beyond, it is left for the reader to show, from Eq. 2.84, that the correlation function is $\Psi(\tau) = E_0(1/\omega)m\tau))\sin(\omega_m\tau)$. In that case, there will be a substantial correlation for values of $\tau < 1/\omega_m$, where $\omega_m = 2\pi v_m$.

impedance $Z(v)$,

$$u(v) = F(v)/Z(v) = F(v)Y(v), \tag{2.89}$$

where $Y = 1/Z$ is the admittance. Actually, for any dynamic variable related to the system there is a corresponding linear relation between the driving force and the response. For example, the complex displacement amplitude $\xi(v) = u(v)/(-i\omega) = F(v)/[(-i\omega)Z(v)]$.

With a terminology borrowed from electrical circuit analysis, if one complex amplitude $x(v)$ is considered to be the input and another, $y(v)$, the output, the linear relation between the two is written

$$y(v) = H(v)x(v), \tag{2.90}$$

where $H(v)$ is the *frequency response function* for the particular quantity involved. If x is the driving force and $y(v)$ the velocity amplitude, H is simply the input admittance $Y = Z^{-1}$. If y is the displacement the frequency response function is $H = [(-i\omega)Z]^{-1}$ which, unlike the admittance, has not been given a generally accepted special name.

When the input and output of the system are converted into electrical signals by means of appropriate transducers, the frequency response function $H(v)$ can be determined and displayed by feeding these signals to a two-channel digital frequency analyzer (Fast Fourier Transform, FFT, analyzer).

If the input is a unit impulse at $t = 0$, $\delta(t)$, so that the Fourier amplitude $x(v)$ is unity, H is called the *impulse (frequency) response function*. The corresponding time dependence is obtained from the Fourier transform equation

$$h(t) = \int_{-\infty}^{\infty} H(v) \exp(-i2\pi v)\, dv. \tag{2.91}$$

With the impulse delivered at $t = t'$ rather than at $t = 0$, the Fourier amplitude is $\exp(i\omega t')$ (see Eq. 2.68) rather than unity. Combined with the factor $\exp(-i\omega t)$ in the Fourier integral leads to $\exp[-i\omega(t-t')]$ so that $h(t)$ will be replaced by $h(t-t')$. The response to an arbitrary input signal can then be calculated as shown in Eq. 2.53.

It is instructive to evaluate the integral in Eq. 2.91 for the special case of the harmonic oscillator, i.e., $H = Y = 1/Z$, to make sure that the result agrees with Eq. 2.52 obtained earlier by an entirely different method.

With $Z = R - i\omega M + iK/\omega$, we have $-i\omega Z = -M(\omega^2 - \omega_0^2 + iR/M) = -M(\omega - \omega_1)(\omega - \omega_2)$ where $\omega_1 = -i\gamma + \omega_0'$ and $\omega_2 = -i\gamma - \omega_0'$, $\gamma = R/2M$ and $\omega_0' = \sqrt{\omega_0^2 - \gamma^2}$. Eq. 2.91 can then be written

$$h(t) = -\frac{1}{(2\pi)^2 M} \int_{-\infty}^{\infty} \frac{\exp(-i2\pi v)}{(v - v_1)(v - v_2)}\, dv, \tag{2.92}$$

where $v = \omega/2\pi$. This integral is evaluated by means of the residue theorem. The poles of the integrand are at v_1 and v_2 and the residues at these poles are $\exp(-i\omega_2)t/[(v_2 - v_1)]$ and $\exp(-i\omega_1)t/(v_1 - v_2)$. The path of integration runs along the real axes and is closed by a semi-circle in the lower half of the complex plane (in this half, the integral along the circular path is zero). The path of integration runs in the clock-wise (negative) direction which is accounted for by a minus sign. The integral is $2\pi i$ times the sum of the residues. Then, with $\exp(i\alpha) - \exp(-i\alpha) = 2i\sin(\alpha)$, we get

$$h(t) = (1/\omega_0')\exp(-\gamma t)\sin(\omega_0' t), \qquad (2.93)$$

where $\omega_o' = 2\pi\nu_0' = \sqrt{\omega_0^2 - \gamma^2}$. If the impulse is delivered at $t = t'$ rather than at $t = 0$, we have to replace t by $t - t'$, as indicated above, in which case the result is identical with Eq. 2.52.

In this particular example, the derivation of the impulse response function in Eq. 2.52 was simpler than that given here. However, the present analysis is general and can be applied to any linear system for which the frequency response function is known.

2.6.7   Cross Correlation, Cross Spectrum Density, and Coherence Function

The quantities referred to in the heading are relations between the input and output signals which can be determined with a two-channel FFT analyzer. With the input and output signals denoted $x(t)$ and $y(t)$, the cross-correlation function is

$$\Psi_{xy}(\tau) = \langle x(t)y(t+\tau)\rangle \qquad (2.94)$$

and, together with the corresponding two-sided cross spectrum density $\hat{s}_{xy}(\nu)$ form the Fourier Transform pair

$$\boxed{\begin{array}{l} \textit{Cross correlation} \leftrightarrow \textit{Cross spectrum density} \\ \hat{s}_{xy}(\nu) = \int_{-\infty}^{\infty} \Psi_{xy}\exp(-i2\pi\nu\tau)\,d\tau \\ \Psi_{xy}(\tau) = \int_{-\infty}^{\infty} \hat{s}_{xy}(\nu)\exp(i2\pi\nu\tau)\,d\nu \end{array}} \qquad (2.95)$$

The derivation is completely analogous to that for the transform pair in Eq. 2.84 for the auto-correlation function and the corresponding two-sided spectrum density. In terms of the present notation, the auto-correlation function for the input signal $x(t)$ is $\Psi_{xx}(\tau)$, and the corresponding two-sided cross spectrum density is $\hat{s}_{xx}$. Analogous expressions apply to the output signal $y(t)$. The ratio of the Fourier amplitudes $y(\nu)$ and $x(\nu)$ of the output and input signals $y(t)$ and $x(t)$ is $H(\nu)$ (see Eq. 2.90) and since the spectrum density is proportional to the squared magnitude of the Fourier amplitudes, it follows from Eq. 2.90 that

$$\hat{s}_{yy}(\nu) = |H(\nu)|^2 \hat{s}_{xx}(\nu). \qquad (2.96)$$

The cross spectrum density is also intimately related to the frequency response function $H(\nu)$, and, as shown below, we have

$$\hat{s}_{xy}(\nu) = H(\nu)\hat{s}_{xx}(\nu), \qquad (2.97)$$

where $\hat{s}_{xx}$ is the two-sided spectrum density of the input function. This relation can be proved as follows. In terms of the impulse response function $h(t - t')$ and the Fourier transform of $H(\nu)$, it follows from Eq. 2.53 that $y(t + \tau) = \int h(t')x(t + \tau - t')\,dt'$ and hence,

$$\Psi_{xy} = \langle x(t)y(t+\tau)\rangle = \int_{-\infty}^{\infty} h(t')\langle x(t)x(t+\tau-t')\rangle\,dt'. \qquad (2.98)$$

Next, we introduce the Fourier integrals for $\Psi_{xy}(\tau)$ and $\Psi_{xx} = (\tau-t') = \langle x(t)x(t+\tau-t')\rangle$ in terms of the density functions $\hat{s}_{xy}$ and $\hat{s}_{xx}$. The integral over t' then yields $H(\nu)$ and by comparing the two sides of the equation, we obtain Eq. 2.97.

The relation between $y(t)$ and $x(t)$ and between s_{xy} and s_{xx} are causal as they involve the output produced by an input. In experiments there are sometimes disturbances that interfere with this relation as the output signal might contain extraneous signals not accounted for in the analysis. A useful diagnostic test for such interferences is the the coherence function γ which is defined by

$$\gamma^2 = |\hat{s}_{xy}(\nu)|^2 / [\hat{s}_{xx}(\nu)\hat{s}_{yy}(\nu)]. \qquad (2.99)$$

It follows from $\hat{s}_{yy} = |H(\nu)|^2 \hat{s}_{xx}(\nu)$ and Eq. 2.97 that under normal conditions, the coherence function is unity. If measurements indicate a deviation from unity, the reason can be: (a) that in addition to the input signal, extraneous signals contribute to the output, (b) that the system is nonlinear, and (c) that the system parameters are time dependent.

2.6.8 Spectrum Analysis

A mechanical vibration (including sound) can be converted into an electrical signal by means of a *transducer*. There are many kinds of transducers based on a variety of physical phenomena such as the induced voltage caused by the motion of a conductor in a magnetic field, the electric effects resulting from the deformation of a piezo-electric or magneto-strictive materials, the variation of the capacity of a condenser resulting from a variation of the separation of the capacitor plates, the velocity dependent cooling and change in electrical resistance of a thin wire, the Doppler effect of light reflected from a vibrating surface, the change in electrical resistance of packed carbon powder or foams, etc. These transducers can be made in such a way that the output current (voltage etc.,) is proportional to sound pressure, displacement, velocity, or acceleration.

The current can be decomposed by means of filters in much the same way as a signal can be decomposed into a sum of harmonic functions, as described in Section 2.6. A frequency analysis can be made not only of a periodic signal but of a signal with a more general time dependence such as a random function and even a pulse. The filters can be analog or digital devices; the latter are now more common. The FFT analyzer (Fast Fourier Transform) yields an almost instantaneous presentation of the spectrum of a signal and can process a signal in many different ways resulting in the rms value, the spectrum, and the correlation function, for example. A two-channel FFT analyzer can produce other useful outputs discussed above such as the frequency response function, the cross correlation function, and the coherence function.

An analyzer filters a signal into frequency bands and gives as an output the rms values in these bands which constitutes the *frequency spectrum* of the signal. Often the bandwidth of the analyzer can be selected. Either the bandwidth Δv itself or the relative bandwidth $\Delta v / v$ can be chosen to be constant over the frequency range under consideration. Normally, analyzers in engineering acoustics cover a range from 16 to 10,000 Hz (compare the range of frequencies on the normal keyboard of a piano, discussed in Chapter 1). The relative bandwidths are generally 1/3 and 1/1 octaves, the first octave being centered at 31.5 Hz. Occasionally, the 1/12 octave is used, corresponding to a semitone on the equally tempered musical scale.

By dividing a narrow band spectrum by the bandwidth Δv, an analyzer can also provide the *spectrum density* $E(v)$, which is the contribution to the rms value per Hz. Formally, it is defined by the relation

$$\langle F^2(t) \rangle \equiv F^2 = \int_0^\infty E(v)\, dv, \tag{2.100}$$

where $\langle F^2(t) \rangle$ is the mean square value, F the rms value, and v the frequency.

Band Spectra; 1/1 OB and 1/3 OB

With the lower and upper frequencies of the band being v_1 and v_2, the meter reading for this band will be determined by the rms value F_b given by

$$F_b^2 = \int_{v_1}^{v_2} W(v)\, dv, \tag{2.101}$$

which is the mean square contribution from this band. If $v_2 = 2v_1$, the bandwidth is one *octave*, and if $v_2 = 2^{1/3} v_1$, it is one-third of an octave.

On a logarithmic scale, the *center frequency* v_{12} of a band is such that $v_2 / v_{12} = v_{12}/v_1$ (i.e., $v_{12} = \sqrt{v_1 v_2}$, $v_1 = v_{12}\sqrt{v_2/v_1}$ and $v_2 = v_{12}\sqrt{v_2/v_1}$). Thus, the bandwidth (in Hz) of an octave band with the center frequency v_{12} can be written $v_2 - v_1 = v_{12}(\sqrt{v_2/v_1} - \sqrt{v_1/v_2}) \approx 0.707\, v_{12}$ (i.e., close to 71 percent of the center frequency).

For a one-third octave band, $v_2/v_1 = 2^{1/3}$ and $v_2 - v_1 = v_{12}(2^{1/6} - 2^{-1/6}) \approx 0.23 v_{12}$. In general, for a bandwidth of $1/n$th octave the result is

$$v_2 - v_1 = v_{12}(2^{1/2n} - 2^{-1/2n}). \tag{2.102}$$

In acoustical engineering practice, the center frequencies of the octave bands have been standardized to the values 16, 31.5, 63, 125, 250, 500, 1000, 2000, 4000, and 8000 Hz and the third octave band center frequencies are 12.5, 16, 20, 25, 31.5, 40, 50, 63, 80, 100, 125, 160, 200, 250, 315, etc.

Frequently, a reference rms value F_r of the quantity involved is used. With reference to it, the level in decibels of the quantity F^2 is expressed as

$$\text{dB} = 10 \log(F^2/F_r^2), \tag{2.103}$$

where F^2 is given by Eq. 2.100.

2.6.9 Problems

1. **Fourier series. Particle in a box**

   Consider the motion of a particle in a box bouncing back and forth between two parallel rigid walls (normal incidence). The speed of the particle is U and the width of the box is $2L$. The collisions with the walls are elastic so the speed of the ball is the same before and after the collision.

   (a) Make a Fourier decomposition of the displacement of the particle. First, let $t = 0$ be at the maximum excursion of the particle so that the function will be symmetric with respect to t. Then, repeat the analysis with $t = 0$ at the time of zero excursion. Use your favorite software and make plots of the sum of 5, 10, and 20 terms of the Fourier series.

   (b) Do the same for the velocity function of the particle.

2. **Fourier expansion of a rectified harmonic function**

   Determine the Fourier series of the function $\xi(t) = |\cos(\omega t)|$.

3. **Fourier series, use of complex variables**

   With reference to the example in Section 2.6, carry out part (b) of the example using complex amplitudes (two sided expansion) by analogy with the expansion in (a).

4. **Correlation function**

   (a) Show that the correlation function of a harmonic function is also harmonic.

   (b) What is the auto-correlation function of $\exp(-\gamma t)\cos(\omega t)$?

5. **Spectrum shape**

   Consider a sound pressure field in which the spectrum density of the sound pressure p is constant, $E(f) = E_0$. Make a sketch of the frequency dependence of the octave band spectrum of the pressure in which the sound pressure level is plotted versus the logarithm of the center frequency.

6. **Oscillator driven by a random force**

   Check the results in Eqs. 2.87 and 2.88 for the correlation function and the mean square displacement of an oscillator driven by a random force.

2.7 The Potential Well and Nonlinear Oscillators

As we have seen, the linear oscillator is characterized by a restoring force proportional to the displacement. It was pointed out, however, that this linearity can be expected to hold only for small displacements from the equilibrium position. The deviation from linearity was illustrated qualitatively for both the coil spring and the air spring. A quantitative study of a nonlinear oscillator requires the solution of nonlinear differential equation which in most cases has to be done numerically, as illustrated in an example at the end of the chapter.

Some aspects of a nonlinear oscillator can be understood from the motion of a particle in a potential well, in which the potential energy of the particle is a known function of the displacement. (The mass-spring oscillator is a special case with the potential energy being proportional to the square of the displacement.)

We denote the potential energy of the particle by $V(\xi)$, where ξ is the displacement in the x-direction from the stable equilibrium position at the bottom of the well ($\xi = 0$) where the potential energy is set equal to zero. As indicated in Fig. 2.8, the

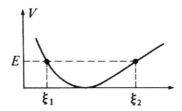

Figure 2.8: Motion of a particle in a one-dimensional potential well. Total energy of the particle is E. The turning points in the oscillatory motion are ξ_1 and ξ_2.

total energy of the particle is E. The kinetic energy is zero where $E = V(\xi)$; this determines the turning points ξ_1 and ξ_2 of the oscillator.

The force on the particle in the x-direction is $-\partial V/\partial \xi \equiv -V'(\xi)$. At the equilibrium position, this force is zero. Furthermore, since we are at a minimum of the potential energy, $V'(0) = 0$ and $V''(0)$ is positive. The Taylor expansion of $V(\xi)$ then becomes

$$V(\xi) = V(0) + \xi V'(0) + (\xi^2/2)\, V''(0) + \ldots = (\xi^2/2)\, V''(0) + \cdots \qquad (2.104)$$

and the force on the particle is $F = -V''(0)\xi - \ldots$ (i.e., proportional to ξ for small ξ); the equivalent spring constant is $K = V''(0)$. For the linear spring, the potential energy is $V = K\xi^2/2$ and $V''(0) = K$, as it should be. Thus, the equation of motion for small oscillations is

$$M\ddot{\xi} = -V''(0)\xi$$

and the solution is a harmonic motion with

$$\omega_0^2 = K/M = V''(0)/M \qquad (2.105)$$

and the corresponding period is $T_0 = 2\pi/\omega_0$.

Sometimes it may not be convenient to place $\xi = 0$ at the equilibrium position; it could equally well have been chosen to be ξ_{st} so that the displacement from equilibrium is $\xi - \xi_{st}$. In the Taylor expansion of the potential in Eq. 2.104, ξ is then replaced by $\xi - \xi_{st}$ and '0' in the argument of the derivatives of V by ξ_{st}.

Example

Consider a nonlinear spring held fixed at its upper end. The force required to change the length by an amount ξ from its relaxed position is $F(\xi)$ (rather than $K\xi$ for a linear spring). A body of mass M is hung from the lower end of the spring. Calculate the frequency of oscillation of the body in small oscillations about the equilibrium position. In particular, let $F(x) = b\xi^3$.

The static displacement ξ_{st} of the spring is determined by the equation $F(\xi_{st}) = Mg$. With $F(\xi) = b\xi^3$, the static displacement becomes $\xi_{st} = (Mg/b)^{1/3}$. The potential energy function is $V = \int F(\xi)d\xi = (b/4)\xi^4 + \text{const}$ and, according to Eq. 2.105, the local spring constant in a small displacement from equilibrium is

$V''(\xi_0) = 3b\xi_0^2 = 3Mg/\xi_{st}$, where we have used $Mg = b\xi_{st}^3$. Thus, the angular frequency of oscillation about the equilibrium position is

$$\omega_0 = \sqrt{V''(\xi_{st})/M} = \sqrt{3g/\xi_{st}}. \qquad (2.106)$$

This should be compared with the result for the linear spring which is $\sqrt{g/\xi_{st}}$.

2.7.1 Period of Oscillation, Large Amplitudes

The period of oscillation for an arbitrary amplitude of motion of a nonlinear oscillator can be expressed in the following manner. Conservation of energy requires that $M\dot{\xi}^2/2 + V(\xi) = \text{constant} = E$, where E is the total mechanical energy of oscillation. Thus,

$$\dot{\xi} \equiv d\xi/dt = \sqrt{(2/M)(E - V(\xi))}. \qquad (2.107)$$

At the turning points ξ_1 and ξ_2 of the oscillation, the kinetic energy is zero and these points are obtained as solutions to $V(\xi) = E$. The period of oscillation is twice the time required to go between the turning points ξ_1 and ξ_2, i.e.,

$$T = 2 \int_{\xi_1}^{\xi_2} \frac{d\xi}{\sqrt{(2/M)(E - V(\xi))}}. \qquad (2.108)$$

For the square law potential, the period is independent of the energy and the spring constant independent of the amplitude. For an oscillator, such as a pendulum, the equivalent spring constant decreases with increasing amplitude and the period increases with amplitude. Such an oscillator is sometimes referred to as 'soft.' For an air spring, on the other hand, the spring constant increases with amplitude and the oscillator is 'hard.'

2.7.2 Pendulum

The pendulum, as in a clock, consists of rigid body of mass M, which can swing freely in a plane about a fixed axis of rotation. If the center of mass is a distance L from the axis and the angle of deflection from the vertical is ϕ, the height of the weight above the equilibrium position $\phi = 0$ is $L(1 - \cos\phi)$ so that the potential energy of the pendulum is

$$V(\phi) = MgL(1 - \cos\phi). \qquad (2.109)$$

If the maximum angle of deflection is ϕ_0, the total energy can be expressed as $E = MgL(1 - \cos\phi_0)$. With the moment of inertia of the pendulum being $I = MR^2$, where R is the radius of gyration, the kinetic energy is $K = (I/2)(d\phi/dt)^2 = E - V(\phi)$ and solving for $\partial\phi/\partial t$ and integrating from $\phi = 0$ to ϕ and from $t = 0$ to $t = t$ yields

$$t = \sqrt{R^2/2gL} \int_0^\phi \frac{d\phi}{\sqrt{(\cos\phi - \cos\phi_0)}}. \qquad (2.110)$$

For the 'simple' pendulum we have $R = L$. With the maximum angle ϕ_0 as the upper limit of integration, $t = T/2$, where T is the period of oscillation.

For small displacements, we can use $\cos\phi \approx 1 - \phi^2/2$, and with $\sqrt{g/L}$ denoted ω_0, it follows that

$$\omega_0 t = \int_0^t \frac{d\phi}{\sqrt{(\phi_0^2 - \phi^2)}} = \arcsin(\phi/\phi_0) \quad \text{or} \quad \phi = \phi_0 \sin(\omega_0 t). \qquad (2.111)$$

In other words, for small amplitudes the pendulum moves like a linear oscillator with the resonance frequency

$$\omega_0 = \sqrt{g/L} \qquad (2.112)$$

independent of the amplitude. The period increases with amplitude which means that the equivalent spring constant is 'soft.'

2.7.3 Oscillator with 'Static' and 'Dynamic' Contact Friction

In the analysis of the motion of a damped mechanical oscillator, the simplest example being the mass-spring variety, it is usually assumed, as we have done earlier, that the damping force is viscous (i.e., proportional to the velocity). This is not always realistic; a typical example is a block sliding on a table and attached to a spring, often used in elementary texts with the tacit assumption of viscous friction.

Actually, the elementary view of the velocity dependence of the contact friction force and the corresponding friction coefficient is simply that one distinguishes between a 'static' and a 'dynamic' coefficient. The former refers to the state in which the body is on the verge of moving under the influence of a horizontal driving force.[10] The latter applies when the body is in motion and the magnitude of the friction force is then usually assumed to be independent of velocity.

There is a fundamental difference between 'dry' and viscous friction. In the case of the dry friction, the driving force must exceed the constant friction force if any motion at all is to occur. By contrast, a viscous friction force allows motion for any magnitude of the driving force. This illustrates in a simple manner the nonlinearity of the oscillator with dry friction.

We shall consider here the impulse excitation of an oscillator in which the damping is due to a combination of a viscous friction force $-Ru$, proportional to the velocity $u = \dot{\xi}$, and a dry contact friction force of magnitude $|F_d|$. If the mass M of the oscillator is sliding on a horizontal plane, the elementary view of the friction force makes it proportional to the normal contact force, and if only gravity is the cause of it, we have $|F_d| = \mu M g$, where μ is the friction coefficient. To include also the direction of the this friction force, we use the expression $F_d = -|F_d| \operatorname{sgn}(u)$, where the sign function $\operatorname{sgn}(u)$, by definition, is $+1$ if u is positive and -1, if u is negative; it can be expressed formally as

$$\operatorname{sgn}(u) = u/|u| = \dot{\xi}/|\dot{\xi}|. \qquad (2.113)$$

---

[10]Actually, when the block is at rest, the friction force increases with the applied horizontal force until it reaches a maximum value F_m which defines the static friction coefficient as the ratio of F_m and the normal contact force.

With the displacement from equilibrium denoted ξ, as before, the equation of free damped motion of the oscillator is the same as Eq. 2.23 except for the addition of the contact friction force F_d.

$$\ddot{\xi} + 2\gamma\dot{\xi} + |F_d|\,\text{sgn}(\dot{\xi}) + \omega_0^2\xi = 0, \tag{2.114}$$

where $\omega_0^2 = K/M$ and $\gamma = R/2M$.

With only the conventional viscous friction the oscillator when started from $\xi = 0$ at $t = 0$ with a velocity $u(0)$, corresponding to an applied impulse of $Mu(0)$, the time dependence of the velocity $u = \dot{\xi}$ in the subsequent motion follows directly from the impulse response function Eq. 2.52 by differentiating with respect to time and multiplying by the impulse Mu_0, since the impulse function refers to a unit impulse. Thus,

$$u(t)/u(0) = \exp(-\gamma t)[\cos(\omega_1 t) - (\gamma/\omega_1)\sin(\omega_1 t)], \tag{2.115}$$

where $\gamma = R/2M$ and $\omega_1^2 = \omega_0^2 - \gamma^2$. The period of free undamped oscillations is $T_0 = 2\pi/\omega_0$. This normalized velocity function will be independent of the magnitude of the initial velocity and the corresponding impulse.

This is not the case when the constant friction force is present, however, since the oscillator is no longer linear. The equation of motion (2.114) now has to be solved numerically and it is convenient in such a computation to use a normalized version of the equation. Thus, if we introduce the normalized time $t' = t/T_0$ and the normalized displacement $\xi' = \xi/u(0)T_0$, the equation takes the form (see Problem 6)

$$\frac{\partial^2\xi'}{\partial t'^2} + 2\gamma T_0\frac{\partial\xi'}{\partial t'} + \beta\,\text{sgn}(\frac{\partial\xi'}{\partial t'}) + (2\pi)^2\xi' = 0, \tag{2.116}$$

where $\beta = F_d T_0/Mu(0)$. The quantity $\partial\xi'/\partial t'$ now becomes the normalized velocity $u/u(0)$ with the value 1 at $t' = 0$. The quantity $\beta = (F_d T_0/Mu(0))$ is the magnitude of the nonlinear term in the equation; it is the ratio of the impulse $F_d T_0$ of the friction force during one period and the external impulse $Mu(0)$ delivered to the oscillator at $t' = 0$. It is a nonlinearity parameter which goes to zero as F_d goes to zero or $Mu(0)$ goes to infinity.

The equation (2.116) is solved numerically and we have used a slightly modified Runge-Kutta fourth order approximation.[11] The accuracy of this procedure is checked by comparing the result obtained for $\beta = 0$ with the known exact solution, Eq. 5.19, for the linear oscillator. The results obtained are illustrated by the examples in Fig. 2.9 where the normalized velocity $u(t)/u(0)$ is plotted as a function of the normalized time $t' = t/T_0$ in the range $t' = 0$ to 10; the linear decay constant is such that $\gamma T_0 = 0.05$ and values of the nonlinearity parameter $\beta = F_d T_0/Mu(0)$ are 0.025, 0.1, 0.2, and 1.0.

With $\gamma T_0 = 0.05$, it will take about 20 periods for the linear oscillator amplitude to decay by a factor of $\exp(-1) \approx 0.37$. For comparison, this linear decay curve (thin line) is shown in each case; then, since $F_d = 0$, it corresponds to a value $\beta = 0$ of the nonlinearity parameter.

---

[11]See standard mathematical texts on differential equations.

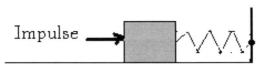

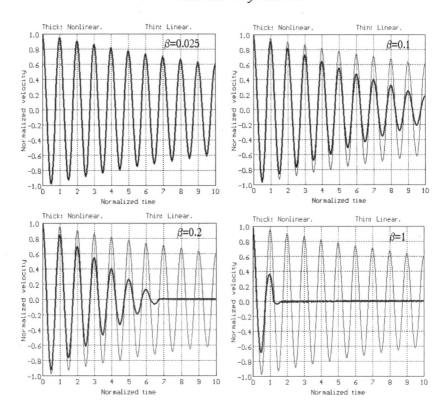

Figure 2.9: Decay of impulse excited oscillator containing both viscous and speed independent contact friction damping; normalized velocity $u(t)/u(0)$ versus normalized time, t/T_0.
Thin lines: Linear oscillator with only viscous damping. Thick lines: Both viscous and speed
independent contact friction force F_d present. 'Nonlinearity' parameter: $\beta = F_d T_0 / M u(0)$.
T_0: Period of undamped oscillations. M: Oscillator mass. $u(0)$: Initial velocity.

The actual decay curves for different degrees of nonlinearity, i.e., different values of β, are shown by the thick lines in the figure. For $\beta = 0.025$, the decay is almost indistinguishable from the linear decay during the first ten periods. With $\beta = 0.1$, there is only a slight difference during the first two periods; the difference increases as the amplitude decreases, however, since the role of the contact friction increases as the momentum of the oscillator decreases. Actually, with $\beta = 0.2$, the oscillator comes to a stop between the sixth and seventh periods, and with $\beta = 1$, only a couple of periods survive.

2.7.4 Problems

1. **Transverse oscillations of mass on a spring**

   Reexamine the example in Chapter 11 when the spring has an initial tension S and the amplitude of oscillation is not necessarily small. Derive an expression for the period of oscillation in terms of the amplitude of oscillation.

2. **Oscillations of a floating body**

   A weight is hung from the vertex of a wooden cone which floats on water. In equilibrium, the cone is submerged a distance y in the water, measured from the vertex. What is the frequency of small vertical oscillations about the equilibrium? What can you say about the frequency of large amplitude oscillations? Density of the cone: 0.5 g/cm^3.

3. **Morse potential**

   The Morse potential (describing the interaction potential in a diatomic molecule) is $V(\xi) = B[\exp(-2b\xi) - 2\exp(-b\xi)]$, where ξ is the displacement from the equilibrium position. A particle of mass m moves under the influence of this potential.
   (a) What is the potential energy in the equilibrium position?
   (b) Show that the angular frequency of small oscillations about the equilibrium is given by $\omega_0^2 = 2b^2 B/m$.

4. **Potential well**

   Consider a particle of mass M oscillating in a well of the one-dimensional periodic potential $V(\xi) = 1 - \cos(k\xi)$, where $k = 2\pi/\lambda$ and λ is the wavelength. Show that the angular frequency of small oscillation is given by $\omega_0^2 = V_0 k^2/M$.

5. **Piston on an air spring**

   Obtain the differential equation for the displacement ξ of the piston of mass M riding on the air column in a vertical tube of length L in the case when the displacement from the static equilibrium position of the piston cannot be regarded as small. Show that in the limit of small displacements, the equation reduces to the linear oscillator equation and what is the spring constant?
   Will the period of oscillation increase or decrease with the amplitude?
   Optional. Explore numerical methods of solving the nonlinear equation of motion.

6. **Oscillator with combined 'static' and 'dynamic' friction**

   Check Eq. 2.116 for the normalized displacement of the nonlinear oscillator with combined dry and viscous friction.

Chapter 3

Sound Waves

Unlike electromagnertic waves, elastic waves require a gas, liquid, or solid for transmission. Both longitudinal and travsevers waves are involved as will be discussed in this chapter.

Classical mechanics is often divided into two major parts, Kinematics and Dynamics. We follow the same major outline in this chapter and start with wave kinematics, examples, and description of waves.

In dynamics, forces enter into the discussion and the elastic properties of the material substance (gas, liquid, solid) that carries the wave need to be discussed. The underlying physics involved are the conservation laws of mass and momentum to which is added the equation of state for the material (the latter contains information, which, in a sense, is analogous to the spring constant in a mass-spring oscillator).

The interaction of waves with boundaries leads to the phenomena of reflection, absorption, transmission, diffraction, and scattering, which will be treated in separate sections.

3.1 Kinematics

3.1.1 Traveling Waves

As a familiar example consider the 'waves' frequently observed amongst the spectators of a football game in the packed stands. The wave can be generated, for example, by repeating the motion of the spectator to the left, lifting an arm, for example. There is some time delay involved in this repeated motion. As a result, a wave traveling to the right is generated. The speed of the wave depends on the reaction time of the individuals and inertia. For the motion suggested, the wave will be transverse. It should be noted that the wave does not carry any mass in the direction of propagation. The shape of the wave depends on the motion that is being repeated.

Similarly, when the end (at $x = 0$) of a stretched rope is suddenly moved sideways, the event of 'moving sideways' travels along the rope (in the x-direction) as a wave with a certain speed v which is known from experiments to depend on the tension in the rope and its mass. The initial displacement will be repeated by the element at x after a travel time x/v. Again, there is no net mass transported by the wave in the direction of wave travel and the wave speed.

In a compressional wave in a fluid or solid, it is the state of being compressed that travels and on an electric transmission line, it is the electromagnetic field which is transmitted with a certain wave speed; it has little to do with the velocity of the electrons which carry the current in the line.

If, in any such example, the time dependence of the displacement at location $x = 0$ is harmonic, $\xi(0, t) = A \cos(\omega t)$, it will be

$$\xi(x, t) = A \cos[\omega(t - x/v)] = A \cos[2\pi(t/T - x/\lambda)] \tag{3.1}$$

at location x. We have here introduced the *period* $T = 2\pi/\omega$ and the *wavelength* $\lambda = vT$, which is the distance traveled by the wave in one period. It is the spatial period of the displacement as obtained in the snapshot referred to in the previous paragraph.

Thus, a traveling harmonic wave can be thought of as a distribution of harmonic oscillators along the x-axis, all with the same amplitude but with a phase lag (phase angle) proportional to x.

In regard to the x-dependence of the wave function, the wavelength λ stands in the same relation to x as the period T does to t. The angular frequency $\omega = 2\pi/T$ has its equivalence in $k = 2\pi/\lambda = \omega/v$, where $v = \lambda/T$ is the wave speed. Quantity k is generally called the *propagation constant*. In terms of these quantities, the wave in Eq. 3.1 can be written $A \cos(\omega t - kx)$ or $A \cos(kx - \omega t)$.

The frequency $f = 1/T$, the number of periods T per second, has its analog in the quantity $1/\lambda$, the number of wavelengths per unit length; it is often called the *wave number*.

If instead of having the time dependence $A \cos(\omega t)$ at $x = 0$ we have the more general harmonic function $A \cos(\omega t - \phi)$, the corresponding wave function will be $\xi(x, t) = A \cos(\omega t - kx - \phi)$, where ϕ is the phase angle or phase lag.

The fact that the wave is traveling in the positive x-direction is expressed by the time delay x/v. If, instead, a wave is traveling from the origin in the negative x-direction, the time delay, being a positive quantity, must be expressed as $-x/v$. Thus, to summarize, harmonic one-dimensional waves are of the form

$$\xi(x, t) = A \cos(\omega t \pm kx - \phi), \tag{3.2}$$

where the minus and plus signs refer to wave travel in the positive and negative x-directions, respectively. The wave can be thought of as a continuous distribution of harmonic oscillators along the x-axis, all with the same amplitude but with a phase difference which is proportional to the distance between the oscillators.

The time dependence of the generator of the wave at $x = 0$ need not be harmonic but can be an arbitrary function $f(t)$. For a wave traveling in the positive x direction, this function is repeated at location x after a time delay x/v, and the wave function then becomes

$$f(x, t) = f(0, t \pm x/v). \tag{3.3}$$

As before, the minus and plus signs correspond to wave travel in the positive and negative x-directions, respectively.

3.1.2 The Complex Wave Amplitude

We have already introduced and used the complex amplitude of a harmonic motion in Section 2.1.1. We refer to it and and also to Appendix B for the use of Euler's identity $\exp(i\alpha) = \cos(\alpha) + i\sin(\alpha)$ to express a harmonic function in terms of an exponential and how to use the resulting complex amplitude in problem solving.

With reference to Section 2.1.1, a displacement wave traveling in the positive x-direction, $\xi(x,t) = A\cos(\omega t - kx)$, at a given x is nothing but a harmonic oscillator with the phase angle $\phi = kx$ and the complex wave amplitude is

$$\xi(\omega) = Ae^{ikx}, \tag{3.4}$$

where $k = \omega/v$, v being the wave speed. Similarly, a harmonic wave traveling in the negative x-direction has the complex amplitude $A\exp(-ikx)$. If we let A be a complex number, $A = |A|\exp(i\phi)$, the corresponding real wave function is $\xi(x,t) = |A|\cos(\omega t - kx - \phi)$.

The frequency ω and the corresponding time factor $\exp(-i\omega t)$ are implied and are not included in the definition of the complex wave amplitude. All we need to know about the motion is contained in the complex amplitude (i.e., the magnitude $|\xi|$ and the phase angle (lag) kx). In regard to notation, we use, as in Chapter 2, $\xi(x,t)$ for the space-time dependence of the real pressure and $\xi(x,\omega)$ for the corresponding complex pressure amplitude. Admittedly, in the course of describing an equation of motion, this kind of careful use of terms often tends to be cumbersome and is often ignored, both $\xi(x,t)$ and $\xi(x,\omega)$ being referred to simply as ξ, and if this is the case, the context will decide which of the quantities is involved. If there is any risk of confusion, we use the full arguments.

3.1.3 Standing Wave

The sum of two waves of the same amplitude but traveling in opposite directions is expected to have no preferred direction of wave travel. This can be seen numerically by a brute force addition of the displacements at different times or, more simply, algebraically, with the use of the trigonometric identity $\cos(a+b) = \cos(a)\cos(b) - \sin(a)\sin(b)$. Thus, if the waves in the positive and negative x-direction are $\xi_+ = A\cos(\omega t - kx)$ and $\xi_- = A\cos(\omega t + kx)$, their sum will be

$$\xi(x,t) = \xi_+ + \xi_- = 2A\cos(kx)\cos(\omega t) \quad \text{(standing wave)}. \tag{3.5}$$

There is no direction of propagation and it is called a *standing wave*. Like the traveling wave, it can be thought of as a continuous distribution of harmonic oscillators, but unlike the traveling wave, the amplitude is not constant but varies with x as expressed by $2A\cos(kx)$. In this case, the amplitude will be zero for $kx = (2n-1)\pi/2$, where n is an integer, and the distance between the zero points or *displacement nodes* will be $\pi/k = \lambda/2$. The maxima of the displacement, the *antinodes*, have the magnitude $2A$ and occur where $kx = n\pi$, i.e., at $x = 0$, $x = \lambda/2$ etc.

The oscillations in a standing wave are either in phase or 180 degrees out of phase. Between two adjacent nodes, the phase is the same and a phase change of π occurs when a node is crossed; this means a change in sign (direction) of the displacement.

If the amplitudes of the traveling waves in the positive and negative directions are different, $A_+ \cos(\omega t - kx)$ and $A_- \cos(\omega t + kx)$, we can always express A_+ as $[A_- + (A_+ - A_-)]$ in which case the sum of the two traveling waves can be written as the sum of a standing wave and a traveling wave (in the positive x-direction) with the amplitude $A_+ - A_-$.

From Section 3.1.2 follows that the complex amplitudes of the two waves involved in the creation of a standing wave are $A \exp(ikx)$ and $A \exp(-ikx)$, and the complex amplitude of the sum

$$\xi(x, \omega) = A(e^{ikx} + e^{-ikx}) = 2A \cos(kx). \tag{3.6}$$

The result follows directly from the Euler identity; the imaginary parts of the exponentials in the sum cancel each other, leaving only the two identical real parts.

If we wish to return to the real displacement, $\xi(x, t)$ we re-attach the time factor $\exp(-i\omega t)$ and take the real part, i.e.,

$$\xi(x, t) = \Re\{\xi(x, \omega)e^{-i\omega t}\} = 2A \cos(kx) \cos(\omega t). \tag{3.7}$$

No further comments are needed in regard to the x-dependence of the displacement amplitude in Eq. 3.6, but how about the phase angle? It is contained in the sign of $\cos(kx)$; if it is positive, i.e., with kx between 0 and $\pi/2$, the phase angle is 0 (or an integral number of 2π) and if it is negative, with kx between $\pi/2$ and π, the phase angle is π, i.e., 180 degrees out of phase. (Remember, $\exp(0) = 1$ and $\exp(i\pi) = -1$.) Thus, crossing a displacement node in the standing wave, for example at $kx = \pi/2$, changes the phase by an amount equal to π.

3.1.4 The Wave Equation

A wave traveling in the positive x-direction is of the form $\xi(x, t) = \xi(0, t - x/v)$, as indicated in Eq. 3.3. Since the time and space dependence is expressed by the combination $t - x/v$ for a wave traveling in the positive x-direction, it follows that there is an intimate relation between the time dependence and space dependence, $\partial\xi/\partial x = -(1/v)\partial\xi/\partial t$. For a wave in the negative x-direction, the corresponding relation is $\partial\xi/\partial x = (1/v)\partial\xi/\partial t$. They are both contained in the equation

$$\partial^2 p/\partial x^2 = (1/v^2)\partial^2 p/\partial t^2, \tag{3.8}$$

which is called the *wave equation*.

In the special case of harmonic time dependence, $\partial^2 p/\partial t^2 = -\omega^2 p$, and with $k = \omega/v$, the *harmonic wave equation* takes the form

$$\frac{\partial^2 \xi}{\partial x^2} + (\omega/v)^2 \xi = 0. \tag{3.9}$$

Thus, if in the study of the dynamics of waves we should encounter an equation of this type, we know that a harmonic wave, either in the positive or negative x-direction or a combination of both, are possible solutions. To find the solution which applies to a particular problem, these possible solutions have to be combined in such a manner as to satisfy boundary conditions, as will be discussed later.

The wave equation is valid if ξ stands for the complex amplitude $\xi(x, \omega)$ of the wave.

3.1.5 Wave Lines

The arrival of the pressure front (or any other part of the wave) at a certain location is represented as a point in the (x, t) plane, and it is referred to as an *'event.'* The event $(0, 0)$ is thus the passage of the pressure through the origin $x = 0$ at time $t = 0$. The collection of events (x, t) define the *wave line* of the wave front. This line goes through the origin and is given by $t - x/c = 0$ or $t = x/c$, where c is the wave speed, a notation that is used in the rest of the book when sound waves in a gas are involved.

The events corresponding to the trailing edge of the pulse are represented by the line $t - x/c = \tau$, where τ is the duration of the pulse. The lines are parallel and have the slope $1/c$ but are separated by the time τ. Along these lines, or any other parallel line, the argument of the wave function and hence the value of the function remain constant. The lines can also properly be called *wave trajectories*. Frequently, the t-axis is replaced by a ct-axis, in which case the slope of a wave line will be 1 for a wave traveling in the positive x-direction and -1 for a wave in the opposite direction.

As a imple illustration of wave lines consider a sound wave incident on the boundary between two regions with different sound speeds, for example air and helium. (The boundary can be considered to be a very thin sound transparent membrane.) The wave speed in helium is about 3 times larger than in air and the slope of the wave line of the transmitted wave will be approximately 1/3 of the slope of the wave line of the incident wave in air.

3.1.6 The Doppler Effect

Moving Source, Stationary Observer on Line of Motion

The source can be considered to emit wave pulses at regular intervals and waves are emitted in both the positive and the negative x-direction. The slopes of the wave lines are determined only by the wave speed c in the surrounding air, and with the speed u of the source smaller than the wave speed, the slope of each line is smaller than the slope of the source trajectory.

From the wave lines, we get an idea of the time dependence of the wave trains recorded by observers at rest ahead of and behind the source. It is clear that the number of wave lines (wave pulses or periods) observed per second ahead of the source, at $x = x_1$, will be greater than behind it, at $x = x_2$. In the case of harmonic time dependence, the wave lines can be thought of as representing the crests of the waves. In that case the number of lines per second will be the observed frequency of the harmonic wave.

These frequencies can be obtained in several ways. Since the source moves with a velocity u in the x-direction as it emits a harmonic wave, the separation of the wave maxima 'imprinted' on the gas and constituting the emitted sound wave will not be the ordinary wavelength. With reference to Fig. 3.1, consider one wave front emitted in

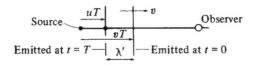

Figure 3.1: Doppler effect. Moving source, stationary observer. Source velocity: u. Sound speed: c. Wave fronts are shown as vertical lines.

the direction of motion at $x = 0$ and $t = 0$ and another at $t = T$, one period later. The first wave front will be at $x = cT$ when the other wave front is emitted, thus located at $x = 0$. The source has then reached the position ut. This means that the separation of the imprinted wave fronts will be $\lambda' = (c-u)T$ which is the wavelength of the wave that travels with the wave speed c. It is shorter than the wavelength $\lambda = c/T$ which would have been obtained if the source had been at rest. For the sound traveling in the opposite direction, the wavelength will be $\lambda' = (c+u)T$.

The wavefronts come closer together in the forward direction and further apart in the opposite direction. The frequency of the emitted wave from the moving source, as observed by a stationary observer ahead of the source, will be $f_1 = c/\lambda' = c/[(c-u)T] = f/(1-m)$, where f is the frequency of the source and $m = u/c$, the Mach number of the source. The corresponding observed frequency f_2 for an observer behind the source is obtained by merely changing the sign of m. Consequently,

$$f_1 = f/(1-m)$$
$$f_2 = f/(1+m), \tag{3.10}$$

where $m = u/c$. These relations express the *Doppler effect*. The difference in frequency $f_1 - f$ (or $f - f_2$) is referred to as the *Doppler shift*. It is important to understand that u is the speed of the source relative to the observer. If the absorber is not located on the line of motion of the source, it is the velocity component of the source in the direction of the observer which counts. Thus, when the sound emitted at an angle ϕ with respect to the direction of motion of the source, the Doppler shift in this direction is determined by the velocity component $u\cos\phi$ so that m in Eq. 3.10 should be replaced by $m\cos\phi$. It is important to realize, however, that when the sound arrives at the observer, the source has moved so that the emission angle is not the same as the *view angle* under which the source is seen at the time of arrival of the Doppler shifted sound. This is explained further in the example given below.

The Doppler effect occurs for all waves. The frequency of the light from a source moving away from us is down shifted (toward the red part of the spectrum) and the shift is usually referred to as the 'red-shift.'

Another way to obtain Eq. 3.10 is geometrical, using a wave diagram. This is done in Example 20 in Chapter 11. The diagram used there looks a bit complicated because of the many lines involved; perhaps you can simplify it.

Eq. 3.10 is valid when the source speed is smaller than the wave speed, i.e., when $m < 1$. For supersonic motion of the source, $m > 1$, we get $f_1 = f/(m-1)$ and $f_2 = f/(m+1)$.

The Doppler effect is important throughout physics. It is used in a wide range of applications both technical and scientific for the measurement of the speed of moving objects ranging from molecules to galaxies.

In the case of sound from a source like an aircraft, the speed of the source can exceed the wave speed. The slope of the trajectory of the source will then be smaller than the slope of the wave lines, and it follows that the wave lines emitted in the positive and negative x-directions will emerge on the same side of the source trajectory and cross each other; this indicates interference between forward and backward running wave.

Observer on Side Line

The observer is now located at a distance h from the line of motion of the source. At time t, the location of the source is at $x_s = ut$. The wave reaching the observer at this time was emitted at an earlier time t_e from the emission point $x_s = ut_e$. The distance from this emission point to the observer can be expressed as $R = c(t - t_e = c(x - x_e)/u$. With the coordinates of the observation point being x, y, R can be calculated from $R^2 = y^2 + (x - x_e)^2$. With $x - x_e = x - x_s + (x_s - x_e) = x - ut + uR/v$ the equation for R can be written $R^2 = y^2 + [x - ut + u(R/v)]^2$ with the solution

$$R = [m(x - ut) \pm R_1/(1 - m^2)$$
$$R_1 = [(x - ut)^2 + (1 - m^2)y^2]^{1/2}, \tag{3.11}$$

where $m = u/c$ is the Mach number of the source. The distance R must be positive, and for subsonic motion only the plus sign corresponds to a physically acceptable solution.

With the *emission angle* between the line of propagation from the emission to the observation point denoted ϕ, the component of the source velocity in this direction will be $u \cos \phi$. The Doppler shifted frequency depends only on this component and is $f' = f/(1 - m \cos \phi)$. This Doppler shifted frequency can be expressed in terms of the observer coordinates and time and we leave it for one of the problems to show that

$$f' = f/(1 = m \cos \phi) = f(R/R_1), \tag{3.12}$$

where R and R_1 are given in Eq. 9.29.

For large negative values of the source location x_s, the component of the source velocity in the direction of the observation point is approximately u, and the corresponding Doppler shifted frequency is the $f/(1 - m)$. Similarly, after the source has passed the observer, the frequency approaches the value $f/(1 + m)$ asymptotically. For example, with a source Mach number of 0.9 the corresponding range in Doppler shifted frequencies goes from $10 f$ to $0.53 f$.

Although the Doppler shift is zero when the emission angle is 90 degrees, there is an upshift in frequency when the source is at $x_s = 0$. The reason is that $x_s = 0$ corresponds to an emission point at an earlier time and the emission angle is less than 90 degrees. For a source with supersonic speed there are two emission points that contribute to the sound pressure at time t, as illustrated in Fig. 3.2. The corresponding travel distances R' and R'' correspond to the two solutions in Eq. 9.29 in which now

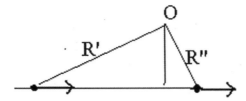

Figure 3.2: For a supersonically moving source, there are two emission points contributing to the sound pressure at the stationary observer O.

both the plus and minus signs are acceptable. There are corresponding emission angles ϕ' and ϕ'' and the Doppler shifted frequency for each can be calculated from Eq. 3.12. It should be noted though, that $f/(1 - m\cos\phi')$ becomes negative. This merely means that the wave fronts emitted from the source arrive in reverse order, the front emitted last arrives first.

Stationary Source, Moving Observer

Consider two successive wave fronts emitted from the stationary source S separated in time by T and in space by λ. These wave fronts travel with the velocity c, the sound speed. The observer O is moving with the velocity u_0. The time it takes for these front to pass the observer is then $T' = \lambda/(c - u_0)$ and the corresponding frequency, $f' = 1/T'$ is

$$f' = f(1 - m_0), \tag{3.13}$$

where $m_0 = u_0/c$ is the Mach number of the observer.

It is instructive to consider the waveline interpretation of this case.

Both Source and Observer Moving

As before, denote by u and u_0 the velocities of the source and the observer along the x-axis. The distance between two wave fronts emitted a time T apart will be $(c-u)T$, where c is the sound speed. The two wave fronts travel with the speed $c - u_0$ with respect to the observer. The time required for the wave fronts to pass by will be $T' = (c - u)T/(c - u_0)$ and the corresponding observed frequency

$$f' = f(1 - m_0)/(1 - m), \tag{3.14}$$

where, as before, m and m_0 are the Mach numbers of the source and receiver. For small values of m and m_0, we get $f' \approx f[1 + (m - m_0)]$ which depends only on the relative velocity of the source and the observer. For electromagnetic waves in vacuum, the Doppler shift depends only on the relative speed under *all conditions*. This is a consequence of the speed of light being the same in all frames of reference.

Source, Observer, and Fluid, All Moving

Finally, we consider source and observer both moving in a moving fluid in arbitrary directions. The corresponding velocities are denoted by the vectors $\boldsymbol{u}$, $\boldsymbol{u}_0$, and $\boldsymbol{U}$. The simplest way of deriving the expression for the corresponding Doppler shifted frequency is probably to consider the motion from a frame of reference in which the fluid is at rest. The velocities of the source and the observer are then $(\boldsymbol{u} - \boldsymbol{U})$ and $(\boldsymbol{u}_0 - \boldsymbol{U})$. The Doppler shift depends on the velocity components in the direction of wave travel and we shall denote by $\hat{\boldsymbol{k}}$ the unit vector for the direction of propagation of the wave from the emission point to the observer. These velocity components are $(\boldsymbol{u} - \boldsymbol{U}) \cdot \hat{\boldsymbol{k}}$ and $(\boldsymbol{u}_0 - \boldsymbol{U}) \cdot \hat{\boldsymbol{k}}$. The Doppler shifted frequency, by analogy with Eq. 3.14, is then

$$f' = f[1 - (\boldsymbol{m}_0 - \boldsymbol{M}) \cdot \hat{\boldsymbol{k}}]/[1 - (\boldsymbol{m} - \boldsymbol{M}) \cdot \hat{\boldsymbol{k}}], \qquad (3.15)$$

where $\boldsymbol{m} = \boldsymbol{u}_0.\boldsymbol{c}$, $\boldsymbol{m}_0 = \boldsymbol{u}_0/\boldsymbol{c}$, and $\boldsymbol{M} = \boldsymbol{U}/\boldsymbol{c}$.

3.1.7   Problems

1. **Sound from a swirling sound source**

   A sound source with a frequency of 100 Hz is located at the end of a string of length $R = 4$ m. It is swirled in air in a horizontal plane with an angular velocity of $\omega = 25 \ \sec^{-1}$.

   (a) What is the range of frequencies observed by a listener in the same plane as the motion but outside the circular path?

   (b) If the listener is at a point on the axis of the circle, what then is the range?

   (c) Comment on the difference between the acoustic Doppler shift and the electromagnetic.

2. **Perception of Doppler shift**

   In the frequency range between 600 and 4000 Hz, the smallest pitch change that can be resolved by a normal human ear corresponds to a relative frequency change $\Delta f/f$ of approximately 0.003.

   A sound source emitting a tone of 1000 Hz moves along a straight line with constant speed. What is the lowest speed of the source that produces an audible pitch change as the source moves by?

3. **Tone from an airplane**

   A propeller plane emits a tone with a frequency f and flies at a constant speed U along a straight line, the x-axis, at a constant elevation H. As the plane crosses the y-axis, an observer at $x = 0$, $y = 0$ receives the Doppler shifted frequency $2f$ and, at a time τ later, the unshifted frequency f. From these data determine

   (a) the Mach number $M = U/c$ of the plane and (b) the elevation H.

4. **Wave diagram**

   A pressure pulse of duration 5 milliseconds is generated at $t = 0$ at the left end of a closed tube. The tube contains air and helium separated by a thin, limp membrane which can be considered transparent to the wave. The total length of the tube is 16 m and the distance to the membrane from the source is 4 m. The tube is closed by rigid walls at both ends.

   (a) Accounting for the reflection at the boundary between the gases and at the end walls, make a wave diagram which covers the first 50 milliseconds.

(b) Indicate in the (t, x)-plane the region where you expect interference to occur between reflected and incident waves.

5. **Emission angle versus view angle**

   A sound source moves with constant speed V along a straight line. An observer is located a distance H from the line. When the observer hears the sound from the source, the line drawn from the source to the observer makes an angle ϕ_v with the line of motion. The corresponding angle at the moment of emission of the sound is denoted ϕ_e. What is the relation between the two angles?

6. **Doppler shift when observer velocity is supersonic**

   By analogy with the discussion of the Doppler shift for a source moving at supersonic speed with the observer stationary, extend the discussion to a stationary source and an observer moving at supersonic velocity. Let the observer be on the line of motion of the source.

3.2   Sound Wave in a Fluid

3.2.1   Compressibility

We now turn to the dynamics of waves and start with some observations regarding the one-dimensional motion of a fluid column (gas, liquid) in a tube when it is driven at one end by a piston, as indicated in Fig. 3.3.

If the fluid is incompressible, it acts like a rigid body and if the tube is closed at the end and held fixed, it would not be possible to move the piston. The same conclusion is reached even with an open tube if is infinitely long since the mass of the fluid column would be infinite, thus preventing a finite force on the piston to accelerate the piston and the fluid.

In reality, we know that the piston indeed can be driven by a force of finite amplitude and that a sound wave can be generated in the tube in this manner. The fallacy of the conclusion that the piston cannot be accelerated lies in the assumption of an incompressible fluid; in order for a sound wave to be produced, compressibility is a necessary requirement and we shall pause here to review this concept.

Compressibility is the measure of the 'ease' with which a fluid can be compressed. It is defined as the relative change in volume per unit change in pressure. If, in a fluid element of volume V, an increase in pressure, ΔP, is associated with a change of volume, ΔV, the average value of the compressibility in this volume range is, by definition, $-(1/V)(\Delta V/\Delta P)$. The minus sign is included since an increase in pressure results in a decrease in volume. The corresponding 'local' value of the compressibility at the volume V is $\kappa = -(1/V)dV/dP$. Alternately, compressibility can be defined as the relative change of the density ρ per unit increase in pressure, $(1/\rho)d\rho/dP$, this time with a positive sign. Thus, the *compressibility* is

$$\kappa = -(1/V)(dV/dP) = (1/\rho)(d\rho/dP). \qquad (3.16)$$

The density ρ is a function of both pressure P and temperature T (or another pair of thermodynamic variables such as pressure and entropy) and the derivative $d\rho/dP$ is ambiguous without specifying the conditions under which the change of state takes place. From the elements of thermodynamics it is known that the equation of state of an ideal gas can be written $P = r\rho T = (R/M)\rho T$ where T is the absolute temperature (in Kelvin, K), $r = R/M$, the gas constant per unit mass, R, the universal gas constant (per mole), ≈ 8.3 joule/K, and M, the molar mass in kg. For air, $M \approx 0.029$ kg (for steam, $M \approx 0.018$ kg).

From the equation of state it follows that $dP/P = d\rho/\rho + dT/T$; for an *isothermal* change of state, the temperature is constant ($dT = 0$) so that $dP/P = d\rho/\rho$. This means that $\kappa = (1/\rho)d\rho/dP = 1/P$. For an *isentropic (adiabatic)* change of state, we have $dP/P = \gamma d\rho/\rho$, where $\gamma = C_p/C_v$ is the specific heat ratio, ≈ 1.4 for air; the compressibility then becomes $\kappa = 1/(\gamma P)$. Thus,

$$\kappa = \begin{cases} 1/P, & \text{isothermal} \\ 1/(\gamma P) & \text{isentropic.} \end{cases} \tag{3.17}$$

At a pressure of 1 atm ($\approx 10^5$ N/m^2), the isothermal value for air is $\kappa \approx 10^{-5}$ m^2/N; the isentropic value is smaller by a factor of γ.

Fluid is the generic term for liquids, gases, and plasmas (ionized gases). The compressibility of a liquid normally is much smaller than for a gas. For water it is about 10^{-5} times the value for normal air, and in the analysis of the dynamics of a liquid, incompressibility (i.e., $\kappa = 0$) is often assumed. On this basis, many important aspects of fluid dynamics can be analyzed and understood, but, as already stated, a compressibility different from zero is required where sound is involved.

3.2.2 Piston Source of Sound

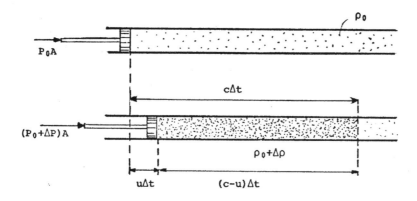

Figure 3.3: Sound generation by a piston in a tube.

With the fluid being compressible, let us consider what happens when the piston in Fig. 3.3 is moved forward with a velocity u during a time Δt. The velocity of a fluid element in contact with the piston will have the same velocity. Through intermolecular collisions this velocity will be transmitted as a wave with a certain wave speed c (as yet unknown) so that at the end of the time interval Δt the wave front has reached $x = c\Delta t$.

Since the piston has moved forward a distance $u\Delta t$ during that time, the length of the wave will be $(c - u)\Delta t$ at the end of the time interval Δt. The fluid velocity throughout this section is the same as that of the piston, i.e., u. Thus, a length $c\Delta t$ of the unperturbed fluid of density ρ_0 has been compressed to the length $(c - u)\Delta t$ with the density ρ. Conservation of mass requires that

$$c\rho_0 = (c - u)\rho. \tag{3.18}$$

For a weak compression, $(\rho - \rho_0)/\rho_0 << 1$, this ratio can be expressed in terms of the compressibility (see Eq. 3.17) as $d\rho/\rho = \kappa p$, where p is the increase in pressure resulting from the compression. Eq. 5.1 then becomes

$$u/c = (\rho - \rho_0)/\rho_0 \approx \kappa p. \tag{3.19}$$

The increase in pressure, p, must equal the force per unit area supplied by the piston. The corresponding impulse $p\Delta t$ delivered by the piston to the fluid column has produced the momentum $\rho u(c - u)\Delta t$ of the fluid column at the end of the time interval Δt, and it follows from Newton's law (in the form of the impulse-momentum relation) that

$$p = \rho u(c - u) \approx (\rho c)u. \tag{3.20}$$

In the last step in this equation, it is assumed that the velocity u of the piston is negligible compared to the velocity c of the wave. As we shall see shortly, this linear approximation is quite good in most of acoustics.

3.2.3 Sound Speed and Wave Impedance

The combination of Eqs. 3.19 and 3.20 yields the following expression for the *speed of sound*

$$\boxed{\begin{array}{c} \textit{Speed of sound; ideal gas} \\ c = \sqrt{1/(\kappa\rho)} = \sqrt{\gamma P/\rho} = \sqrt{\gamma RT/M} \end{array}} \tag{3.21}$$

[κ: *Compressibility.* ρ: *Density.* γ: *Specific heat ratio,* ≈ 1.4 *for air.* $R \approx 8.3$ *joule/K, universal gas constant (per mol). T: Absolute temperature (K). M: Molar mass,* ≈ 0.029 *kg for air. At* $20°C (T = 293 K)$, $c \approx 342.6$ *m/s for air.*]

We have assumed isentropic compression which turns out to be appropriate in most cases of sound propagation in free field. Within a porous material, on the other hand, the large heat conduction and heat capacity of the solid material prevent temperature fluctuations from occurring, and the compressibility becomes closer to isothermal, at least at sufficiently low frequencies.

The last step in Eq. 3.21 refers to an ideal gas. In that case, the *sound speed depends only on temperature.* This is consistent with the molecular model of sound propagation according to which the sound speed is expected to be approximately equal to the average thermal speed of the molecules which is known to be proportional to $\sqrt{T}$.

With $R = 8.3$ joule$\cdot K^{-1}$, $M = 0.029$ kg, $\gamma = 1.4$ and a temperature of 20°C (68°F, $T = 293K$) the sound speed in air becomes 343 m/sec (1125 ft/sec). At a temperature of 1000 °F, it is 570.6 m/sec (1872.5 ft/sec). The isothermal wave speed in normal air is smaller than the isentropic by a factor of $1/\sqrt{\gamma} \approx 0.845$ and is ≈ 290 m/sec. The experimental evidence for sound waves over a wide range of frequencies is in overwhelming favor of the isentropic value.

In the linear approximation, $u \ll c$, the sound speed is independent of the strength of the wave, i.e., independent of the fluid velocity u. However, had we not assumed $u \ll c$ in Eqs. 3.19 and 3.20, we would have found the wave speed to be $c + u$, where c is the sound speed at the slightly elevated temperature in the wave due to the compression.

It is sometimes convenient to express the compressibility in terms of the sound speed. Thus, with $c = \sqrt{1/\kappa\rho}$, we get

$$\kappa = 1/(\gamma P) = 1/(\rho c^2). \tag{3.22}$$

The motion thus described is a sound wave. In the linear approximation, $u \ll c$, the sound pressure p is proportional to the fluid velocity u, $p = (\rho c)u$, as obtained from Eq. 3.20. Actually, as derived in Eqs. 3.19 and 3.20, this relation is valid for a wave traveling in the *positive* x-direction, with u counted positive in this direction. The pressure p does not depend on the direction, and for a wave traveling in the negative x-direction, the fluid velocity becomes negative, and we have to put $p = -(\rho c)u$. Thus, for a plane traveling wave, the relation between sound pressure and fluid velocity is

$$
\boxed{
\begin{array}{ll}
\textit{Pressure–velocity relation in plane wave} & \\
p = (\rho c)u & \text{for wave in positive } x\text{-direction} \\
p = -(\rho c)u & \text{for wave in negative } x\text{-direction}
\end{array}
} \tag{3.23}
$$

The constant of proportionality ρc is called the *wave impedance* of the fluid,

$$
\boxed{
\begin{array}{l}
\textit{Wave impedance} \\
\rho c = \sqrt{\rho/\kappa}
\end{array}
} \tag{3.24}
$$

[ρ: *Density. c: Sound speed. For air at 1 atm and 20°C($T = 293$ K), $\rho \approx 1.27$ kg/m^3, $c \approx 342.6$ m/s, $\rho c \approx 435$ MKS. At 0 °C(273 K) the value is ≈ 420 MKS.*]

For water vapor at 1000°F (811 K) and a pressure of 1000 psi ($\approx 6.8 \cdot 10^6$ N/m^2), $\rho c \approx 12700$ MKS (i.e., about 32 times greater than for normal air). This latter condition is typical for the steam in a nuclear power plant.

The derivation of the relation between p and u was based on the analysis of a positive displacement of the piston of duration Δt generating a *compressional* wave. If the piston is moved in the negative direction, a *rarefaction (expansion)* wave is generated in which the perturbations of density, pressure, and fluid velocity will be negative. A succession of positive and negative pulses can be used to build up an arbitrary time dependence. Thus, for an infinitely extended tube or a tube with an absorber at the end so that no reflected sound is present, the relation $p = \rho cu$ is valid for *any time dependence* of a wave traveling in the positive x-direction.

If the tube extends to the left of the piston along the negative x-direction, a negative displacement (and velocity) of the piston gives rise to a compression wave in the negative x-direction but the velocity in this wave is in the negative x-direction and the relation between pressure and velocity is still $p = -\rho c u$. In the rarefaction wave generated in the positive x-direction, both pressure and velocity are negative so that $p = \rho c u$ is still valid.

Of particular interest is the harmonic time dependence. Then, for a wave traveling in the positive x-direction the pressure and velocity waves take the form

$$p(x, t) = |p| \cos(\omega t - kx)$$
$$u(x, t) = (|p|/\rho c) \cos(\omega t - kx), \tag{3.25}$$

where $k = \omega/c = 2\pi/\lambda$, as mentioned in Eq. 3.1. For a wave traveling in the negative x-direction, $-kx$ is replaced by kx and u by $-u$. Quantity $|p|$ is the pressure amplitude.

If the piston is located at $x = x'$ rather than at $x = 0$, the time of wave travel to the observation point x will be $(x - x')/c$ so that kx in Eq. 3.25 will be replaced $k(x - x')$. We can incorporate both directions of wave travel by replacing $\omega t = k(x - x')$ by $\omega t - k|x - x'|$.

Rms value.

The *mean square value* of the sound pressure is

$$\langle p^2(t) \rangle = (1/T) \int_0^T p^2(t)dt \tag{3.26}$$

and for a harmonic pressure wave, this becomes $|p|^2/2$. The square root of this quantity is the *rms-value* of the pressure which for the harmonic wave is

$$p_{rms} = |p|/\sqrt{2} \quad \text{(harmonic wave)}. \tag{3.27}$$

It is this value that is usually indicated on instruments that measure sound pressure and we shall often use the symbol p for it if there is no risk of misunderstanding.

Density and Temperature Fluctuation in Sound

With an isentropic compressibility in the change of state that occurs in a sound wave we have $dP/P = \gamma d\rho/\rho$. Then, with $dP = p$ follows that the density fluctuation that is caused by a sound pressure $dP = p$ becomes $d\rho = p/(\gamma P/\rho) = p/c^2$.

There is also a temperature fluctuation. From the equation of state $P = r\rho T$ follows $dP/P = d\rho/\rho + dT/T$, or $dT/T = (\gamma - 1)d\rho/\rho = (\gamma - 1)p/c^2$. The acoustic perturbation in temperature is then

$$dT = (\gamma - 1)T \, p/c^2 = \frac{\gamma - 1}{\gamma}(p/P)T. \tag{3.28}$$

In a plane wave, $u = p/\rho c$ so that $dT = T(\gamma - 1)(u/c)$.

Intensity

The work per unit area done by the piston in Fig. 3.3 as it moves forward during the time Δt is $pu\,\Delta t$; it is pu per unit of time. Conservation of energy requires that this energy must be carried by the wave. Thus, the wave energy per second and unit area is $I = pu$; it is called the *acoustic intensity*. Since $p = \rho cu$, it follows that

$$I(x,t) = p(x,t)u(x,t) = \rho cu^2(x,t) = (p^2(x,t)/\rho c). \qquad (3.29)$$

In the case of a single traveling wave with harmonic time dependence, $p(x,t) = |p|\cos(\omega t - kx)$, the intensity is $I(x,t) = \rho c|u|^2\cos^2(\omega t - kx) = |p||u|\cos^2(\omega t - kx)$. We are generally interested in the time average of the intensity which is $I = |p||u|/2$. The same notation, I, will be used for both, but when time is involved, it is shown explicitly as an argument, $I(t)$; without this argument, time average is implied. If rms values p and u are used for the amplitudes, the intensity is simply $I = pu = \rho cu^2 = p^2/\rho c$. For the traveling wave, it is independent of x.

At the *threshold of hearing*, with $p_r = 2 \times 10^{-5}$ N/m^2, the *threshold intensity* is $I_r \approx 10^{-12}$ w/m^2.

Acoustic Energy Density

The *energy density* in a wave is the sum of the kinetic energy density $\rho u^2(t)/2$ and the potential or compressional energy density which can be expressed as $\kappa p^2(t)/2$, where $\kappa = 1/(\gamma P) = 1/\rho c^2$ is the compressibility. In a single traveling wave, with $p = \rho cu$, these quantities are the same and if the total energy density is denoted $W = \rho u^2/2 + \kappa p^2/2$, it follows that

$$I = Wc$$
$$W = \rho u^2/2 + \kappa p^2/2. \qquad (3.30)$$

In a single traveling wave, the pressure is ρcu, and the corresponding reaction force on the piston that drives the wave is $A\rho cu$, proportional to the velocity like a viscous friction force. If the piston is part of a harmonic oscillator, the power transferred to the wave results in damping of the oscillator, usually referred to as *radiation damping* and the wave impedance ρc is often called *wave resistance*.

Complex Amplitude Description

Suppose a problem has been solved for the complex pressure amplitude $p(\omega)$ and the corresponding velocity $u(\omega)$. How do we use these amplitudes to express the intensity in the sound field? To find out, we go back to the corresponding real quantities, $p(t)$ and $u(t)$, and express these quantities in terms of the complex amplitudes. This is facilitated with the aid of *complex conjugate* quantities. With reference to Appendix B, we are reminded that the complex conjugate of a complex number $z = r + ix$ is $z^* = r - ix$, so that $r = (z + z^*)/2$.

Thus, we express $p(t)$ as $p(t) = (1/2)(p(\omega)\exp(-i\omega t) + p^*(\omega)\exp(i\omega t)$ and $u(t) = (1/2)(u(\omega)\exp(-i\omega t) + u^*(\omega)\exp(i\omega t)$. The time average intensity

$I = \langle p(t)u(t) \rangle$ then becomes $I = (pu^* + p^*u)/4$. We note that pu^* is the complex conjugate of p^*u so that the sum is twice the real part of pu^*. Thus, with $p = |p|\exp(i\phi_1)$ and $u = |u|\exp(i\phi_2)$, we have $u^* = |u|\exp(-i\phi_2$ and $pu^* = |p||u|\exp(i\phi)$, where $\phi = \phi_1 - \phi_2$. Thus,

$$I = (1/2)\Re\{pu^*\} = (1/2)|p||u|\cos(\phi), \qquad (3.31)$$

where ϕ is the phase difference between pressure and velocity. If the amplitudes are rms values, the factor 1/2 has to be eliminated.

Intensity Probe

An intensity probe consists of two closely spaced microphones in combination with a two-channel FFT (Fast Fourier Transform) analyzer. The sum of the output signals is the average sound pressure between the microphones and the difference represents the gradient of the pressure, respectively. The particle velocity is proportional to the gradient and the product of these quantities yields the intensity. In terms of the signals from the two microphones, this turns out to proportional to the cross spectrum density of these signals, which is automatically determined by the analyzer. All that remains is a constant of proportionality which can be incorporated in the signal processing program.

The formal derivation of this result is given below. It is based on the Fourier transforms of the pressure and the velocity and is quite similar to the derivation of the energy spectrum density discussed in Chapter 2.

Derivation

The sound pressure $p(x, t)$ is expressed in terms of its Fourier amplitude $p(\nu)$, i.e.,

$$p(x, t) = \S p(x, \nu)e^{-i2\pi\nu t}d\nu \qquad (3.32)$$

and the particle velocity $u(x, t)$ in the x-direction in terms of its Fourier amplitude is $u(x, \nu)$. Then, from the momentum equation $\rho du/dt = -\partial p/\partial x$ it follows that $u(x, \nu) = (1/i\omega\rho)\partial p/\partial x$.

The intensity in the x-direction is

$$I(t) = p(x, t)u(x, t) = \S p(x, \nu)e^{-i2\pi\nu t}d\nu\S(1/i\omega\rho)\partial p(x, \nu')/\partial x e^{-i2\pi\nu' t}d\nu'$$
$$= (1/i\omega\rho)\S\S e^{-i2\pi(\nu+\nu')t}d\nu d\nu'. \qquad (3.33)$$

Integrating $I(t)$ over all times produces $\delta(\nu+\nu')$ and integration over ν' yields a contribution only if $\nu' = -\nu$ and we obtain

$$\S I(t)dt = (1/i\omega\rho)\S p(x, \nu)\partial p(x, -\nu)/\partial x \, d\nu. \qquad (3.34)$$

The microphones are located at $x - d/2$ and $x + d/2$ at which points the pressures are p_1 and p_2. We put $p(x) = (p_1 + p_2)/2$ and express the gradient as $\partial p(x)/\partial x = (p_2 - p_1)/d$. With $p(-\nu) = p^*(\nu)$, the integrand in Eq. A.9 becomes $(p_1 + p_2)(p_2^* - p_1^*)$. Neglecting the term $|p_2|^2 - |p_1|^2$ and realizing that $p_2 p_1^*$ is the complex conjugate of $p_1 p_2^*$, the remaining $p_1 p_2^* - p_2 p_1^*$ is twice the imaginary part of $p_1 p_2^*$. Thus, we obtain

$$\S I(t)dt = (1/i\omega\rho d)\S 2\Im\{p_1(\nu)p_2^*(\nu)\}\,d\nu \equiv \S I(\nu)d\nu, \qquad (3.35)$$

where the intensity spectrum is

$$I(\nu) = (2/\omega\rho d)\Im\{p_1(\nu)p_2^*(\nu)\}. \tag{3.36}$$

With the signals from the two microphones analyzed with a two-channel analyzer, the quantity $p_1(\nu)p_2^*(\nu)$, the cross spectrum density, is obtained directly from the two-channel FFT analyzer.

3.2.4 Acoustic Levels. Loudness

Sound Pressure Level

The sound pressures normally encountered in practice cover a wide range, from the threshold value of hearing, $\approx 10^{-5}$ N/m^2 up to pressures of the order of 1 atm, $\approx 10^5$ N/m^2. This represents a range of about 10^{10}; the range for the corresponding intensities and powers then will be about 10^{20}. Under those conditions, it is convenient to introduce a logarithmic scale for sound intensity such that the ratio of two intensities is expressed as $I_1/I_2 = p_1^2/p_2^2 = 10^B$, where $B = \log_{10}(I_1/I_2)$ is the intensity ratio expressed in Bel. Actually, a unit *decibel, dB*, which is 10 times smaller is generally used, so that

$$dB = 10\log_{10}(I_1/I_2)^2 = 20\log_{10}(p_1/p_2). \tag{3.37}$$

If the rms value of p_2 is taken to be the hearing threshold $p_r = 2 \times 10^{-5}$ N/m^2, the dB-value is referred to as the *sound pressure level*, SPL. For example, at the threshold value, the SPL is 0 and if $p_1 = 2$ N/m^2, the SPL is 100 dB. The threshold pressure is approximately the threshold of hearing of the average human of a pure tone at 1000 Hz. Similarly, if I_2 is taken to be the corresponding reference intensity 10^{-12} w/m^2, the corresponding dB value is called the *intensity level*.

The *acoustic power* going through an area A is $\Pi = IA$, where I is the average intensity I over the area. The power that corresponds to the reference intensity $I_r = 10^{-12}$ w/m^2 and an area of 1 m^2 is the reference power $\Pi_r = 10^{-12}$ w. The power level of an acoustic power Π expressed in dB.

$$PWL = 10\log(\Pi/\Pi_r) \tag{3.38}$$

The acoustic power of a source can be measured by means of an intensity probe by integrating the normal component of the intensity over a control surface surrounding the source. The accuracy of this procedure is best in a free field environment. Another method is to place the source in a reveruberation room and measure the average sound pressure level in the room. Then, from the measured reverberation time of the room, the power output of the source can be determined as described in Chapter 6.

Loudness and Equal Loudness Contoers

Loudness is the subjective measure of the 'strength' of a sound. The threshold of hearing depends on frequency as indicated by the bottom curve in Fig. 3.4. At 1000 Hz, the threshold sound pressure level is set equal to 0. At frequencies below 1000 Hz, the threshold value of the sound pressure level is higher; at 100 Hz, for example,

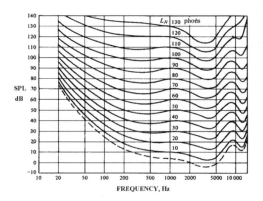

Figure 3.4: Equal loudness contours according to international standards, (ISO).

it is about 23 dB. With the sound pressure level of the 1000 Hz tone of 10 dB, for example, the sound is, of course, audible with a certain loudness. The *loudness level* by definition is the same as the sound pressure level of the tone at 1000 Hz. The sound pressure level required at another frequency to make it sound as loud as the 1000 Hz tone can readily be determined experimentally and the results obtained over a frequency range from 20 to 10000 Hz are indicated by the curves in Fig. 3.4. They are referred to as the *equal loudness contour* for the *loudness level, phons*. This value of the loudness level L_N in phons, by definition, is the same as the sound pressure level of the 1000 Hz reference tone. The loudness level is often referred to simply as the *sound level*.

In a similar manner the contours at other values of the loudness level can be obtained with the results shown in Fig. 3.4. It is significant that the frequency dependence of the contours depends on the loudness level L_N; the increase of the counters at low frequencies becomes less pronounced with an increasing loudness level.

The loudness level of a complex tone containing a band of frequencies can be obtained experimentally in an analogous manner by comparing its loudness with the reference tone at 1000 Hz.

An instrument designed to measure the loudness level, a sound level meter, contains a standardized frequency weighting network based on the equal loudness contours. With the frequency weighting factor denoted $A(f)$, the output of the meter, the loudness level or sound level in dBA, is

$$dBA = 10\log\left[\int A(f)E(f)df/p_r^2\right], \qquad (3.39)$$

where $E(f)$ is the spectrum density and p_r the rms value of the reference sound pressure at the hearing threshold at 1000 Hz. For all the contours, $A(1000) = 1$, by definition. For the zero phon contour, $A(f) < 1$ for $f < 1000$ and decreases with decreasing frequency with a corresponding difference between the loudness level and the SPL. The difference increases with decreasing frequency; for example, at 500 Hz, it is about 2.5 dB and at 100 Hz, close to 23 dB. In the range between 1000 and 5000 Hz, $A(f)$ is somewhat larger than 1 with a maximum close to 4000 Hz, indicating maximum sensitivity of the ear. In Section 3.2.5 we have attempted to understand the weighting function $A(f)$ in terms of the frequency response of the ear drum to an incoming wave. The result is summarized in Fig. 3.6, where also $A(f)$ is shown.

Hearing Damage Risk. Annoyance

The loudness level (sound level) is commonly used in efforts to correlate the effects of noise on man with some physical measure of the sound. For example, the sound level that is considered to present risk for hearing damage in industry is usually considered to be 90 dBA for an 8 hour daily exposure. The risk level increases with decreasing time of exposure; thus, it is set to be 95 dBA for 4 hrs, 100 dBA for 2 hrs, 105 dBA for 1 hr, 110 dBA for 1/2 hr, and 115 dBA for 1/4 hr and below. At higher exposures, hearing protection devices should be used. Federal legislation concerning industrial noise exposure covers this subject in great detail.

Criteria regarding the annoyance of noise can be found in local community ordinances. Typically, a night time criterion level is 40 dBA.

Loudness, Sones

Loudness N is the quantity used to subjectively rank sounds of different loudness levels. It has been found experimentally that a sound which is judged to have the same loudness as a 1000 Hz (reference) tone with a sound level L_N is N times louder (subjectively) than a reference tone of 40 dBA, where

$$N \approx 2^{(L_N-40)/10}. \tag{3.40}$$

The scale thus defined expresses loudness in *sones* and is valid in the range 20 to 120 dBA for L_n.

It follows from this relation that a doubling of the loudness corresponds to an increase in the sound level by 10 dBA.

The smallest detectable change in loudness (difference limen, loudness) as been found to correspond to a change in the sound level of 2 to 3 dBA.

The effect of multiple sources on loudness depends on their correlation. The mean square value of the sum of two pressures $p = p_1(t) + p_2(t)$ is $\langle p^2 \rangle = \langle (p_1 + p_2)^2 \rangle = \langle p_1{}^2 \rangle + \langle p_2{}^2 \rangle + 2\langle p_1 p_2 \rangle$. If the pressures are uncorrelated, $\langle p_1 p_2 \rangle = 0$, and the mean square value of p is the sum of the individual mean square values. Then, if $\langle p_1{}^2 \rangle = \langle p_2{}^2 \rangle$, the sound pressure level of the sum of the two pressures will be $10 \log(2\langle p_1^2 \rangle / p_0^2) = L + 10 \log(2) \approx L + 3$ dB, where L is the sound pressure level of each of the two pressures. Thus, a doubling of the intensity or power increases the sound pressure level by 3 dB. The same holds true for the dBA value if the frequency spectra of the sources are the same. The corresponding change in the loudness is then barely noticeable.

If the pressures are perfectly correlated, as is the case if $p_2 = p_1$, the sum of the two pressures leads to an increase of the sound pressure level of $10 \log(4) \approx 6$ dB. In this case the change in loudness is quite apparent.

3.2.5 Hearing Sensitivity and Ear Drum Response

With reference to the discussion in Section 3.2.4, we present here an attempt to understand the dBA weighting function $A(f)$ in Eq. 3.39 in terms of the frequency response of the ear drum to an incoming sound wave.

The ear canal is terminated by the ear drum which is connected to the bones in the middle ear. They transmit the sound-induced motion of the ear drum to the fluid filled inner ear, where the fluid motion is sensed and converted into electrical impulses which are carried by the auditory nerves and then decoded in the brain.

The frequency dependence of the 'sensitivity' of the ear was discussed in connection with the well-known contours in Fig. 3.4 upon which the weighting function $A(f)$ is based. How is the sensitivity related to the motion of the ear drum? Is it determined by the frequency dependence of the displacement, velocity, or acceleration of the ear drum? Or is it the sound pressure spectrum at the ear drum that is essential? We will try to answer this question by using a simple model of the ear canal and a knowledge of the input impedance of the ear drum.

Measurements of this impedance have indeed been carried out, see for example A. R. Möller, *J. Acous, Soc. Am.* **32**, 250-257, (1960), and we shall use these data here. They cover a frequency range from 200 to 2000 Hz and represent the average of the results obtained from measurements on ten different ears. In Fig. 3.5 are shown smoothened versions of the frequency dependence of the normalized resistance and the magnitude of the reactance; the data have been extrapolated down to 100 and up to 10000 Hz. We treat the ear canal as a straight, uniform tube which is terminated by the ear drum. The acoustic field variables at the entrance and the end of the tube are labeled by the subscripts 1 and 2. Then, with reference to Section 4.4.5 and with the transmission matrix elements of the ear canal denoted T_{ij}, we have

$$p_1 = T_{11}p_2 + T_{12}\rho c u_2$$
$$\rho c u_1 = T_{21}p_2 + T_{22}\rho c u_2. \tag{3.41}$$

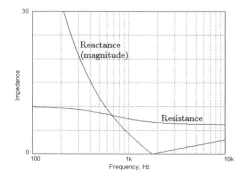

Figure 3.5: Smoothened curves for the frequency dependence of the normalized input resistance and reactance of the ear drum, based on data from Möller referenced in the text. The experimental data, covering the range 200 Hz to 2000 Hz, to have been extrapolated to the range 100 to 10000 Hz.

Furthermore, with the normalized impedance of the eardrum denoted ζ_2, it follows that $p_2 = \rho c u_2 \zeta_2$ and if this is used in the first of these equations, the pressure response p_2/p_1 and the velocity response $\rho c u_2/p_1$ can be obtained directly.

However, as indicated above, it is the *incident* sound pressure p_i rather than the pressure at the entrance of the ear which is involved in the experimental data on the hearing threshold, and p_1 has to be expressed in terms of p_i. This can be done as follows. The scattered pressure at the ear can be expressed as $p_s = -\zeta_r \rho c u_1$, where ζ_r is the normalized radiation impedance of the ear (a negative sign has to be used because the definition of ζ_r refers to a velocity in the *outward* direction and not the inward direction which is implied in the definition of u_1). We also introduce the normalized input impedance of the ear which follows directly by dividing the relations in Eq. 3.41,

$$\zeta_i = \frac{p_1}{\rho c u_1} = \frac{T_{11}\zeta_2 + T12}{T21\zeta_2 + T_{22}}. \tag{3.42}$$

Then, with $p_1 = p_i - \rho c u_1 \zeta_r$ and $\zeta_i \rho c u_1 = p_1$, we get

$$p_1 = \frac{\zeta_i}{\zeta_i + \zeta_r}\, p_i. \tag{3.43}$$

Frequency Responses of the Ear Drum

Using this expression for p_1 in Eq. 3.41, we can express the velocity of the ear drum in terms of the incident sound pressure p_i,

$$\frac{\rho c u_2}{p_i} = \frac{\zeta_i}{\zeta_i + \zeta_r} \frac{1}{T_{11}\zeta_2 + T_{12}}. \tag{3.44}$$

For ζ_r we could use the radiation impedance of a piston in a rigid sphere, as can be found in acoustics texts,[1] but, for the present purpose, it is sufficient to a simple approximate expression

$$\zeta_r \approx 0.25(ka)^2/(1 + 0.25(ka)^2) - ika/[1 + (ka)^2], \tag{3.45}$$

where a is the radius of the ear canal and $k = \omega/c$.

It remains to discuss the transmission matrix elements. For a loss-free tube we have to anticipate the results in Eq. 4.116. To account for the flexibility of the tube walls and a corresponding wave attenuation in the tube, the matrix elements have to be modified to

$$T_{11} = T_{22} = \cos(k_x d)$$
$$T_{12} = -i(k_x/k)\sin(k_x d), \quad T_{21} = (-i(k/k_x)\sin(k_x d), \tag{3.46}$$

where $k_x \approx k\sqrt{1 - \eta/kd_c}$ and $k = \omega/c$. The quantity η is the normalized admittance of the wall and d_c is the diameter of the ear canal. The complex rather than the

---

[1] See, for example, Morse and Ingard, Theoretical Acoustics, (1968), p 343.

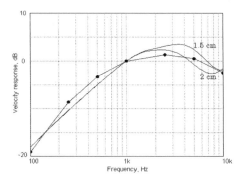

Figure 3.6: Frequency response in dB of velocity amplitude of the ear drum using the value at 1000 Hz as a reference. The computed curves obtained for two different lengths of the ear canal, 1.5 and 2 cm, are shown. The standard experimentally based A-weight function for the frequency dependence of the hearing threshold is outlined by the plot markers.

approximate expression for k_x cannot be expressed in closed form and we refer to the quoted reference for further details.

To obtain the numerical result for the velocity response $\rho c u_2 / p_i$ in Fig. 3.6, we have used the approximate expression for k_x and a frequency independent value of 1.5 for η. The velocity response is expressed as $20 \log[u_2(f)/u_2(1000)]$. The radius of the ear canal has been set equal to 0.5 cm and two lengths of the canal, 1.5 cm and 2.0 cm, have been used as indicated in the figure. Decreasing η ($\eta = 0$ corresponds to the hard-walled tube) produces an increase in the maximum of the response curve but does not significantly change the response below 1000 Hz. The computed response curve thus obtained has the same general shape as the accepted standardized experimentally based 'A-weighting' function which is outlined in the figure by the plot markers.

The calculated *displacement* response of the ear drum has an entirely different shape; it is almost a constant at frequencies below 1000 Hz and is about 20 dB higher than the velocity response at 100 Hz. Similarly, the corresponding acceleration response curve starts out about 20 dB lower than the velocity response at 100 Hz. Finally, the pressure response is almost constant below 1000 Hz and is about 20 dB higher than the velocity response at 100 Hz.

Consequently, according to this analysis, there is good reason to believe that it is the *velocity amplitude* of the ear drum that is directly correlated with the frequency dependence of the sensitivity of the ear as expressed by the A-weighting function.

Another point to observe is that the maximum sensitivity is not solely a result of a 'quarter wavelength' resonance of the ear canal, as is sometimes stated. The frequency dependence of the ear drum impedance is an equally important factor. The experimental results indicate (see Fig. 3.5 that the reactance becomes zero which means resonance of input impedance of the ear drum) at a frequency of about 1740 Hz. The maximum sensitivity in our response curve occur at a higher frequency which is true also for the threshold curve in Fig. 3.4, close to 4000 Hz.

3.2.6 Problems

1. **Sound speed in high temperature steam**

   The steam (use $M = 0.018$ kg) in a power plant generally has a temperature of $1000°$F and a pressure of 1000 psi.
   (a) What is the speed of a sound wave in the steam? How does the pressure of the steam affect the result?
   (b) What is the wave impedance?

2. **Sound pressure level**

   Consider the sum of two harmonic sound pressures, $p = A \cos(\omega t) + B \cos(\omega t - \phi)$.
   (a) What is the resulting sound pressure level as a function of ϕ if $A = B = 1$ N/m^2?
   (b) If the sound pressure levels of the individual sound pressures are $L_1 = 80$ dB and $L_2 = 85$ dB, what is the resulting sound pressure level if $\phi = 0$?

3. **Sound pressure, particle velocity, displacement, and temperature fluctuation**

   Determine the amplitudes (rms) of sound pressure, particle velocity, particle displacement, and temperature fluctuation in $°$C in a plane sound wave with a frequency of 1000 Hz and a sound pressure level of 100 dB.

4. **Examples of power levels**

   Typical values of the total acoustic power outputs from a jet engine, a pneumatic hammer, and ordinary conversational speech are 10 kw, 1 w, and 20 microwatt, respectively.
   (a) What are the corresponding power levels?
   (b) What is the sound pressure level of a sound with an rms value equal to 1 atm?

5. **Density fluctuations and laser performance**

   In a pulsed laser, the performance of the laser was found to be affected by the density variations produced by the ignition pulses. The reason is that the index of refraction and hence the speed of light depends on the gas density and spatial variations in it will then distort the optical wave fronts in the laser. If the distortion is sufficiently high, lasing cannot be achieved. In a particular installation it was estimated that the density fluctuation amplitude should be less than 10^{-5} of the static value for normal operation of the laser. What is then the highest permissible sound pressure level in dB in the laser cavity if the static pressure is 1 atm?

3.3 Waves on Bars, Springs, and Strings

3.3.1 Longitudinal Wave on a Bar or Spring

Wave motion on bars, springs, and strings is analogous to that on a fluid column, considered in Section 3.2. In a solid bar and one-dimensional motion, the quantity that corresponds to the compressibility of a fluid is $1/Y$, where Y is the Young's modulus. The wave speed becomes

$$v = \sqrt{Y/\rho} \qquad (3.47)$$

and the wave impedance, as before, is ρv, where ρ is the density. The pressure p in the sound wave is now replaced by the stress σ in the material, i.e., the force per unit area of the rod.

The wave power and intensity has the same form as for the sound wave with the sound pressure replaced by the stress.

The Young's modulus in N/m^2 and density in kg/m^3 are for

Steel: $Y \approx 20 \times 10^{10}$, $\rho = 7.8 \times 10^3$

Aluminum, rolled: $Y = 6.9 \times 10^{10}$, $\rho = 2.7 \times 10^3$

Tungsten, drawn: $Y = 35.5 \times 10^{10}$, $\rho = 14 \times 10^3$.

For a (coil) spring, the situation is not much different; the Young's modulus is replaced by the compliance per unit length. With the spring constant of a spring of length L being K, the compliance is $C = 1/K$ and the compliance per unit length, $\kappa = C/L$. The mass per unit length μ of the spring takes the place of the mass density in the gas. Thus, the longitudinal wave speed on the spring will be

$$v = \sqrt{1/(\kappa\mu)} = \sqrt{KL/\mu}. \tag{3.48}$$

Example

A longitudinal harmonic wave with a frequency of $f = 10^6$ Hz is generated in a steel bar by a piezo-electric crystal mounted at the end of the bar. What should be the displacement amplitude of the bar in order that the energy flux (intensity) of the wave be 10 w/cm^2? Density $\rho = 7.8$ g/cm^2. What speed $v = 5300$ m/sec?

With reference to the text, the ratio of the force per unit area and the particle in a single wave on the bar is equal to the wave impedance $Z = \rho v$, where ρ is the mass density and v the longitudinal wave speed. The wave power per unit area, the intensity I, is then obtained as the time average

$$I = Z\langle u^2 \rangle = \rho v \langle u^2 \rangle = (1/2)\rho v \omega^2 \langle \xi_0^2 \rangle,$$

where the angle brackets indicate time average. In the last step, we have accounted for the harmonic time dependence in relating the velocity u and the displacement ξ; the factor $1/2$ is due to time averaging.

Expressing the displacement amplitude in terms of intensity, we obtain the displacement amplitude $\xi_0 = \sqrt{2I/(\rho v \omega^2)}$. Numerically, with $\rho = 7800$ kg/m^3, $v = 5300$ m/sec, and $\omega = 2\pi 10^6$, we obtain $\xi_0 = 9.8 \times 10^{-9}$ m, i.e., about 100 Ångström.

3.3.2 Torsional Waves

A rod can carry not only a longitudinal but also a torsional wave. To study torsional wave motion, we proceed by analogy with the discussion of the longitudinal wave in connection with Fig. 3.3. Thus, during a time Δt, the end of the rod is twisted with an angular velocity $\dot{\theta}$. The angular displacement travels along the rod as a wave and at the end of the time interval Δt, the front of the angular velocity wave has reached the position $x = v\Delta t$, where v is the wave velocity, yet to be determined; beyond the front, there is no angular displacement. We have already calculated the torque in the discussion of the torsional pendulum in Eq. 2.47; the length of the rod used there has to be replaced by the length $v\Delta t$ of the activated region here. Since there is no displacement beyond the wave front, the response of the rod is the same as if the rod had been clamped, as was the case in Eq. 2.47. Thus,

$$\tau = (G\pi a^4/2)\theta/(v\Delta t) = (GI/v)\dot{\theta}$$
$$I = \pi a^4/2, \tag{3.49}$$

where G is shear modulus and ρI is the moment of inertia per unit length.

After the time interval Δt, the bar contains the angular momentum $\rho I \dot{\theta}(v\Delta t)$ and, from conservation of angular momentum, this must equal the angular impulse $\tau \Delta t$, i.e.,

$$\tau = \rho I v \dot{\theta}. \tag{3.50}$$

Combining these equations leads to the wave speed

$$v = \sqrt{G/\rho}. \tag{3.51}$$

The driving torque τ, (Eq. 3.50), is proportional to the angular displacement velocity $\dot{\theta}$; the constant of proportionality $\rho I v$ is analogous to the wave impedance ρv for the longitudinal wave.

Wave ladder demonstration. A 'wave ladder' consists of a long vertical torsion wire or metal band which is held fixed at its upper end and with equally (and closely) spaced bars or dumbbells mounted along its entire length. If the lowest dumbbell is given an angular displacement, a torsion wave is produced which travels up on the ladder. The speed of this wave is determined by the torsion constant and the moment of inertia of the dumbbells and can be made very low so that the wave motion can be easily observed. The excitation can be a pulse or a harmonic motion of the lowest rod. The wave speed and the wavelength can be measured using simply a ruler and a stop watch. Various wave phenomena, such as reflection and transmission at the junction of two ladders with different wave speeds, can be demonstrated.

Problem

1. **Wave damper**

   Discuss the feasibility of an electromagnetic damping device at the end of the ladder consisting of a conducting disk oscillating in the field of a magnet. The damping is provided by the induced current in the disk and the electrical resistance of the disk.

3.3.3 Transverse Wave on a String. Polarization

The string considered here is limp (i.e., it has no bending stiffness). This is an idealization which is satisfactory in most cases. In order to have wave motion, some form of restoring force is required, and for the limp string, it is provided by a static tension S (the symbol T would have been better, but it is reserved for the period of oscillation).

The wave motion to be studied involves a transverse displacement, but otherwise the arguments given for the sound wave on an air column still apply in principle (Fig. 3.3). Instead of an axial velocity on an air column, we now generate a transverse velocity on a string by a driving force F at the end of the string during a time Δt (Fig. 3.7). This disturbance travels out on the string as a velocity wave. At the end of

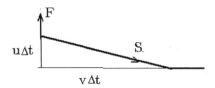

Figure 3.7: Transverse wave displacement on a string under tension. The displaced portion of the string has a transverse velocity u, the same as the velocity of the driving point.

the time interval Δt, the driving point of the string has moved sideways a distance $u\Delta t$ whereas the front of the wave has reached the position $v\Delta t$ along the string, where v is the wave speed, yet to be determined. The displacement of the string at time Δt is then as shown in Fig. 3.7. If the transverse velocity rather than the displacement had been plotted, it would have been a square wave with a velocity u in the displaced portion (the same as at the driving point) and zero elsewhere.

The slope of the displacement is $\tan(u/v) \approx u/v$ with the approximation u/v valid for small displacements. The component of the tension force S in the transverse direction has to be matched by the driving force, i.e.,

$$F = S(u/v). \tag{3.52}$$

The impulse delivered by the force during the time Δt is $F\Delta t$ which must equal the momentum of the activated portion $v\Delta t$ of the string, so that $F\Delta t = u\mu v\Delta t$ or

$$F = (\mu v)u, \tag{3.53}$$

where μ is the mass per unit length. This has the same form as the relation between p and u in the sound wave, the *wave impedance* now being μv.

Combining these two equations yields the *wave speed* on the string

$$v = \sqrt{S/\mu}. \tag{3.54}$$

This has the same form as the wave on a rod with S and μ taking the places of Y and ρ.

Wave power. If the transverse velocity of the string is u, the power transferred from the driving force to the string is $W = Fu = (v\mu)u^2 = F^2/(\mu v)$; in a traveling wave, this power is carried along the string. As for the sound wave, there is a corresponding energy density E such that $W = Ev$, where v is the wave speed; i.e., $E = \mu u^2$. The kinetic energy density is $\mu u^2/2$ which makes up half of the total. The remaining half, the potential energy density, has the same value since the total is μu^2 and can be expressed as $(1/S)F^2/2$ with $1/S$ taking the place of the compressibility in the corresponding expression $\kappa p^2/2$ for the potential energy density in a sound wave in a fluid.

Polarizer for Waves on a String

The wave on a string considered above involved a transverse displacement η in the y-direction for a wave traveling in the x-direction, $\eta = |\eta| \cos(\omega t - kx)$. In this respect, it resembles more an electromagnetic wave than does a sound wave, and, as for the electromagnetic wave, the concepts of a plane of polarization and linearly polarized wave have meaning.   In our case, the plane of polarization of the string wave is the xy-plane. If another similar wave with a displacement in the xz-plane, $\zeta = |\zeta| \cos(\omega t - kx)$ is superimposed on the first, and the resulting wave will be linearly polarized in a plane inclined at an angle $\psi = \arctan(|\eta|/|\zeta|)$ with respect to the z-axis.

If this wave is incident on a 'polarizer,' consisting of a rigid screen with a horizontal, frictionless slot in the y-direction, the y-component of the wave goes through the screen unperturbed but the z-component will be totally reflected since the slot forces the amplitude to be zero at the screen. The corresponding reflection coefficients for the velocity and displacement are then -1. In other words, the polarizer breaks up the wave into two linearly polarized components, one transmitted and the other reflected.

If the z-component of the incident displacement wave lags the y-component by a phase angle $\pi/2$ so that it is $\zeta = \zeta_0 \sin(\omega t - kx)$, the wave represents a counter-clockwise swirling motion, a *circularly polarized* wave. Again, the polarizer lets through the y-component and reflects the z-component with a reflection coefficient for displacement of -1. This reflected wave component combines with the incident wave to form a standing wave but superimposed on it is the traveling y-component of the displacement.

3.3.4   Problems

1. **Radiation load on an oscillator**

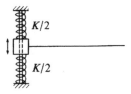

Figure 3.8: Damping of oscillator by a string wave.

A mass spring oscillator ($M = 2\,\text{kg}$, $K = 32\,\text{N/m}$) is connected to a long string (tension $S = 100\,\text{N}$, mass per unit length $\mu = 0.25\,\text{kg/m}$). The mass element is sliding on a horizontal frictionless guide bar, as shown in Fig. 3.8.

(a) What is the nature of the effect of the string on the oscillator, i.e., is it equivalent to a mass-, stiffness-, or resistive load?

(b) What is the Q-value of the oscillator, accounting for the effect of the string?

(c) The oscillator is started from an initial displacement $A = 5\,\text{cm}$. Indicate the shape and length of the wave on the string at the time when the amplitude of the oscillator has decreased to the value $1/e$ of the initial displacement.

2. **Maximum wave speed on a string**

Consider a string of length L clamped on both ends. To make the wave speed as high as possible the tension is brought up to correspond to a stress (force per unit area) equal to the tensile strength of the material. For steel, the tensile strength is 3.2×10^{10} and for aluminum, $.8 \times 10^9$ dyne/cm2. The mass density of steel is 7.8 and for aluminum, 2.7 g/cm^2. Will the wave speed exceed the speed of sound in air, 340 m/s?

Are the following answers correct? Steel: 640 m/s. Aluminum: 320 m/s.

3. **Wave energy on a string**

The end of a string is driven at $x = 0$ in harmonic motion with a frequency 10 Hz and a displacement amplitude 0.2 m. The wave speed on the string is 10 m/sec and the mass per unit length 0.001 kg/m.
(a) Calculate the time average power in watts by the wave.
(b) What is the average wave energy per unit length of the string?
(c) What is the change in power and wave energy density if the tension of the string is doubled?

Are the following answers correct? (a): 7.9 w. (b): 0.79 joule/m. (c): Power increase by a factor of $\sqrt{2}$. Energy density remains the same.

4. **The complex Young's modulus**

A long glass rod is driven at one end with a transducer producing a longitudinal wave in the bar. The mass density is $\rho_p = 2.5$ g/cm^3 and the Young's modulus is $Y_0 = 6 \times 10^{11}$ dyne/cm^2. The loss factor is $\epsilon = 0.1$ so that the complex modulus is $Y = Y_0(1 - i\epsilon)$.
(a) What is the wave speed in the rod (neglect losses)?
(b) Accounting for the loss factor, what is the expression for the x-dependence of the complex amplitude of displacement in the rod? Determine the attenuation in dB in a distance of 10 m.
(c) Explain why the expression for the complex Young's modulus has to be $Y = Y_0(1 - i\epsilon)$ rather than $Y = Y_0(1 + i\epsilon)$ using our sign conventions in the definition of a complex amplitude according to which the x-dependence of the complex amplitude of a wave traveling in the positive x-direction is $\exp(ikx)$, where $k = \omega/v$ and v the wave speed.

3.4   Normal Modes and Resonances

3.4.1   Normal Modes and Fourier Series

The standing wave in Eq. 3.6 was a special case of a one-dimensional wave field involving waves traveling in both the positive and negative x-directions. In that case the amplitudes of the waves were the same. In a more general wave field the amplitudes are different both in magnitude and phase and a complex amplitude description of a harmonic one-dimensional pressure field takes the form

$$p(x, \omega) = A \exp(ikx) + B \exp(-ikx), \tag{3.55}$$

where A and B are complex constants and $k = \omega/c$. Until these constants are specified, the wave field applies to any one-dimensional problem. The corresponding velocity field follows from the equation of motion, $-i\omega\rho u = -\partial p/\partial x$,

$$u(x, \omega) = (1/\rho c)[A \exp(ikx) - B \exp(-ikx)]. \tag{3.56}$$

The constants A and B are now chosen in such a manner that the fields apply to a pipe of length L closed at both ends with rigid acoustically hard walls. This means that the velocity amplitude at both ends $x = 0$ and $x = L$ must be zero, i.e.,

$$A - B = 0$$
$$A \exp(ikL) - B \exp(-ikL) = 0. \qquad (3.57)$$

The first of these refer to $x = 0$ and yields $B = A$. In order that the condition at $x = L$ be satisfied for a value of A different from zero we must have $\sin(kL) = 0$ (recall $[\exp(ikL) - \exp(-ikL)] = 2i \sin(ikL)$]). Thus, only the frequencies satisfying this condition are possible, i.e., $\omega_n L/c = n\pi$ or

$$v_n = \omega_n/2\pi = nc/2L \quad \text{or} \quad \lambda_n = 2L/n. \qquad (3.58)$$

These frequencies are called the *characteristic, 'eigen' or normal mode* frequencies of the pipe and the corresponding wave fields the *normal or 'eigen' modes*

$$p_n(x) = 2A_n \cos(k_n x) = 2A \cos(n\pi x/L)$$
$$u_n(x) = i(1/\rho c)2A_n \sin(n\pi x/L), \qquad (3.59)$$

where we have used $(1/2i)[\exp(ikL) - \exp(-ikL)] = \sin(kL)$. The factor $i = \exp(i\pi/2)$ in the expression for the velocity merely means that it lags behind the pressure by an angle $\pi/2$. To obtain the time dependence, we have to multiply by $\exp(-i\omega_n t)$ and take the real part of the product. For the pressure mode, the resulting time function is $\cos(\omega_n t - \phi_n)$, where ϕ_n is the phase angle of A_n. The corresponding function for the velocity mode is $\cos(\omega_n t - \phi_n - \pi/2) = \sin(\omega_n t - \phi_n)$.

The modes are called 'orthogonal' because the integral of the product of two different modes over the length L is zero,

$$\int_0^L \sin(n\pi x/L) \sin(m\pi x/L)$$
$$= (1/2) \int_0^L (\cos[(m-n)\pi x/L] - \cos[(m+n)\pi x/L] = 0 \quad (m \neq n). \qquad (3.60)$$

If $m = n$ the result is $(L/2)$. Sometimes the normal mode wave function is normalized to make this integral equal to unity. This wave function is $\Psi_n(x) = \sqrt{2/L} \sin(n\pi x/L)$.

An arbitrary function in this region can be expanded in a series of normal modes in the same manner as in the Fourier expansion in Chapter 2. Thus, if at $t = 0$ there is a pressure distribution $p(x, 0)$, it can be expressed as

$$p(x, 0) = \sum_0^\infty P_n \cos(k_n x). \qquad (3.61)$$

The coefficients P_n are obtained by multiplying both sides with $\cos(k_n x)$ and integrating over L to yield

$$P_n = (2/L) \int_0^L p(x, 0)\, dx, \qquad (3.62)$$

for $n > 0$ and $P_0 = (1/L) \int_0^L p(x, 0) \, dx$. The time dependence of each mode will be $P_n \cos(k_n x) \cos(\omega_n t - \phi_n)$. In other words, all the modes and corresponding frequencies will generally be excited. We shall pursue this question in more detail later in connection with the motion of a string.

3.4.2 The 'Real' Mass-Spring Oscillator

We are now prepared to reexamine the motion of a mass-spring oscillator accounting for the mass of the spring which was ignored in the analysis in Chapter 2 although the shortcoming of this omission was discussed.

As we have seen above, the longitudinal wave motion on a coil spring is similar to that of an air column with the air density replaced by the mass μ per unit length and the compressibility by the compliance $1/KL$ per unit length, where K is the spring constant. The wave speed on the spring is then $v = \sqrt{KL/\mu}$.

The spring is anchored at $x = L$ and the other end is connected to a mass M which is driven in harmonic motion by a force $F(0, t) = |F| \cos(\omega t)$ at $x = 0$. The corresponding complex amplitude is $F(0, \omega)$. Proceeding by analogy with the sound wave in the pipe above, the force wave and the corresponding velocity wave on the spring will have the complex amplitudes

$$F(x, \omega) = A \exp(ik_x) + B \exp(-ikx)$$
$$u(x, \omega) = (1/\mu v)[A \exp(ikx) - B \exp(-ikx)], \tag{3.63}$$

where $k = \omega/v$. The velocity must be zero at $x = L$, which yields $A \exp(ikL) = B \exp(-ikL)$. The force at the beginning of the spring is then $A + B$ and the velocity $(1/\mu v)(A - B)$, where μv is the wave impedance, corresponding to ρc for the air wave. The input impedance of the spring is then

$$Z_s = F(0, \omega)/u(0, \omega) = (A + B)/(A - B) = i\mu v \cot(kL), \tag{3.64}$$

where we have use $2 \cos(kL) = \exp(ikL) + \exp(-ikL)$ and $2i \sin(kL) = \exp(ikL) - \exp(-ikL)$. This expression can be applied to an air layer in a tube closed at one end by replacing the wave impedance μv by ρc.

As a check on the input impedance we consider very low frequencies for which $\cot(kL) \approx 1/kL = v/\omega L$. The impedance is then $Z_s = i\mu v^2/\omega L = iK/\omega$, where we have used the expression for the wave speed above, $v = \sqrt{KL/\mu}$. This result is familiar from Chapter 2.

The total input impedance of the oscillator, including the mass, is then $Z(\omega) = -i\omega M + Z_s$ and with the complex amplitude of the driving force on the oscillator being $F(\omega)$, we get for the velocity of M

$$u(\omega) = F(\omega)/Z(\omega) = F(\omega)/[-i\omega M + i\mu v \cot(kL)]. \tag{3.65}$$

The resonance frequencies of the system are obtained from $Z(\omega) = 0$ which can be written $(\omega M/\mu v) \tan(kL) = 0$, or, with $k = \omega/v$,

$$kL \tan(kL) = m/M, \tag{3.66}$$

where we have introduced the mass of the spring, $m = \mu L$.

If the wavelength is much greater than the length of the spring, so that $kL \ll 1$, we get $\tan(kL \approx kL$, and $(kL)^2 \approx m/M$. It follows then that $\omega_0^2 = v^2 m/ML^2$ or, with $m = \mu L$ and $v^2 = KL/\mu$, $\omega_0 \approx \sqrt{K/M}$, as in Chapter 2. To get an improved approximation of the lowest resonance frequency, we include one more term in the expansion of $\tan(kL)$, i.e., $\tan(kL) \approx kL + (kL)^3/3 \approx kL(1 + m/3M)$. Using this expression in Eq. 3.66, we get for the corresponding resonance frequency

$$\omega_0 \approx \sqrt{K/(M + m/3)}. \tag{3.67}$$

In effect, one-third of the spring mass should be added to M to get the influence of the spring mass on the lowest frequency.

Higher mode frequencies are obtained by solving the equation numerically. They can also be deduced by examining the frequency dependence of the velocity or displacement, as obtained from Eq. 3.65. It is instructive to study the frequency response of the displacement for different values of m/M, where m is the spring mass and M the load mass. (Normalize the frequency with respect to $\omega_0 = \sqrt{K/M}$ corresponding to zero spring mass.) The idealized mass-spring oscillator, with $m = 0$, has only one resonance, but the real oscillator has, theoretically, an infinite number.

If $M = \infty$, the resonance frequencies are given by $\tan(kL) = 0$, i.e., $kL = n\pi$, where $n = 1, 2, \ldots$. The resonances then occur when the length L is an integer number of half wavelengths, analogous to the pipe closed at both ends. $M = 0$, the resonance frequencies are obtained from $\cot(kL) = 0$, i.e., $kL = (2n - 1)\pi/4$, in which case the length is an odd number of quarter wavelengths. It corresponds to an open pipe closed at one end.

Oscillator Response; Analysis without the Use of Complex Amplitudes

For comparison, it is instructive to reconsider the problem above and indicate how it should be handled without the use of complex amplitudes.

As before, the force wave on the spring will be a superposition of waves in the positive and negative x-direction, i.e.,

$$F(x, t) = A \cos(\omega t - \phi_a - kx) + B \cos(\omega t - \phi_b + kx), \tag{3.68}$$

where $k = \omega/v = 2\pi/\lambda$ and ϕ_a and ϕ_b are phase angles, as yet unknown. The corresponding velocity field, according to Eq. 3.25, is

$$u(x, t) = (A/\mu v) \cos(\omega t - \phi_a - kx) - (B/\mu v) \cos(\omega t - \phi_b + kx). \tag{3.69}$$

The boundary condition at the end of the spring is $u(L, t) = 0$ from which it follows

$$A \cos(\omega t - \phi_a - kL) - B \cos(\omega t - \phi_b + kL) = 0.$$

We expand both the terms using the identity $\cos(\omega t - \alpha) = \cos(\omega t) \cos(\alpha) + \sin(\omega t) \sin(\alpha)$. Then, if the condition is to be satisfied at all times, the coefficients for the $\cos(\omega t)$ and $\sin(\omega t)$-terms must be zero individually so that two equations are obtained.

At the driving end $x = 0$ end, the equation of motion of M provides the condition

$$F\cos(\omega t) - F(0, t) = M du(0, t)/dt.$$

Also here we express the time dependence in terms of $\cos \omega t)$ and $\sin(\omega t)$-terms and require that the coefficients for these terms be the same on the two sides of the equation. This yield an additional two equations. Thus, in all we have four equations for the determination of the amplitudes A and B and the phase angles ϕ_a and ϕ_b.

Although straight-forward, this procedure is quite cumbersome and unattractive; nevertheless, it might be useful to the reader to carry it out for comparison. With increasing complexity of the problem, this approach becomes even more intractable.

3.4.3 Effect of Source Impedance

The analysis of the frequency response of the mass-spring oscillator in the previous section is directly applicable to the forced motion of a piston at the entrance of a tube of length L producing a sound wave in the tube. We then replace the complex amplitude of the driving force $F(0, \omega)$ by the sound pressure $p(0, \omega)$. If $M = 0$, the solution applies to the forced motion of an acoustic tube resonator. If the driving pressure $p(0, \omega)$ at the tube entrance is constant, independent of frequency, the frequency response is expressed by the factor $1/\cos(kL)$, where $k = \omega/c$. If, instead, the velocity amplitude at the entrance to the tube is kept constant, this factor will be replaced by $1/\sin(kL)$ (see Problem 3).

In other words, the response depends on the character of the source, whether it provides a frequency independent velocity or pressure amplitude at the entrance. This property of the source is often described in terms of the *internal impedance* of the source. If this impedance is very high, the velocity will be essentially independent of the load impedance and if the source impedance is very low, it will be the driving pressure that will be frequency independent. These two types of sources are referred to as 'constant velocity' and 'constant pressure' sources. If the tube is driven by a piston of mass M and a frequency independent harmonic force, the source impedance becomes the inertial reactance of the piston; the constant velocity source then corresponds to a very heavy piston and the constant pressure source, to a very light piston.

In electrical circuits, another property of a source is the 'electromotive force.' The analogous quantity for an acoustic source would be an 'internal pressure,' as illustrated in the following example.

In Fig. 3.9 is shown schematically an electrodynamic loudspeaker. It consists of a coil placed between the poles of a magnet. A cone-shaped piston is attached to the coil and is set in motion when a time dependent current is sent to the coil. In this example, we assume that the current $I(t)$ is harmonic with a frequency independent amplitude. The force on the coil will be BIL, where L is the length of the wire in the coil and B the magnetic field. The sound pressure difference on the two sides of the cone is Δp and if the coil-cone assembly is described as a harmonic oscillator, mass M, spring constant K, and resistance R, the equation of motion of the assembly becomes

$$IBL = \Delta p A + M \partial u/\partial t + K \int u dt + Ru, \tag{3.70}$$

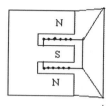

Figure 3.9: Schematic of an electrodynamic loudspeaker.

where A is the equivalent piston area of the cone.

For harmonic time dependence, we get

$$\Delta p = (IBL/A) - z_i u, \tag{3.71}$$

where $z_i = (1/A)(-i\omega M + iK/\omega + R)$ is the equivalent internal impedance of the source and $p_i = IBL/A$, the 'internal pressure,' both per unit area.

In the design of an efficient loudspeaker we are interested in a 'smooth' response of the sound pressure on the outside of the cone and irregularities in the inside pressure contribution to the pressure difference Δp due to frequency response of the air in the speaker cabinet must be considered carefully and should be eliminated or used appropriately for best performance.

3.4.4 Free Motion of a String. Normal Modes

As we have seen in Section 3.3, the properties of one-dimensional sound waves are directly applicable to other waves, including the transverse waves on a string. All that is needed is to replace the sound pressure by the transverse force, the sound speed c by the wave speed $v = \sqrt{S/\mu}$, and the wave impedance ρc by μv, where S is the string tension and μ the mass per unit length.

By analogy with the normal modes of sound in a pipe closed at both ends, Eq. 3.59, the normal modes of displacement of a string of length L clamped at both ends are given by

$$\eta_n(x, t) = A_n \sin(n\pi x/L) \cos(\omega_n t - \phi_n), \tag{3.72}$$

where $(\omega_n/v)L = n\pi$, i.e., $\nu_n = \omega_n/2\pi = nv/2L$ where ϕ_n depends on the initial condition of the spring. If it is started from rest, $\phi_n = 0$.

3.4.5 Forced Harmonic Motion of a String

The unperturbed string is along the x-axis. The displacement in the y-direction is $\eta(x, t)$. An external force $f(x, t)$ is acting on it per unit length. The tension in the string is S. We isolate an element of length Δx and apply Newton's law to it. First we have the external force $f(x, t)\Delta x$ in the y-direction. Second, there is the tension acting on the two sides of the element from the rest of the string. If the displacements of the string is $\eta(x)$ the slope is $\partial\eta/\partial x$ and the y-component of the tension S acting on the element from the left $F(x) = -S\partial\eta/\partial x$ (with a positive slope of the string, the tension force acting on the element has a downwards component, hence the minus sign). Accounting for both ends of the element, the net transverse force from the tension is $F(x + \Delta x) - F(x) = -(\partial F/\partial x)\Delta x$. Newton's law then takes the form

$$\mu\partial u/\partial t = -\partial F/\partial x + f(x, t). \tag{3.73}$$

It is combined with the expression for $F = -S\partial\eta/\partial x$. Since the equation of motion contains the velocity u rather than the displacement, we express F also in terms of it and obtain after time differentiation and with $\kappa = 1/S$

$$\kappa\partial F/\partial t = -\partial u/\partial x, \tag{3.74}$$

where $\kappa = 1/S$.

Differentiating the first of these equations with respect to t and the second with respect to x, we can eliminate F and obtain

$$\partial^2 u/\partial x^2 - k^2\partial^2 u/\partial t^2 = -\kappa\partial f/\partial t, \tag{3.75}$$

where $k = \omega/v$ and $v^2 = 1/\kappa\mu$.

Since we are interested in harmonic time dependence, we introduce the complex amplitudes $u(x, \omega)$ and $f(x, \omega)$. Then, with $\partial/\partial t \rightarrow -i\omega$, where ω is the angular frequency of the driving force, we get

$$d^2 u(x, \omega)/dx^2 + (\omega/v)^2 u(x, \omega) = i\omega\kappa f(x, \omega). \tag{3.76}$$

Next, the functions u and f are expanded in terms of the normal modes of the spring,

$$u(x) = \sum u_n \sin(k_n x), \quad f(x) = \sum f_n \sin(k_n x), \tag{3.77}$$

where $k_n L = n\pi$ and $k_n = \omega_n/v$. With $d^2 u/dx^2 = -\sum k_n^2 u_n \sin(k_n x)$, this equation reduces to

$$\sum (k^2 - k_n^2)u_n \sin(k_n x) = i\omega\kappa \sum f_n \sin(k_n x). \tag{3.78}$$

This equation is satisfied for all values of x only if

$$u_n(\omega) = i\omega\kappa f_n/(k^2 - k_n^2) = -i\omega\kappa(f_n/k_n^2)/(1 - \Omega_n^2), \tag{3.79}$$

where $\Omega_n = \omega/\omega_n$ is the normalized frequency. With $u_n = -i\omega\xi_n$, the corresponding displacement amplitude of the nth mode is

$$\xi_n(\omega) = A_n/(1 - \Omega^2), \tag{3.80}$$

where $A_n = \kappa f_n/k_n^2 = (1/n\pi)^2(f_n l/S)L$. The function ξ_n has the same form as the frequency response of the harmonic oscillator. In this expression A_n is the 'static' displacement of the nth mode of the string, corresponding to $\Omega = 0$.

This analysis is another example of the considerable importance of the harmonic oscillator to which we devoted considerable time in Chapter 2. Thus, by decomposing the displacement of a continuous system into its normal modes, the response to an external force can be described in terms of harmonic oscillator response functions, one for each mode.

Example

Consider a force density distribution with harmonic time dependence (frequency ω) concentrated at the location x' of the string and described by the delta function $f(x) = F(x)\delta(x - x')$. Fourier expansion of this function, $\sum f_n \sin(k_n x)$, has the coefficients $f_n = (2/L)\int(\delta(x - x')F(x)\sin(k_n x) = (2/L)F(x')\sin(k_n x')$. Then, according to Eq. 3.79, the complex velocity amplitude of the string becomes

$$u(x) = (-i\omega\kappa)(2/Lk_n^2)F(x')\sum \sin(k_n x')\sin(k_n x)/(1 - \Omega_n^2). \qquad (3.81)$$

The corresponding displacement function $\eta(x)$ is obtained by dividing by $-i\omega$). With $\kappa = 1/S$ and $k_n L = n\pi$, the displacement function then can be written

$$\eta(x, x') = L(2/n\pi)^2[F(x')/S]\sin(k_n x')\sin(k_n x)/(1 - \Omega^2). \qquad (3.82)$$

It is the displacement amplitude at x caused by a harmonic force $F(x')$ at x' at x'. If this force has unit magnitude, the function is often referred to as the harmonic Green's function. The displacement at x produced by a uniform force distribution is obtained by integrating over x'.

It is important to note that the function $\sin(k_n x')$ expresses the 'coupling' of the driving force to the nth normal mode. If the location is such that $\sin(k_n x') = 0$, the nth mode will not be excited. The tonal quality, harmonic composition of the sound produced by plucking a string depends on where it is plucked.

3.4.6   Rectangular Membrane

The derivation of the equation of motion of a membrane is analogous to that of a string, as described by Eqs. 3.73 and 3.74. The unperturbed membrane is in the xy-plane and the displacement ζ is in the z-direction. The tension S in the membrane is the force per unit length in a cut of the membrane. Then, by analogy with Eq. 3.74 for the string, the z-component of tension force along the edge of length Δy at x is $F(x, y) = -S[\partial\zeta(x, y)/\partial x]\Delta y$. The corresponding component acting on the element along the edge at $x + \Delta x$ is $S[\partial\zeta(x + \Delta x, y)/\partial x]\Delta y$ and the combination of the two is $S\partial^2\zeta/\partial x^2 \Delta x\Delta y$. There is a similar force from the two Δx edges so that the total force on the element will be $S[\partial^2\zeta/\partial x^2 + \partial^2\zeta/\partial y^2]$.

Then, with the mass per unit area of the membrane denoted μ (not to be confused with the mass per unit length of a string), the equation of motion becomes

$$\mu \, \partial^2\zeta/\partial t^2 = S[\partial^2\zeta/\partial x^2 + \partial^2\zeta/\partial y^2], \qquad (3.83)$$

which is the wave equation for the displacement. With harmonic time dependence, $\partial/\partial t \to -i\omega$ and $\partial^2/\partial t^2 \to -\omega^2$ so that the corresponding equation for the complex displacement amplitude $\zeta(x, y, \omega)$ becomes

$$\partial^2\zeta/\partial x^2 + \partial^2\zeta/\partial y^2 + (\omega/v)^2\zeta = 0, \qquad (3.84)$$

where $v = \sqrt{S/\mu}$ is the wave speed on the membrane.

This is the (wave) equation for the free motion of the membrane. (In this form it is sometimes called the Helmholtz equation.)

We consider the case when the membrane is clamped along its edges at $x = 0$, $x = L_1$, and $y = 0$, $y = L_2$. The equation and the boundary conditions of zero displacement are then satisfied by a normal mode function

$$\zeta_{mn}(x, y) = A \sin(k_m x) \sin(k_n y), \qquad (3.85)$$

where $k_m L_1 = m\pi$ and $k_n L_2 = n\pi$, where m and n are integers.

Insertion of this displacement in Eq. 3.84 yields the expression for the corresponding normal mode frequency

$$\omega_{mn} = v\sqrt{k_m^2 + k_n^2} = v\sqrt{(m\pi/L_1)^2 + (n\pi/L_2)^2}. \qquad (3.86)$$

The mnth mode has $m - 1$ nodal lines perpendicular to the x-axis and $n - 1$ nodal lines perpendicular to the y-axis. The normal mode can be regarded as a standing wave and crossing a nodal line results in a change of sign of the function.

3.4.7 Rectangular Cavity

The normal modes of sound in a rectangular room will be discussed separately in Chapter 6 and we refer to this and the next three chapters for details. We present here merely the expression for the normal mode functions and the corresponding normal mode frequencies. Thus, for a sound pressure field with harmonic time dependence, with $p(x, y, z, t) = \Re\{p(x, y, z, \omega)\exp(-i\omega t)\}$, the three-dimensional wave equation becomes

$$\partial^2 p/\partial x^2 + \partial^2 p/\partial y^2 + \partial^2 p/\partial z^2 + (\omega/c)^2 p = 0, \qquad (3.87)$$

where $c = \sqrt{1/\kappa\rho}$ is the sound speed, κ, the compressibility and ρ, the density.

This equation describes the free acoustic oscillations of the air in the room. With the walls of the room acoustically hard, so that the normal particle velocity is zero at the walls, and with the dimensions of the room L_1, L_2, L_3 with the origin at one of the corners, the normal modes are

$$p_{\ell mn} = A\cos(k_1 L_1)\cos(k_2 L_2)\cos(k_3 L_3), \qquad (3.88)$$

where $k_1 = \ell\pi/L_1$, $k_2 = m\pi/L_2$, and $k_3 = n\pi/L_3$. Then, from Eq. 3.87 follows the corresponding normal mode frequencies

$$\omega_{\ell mn} = c\sqrt{k_1^2 + k_2^2 + k_3^2}. \qquad (3.89)$$

3.4.8 Modal Densities

The number of normal modes with frequencies below a specified frequency and the number of modes in a given frequency range play an important role in many areas of physics and engineering from the theory of specific heat to the acoustics of concert halls.

The frequency of the nth mode on a string clamped at $x = 0$ and $x = L$ is $v_n = nv/2L$, where v is the wave speed. The number of modes with frequencies below a value v is $N(v) = 2Lv/v$. We define the modal density in 'frequency space' as $n(v) = daN(v)v$, i.e.,

$$N(v) = 2Lv/v, \qquad n(v) = dN(v)/dv = 2(L/v) \quad \text{(one-dimensional)}. \qquad (3.90)$$

In 'k-space,' with $k_n = n\pi/L$, we get $N(k) = (L/\pi)k$ and $n(k) = L/\pi$.

For two-dimensional waves, we consider as an example the modes of a rectangular membrane in Section 3.4.6. We have $k_m = m\pi/L_1$ and $k_n = n\pi/L_2$. The modes are identified by points in a two-dimensional k-space in which the axes are k_m and k_n. The spacing between adjacent points on the two axes are π/L_1 and π/L_2 and the average 'area' in k-space occupied by one mode is $\pi^2/(L_1 L_2)$. For a sufficiently large value of k, the number of normal modes with k_{mn} less than k can be expressed as $N(k) = (\pi k^2/4)/(\pi^2/(L_1 L_2))$, where $\pi k^2/4$ is the area in k-space enclosed by the circle of radius k in the quadrant between the positive axes k_m and k_n. With $k = 2\pi v/v$, the corresponding expression for $N(v) = (L_1 L_2/v^2)\pi v^2$. Thus,

$$n(v) = dN(v)/dv = (L_1 L_2/v^2)2\pi v \quad \text{(two-dimensional)}$$
$$N(k) = (k^2 L_1 L_2/4\pi) \qquad n(k) = \partial N(k)/\partial k = (L_1 L_2/2\pi)k. \qquad (3.91)$$

In a completely analogous manner we obtain for the density of the acoustic modes in the rectangular cavity

$$n(v) = (L_1 L_2 L_3/c^3)4\pi v^2 \quad \text{(three-dimensional)}. \qquad (3.92)$$

It should be noted that the modal density increases with the size of the system, the length L of the string, the area $L_1 L_2$ of the membrane, and the volume $L_1 L_2 L_3$ of the cavity. In many engineering problems, this size effect can be of considerable importance in regard to the risk of exciting resonances and generating instabilities of oscillation of structures due to the interaction of structural and acoustic modes, as discussed in Chapter 7.

3.4.9 Problems

1. **Response of a tube resonator**

   (a) Determine the input impedance of an air column in a tube of length L and closed at the end with an acoustically hard wall.

   (b) What is the frequency dependence of the maximum pressure and velocity in the tube if the velocity amplitude of the driven end is independent of frequency? Do the same if the driving *pressure* is independent of frequency. In each case determine the frequencies at which the sound pressure at the end of the tube will be a maximum.

2. **Orthogonality of normal modes**

   (a) The normal modes in a tube resonator of length L, open at one end and closed at the other, are such that $kL = (2n - 1)\pi/2$, where $k = 2\pi/\lambda = \omega/c$. Show that the integral over L of the product of two normal mode wave functions with different values of n is zero.

(b) What are the normal mode wave functions for the pressure in an organ pipe of length L which is open at both ends (assume sound pressure is zero at the ends)? Also show that the modes are orthogonal.

3. **Resonance frequencies of a tube resonator**

A piston of mass M rides on the air column in a vertical tube of length L, closed at the bottom with a rigid wall.

(a) By analogy with the analysis of the forced harmonic motion of the 'real' mass-spring oscillator in the text determine the frequency response of the sound pressure in the tube.

(b) Obtain an equation for the resonance frequencies of this oscillator and indicate how the equation can be solved graphically.

4. **Effect of spring mass in the mass-spring oscillator**

Following the outline in the text, prove the approximate expression (3.67) for the fundamental frequency of a mass-spring oscillator accounting for the mass of the spring.

5. **Equivalent source characteristics**

A loudspeaker is mounted on the side of a tube a distance L from the closed end of the tube. The tube is so long that reflections from the other end can be ignored. The loudspeaker has an internal impedance z_s and an equivalent internal pressure p_s. As far as sound generation in the tube is concerned, regard the speaker in combination with the closed end tube section as an equivalent source placed at the location of the speaker and determine the internal impedance and the internal pressure of this source. The area of the tube is A and the area of the speaker A_s.

6. **Maximum frequency of a string**

What is the maximum frequency of the fundamental mode that can be obtained with of a 1 m long string, (a) of steel, (b) of aluminum? Look up the tensile strengths and the density of these materials in an appropriate handbook. Does the result depend on the diameter of the string (neglecting bending stiffness)?

7. **Forced harmonic motion of a string**

A string of length L and clamped at both ends is driven by a harmonic force with a frequency one-tenth of the fundamental normal mode frequency. What is the (relative) amplitudes of the first five modes of the string if the force is applied (a), at the center of the string, $x' = L/2$, and (b), at $x' = 3L/4$?

3.5   The Flow Strength of a Sound Source

In Fig. 3.10 is shown a more general piston source than the one considered in Fig. 3.3; it is used here to introduce the concepts of the *flow strength* of an acoustic source. The source can be regarded as a thin 'pill box' with moving side walls representing pistons. The velocities of the pistons are the same in magnitude but opposite in direction so that the box pumps air in and out of the source region. Thus, if the velocity of the piston on the right-hand side is $u(t)$, it is $-u(t)$ on the left. The mass flow rate out of the source region per unit area is then $2\rho u(t)$. From what we have seen so far,

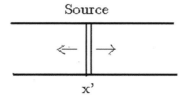

Figure 3.10: Acoustic piston source pair as a basis for the one-dimensional flow strength of an acoustic source.

it is clear that this source will generate a plane wave in both the positive and negative x-directions.

An equivalent mass flow rate out of the source region can be obtained also if we had a mass injection or creation in the gas at a rate Q_f per unit volume. (It can be shown that *heat generation* at a rate H per unit volume is acoustically equivalent to a mass flow rate injection $Q_f = (1/c^2)(\gamma - 1)H$, where γ is the specific heat ratio and c the sound speed.)

Then, if the thickness of the box located at $x = x'$ is $\Delta x'$, this equivalence requires that $Q_f(x', t) \Delta x' = 2\rho u(x', t)$.

From the relation between the plane wave pressure and the velocity, $p = \rho c u$ (see Eq. 3.20), it follows that the pressure field contribution at a location x from a source at x' becomes $\Delta p(x, x', t) = (c/2)Q_f(x', t - |x - x'|/c)$, where we have used $\rho u = \Delta x' Q_f/2$ and the fact that there is a time delay $|x - x'|/c$ (i.e., $x - x'/c$ for $x > x'$ and $-(x - x')/c$ for $x < x'$) between the emission at the source and the arrival of the emitted sound at x. Integrating over the source region, we then obtain the pressure field from a finite source distribution

$$p(x, t) = (c/2) \int Q_f(x', t - |x - x'|/c)dx', \qquad (3.93)$$

where the integral extends over the source region.

As an example, consider harmonic time dependence, $Q_f(x', t) = |Q(x')| \cos(\omega t)$ and an amplitude $|Q_f(x')| = Q_f$ independent of x' in the region between $x = -L$ and $x = L$ and zero outside. At a point of observation to the right of the source region, i.e., $x > L$, we have $|x - x'| = x - x'$ and the integral becomes

$$p(x, t) = (Q_f c/2) \int_{-L}^{L} \cos[\omega t - k(x - x')] \, dx' = (Q_f L)\frac{\sin(kL)}{kL} \cos(\omega t - kx), \qquad (3.94)$$

where we have made use of $\sin(A + B) = \sin(A)\cos(B) + \cos(A)\sin(B)$.

Let us see if this result makes sense. First we note that if the source region is small compared to the wavelength, i.e., $kL \ll 1$, the factor $\sin(kL)/kL \approx 1$, and the result is $Q_f L \cos(\omega t - kx)$, i.e., a traveling wave with an amplitude equal $Q_f L$, i.e., half of the total source strength $2LQ_f$. This is as it should be since there is a wave also in the negative x-direction with the same amplitude. If $kL = \pi$, i.e., $2L = \lambda$, the amplitude will be zero. This is also as expected, since for each source element on the right of center there is an element on the left which is half a wavelength removed so that their pressure contribution arrive at x out of phase by 180 degrees and thus cancel each other.

It is left as a problem to calculate the pressure field inside the source region. The integral now has to be broken up into two parts, one for $x' < x$ and one for $x' > x$.

From a physical standpoint, the equivalent source distribution introduced here is not very satisfactory if depicted as the rate of 'mass creation' or mass injection per unit volume. In a one-component fluid, such as a neutral gas, there is no mass creation and mass injection from a foreign object (a tube or the like) is not properly a volume source and should be treated as a boundary condition. In a multi-component gas, however, such as weakly ionized gas, there are three components, the neutrals, the electrons, and the ions. Recombination of electrons and ions leads to the creation of neutrals so that in the equation for sound generation in the neutral gas component, there is indeed a mass creation source of sound.

A source distribution due to a heat source with a heat transfer H' per unit volume is shown in Section 7.9 to be equivalent to a flow source $Q_f (\gamma - 1)/c^2)H'$, where Q_f is the mass transfer rate per unit volume. It can be realized in practice either through combustion or absorption of radiation. Its effect is equivalent to that of a flow source. Thus, sound can be generated by a modulated laser beam in a gas if it contains molecules with an absorption line at the laser frequency. In fact, this effect has been used as a tool in gas analysis.

3.5.1 Problems

1. **Field inside a uniform source distribution**

   With reference to Section 3.5 and the uniform source distribution in a tube, calculate the sound field inside the source region, following the suggestion made in that section.

2. **Nonuniform source distribution**

   Instead of the uniform source distribution leading to the field in Eq. 3.94 use a source distribution given by $Q = |Q| \cos(kx'/4L)$ and determine the pressure field outside the source region.

3.6 Sound on the Molecular Level

Sound, unlike light, requires matter for its existence and can be regarded as a molecular interaction or collision process.

In a naive one-dimensional model, the molecules in a gas may be pictured as identical billiard balls arranged along a straight line. We assume that these balls are initially at rest. If the ball on the left end of the line is given an impulse in the direction of the line, the first ball will collide with the second, the second with the third, and so on, so that a wave disturbance will travel along the line. The speed of propagation of the wave will increase with the strength of the impulse. This, however, is not in agreement with the normal behavior of sound for which the speed of propagation is essentially the same independent of the strength. Thus, the model is not very good in this respect.

Another flaw of the model is that if the ball at the end of the line is given an impulse in the opposite direction, there will be no collisions and no wave motion. A gas, however, can support both a compression and rarefaction wave.

Thus, the model has to be modified to be consistent with these experimental facts. The modification involved is to account for the inherent thermal random motion of the molecules in the gas. Through this motion, the molecules collide with each other even when the gas is undisturbed (thermal equilibrium). If the thermal speed of the molecules is much greater than the additional speed acquired through an external impulse, the time between collisions and hence the time of communication between the molecules will be almost independent of the impulse strength under normal conditions. Through collision with its neighbor to the left and then with the neighbor to the right, a molecule can probe the state of motion to the left and then 'report' it to the right, thus producing a wave that travels to the right which involves a transfer of a perturbation of molecular momentum.

The speed of propagation of this wave, a sound wave, for all practical purposes will be essentially the thermal speed since the perturbation in molecular velocity typically is only one-millionth of the thermal speed. Only for unusually large amplitudes, sometimes encountered in explosive events, will there be a significant amplitude dependence of the wave speed as demonstrated in Chapter 10. Thus, like the thermal speed, the wave speed (sound speed) will be proportional to $\sqrt{T}$, where T is the absolute temperature.

Chapter 4

Sound Reflection, Absorption, and Transmission

4.1 Introduction

In Chapters. 2 and 3, complex amplitudes were gradually introduced in the analysis of simple problems and it was mentioned that with increased problem complexity, the advantage of complex variables becomes more apparent. This will be further illustrated in this and subsequent chapters. Actually, after having become used to solving problems in this manner, it often becomes difficult to do it any other way.

4.1.1 Reflection, an Elastic Particle Collision Analogy

A ball thrown against a rigid wall bounces back with the same speed as the incident if the collision is elastic. This 'reflection' is not unlike what happens when a sound wave strikes a rigid, impervious wall; it is reflected with no change in strength.

Consider next a head-on collision between an incident ball, the projectile, and a stationary ball, the target. The masses and initial velocities of these balls are M_1, M_2, and U_1, U_2. It is well known (from billiards, for example) that if the masses are the same, the projectile comes to rest after the collision and the target acquires the velocity of the projectile.[1] If the masses are not the same, we find the velocities U_1' and U_2' of the projectile and the target after the collision to be such that

$$T_U = U_2'/U_1 = 2M_1/(M_1 + M_2). \tag{4.1}$$

These results follow from the equations for conservation of momentum and energy (see Problem 1). The quantities R_U and T_U can be considered to be reflection and transmission coefficients for velocity.

As we shall find shortly, the expressions for the coefficients of reflection and transmission of a wave at the junction between two transmission lines have the same form if the masses are replaced by the wave impedances Z_1 and Z_2 of the lines. This does not mean that the physics involved is identical in the two cases but in a loose sense, the analogy is intuitively helpful.

---

[1] To give the game of billiards an additional dimension, let one ball (or more) be heavier than the others but the same size.

4.1.2 Gaseous Interface

Let us reexamine the example of the reflection of a sound at the boundary between air
and helium columns separated by a limp membrane of negligible mass. We choose
$x = 0$ at the boundary. The fluid velocities of the incident, reflected, and transmitted
waves at $x = 0$ are u_1, u_1', and u_2'. Thus, the total velocity in the air at $x = 0$ is $u_1 + u_1'$
and in the helium, it is u_2'. We have then assumed that the helium column is infinitely
long so that we need not be concerned with any reflected wave in it.

According to Eq. 3.23, the sound pressure is $p = Zu$ for wave travel in the positive
and $p = -Zu$ for travel in the negative x-direction, where $Z = \rho c$ is the wave
impedance of the material that is carrying the wave.

Thus, the total pressure at $x = 0$ can be written $Z_1(u_1 - u_1')$ in the air and $Z_2 u_2'$
in the helium. The boundary conditions of continuity of velocity and of pressure at
$x = 0$ are then expressed by

$$u_1 + u_1' = u_2'$$
$$Z_1(u_1 - u_1') = Z_2 u_2' \tag{4.2}$$

from which follows the *reflection and transmission coefficients* for velocity,

$$R_u = u_1'/u_1 = (Z_1 - Z_2)/(Z_1 + Z_2)$$
$$T_u = u_2'/u_1 = 2Z_1/(Z_1 + Z_2). \tag{4.3}$$

These expressions have the same form as Eq. 4.1 for elastic collisions with Z taking
the place of M. The corresponding coefficients for the pressure are

$$R_p = p_1'/p_1 = (Z_2 - Z_1)/(Z_1 + Z_2)$$
$$T_p = p_2'/p_1 = 2Z_2/(Z_1 + Z_2). \tag{4.4}$$

The *power transmission coefficient* τ, the ratio $I_t/I_i = Z_2 u_2'^2/Z_1 u_1^2$ of the transmitted and incident intensities, becomes

$$\tau = \frac{I_t}{I_i} = \frac{4Z_1 Z_2}{(Z_1 + Z_2)^2} = \frac{4Z_1/Z_2}{[1 + (Z_1/Z_2)]^2}. \tag{4.5}$$

If the helium column is replaced by a solid bar, the impedance Z_2 will be much
larger than Z_1, so that we may set $Z_2/Z_1 \approx \infty$. In that case $R_p \approx 1$ so that
the reflected pressure wave has about the same amplitude as the incident. The
pressure amplitude at the boundary is then $\approx 2p_1$ (i.e., pressure doubling occurs). If
we have harmonic time dependence, with angular frequency ω, the incident pressure
wave will be $p_1 = A\cos(\omega t - kx)$ and the reflected wave, $p_1' = A\cos(\omega t + kx)$,
where $k = \omega/c = 2\pi/\lambda$. The addition of the two yields the standing wave

$$p(x, t) = 2A\cos(kx)\cos(\omega t) \tag{4.6}$$

as explained in Chapter 3.

The expressions for the reflection and transmission coefficient apply also to the
various waves on bars, springs, and strings considered in Chapter 3 for the field
variables that correspond to the present ones, velocity and pressure. In each case,
the analogous wave impedance must be used, of course.

Effect of Membrane Mass

The membrane at the interface between the two gases was assumed to be mass-less in the analysis above. The result can readily be extended to include the mass as follows. With the membrane assumed to be limp, its impedance is simply $-i\omega m$, where m is the mass per unit area. The impedance at the end of the air column is the sum of the membrane impedance and the impedance Z_2 of the Helium column. Thus, to account for the membrane mass we have to replace Z_2 in Eq. 4.4 by $Z_2' = Z_2 - i\omega m$. The pressure reflection coefficient then becomes

$$R_p = (Z_2' - Z_1)/(Z_2' + Z_1). \qquad (4.7)$$

If $m = 0$ we obtain the previous result, of course, and with $m = \infty$, the membrane acts like a rigid wall and the pressure reflection coefficient becomes $R_p = 1$. If the Helium column is finite and closed at the end, the membrane and the column become an acoustic tube resonator, which will be discussed later.

With the impedances normalized with respect to $Z_1 = \rho c_1$ and with $\zeta_2 = Z_2'/Z_1$, the reflection coefficient in Eq. 4.7 takes the general form

$$R_p = (\zeta_2 - 1)/(\zeta_2 + 1). \qquad (4.8)$$

4.1.3   Reflection from an Area Discontinuity in a Duct

The reflection from the interface between two gases discussed above was due to the discontinuity of the wave impedance at the interface. The membrane interface added a mass reactance to the impedance discontinuity and in terms of the effect on the reflection it is similar to that from an area discontinuity in a duct, shown in Fig. 4.1. A plane harmonic wave is incident from the left. As it encounters the discontinuity

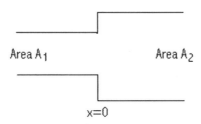

Figure 4.1: Reflection from an area discontinuity in a duct.

in area it is partially reflected. To calculate the reflection coefficient we assume the velocity distribution at the discontinuity to be uniform. In that case, the wave transmitted to the right will be the same as that produced by a plane piston radiating into a tube, discussed in Section 6.2.3. It is shown there that the piston generates a plane wave as well as higher modes. If the frequency is below the cut-off frequency of the duct, the higher modes will be evanescent and contribute a mass reactive load on the piston. The corresponding mass per unit area can be expressed as $\delta\rho$, where

δ is the mass end correction given in Fig. 6.5. This takes the place of the membrane mass m in Eq. 4.7.

The corresponding contribution of the higher modes to the radiation impedance of the piston is then $-i\omega(\delta\rho)$ with the normalized value $-ik\delta$. The plane wave contribution is a resistance $\rho c u_2$, where u_2 is the axial velocity the downstream duct. Continuity of mass flow requires that the velocity in to the left of the piston at the piston be $u_1 = (A_2/A_1)u_2$. This means that the resistive part of the impedance will be $\rho c u_2 = \rho c(A_1/A_2)u_1$ with the normalized value A_1/A_2. The total equivalent impedance of the area discontinuity is then

$$\zeta_2 = (A_1/A_2) - ik\delta. \tag{4.9}$$

The corresponding pressure reflection coefficient is then obtained from Eq. 11.69.

Reflection from the End of a Duct

The assumption of a uniform velocity distribution at the area discontinuity in Section 4.1.3 is an approximation. The true velocity distribution deviates from it, particularly in the vicinity of the edges of the discontinuity; a rigorous analysis is beyond the present scope.

An exact solution of the related problem of the reflection from the open end of a long circular pipe has been given by Levine and Schwinger in a classic paper.[2] Their calculated frequency dependence of the magnitude of the pressure reflection coefficient at the end of the duct for an incident plane wave is shown in Fig. 4.2. In this figure is shown also the mass end correction Δ, corresponding to the δ in the radiation from the piston in an infinite baffle in Section 5.3.5. The low frequency values

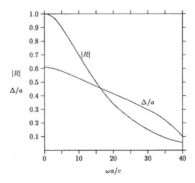

Figure 4.2: Magnitude R of the pressure reflection coefficient and the mass end correction Δ at the open end of an unflanged pipe of radius a. [From H. Levine and J. Schwinger, *Phys. Rev.* 73, 383, (1948)].

---

[2]H. Levine and J. Schwinger, *Phys. Rev.* 73, 383, (1948).

of the end correction and the pressure reflection coefficient at the end of the pipe were found to be

$$\Delta = 0.6133\,a$$
$$|R| = e^{-(ka)^2/2}. \tag{4.10}$$

The value for Δ should be compared with $\delta = (8/3\pi)\,a \approx 0.85\,a$ for the uniform piston in an infinite baffle.

The average normalized radiation impedance at the end of the pipe is

$$\zeta_r = (1 + R)/(1 - R). \tag{4.11}$$

4.1.4 Problems

1. **Review: Elastic collisions**

   Derive the expressions for the reflection and transmission coefficient in Eq. 4.1.

2. **Tennis, anyone?**

   As a refresher of elementary mechanics, consider the following. The mass M of a tennis racket is about 5 times the mass m of a tennis ball. The ball initially has no forward motion, as in a serve, and the speed of the racket as it hits the ball is U.

   (a) What will be the speed of the ball after the serve? Treat the problem as a one-dimensional elastic collision between two bodies. Assume also that the ball is hit at the center (of percussion) of the racket.

   (b) If the mass of the racket is increased by 20 percent (the mass of the ball is kept the same), what is the percentage increase of the ball speed, assuming the racket speed to be the same as before (discuss the validity of this assumption)?

   (c) What effect does an off-center hit have on the ball speed? Let the radius of gyration of the racket be R and the distance of the impact point from the center be r.

   (d) Repeat the calculation with the ball having a speed U_0 toward the racket before it is hit. What now is the ball speed after a centered hit with a racket speed U? If the incident ball speed is the same as the racket speed, what is the percentage increase in the ball speed after the hit if the racket mass is increased by 20 percent?

   (e) With what speed should the racket be moved backwards in (d) to make the ball come to rest after the impact (stop volley)?

   ANSWERS: (a) Ball speed: $U_b = \frac{2U}{1+(m/M)}$.

   (b) Increase in ball speed: 2.4 percent.

   (c) Ball speed: $U_b = \frac{2U}{1+(m/M)(1+r^2/R^2)}$.

   (d) Ball speed after hit: $U_b = \frac{2U}{1+m/M} + U_0\frac{1-m/M}{1+m/M}$

   Increase in ball speed for 20% increase in racket weight: 3.8 percent.

   (e) Backwards speed of racket for stop volley: $U = \frac{U_0/2}{1+(m/M)} = 0.42\,U_0$

3. **Coil spring with resistive termination**

   A coil spring is terminated by a dashpot damper so that the ratio of the driving force and the velocity at the end is real and equal to R. The wave impedance of the coil spring is Z. Plot the velocity reflection coefficient as a function of R/Z.

4. **Sound reflections in a duct**

(a) Determine the pressure reflection coefficient (magnitude and phase) at an area transition in a circular duct of radius a at which the area is doubled. What fraction of the acoustic power is reflected and transmitted?

(b) What is the fraction of power transmission coefficient at the open end of a pipe when $ka = 10$?

(c) What is the ratio of the maximum and minimum sound pressures in the standing wave in the pipe in (b)?

4.2 Sound Absorption

4.2.1 Mechanisms

Sound absorption is the conversion of acoustic energy into heat through the effects of viscosity and heat conduction. These effects increase with the gradients of fluid velocity and temperature in the sound field. In free field, far from boundaries, the characteristics length of spatial variation is the wavelength and the gradients are proportional to the frequency.

The interaction of sound with solid boundaries gives rise to acoustic boundary layers in which the gradients and the corresponding viscous and thermal effects are much larger than in free field. The sound absorption can be considerable, particularly when porous materials are involved. The 'contact' or 'sonified' area is then large and if the material is chosen properly, efficient absorption will result. This requires the width of the pores or channels in the material to be quite small, typically of the order of a thousandth of an inch.

There are other mechanisms of sound absorption. One involves the separation of the oscillatory flow in a (large amplitude) sound wave at sharp corners and in orifices; acoustic energy is then converted into vorticity and then ultimately into heat as a result of the decay of the vorticity.

A related effect involves the interaction of sound with turbulent flow. The acoustic modulation of such a flow results in a conversion of sound energy into vorticity. This effect can be of considerable importance in acoustic resonators. Both of these mechanisms will be discussed in Chapter 10.

Another effect, normally less significant, involves nonlinear distortion in which a sound wave at one frequency generates waves at other frequencies. Although the total energy is conserved, the energy transfer results in attenuation of the primary wave.

The absorbed energy in a porous material is proportional to the product of the squared velocity amplitude within the material and the contact area referred to above. If this area is increased by making the fibers and pores smaller and more numerous, the density of the material increases. This has the effect of preventing the sound from effectively penetrating into the material and the average velocity amplitude within the material decreases. Most of the incoming acoustic energy is then reflected from the surface and little absorption results. Similarly, if the contact area (and hence the density of the material) is very small, the sound goes through the material practically unimpeded and, again, little energy is absorbed. Consequently, there is an optimum design of any given absorber configuration where a compromise is struck between contact area (density) and velocity amplitude. Thus, for each frequency and thickness of a porous layer there exists an optimum density of the material for maximum sound absorption as will be discussed further in Section 4.2.8.

4.2.2 The Viscous Boundary Layer

As in steady flow, there is a viscous boundary layer also in oscillatory flow, i.e., in a sound field. In addition, there is a thermal boundary layer; both play important roles in sound absorption.

First, let us discuss the viscous boundary layer. To illustrate it, we consider the shear flow generated by a flat infinite plate in the plane $y = 0$. It oscillates in harmonic motion in the x-direction. Due to friction, this induces a harmonic (shear) in the surrounding fluid in which the velocity in the air is the same as that of the plate at $y = 0$. In the following discussion, the field variables are complex amplitudes. With reference to derivation following this section, the complex velocity amplitude is found to decrease exponentially with the distance y from the plate,

$$
\boxed{
\begin{array}{c}
\textit{Viscous boundary layer} \\
u_x = u(0)e^{i(1+i)y/d_v} = u(0)e^{-y/d_v}e^{iy/d_v} \\
d_v = \sqrt{2v/\omega} \approx 0.22/\sqrt{f}
\end{array}
} \tag{4.12}
$$

[$v = \mu/\rho$: *Kinematic viscosity* (≈ 0.15 *for air at* $20°C$). μ: *Coefficient of shear viscosity.* ρ: *Density.* $f = \omega/2\pi$: *Frequency in Hz. In the numerical expression,* d_v *is in cm and* f *in Hz, and it refers to air at* $20°C$].

The characteristic length d_v, at which the velocity amplitude has decreased to $1/e \approx .37$ of the amplitude of the plate, is called the *viscous boundary layer thickness*. The 'transmission' of the motion from the plate out into the fluid is a diffusion process and the quantity that 'diffuses' is the vorticity in the fluid. Associated with this process is also a phase lag y/d_v of the velocity at y with respect to the velocity at $y = 0$. With $f = 100$ Hz, the boundary layer thickness is ≈ 0.022 cm in air at room temperature.

Viscous Boundary Layer Derivation

We derive here the expression for the (shear) velocity distribution (Eq. 4.12) in the air above a plane boundary which oscillates in harmonic motion in a direction parallel to the plane. The corresponding velocity distribution resulting from the interaction between a sound wave and a stationary plane boundary is also considered. In the process, the expressions for the surface impedance in Eq. 4.20 and the viscous dissipation per unit area are obtained.

A flat plate oscillates in harmonic motion with the velocity $u_0 \cos(\omega t)$ in the x-direction (in the plane of the plate). With the y-direction chosen normal to the plate, the rate of momentum flux (shear stress) in the y-direction is $\tau(y) = -\mu \partial u/\partial y$ so that the net force (in the x-direction) per unit area on a fluid element of thickness dy is $\tau(y) - \tau(y + dy) = -(\partial \tau/\partial y)dy$. The equation for the x-component of the fluid velocity is then

$$
\rho \frac{\partial u}{\partial t} = \mu \frac{\partial^2 u}{\partial y^2}. \tag{4.13}
$$

The corresponding equation for the complex velocity amplitude $u(\omega)$ is[3]

$$\frac{\partial^2 u}{\partial y^2} + (i\omega\rho/\mu)u = 0. \qquad (4.14)$$

With $k_v^2 = i\omega\rho/\mu$, this equation is of the same form as the ordinary wave equation and the relevant solution is

$$u = u_0 e^{ik_v y} = u_0 e^{-y/d_v} e^{iy/d_v}$$
$$k_v = (1+i)\sqrt{\rho\omega/2\mu} \equiv (1+i)/d_v, \qquad (4.15)$$

where we have used $i = \exp(i\pi/2)$ and $\sqrt{i} = \exp(i\pi/4) = \cos(\pi/4) + i\sin(\pi/4) = (1+i)/\sqrt{2}$.

There is a second solution $\propto \exp(-ik_v y)$ but it grows with y and does not fit the 'boundary condition' at $y = \infty$. Such a solution would have to be included, however, if a second plane boundary were involved above the first.

The velocity amplitude decreases exponentially with y and is reduced by a factor e at the distance d_v above the plate which defines the boundary layer thickness,

$$d_v = \sqrt{2\mu/\rho\omega} \approx \frac{0.22}{\sqrt{f}} \text{ cm.} \quad \text{(normal air)}, \qquad (4.16)$$

where f is the frequency in Hz.

The complex amplitude of the shear stress $\mu\partial u/\partial y$ on the plate $(y = 0)$ is $F = -\mu i k_v u_0 = u_0(1-i)\mu/d_v$ and the corresponding shear impedance per unit area is

$$Z_s \equiv R_s + iX_s = F/u_0 = (1-i)\sqrt{\mu\rho\omega/2} = (1/2)(1-i)(kd_v)\rho c, \qquad (4.17)$$

where $k = \omega/c$.

In the reverse situation, when the plate is stationary and the velocity of the fluid in the *free stream* far away from the plate is $u_0\cos(\omega t)$, the corresponding complex amplitude equation of motion in the free stream is $-i\rho\omega u_0 = -\partial p/\partial x$, where the right-hand side is the pressure gradient required to maintain the oscillatory flow. If we assume that this pressure gradient is independent of y, the equation of motion in the boundary layer will be $-i\omega\rho u(y) = \mu\partial^2 u/\partial y^2 - i\omega\rho u_0$, where we have replaced $-\partial p/\partial x$ by $-i\omega\rho u_0$, as given above. The solution is $u = u_0[1 - \exp(ik_v y)]$. Thus, the velocity increases exponentially with y from 0 to the free stream value u_0, and we can use the same definition for the boundary layer thickness as in Eq. 4.16. The viscous stress on the plate will be the same as before and the real part represents the resistive friction force per unit area of the plate and is responsible for the viscous boundary losses in the interaction of sound with the boundary.

The time average power dissipation per unit area in the shear flow at the boundary is simply $R_s|u_0|^2$, i.e.,

$$L_v = R_s|u_0|^2 = (1/2)kd_v\rho c\,|u_0|^2$$
$$R_s = (1/2)kd_v\rho c = \rho c\sqrt{v\omega/2c^2} \approx 2 \times 10^{-5}\sqrt{f}\rho c \quad \text{(normal air)}, \qquad (4.18)$$

where $|u_0|$ is the rms magnitude of the tangential velocity outside the boundary layer and R_s the surface resistance. In the numerical approximation for normal air, f is the frequency in Hz.

The viscous losses per unit area in the shear flow can be obtained also by direct integration of the viscous dissipation function over the boundary layer, as follows. Consider an element of thickness Δy. In a frame of reference moving with the fluid with its origin at the center of Δy,

---

[3]Recall that the complex velocity amplitude is defined by $u(t) = \Re\{u(\omega)\exp(-i\omega t)\}$.

the velocity at the top surface is $(\partial u_x/\partial y)\,\Delta y/2$. The shear stress is $-\mu \partial u_x/\partial y$ (in the positive y-direction) and the power transfer to the element through the top surface (which is in the negative y-direction) is then $\mu \partial u_x/\partial y\,(\Delta y/2)(\partial u_x/\partial y)$. There is a similar transfer from the bottom surface, so that the total power transfer per unit volume will be $\mu(\partial u_x/\partial y)(\partial u_x/\partial y)$.

In harmonic time dependence, the time average of this quantity will be $L_v = \mu\Re\{(\partial u_x/\partial y)$ $(\partial u_x^*/\partial y)\}$, where u_x is the rms value and u_x^* the complex conjugate of u_x. With $u_x = u_0\exp(ik_v y)$, integration from $y=0$ to $y=\infty$ yields the result in Eq. 4.18.

$F = -\mu\partial u/\partial y$ is the viscous stress on the surface which can be used to obtain an approximate value for the impedance per unit length of a channel of arbitrary cross section as long as its transverse dimensions are large compared to the boundary layer thickness. The flow velocity in the center of the channel can then be considered to be the free stream velocity. With the perimeter of the channel denoted S and the area by A, the total viscous stress per unit length of the channel is $SF = SZ_s u_0$, where $F = -\mu\partial u/\partial y$ and $Z_s = F/u_0$. The reaction force on the fluid will be the same but with opposite sign and the equation of motion for a fluid element of unit length is $-iA\omega\rho u_0 = -A\partial p/\partial x - SZ_s u_0$. With $Z_s = R_s + iX_s$, the corresponding impedance per unit length of the channel is then

$$z_{1c} = r_{1c} + ix_{1c} = (1/u_0)\left(-\frac{\partial u}{\partial x}\right) = (S/A)[R_s - i\omega\rho + iX_s]. \tag{4.19}$$

Since X_s represents a mass reactance (i.e., it is negative), as explained earlier, the total reactance can be written $-i\omega\rho_e$, where $\rho_e = \rho + |X|/\omega$ is an equivalent mass density. The impedance per unit length is then $z_{1c} = (S/A)[R_s - i\omega\rho_e]$.

Surface Impedance for Shear Flow

From the velocity field in the fluid, we can determine, at $y=0$, the (shear) stress $-\mu\partial u_x/\partial y$ per unit area of the plate that is required to drive the oscillating flow. The ratio of the complex amplitudes of this stress and the velocity of the plate is defined as a *surface impedance* per unit area. The resistive and reactive parts of the impedance turn out to be equal and the magnitude of each can be expressed as $(kd_v)\rho c/2$, where $k = \omega/c$. From the frequency dependence of d_v (Eq. 4.12) and with reference to Section 4.2.2, it follows that the surface impedance is proportional to the square root of frequency,

$$Z_s \equiv R_s + iX_s = F/u_0 = (1-i)\sqrt{\mu\rho\omega/2} = (1/2)(1-i)(kd_v)\rho c \tag{4.20}$$

We can interpret the mass reactance in terms of the total kinetic energy of the oscillatory flow in the boundary layer. Integrating the kinetic energy density from 0 to ∞, and expressing the result as $(1/2)m|u_0|^2$, we find that the corresponding normalized mass reactance $\omega m/\rho c$ agrees with the expression $kd_v/2$ given in Eq. 4.20.

The reverse situation, when the plate is stationary and the velocity of the fluid far away from the plate has harmonic time dependence, the fluid velocity goes to zero at the plate. With reference to Section 4.2.2, the velocity distribution is now

$$u_x(y) = u_0(1 - e^{ik_v y}), \tag{4.21}$$

where $k_v = (1+i)/d_v$. The transition from the 'free stream' velocity u_0 to zero occurs in a boundary layer which has the same form as above. This is not surprising since it is only the relative motion of the fluid and the plate that should matter. There will be an oscillatory force on the plate and a corresponding surface impedance Z_s per unit area with a resistive and a mass reactive part, the same as before. This means that the viscous interaction force on the boundary by the sound field yields not only a force proportional to the velocity but also a component that is proportional to the *acceleration* of the fluid with respect to the boundary.

We can use this impedance as a good approximation also for sound interacting with curved boundary as long as the radius of curvature of the surface is much larger than the acoustic boundary layer thickness; the surface can be treated locally as plane. Using this approximation, we can determine the total surface impedance for oscillatory flow in a channel of arbitrary cross section as long as the transverse dimensions are large compared to the boundary layer thickness. Thus, if the perimeter of the channel is S and its area A, the total surface impedance per unit length of the channel will be $(S/A)Z_s$. In addition, there is the mass reactance $\omega\rho$ of the air itself. When combined with the reactive part of the surface impedance, the total reactance can be expressed as $\omega\rho_e$, where ρ_e is an equivalent mass density.

4.2.3 The Thermal Boundary Layer

By analogy with the discussion of the viscous boundary layer, we consider next the temperature field in a fluid above a plane boundary produced as a result of a harmonic temperature variation of the boundary about its mean value. Temperature rather than vorticity is now diffused into the fluid, and the temperature field takes the place of the velocity field in the shear motion discussed above. With reference to the derivation following this section, the temperature field is described by an equation similar to that for the diffusion of vorticity and the y-dependence of the temperature is found to be

$$\boxed{\begin{array}{c} \textit{Thermal boundary layer} \\ \tau(y) = \tau(0)e^{i(1+i)y/d_h} = \tau(0)e^{-y/d_h}e^{iy/d_h} \\ d_h = \sqrt{2K/\rho c_p \omega} = \sqrt{K/\mu c_p}\, d_v \approx 0.25/\sqrt{f} \end{array}} \qquad (4.22)$$

[d_h: Thermal boundary layer thickness. $\tau(y)$: Temperature amplitude. ρ: Density. K: Heat conduction coefficient. c_p: Specific heat per unit mass at constant pressure. In the numerical result, d_h is expressed in cm and f in Hz].

The thermal boundary layer thickness is slightly larger than the viscous (by about 10 percent). The expression for d_h can be obtained from the viscous boundary layer thickness given above by replacing the coefficient of shear viscosity μ by K/c_p.

The example above with a boundary with a harmonic temperature dependence was used merely to introduce the idea of the thermal boundary layer. Of more interest here, of course, is the case of the interaction of a sound wave with an isothermal boundary. Sufficiently far from the boundary, the conditions are the same as in free field which means that the change of state is isentropic (adiabatic) and the pressure fluctuations in the sound field produce a temperature fluctuation. The boundary has a much greater heat conduction and heat capacity than the air above so that the conditions at the boundary can be considered to be isothermal, i.e., there is no temperature fluctuation at $y = 0$. Thus, there is a transition from isentropic to isothermal conditions as the boundary is approached, and the transition, as we shall see, occurs in a *thermal boundary layer*. Unlike the example above, the temperature now goes from its maximum value outside the boundary layer to zero at the boundary. With reference to the discussion earlier in this section, the complex amplitude of the temperature is given by

$$\tau(y, \omega) = \tau_0[1 - e^{i(1+i)y/d_h}] = \tau_0[1 - e^{-y/d_h}e^{iy/d_h}], \qquad (4.23)$$

where τ_0 is the amplitude of the temperature fluctuation in the field far away from the boundary and d_h is the boundary layer thickness given in Eq. A.63.

In the thermal boundary layer the compressibility varies from the isentropic value in free field, $1/(\gamma P)$, to the isothermal value, $1/P$, at the boundary (γ is the specific heat ratio, ≈ 1.4 for air, and P the static pressure). In both these regions, a change of state is reversible and a compression of a fluid element is in phase with the pressure increase. This means that the rate of compression of a volume element will be 90 degrees out of phase so that there will be no net work done on the element in one period of harmonic motion. (If the time dependence of pressure is $= \cos(\omega t)$ it will be $\sin(\omega t)$ for the velocity and the time average of the product will be zero.)

The situation is different within the boundary layer where the conditions are neither isothermal, nor isentropic. A compression leads to a delayed leakage of heat out of the compressed region and the build-up of temperature and pressure will be delayed accordingly. The pressure no longer will be 90 degrees out of phase with the rate of compression and there will be a net energy transfer from the sound field into the gas and then, via conduction, into the boundary. The maximum transfer per unit volume of the gas is found to occur at a distance from the boundary approximately equal to the boundary layer thickness.

This is the nature of the acoustic losses caused by heat conduction. Formally, it can be accounted for by means of a *complex compressibility* $\tilde{\kappa}$ in the thermal boundary layer. The loss rate per unit volume is then proportional to the imaginary part of $\tilde{\kappa}$. (Compare the complex spring constant in a damped harmonic oscillator, Section 2.4.2.)

There is some heat conduction also in the free field, far away from the plate, which leads to a slight deviation from purely isentropic conditions. However, the heat flow is now a result of a gradient in which the characteristic length is the wavelength λ rather than the boundary layer thickness d_h; with $\lambda >> d_h$, this effect can be neglected in the present discussion.

Thermal Boundary Layer Derivation

We derive here the temperature amplitude distribution over a boundary with a periodic variation in temperature, the temperature distribution (4.23) resulting from the interaction of a sound wave with an isothermal boundary. The corresponding complex compressibility in the boundary layer and the acoustic power loss per unit area of the boundary due to heat conduction and the total visco-thermal loss are also considered.

By analogy with the discussion of the viscous boundary layer, we consider now the temperature field produced by a plane boundary with a temperature which varies harmonically with time about its mean value, the variation being $\tau(t) = \tau_0 \cos(\omega t)$. The temperature away from the boundary is obtained from the diffusion equation

$$\frac{\partial \tau}{\partial t} = (K/C_p\rho)\,\frac{\partial^2 \tau}{\partial y^2}, \tag{4.24}$$

where K, C_p, and ρ are the heat conduction coefficient, the specific heat at constant pressure and unit mass, and the density, respectively.

For harmonic time dependence ($\partial/\partial t \to -i\omega$) and with the y-dependence expressed as

$$\tau(y, \omega) = \tau_0 \exp(ik_h y) \tag{4.25}$$

it follows from Eq. 4.24 that

$$k_h^2 = i(\omega\rho C_p)/K$$

i.e.,

$$k_h = (1+i)/d_h$$
$$d_h = \sqrt{2K/\rho C_p\omega} = \sqrt{\frac{K}{\mu C_p}}\,d_v \approx 0.25/\sqrt{f}\ \text{cm.} \quad \text{(normal air)} \tag{4.26}$$

where d_h is the thermal boundary layer thickness and d_v, the viscous. In the numerical approximation, f is the frequency in Hz.

From Eq. 4.25, the complex amplitude of the temperature is

$$\tau(y, \omega) = \tau_0 e^{-y/d_h}\, e^{iy/d_h} \tag{4.27}$$

so that at a distance from the plate equal to the thermal boundary layer thickness, $y = d_h$, and the magnitude of the temperature is $1/e$ of the value at the plate at $y = 0$.

The ratio of the viscous and thermal boundary layer thicknesses is

$$d_v/d_h = \sqrt{\mu C_p/K} = \sqrt{P_r}$$
$$P_r = \mu C_p/K, \tag{4.28}$$

where P_r is the *Prandtl* number. For air at 1 atm and 20 degree centigrade, $\mu \approx 1.83 \times 10^{-4}$ CGS (poise), $C_p \approx 0.24$ cal/gram/degree, and $K \approx 5.68 \times 10^{-5}$ cal cm/degree, so that $P_r \approx 0.77$, $d_v/d_h \approx 0.88$ and $k_h \approx 0.88 k_v$.

The reverse situation, when the temperature fluctuation in a sound wave far from the plate is equal to τ_0 and the temperature fluctuation at the plate is zero (due to a heat conduction coefficient and a heat capacity of the solid is much larger than those for air), the appropriate solution to Eq. 4.24 is

$$\tau(y, \omega) = \tau_0[1 - e^{ik_h y}]. \tag{4.29}$$

This solution is applicable to the case when a harmonic sound wave is incident on the plate. Far away from the plate, $y >> d_h$, the conditions in the fluid are isentropic and the compressions and rarefactions in the sound wave produce a harmonic temperature fluctuation with the amplitude (see Section 3.2.3, Eq. 3.28)

$$\tau_0 = \frac{\gamma - 1}{\gamma}\frac{p}{P}T. \tag{4.30}$$

Quantity p is the sound pressure amplitude, $\gamma = C_p/C_v$, the specific heat ratio, P the ambient pressure, and T the absolute temperature. The acoustic wavelength of interest is large compared to the boundary layer thickness so that we need not be concerned about any change of the sound pressure with position across the boundary layer. However, the compressibility varies, going from the isentropic value $1/\gamma P$ to the isothermal, $1/P$, as the boundary is approached. These values refer to an ideal gas.

To determine the complex compressibility throughout the boundary layer, we start with the density $\rho(P, T)$ being a function of both pressure P and temperature T (not only of pressure alone as in the isentropic case) so that

$$d\rho = \left(\frac{\partial\rho}{\partial P}\right)_T dP + \left(\frac{\partial\rho}{\partial T}\right)_P dT. \tag{4.31}$$

From the gas law, $P = r\rho T$, we have $(\partial\rho/\partial P)_T = \rho/P$ and $(\partial\rho/\partial T)_P = -\rho/T$. Then, the quantities $dP = p$, $d\rho$ and $dT = \tau(y, \omega)$ are treated as complex amplitudes, where τ is given in Eqs. 4.29 and 4.30 in terms of the sound pressure amplitude $dP = p$. The compressibility then follows from Eq. 4.31

$$\tilde{\kappa} = (1/\rho)(d\rho/dP) = \frac{1}{\gamma P}[1 + (\gamma - 1)e^{-k_h y}e^{ik_h y}]. \tag{4.32}$$

The tilde symbol is used to indicate that the compressibility is complex and different from the normal isentropic compressibility $\kappa = 1/\gamma P = 1/\rho c^2$.

For $y = 0$, $\tilde{\kappa} = 1/P$ equals the isothermal value, and for $y = \infty$, $\tilde{\kappa} = 1/\gamma P$, the isentropic value; in the transition region, $\tilde{\kappa}$ is complex. The imaginary part can be written

$$\kappa_i = \kappa(\gamma - 1)e^{-y/d_h}\sin(y/d_h). \tag{4.33}$$

It has a maximum 0.321κ at $y/d_h = \pi/4$.

The power dissipation per unit area due to viscosity in the acoustically driven oscillatory shear flow over a solid wall has already been expressed in Eq. 4.18.

To determine the dissipation due to heat conduction, we start from the conservation of mass equation for the fluid $\partial\rho/\partial t + \rho\,\mathrm{div}\,\mathbf{u} = 0$. For harmonic time dependence and with the relation between the complex amplitudes of density and pressure (δ and p) expressed as $\delta = \rho\tilde{\kappa}p$ in terms of a complex compressibility $\tilde{\kappa}$, this equation becomes $-i\omega\tilde{\kappa}p + \mathrm{div}\,\mathbf{u} = 0$.

After integration of this equation over a small volume V with surface area A, and replacing the volume integral of $\mathrm{div}\,\mathbf{u}$ by a surface integral over A, we can express the time average power $\mathfrak{R}\{u_n p*\}A$ transmitted through A into the volume element as $\mathfrak{R}\{(-i\omega)\tilde{\kappa}|p|^2\}V$, where u_n is the inward normal velocity component of the velocity at the surface, $|u_n|$ and $|p|$ being rms values to avoid an additional factor of 1/2. Thus, the corresponding power dissipation per unit volume becomes

$$D_h = \omega\kappa_i|p|^2. \tag{4.34}$$

The integral of this expression over the boundary layer yields the corresponding acoustic power loss per unit area of the wall. The integration can be taken from 0 clear out to infinity. The contribution to the integral comes mainly from y-values less than a couple of boundary layer thicknesses and quickly goes to zero with increasing y outside the boundary layer. The pressure amplitude $|p|$ can be taken to be constant throughout the layer since the wavelength of interest is much larger than the boundary layer thickness. After insertion of the expression for the compressibility in Eq. 4.32, the loss due to heat conduction per unit area of the wall can be expressed as

$$L_h = (1/2)(\gamma - 1)kd_h|p|^2, \tag{4.35}$$

which is the counterpart of the expression for the viscous power dissipation L_v in Eq. 4.18.

The total visco-thermal power dissipation per unit area of the wall then becomes

$$L_s = L_v + L_h = (k/2)[d_v|u|^2\rho c + (\gamma - 1)d_h|p|^2/\rho c]$$
$$\approx 2 \times 10^{-5}\sqrt{f}[|u|^2\rho c + 0.45|p|^2/\rho c], \tag{4.36}$$

where $|u|$ is the tangential velocity outside the boundary layer and $|p|$ the pressure amplitude at the wall, both rms magnitudes.

4.2.4 Power Dissipation in the Acoustic Boundary Layer

We summarize the result presented in Eq. 4.36 as follows: The acoustic power dissipation at a boundary is the sum of two contributions. The first is due to the shear stresses in the viscous boundary layer and is proportional to the squared tangential velocity amplitude just outside the boundary layer. The second is due to the heat conduction in the thermal boundary layer and is proportional to the squared sound pressure amplitude at the boundary. The dissipation per unit area of the boundary is obtained by integrating the viscous and thermal losses per unit volume in the boundary layers as shown above with the result

$$\boxed{\begin{array}{l} \text{Power dissipation per unit area in acoustic boundary layer} \\ L_s = L_v + L_h = (k/2)[d_v|u|^2\rho c + (\gamma - 1)d_h|p|^2/\rho c] \\ \qquad \approx 2 \times 10^{-5}f^{1/2}[|u|^2\rho c + 0.45|p|^2/\rho c] \end{array}} \tag{4.37}$$

[L_v, L_h: *Viscous and heat conduction contributions.* d_v, d_h: *Viscous and thermal boundary layer thicknesses (Eqs. 4.12 and A.63).* $|p|$: *Sound pressure amplitude (rms) at the boundary.* $|u|$: *Tangential velocity amplitude (rms) outside the boundary layer. The numerical coefficient refers to air at* $20°C$].

Since the velocity and pressure amplitudes are simply related, the total visco-thermal power dissipation per unit area at the boundary can be expressed in terms of either the pressure amplitude or the velocity amplitude.

The result obtained for a plane boundary can be used also for a curved boundary, if the local radius of curvature is much larger than the thermal boundary layer thickness.

Example. The Q-value of a tube resonator

For a simple mass-spring oscillator with relatively small damping, the sharpness of its resonance is usually expressed as $1/(2\pi)$ times the ratio of the total energy of oscillation (twice the kinetic energy) and the power dissipated in one period. This relation is valid also for an acoustic cavity resonator. The total energy of oscillation is now obtained from the known pressure and velocity fields in the resonator and by dividing it with the total visco-thermal losses at the boundaries. The Q-value can be determined since both quantities are proportional to the maximum pressure amplitude in the resonator.

The constant of proportionality for the total losses contains a visco-thermal boundary layer thickness $d_{vh} = d_v + (\gamma - 1)d_h$, where d_v and d_h are the viscous and thermal boundary layer thicknesses and $\gamma = C_p/C_v \approx 1.4$ (for air) is the specific heat ratio. If this scheme is used for a circular tube (quarter wavelength resonator), the Q-value turns out to be simply

$$Q \approx a/d_{vh}, \tag{4.38}$$

where a is the radius of the tube. By introducing the frequency dependence of the boundary layer thickness, this can be expressed as $\approx 3.11\, a\sqrt{f}$, where a is expressed in cm and f is the frequency in Hz. (The expression for a parallel plate cavity is the same if a stands for the separation of the plates). Thus, a circular resonator with a radius of 1 cm and a resonance frequency of 100 Hz has a Q-value of 31.1.

In this context, we should be aware of the fact that the boundary layer thickness depends on the kinematic viscosity $\nu = \mu/\rho$ and will decrease with increasing static pressure at a given temperature (μ is essentially independent of density). Thus, if a very high Q-value is desired in an experiment, a high pressure and a high density gas, or both, should be used.

In a nuclear power plant, the static pressure of the steam typically is of the order of 1000 atmospheres and the Q-value of acoustic resonances typically will be very high (damping low). This has a bearing on the problem of acoustically induced flow instabilities and their impact on key components in such plants, for example, control valves and related structures.

4.2.5 Resonator Absorber

In the example about sound transmitted into the (infinitely extended) Helium column, the sound was absorbed in the sense that it did not return, but not in the sense that it was converted into heat through friction. When we talk about sound absorption and sound absorbers in general, it is this latter mechanism which is implied. The study of sound absorption then involves an identification of the mechanisms involved and their dependence.[4]

Rather than terminating the air column by a Helium column, as in the example referred to above, we now let the termination be a piston which forms the mass in a damped mass-spring oscillator. We wish to determine the amplitude of the reflected wave and from it the energy absorbed by the resonator. The resonator is described by a mass M, a spring constant K, and a dashpot resistance R, all per unit area. The resonance frequency of the undamped resonator is $\omega_0 = \sqrt{K/M}$, as discussed in Chapter 2.

In this section we analyze this problem without the use of complex variables. It is generalized in the next section to a boundary with a given normal impedance and to oblique angles of incidence of the sound and complex amplitudes are then used exclusively.

The termination is placed at $x = 0$; the incident and reflected sound pressures at this location are denoted $p_i(t)$ and $p_r(t)$. The velocity of the piston is $u(t)$. The total sound pressure driving the piston is then $p_i + p_r$ and the equation of motion of the piston is then

$$p_i + p_r = ru + m\frac{du}{dt} + K \int u\, dt = Z(t)u, \tag{4.39}$$

---

[4]An extensive study of absorption is given later.

where $Z(t)$ is short for $r + m(d/dt) + K \int dt$. The total velocity in the sound field at the piston is the sum of the incident and reflected wave contributions, $p_i/\rho c$ and $-p_r/\rho c$, which must equal the velocity u of the piston, i.e.,

$$p_i - p_r = \rho c\, u. \tag{4.40}$$

Addition and subtraction of these equations yields

$$2p_i = [Z(t) + \rho c]u$$
$$2p_r = [Z(t) - \rho c]u \tag{4.41}$$

which establishes the relation between the reflected and incident waves. The time dependence is harmonic, and the velocity of the piston is put equal to $u = |u|\cos(\omega t)$ with a phase angle chosen to be zero. Using this expression in Eq. 4.41 yields $2p_i = |u|[(r+\rho c)\cos(\omega t) - m\omega\sin(\omega t) + (K/\omega)\sin(\omega t)]$ which can be written $2|p_i|\cos(\omega t - \phi_i)$, where $2|p_i| = \sqrt{(r + \rho c)^2 + (\omega m - K/\omega)^2}$. The expression for $2|p_r|$ is the same except for a change in sign of ρc. Thus, if we introduce the notation $X = \omega m - K/\omega$ (the reactance), it follows that the reflection coefficient for intensity is

$$R_I = |p_r|^2/|p_i|_2 = [(r - \rho c)^2 + X^2]/[(r + \rho c)^2 + X^2]. \tag{4.42}$$

Conservation of acoustic energy requires that the absorbed intensity I_a by the termination is the difference between the incident and reflected intensities, $I_a = I_i - I_r$. Then, if the *absorption coefficient* is defined as $\alpha = I_a/I_i$ it follows that $\alpha = 1 - R_I$. It is frequently convenient to normalize the resistance and the reactance with respect to the wave impedance ρc. Then, if we introduce the $\theta = r/\rho c$ and the reactance $\chi = x/\rho c$, we get

$$R_I = [(1 - \theta)^2 + \chi^2]/[(1 + \theta)^2 + \chi^2]$$
$$\alpha = 1 - R_I = 4\theta/[(1 + \theta)^2 + \chi^2]. \tag{4.43}$$

At resonance, $\chi = 0$ and $\alpha = 4\phi/(1 + \theta)^2$ and if $\theta = 1$, 'impedance matching,' 100 percent absorption results, $\alpha = 1$.

The absorbed acoustic power per unit area can be expressed in terms of the normal velocity amplitude at the boundary as $W = u^2\rho c\theta$ or, with $\rho c u^2 = (p^2/\rho c)(\theta^2 + \chi^2)$ as

$$W = (p^2/\rho c)[\theta/(\theta^2 + \chi^2)] \equiv (p^2/\rho c)\,\mu, \tag{4.44}$$

where p is the rms value of the sound pressure at the surface and μ the *conductance* of the boundary (real part of the admittance). Sound absorption is of obvious importance in noise control engineering as a means of reducing noise (unwanted sound) and to modify the acoustics of enclosed spaces. It is usually achieved by applying sound absorptive material on interior walls but free hanging absorbers, functional absorbers, are sometimes used.

4.2.6 Generalization; Impedance Boundary Condition

The analysis of reflection and absorption will now be generalized to a boundary which is specified acoustically by a complex normal impedance $z(\omega)$, i.e., the ratio of the complex amplitudes of sound pressure and the normal component of velocity at the boundary. It is assumed that this impedance is known from experiments or has been calculated from known properties of the boundary, as was the case for the resonator example given in the previous section. The generalization also involves considering

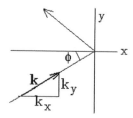

Figure 4.3: Obliquely incident wave on a boundary.

sound at oblique incidence and we start by discussing the description of such a wave, shown schematically in Fig. 4.3.

The wave is incident on the plane boundary at $x = 0$ (yz-plane) under an angle ϕ with the x-axis which is normal to the boundary. Let the coordinate along the direction of propagation be r. The corresponding vector is r. We also introduce the propagation vector k with the magnitude $k = \omega/c = 2\pi/\lambda$ and direction along the line of propagation, i.e., along r. Thus, $kr = k \cdot r = k_x x + k_y y$, where we have expressed the scalar product $k \cdot r$ in terms of the components k_x, k_y and x, y of k and r.

The complex amplitude of the incident plane wave can then be expressed as

$$p_i(\omega) = |p_i| e^{ikr} = |p_i| e^{ik_x x} e^{ik_y y}, \qquad (4.45)$$

where $k_x = k \cos(\phi)$ and $k_y = k \sin \phi$. With $k = 2\pi/\lambda$ it follows that $k_x = 2\pi/\lambda_x$, where the geometrical meaning of $\lambda_x = \lambda/\cos \phi$ is shown in the figure. It is the spatial period of the wave in the x-direction (i.e., the distance between two adjacent wave crests).

The reflected wave from the boundary has a propagation vector with the components $-k_x$ and k_y so that the complex amplitude of the reflected wave is

$$p_r(\omega) = |p_r| e^{-ik_x x} e^{ik_y y}. \qquad (4.46)$$

The factor $\exp(ik_y y)$ is of little interest in what we are going to do here so that in what follows it can be considered included in $|p_i|$ and $|p_r|$.

The total sound pressure field on the left side of the boundary is then

$$p(x) = p_i(x) + p_r(x), \qquad (4.47)$$

where p stands for the total complex pressure amplitude.

As discussed in detail in Chapter 5, the velocity field follows from the equation of motion $\rho \partial u_x / \partial t = -\partial p / \partial x$ with the complex amplitude version $-i\omega \rho u_x(\omega) = -\partial p(\omega)/\partial x$.[5]

We have $-i\omega \rho u_i = -ik_x \, p_i = -ik \cos \phi p_i$ with a similar expression for the reflected wave. Thus, the total velocity field that corresponds to the pressure field in Eq. 4.47 is

$$\rho c \, u_x = \cos \phi [p_i(x) - p_r(x)], \tag{4.48}$$

where we have made use of $k_x = k \cos \phi$ and $k = \omega/c$.

The complex normal impedance of the boundary (at $x = 0$) is z, and this condition requires that $p(0)/u(0) = z$, i.e.,

$$\frac{p_i(0) + p_r(0)}{p_i(0) - p_r(0)} = \zeta \cos \phi, \tag{4.49}$$

where $\zeta = z/\rho c$ is the normalized impedance. It follows then that the pressure reflection coefficient is

$$\boxed{\begin{array}{c} \textit{Pressure reflection coefficient} \\ R_p(\omega) = p_r(0)/p_i(0) = (\zeta \cos \phi - 1)/(\zeta \cos \phi + 1) \end{array}} \tag{4.50}$$

[$p_i(0)$, $p_r(0)$: *Incident and reflected complex pressure amplitudes at the boundary. ζ: Normalized impedance of the boundary. ϕ: Angle of incidence*].

The ratio of the incident and reflected intensities is $I_i/I_r = |R|^2$ and the ratio of the absorbed and incident intensity is $I_a/I_i = (I_i - I_r)/I_i = 1 - |R|^2$. In other words, the absorption coefficient I_a/I_i is

$$\alpha = 1 - |R|^2. \tag{4.51}$$

If the impedance is expressed in terms of a real and imaginary part, $\zeta = \theta + i\chi$, it follows from Eq. 4.50 that

$$\boxed{\begin{array}{c} \textit{Absorption coefficient} \\ \alpha(\phi) = 4\theta \cos \phi / [(1 + \theta \cos \phi)^2 + (\chi \cos \phi)^2] \end{array}} \tag{4.52}$$

[θ, χ: *Real and imaginary parts of the normalized boundary impedance. ϕ: Angle of incidence*].

As will be discussed in Chapter 6, the sound field in a room often can be approximated as diffuse, which means that if the field is regarded as a superposition of plane waves traveling in different directions, the probability of wave travel is the same in all directions. In regard to the absorption by a plane boundary in such a field, we have to average the absorption coefficient in Eq. 4.52 over all angles of incidence.

There are many directions of propagation that correspond to an angle of incidence ϕ and these directions are accounted for in the following way. The probability that acoustic intensity will strike an element of the wall at an angle between ϕ and $\phi + d\phi$ is proportional to the solid angle $2\pi \sin \phi \, d\phi$, i.e., the ring-like surface element on a

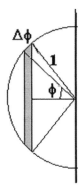

Figure 4.4: The probability of a wave having an angle of incidence ϕ in a diffuse field is proportional to the solid angle element (shaded) $2\pi \sin \phi \, d\phi$.

unit sphere centered at the wall element, as indicated schematically in Fig. 4.4. The power that strikes a wall element of unit area is the product of the intensity $i(\phi) = I$ and the projection $\cos\phi$ of this area is perpendicular to the incident wave direction. Thus, a factor $\cos\phi$ has to be included in calculating the average absorption coefficient which then becomes

$$\alpha_d = \frac{\int_0^{\pi/2} \alpha(\phi) \, 2\pi \, \sin\phi \cos\phi \, , d\phi}{\int_0^{\pi/2} 2\pi \, \sin\phi \cos\phi d\phi} = 2 \int_0^{\pi/2} \alpha(\phi) \sin\phi \cos\phi d\phi, \qquad (4.53)$$

where $\alpha(\phi)$ is obtained from Eq. 4.52. The denominator expresses the total intensity striking the wall element. The coefficient α_d will be called the *diffuse field* absorption coefficient, sometimes also called the *statistical average*. The results in Eqs. 4.50 and 4.52 are valid even if the impedance ζ depends on the angle of incidence. For some boundaries, called *locally reacting*, the impedance is independent of the angle and thus equals the value for normal incidence. The impedance can then be measured with relatively simple experiments in which the sample is placed at the end of a tube and exposed to a plane wave of sound, as described in Section 4.2.7. An example of a locally reacting boundary is a honeycomb structure backed by a rigid wall, in which the cell size is much smaller than a wavelength. The oscillatory velocity in each of the cells then depends only on the local pressure at the entrance to the cell and there is no coupling between the cells, preventing wave propagation along the boundary within the absorber.

With ζ independent of ϕ, the diffuse field absorption coefficient in Eq. 4.53 can be expressed in closed form (see Problem 8).

For a *nonlocally reacting boundary* or boundary with an extended reaction, the impedance is angle dependent and the experimental data of it are normally not avail-

[5]An element of thickness Δx has the mass $\rho \Delta x$. With the pressure being a function of x, the pressures at the two surfaces of the elements are $p(x)$ and $p(x + \Delta x)$ so that the net force on the element in the x-direction is $p(x) - p(x + \Delta x) = -\partial p/\partial x \, \Delta x$ and the equation of motion, Newton's law, is $\rho \partial u_x/\partial t = -\partial p/\partial x$. For further details, see Chapter 5

able. For relatively simply types of boundaries, however, the impedance can be calculated, but α_d generally has to be determined by numerical integration in Eq. 4.52. An example of a nonlocally reacting boundary is a uniform porous layer backed by a rigid wall.

Sheet Absorber

As an example of a resonator absorber, we have chosen to analyze an absorber which is frequently used in practice. It consists of a porous sheet or wire mesh screen backed by an air layer and a rigid wall, as illustrated schematically in Fig. 4.5. Two configurations are shown, one with and the other without a honeycomb structure in the air layer. The honeycomb has a cell size assumed to be much smaller than a wavelength and it forces the fluid velocity in the air layer to be normal to the wall, regardless of the angle of incidence of the sound. The first configuration is a *locally* and the second a *nonlocally* reacting absorber, as indicated.

As we shall see, either configuration can be considered to be a form of acoustic resonator but unlike the resonator absorber in the previous example, it has multiple resonances. In the present context, the relevant property of a sheet or screen that

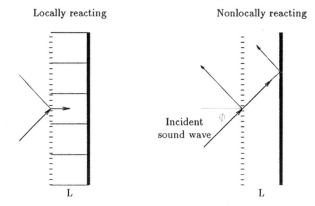

Figure 4.5: Porous sheet-cavity absorber. Left: Locally reacting. Right: Nonlocally reacting.

can readily be measured is the steady *flow resistance*. A pressure drop ΔP across the sheet produces a velocity U through the sheet and the flow resistance is defined as $r = \Delta P/U$. The same resistance is approximately valid also for the oscillatory flow in a sound wave.

In the locally reacting absorber, the fluid velocity in the air layer is forced by the partitions to be in the x-direction, normal to the boundary, so that $k_x = k$. The normal impedance is simply the sum of the sheet resistance θ and the impedance of the air column in a cell which we have found earlier to be $i \cot(kL)$ (see Eq. 3.64), both normalized with respect to ρc, where $k = \omega/c$ and L the thickness of the air layer. Thus, the absorption coefficient is obtained by inserting the impedance

$\zeta = \theta + i \cot(kL)$ into Eq. 4.52.

$$\alpha(\phi) = \frac{4\theta \cos \phi}{(1 + \theta \cos \phi)^2 + \cos^2(\phi) \cot^2(kL)} \tag{4.54}$$

For the sheet absorber without partitions, the fluid velocity in the air layer is no longer forced to be in the x-direction and the normal impedance of the air layer has to be modified. One obvious modification is that we have to use $k_x L = kL \cos \phi$ rather than kL. Furthermore, since the normal impedance is the ratio of the complex amplitude of the pressure and the *normal component* $u_x = u \cos \phi$ of the fluid velocity, the normalized impedance of the air layer will be $i(1/\cos \phi) \cos(kL \cos \phi)$. Thus, the absorption coefficient for the nonlocally reacting sheet absorber becomes (see Eq. 4.52)

$$\alpha(\phi) = \frac{4\theta \cos \phi}{(1 + \theta \cos \phi)^2 + \cot^2(kL \cos \phi)}. \tag{4.55}$$

The corresponding diffuse field absorption coefficient is obtained from Eq. 4.53.

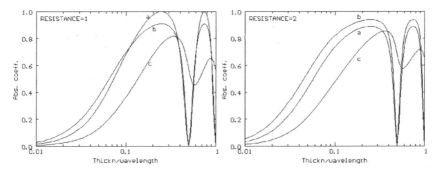

Figure 4.6: Absorption spectra of sheet absorber. (a) Normal incidence. (b) Diffuse field, locally reacting. (c) Diffuse field, nonlocally reacting.

Fig. 4.6 shows the computed frequency dependence of the absorption coefficient of a sheet absorber in which the frequency parameter is the ratio of the thickness L and the wavelength λ. On the left, the flow resistance of the sheet is $r = \rho c$, i.e., $\theta = 1$, and on the right, $\theta = 2$. In each graph, three curves are shown; one for normal incidence and two for diffuse fields corresponding to an air backing with and without a honeycomb.

When the thickness is an odd number of quarter wavelengths in the locally reacting absorber, the absorption coefficient will have a maximum. The standing wave in the air cavity then has a pressure anti-node at the sheet so that there will be no back pressure on it. Then, if $r = \rho c$, the impedance of the absorber is matched to the wave impedance so that no reflection occurs and all incident sound is absorbed. On the other hand, when the thickness is an integer number of half wavelengths, the velocity is zero at the screen so that there will be no absorption (anti-resonance). This is true for both normal incidence and diffuse field. For nonlocal reaction, however, the standing wave pattern depends on the angle of incidence and zero absorption cannot be obtained at all angles of incidence at a given frequency.

4.2.7 Measurement of Normal Incidence Impedance and Absorption Coefficient

Before the advent of digital instrumentation and FFT analyzers, the impedance usually was determined by the *standing wave method*. The sample was then placed at the end of a tube and exposed to a pure tone. Then, from the measured ratio of the maximum and minimum sound pressures in the wave in the tube and the distance to the first minimum from the sample, the impedance and the absorption coefficient could be determined. This measurement had to be repeated at every frequency over the range of interest and was quite time consuming.

Now, with digital instrumentation and the use of a two-channel FFT analyzer in the so-called *two microphone method*, the impedance can be determined directly as a function of frequency from a single measurement with random noise.

Again, the sample is placed at the end of a tube which is now driven at the other end by a source of random noise, $p(t) = \int p(\nu) \exp(-i2\pi \nu t) d\nu$. The signals from two microphones in the tube, separated a distance d, are the inputs to a two-channel FFT analyzer operating in the transfer impedance mode, which means that the ratio of the Fourier amplitudes of the two signals, magnitude and phase, is determined directly by the instrument.

The complex amplitude of the pressure field in the tube at a given frequency is expressed as

$$p(x, \nu) = A(e^{ikx} + Re^{-ikx}), \tag{4.56}$$

where $k = \omega/c = 2\pi \nu/c$ and R the complex reflection coefficient from the sample at the end of the tube.

Let the microphones be located at $x = 0$ and $x = d$. The ratio of the corresponding Fourier amplitudes which is determined by the analyzer is then

$$p(0)/p(d) = \frac{1 + R}{\exp(ikd) + R\exp(-ikd)} = H(\nu), \tag{4.57}$$

where $H(\nu)$ is determined by the analyzer over the entire frequency range. It follows that

$$R = [H\exp(ikd) - 1]/[1 - H\exp(-ikd)]. \tag{4.58}$$

From the relation between the reflection coefficient and the impedance of the boundary in Eq. 4.50, the normalized impedance of the boundary is

$$\zeta = (1 + R)/(1 - R). \tag{4.59}$$

With the output $H(\nu)$ from the analyzer combined with a simple computer program, the impedance can be obtained as a function of frequency. The corresponding normal incidence absorption coefficient is $\alpha = 1 - |R|^2$.

The method assumes a plane wave field in the tube. This means that at frequencies above the cut-on frequency of the lowest higher order mode in the tube (see Chapter 6), this assumption may not be valid. With a circular tube diameter of D, the cut-on frequency is $\approx c/1.7D$. Then, with a diameter of 5 cm, the measurements are limited to the portion of the spectrum below 4000 Hz.

4.2.8 Uniform Porous Absorber

The most common acoustic absorber used as a wall treatment is a uniform porous layer of fiber glass, foam, porous metals, etc. Acoustically, it is equivalent to a collection of closely spaced porous screens; the absorption spectrum is broader than for the single screen absorber and the anti-resonances are absent. The flow resistance per unit thickness is obtained in the same way as for the resistive sheet. It is the most important material property as far as sound absorption is concerned. In this analysis, the porous frame will be assumed to be rigid.

The oscillatory air flow of the sound within the porous material is forced to follow an irregular path by the randomly oriented fibers and pores in the material. The corresponding repeated local changes in direction and speed of the flow results in a force on the porous material and a corresponding reaction force on the fluid which is proportional to acceleration and can be accounted for in terms of an *induced mass density*. In the mathematical analysis of sound absorption, only an average velocity is used and the irregular motion and the corresponding inertial reaction force on the material is accounted for by assigning a higher inertial mass density to the air, the sum of ordinary mass density and the induced mass density. The effect is analogous to the apparent increase of mass we experience when accelerating a body in water such as a leg or an arm. The (empirically determined) factor used to express the apparent increase in density of the air in a porous material is called the *structure factor* Γ. It is typically 1.5-2.

The heat conduction and heat capacity of a solid material is much larger than for a gas. As was the case in the thermal boundary layer in Section 4.2.3 this makes the compressibility different than in free field, far from boundaries. The effect of heat conduction could be accounted for by means of a complex compressibility and the same is true in a porous material. Associated with it is a thermal relaxation time which expresses the time delay between the change in pressure caused by a change of volume. In harmonic time dependence this means a phase difference between the two. In a porous material, the relaxation time is related to the pore size which influences the flow resistance of the material. Consequently, there is a relation between the thermal relaxation time and the flow resistance and between the complex compressibility and the flow resistance. If the flow resistance per unit length in the material is denoted r, it is left as a problem to show that it is a good approximation to express the complex compressibility as

$$\tilde{\kappa} \approx \kappa \left[\gamma + (\gamma - 1)[\frac{\Omega^2}{1 + \Omega^2} + i \frac{\Omega}{1 + \Omega^2}] \right], \qquad (4.60)$$

where κ is the isentropic (free field) compressibility, $\Omega = \omega/\omega_v$, and $\omega_v = r/\rho$. As the frequency goes to zero, $\tilde{\kappa}$ goes to the isothermal value $\gamma\kappa$ and in the high frequency limit it is κ. The transition frequency between the two regions is $f_v = \omega_v/(2\pi) = r/(2\pi\rho)$. In terms of the normalized resistance $\theta = r/\rho c$, we get $f_v = c\theta/2\pi$. Thus, for a (typical) material, $\theta = 0.5$ per inch and with $c \approx 1120 \cdot 12$ inch/s, we get $f_v \approx 1070$ Hz. In other words, the compressibility will be approximately isothermal over a substantial frequency range.

Equations of Motion

In this section we shall outline how the absorption spectra of a uniform porous material can be calculated in terms of the physical properties of the material. It also serves as a practical example of the utility of complex amplitudes.

With H being the porosity, the amount of air per unit volume of the porous material is $H\rho$. We define the average fluid velocity in the sound field in such a way that ρu (rather than $H\rho u$). We choose this definition since it will make the equations and boundary conditions simpler. Under isentropic conditions, neglecting heat conduction, the relation between the density and pressure perturbations δ and p is $\delta/\rho = \kappa p = 1/\rho c^2$, where κ ($= (1/\rho)\partial\rho/\partial P$) is the compressibility of the fluid involved and c the ordinary (isentropic) speed of sound. The first term in the mass conservation equation $\partial(H\rho)/\partial t + \operatorname{div} u = 0$ can then be written $\rho\kappa\,\partial p/\partial t$ and we get

$$H\kappa\frac{\partial p}{\partial t} = -\operatorname{div} \boldsymbol{u}. \tag{4.61}$$

In the momentum equation we have to account for both the flow resistance and the induced mass. Thus, with the flow resistance per unit length of the material denoted r and the equivalent mass density $\Gamma\rho$, where Γ is the structure factor defined above, accounting for the induced mass, the momentum equation becomes

$$\frac{\partial\Gamma\rho\boldsymbol{u}}{\partial t} + r\boldsymbol{u} = -\operatorname{grad} p. \tag{4.62}$$

The velocity can be eliminated between these equations by differentiating the first with respect to time and taking the divergence of the second. With div grad $p = \nabla^2 p$, we then get

$$\nabla^2 p - (H\Gamma/c^2)\frac{\partial^2 p}{\partial t^2} - (\kappa r H)\frac{\partial p}{\partial t} = 0. \tag{4.63}$$

If the flow resistance is small so that the third term can be neglected, we get an ordinary wave equation with a wave speed $c/\sqrt{H\Gamma}$. If the flow resistance is large so that the second term, representing inertia, can be neglected, we get instead a diffusion equation.

The assumption of an isentropic compressibility in the porous material is unrealistic because of the narrow channels in the material and the high heat conductivity of the solid material (compared to air). In harmonic time dependence we can account for heat conduction by using the making the compressibility complex and, as in Section 4.2.3, we denote it $\tilde{\kappa}$. Furthermore, in the momentum equation we combine the first and second term into one, $(-i\omega\rho\Gamma + r)\boldsymbol{u} \equiv \tilde{\rho}\boldsymbol{u}$, where $\tilde{\rho}$ is a complex density. The complex amplitude versions of Eq. 4.61 and 4.62 can then be expressed as

$$-i\omega\tilde{\kappa}' p = -\operatorname{div} \boldsymbol{u}$$
$$-i\omega\tilde{\rho}\boldsymbol{u} = -\operatorname{grad} p, \tag{4.64}$$

where $\tilde{\rho} = \rho(\Gamma + ir/\omega\rho)$ and $\kappa' = H\kappa$. Incidentally, the complex compressibility is analogous to the inverse of the complex spring constant which is used to account for compressional losses in a spring in parallel with a dashpot damper. Similarly, the complex density corresponds to the complex mass in a mass-spring oscillator in which the forces due to inertia and friction are combined into one.

The complex density contains the flow resistance and the structure factor and on the basis of the results obtained from this analysis, experiments can be devised for the measurement of these quantities. For example, they can be obtained from the measurement of the phase velocity and the spatial decay rate of a sound wave in a porous material, assuming that the porosity has been determined from another experiment.

Propagation Constant and Wave Impedance

Eliminating **u** between the equations in Eq. 4.64, we obtain

$$\nabla^2 p + \tilde{\rho}\tilde{\kappa}' \, p = 0. \tag{4.65}$$

With a space dependence of the complex sound pressure amplitude $\propto \exp(iq_x x + iq_y y + iq_z z)$, we obtain from Eq. 4.65,

$$q^2 = q_x^2 + q_y^2 + q_z^2 = k^2 \, (\tilde{\rho}/\rho)(\tilde{\kappa}'/\kappa), \tag{4.66}$$

where we have used for normalization the isentropic compressibility $\kappa = 1/\rho c^2$ and $k = \omega/c$.

The corresponding normalized propagation constant is

$$Q \equiv q/k \equiv Q_r + i Q_i = \sqrt{(\tilde{\rho}/\rho)(\tilde{\kappa}'/\kappa)}, \tag{4.67}$$

where $\tilde{\rho}$ and $\tilde{\kappa}'$ are given in Eqs. 4.64.

The front surface of the porous material is located in the yz-plane at $x = 0$ and a plane sound wave is incident on it. The complex pressure amplitude is expressed as $p(x, y, z, \omega) = A \exp(ik_x x + ik_y y + ik_z z)$, where, from the wave equation in free field, we get $k_x^2 + k_y^2 + k_z^2 = k^2 \equiv (\omega/c)^2$. The direction of the wave is specified by the polar angle ϕ with respect to the x-axis and the azimuthal angle ψ, measured from the z-axis. In other words, the projection of the propagation vector on the yz-plane has the magnitude $k \sin\phi$ and we have $k_y = k \sin\phi \sin\psi$ and $k_z = k \sin\phi \cos\psi$.

Similarly, the wave function inside the material is $\exp(iq_x x + iq_y y + iq_z z)$, where $q^2 = q_x^2 + q_y^2 + q_z^2$. The wave vector components in the y- and z-direction are continuous across the surface of the absorber so that $q_y = k_y = k \sin\phi \sin\psi$ and $q_z = k_z = k \sin\phi \cos\psi$. This is equivalent to saying that the intersection of the incident wave front with the boundary and the corresponding intersection of the wave front in the porous material are always the same.

It follows then that

$$q_x \equiv (\omega/c)Q_x = \sqrt{q^2 - q_y^2 - q_z^2} = \sqrt{q^2 - k^2 \sin^2\phi} = (\omega/c)\sqrt{Q^2 - \sin^2\phi}. \tag{4.68}$$

where $Q = q/k$, $k = \omega/c$, and ϕ the angle of incidence. The velocity component in the x-direction is obtained from

$$u_x = (1/i\omega\tilde{\rho})\frac{\partial p}{\partial x}, \qquad (4.69)$$

where $\tilde{\rho}/\rho = \Gamma_s + iz_v/\omega\rho$.

The *wave admittance* in the x-direction is the ratio u_x/p for a *traveling* wave in the x-direction for which $\partial p/\partial x = iq_x p$. It follows from the equations above that the normalized value of the wave admittance and the corresponding impedance are given by

$$\eta_w = 1/\zeta_w = \rho c u_x/p = \frac{Q_x}{\tilde{\rho}/\rho}, \qquad (4.70)$$

where Q_x is given in Eq. 4.68 and $\tilde{\rho}$ in Eq. 4.64.

We recall that the input impedance of an air layer of thickness L is $z = i\,(\rho c)$ $\cot(kL)$. The impedance of a uniform porous layer has the same general form but with ρc replaced by a complex wave impedance and k by a complex propagation constant q_x, both containing the flow resistance per unit length and the structure factor. Once the input impedance of the layer has been expressed in this manner, the absorption coefficient can be computed from Eqs. 4.52 and 4.53. Examples of

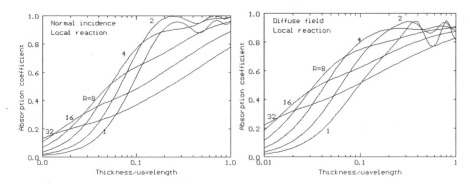

Figure 4.7: Sound absorption spectra of a uniform porous layer of thickness L backed by a rigid wall. The frequency variable is L/λ and the parameter R, ranging from 1 to 32, is the normalized total flow resistance of the layer. λ is the free field wavelength. Left: Normal incidence. Right: Diffuse field.

computed absorption spectra thus obtained are shown in Fig. 4.7. The graphs on the left and on the right refer to normal incidence and diffuse field, respectively. The parameter which ranges from 1 to 32 is the normalized total flow resistance R of the layer. The absorber is assumed locally reacting. Under these conditions, R-values less than 1 yield a lower absorption than for $R = 1$ and generally are of little interest.

With the use of normalized values of the frequency parameter and the flow resistance in this figure, different curves for different layer thicknesses are not needed. However, in practice, it is more convenient to have the frequency in Hz as a variable.

We leave it for Problem 9 to plot a spectrum or two in this manner, using the universal curves in the figure as sources of data. For an R-value of 4 and a layer thickness larger than one-tenth of a wavelength, the absorption coefficient exceeds 80 percent for both normal incidence and diffuse fields.

4.2.9 Problems

1. **Surface impedance for shear flow**

   Following the computational outline in the text, prove that the mass reactance of the surface impedance in Eq. 4.20 is consistent with the kinetic energy in the oscillatory flow in the boundary layer.

2. **Reflection–absorption, anechoic room**

   In an anechoic room, the walls are usually treated with wedges of porous material, typically 2-3 feet deep to provide absorption over a wide range of frequencies, from 100 Hz (or lower) to 8000 Hz. In many measurements in such a room, it is desirable that the reflected amplitude be 20 dB lower than the incident. Show that in order to achieve this, the absorption coefficient must be at least 0.99.

3. **Intensity of reflected wave**

   The absorption coefficient of a 4 inch thick layer of fiberglass mounted on a rigid wall is 0.8 at 500 Hz. An incident wave with a sound pressure level of 120 dB is incident on the absorber. Determine
   (a) the magnitude of the pressure reflection coefficient.
   (b) the intensity of the incident and reflected waves in watts/m^2.
   (c) the reduction in sound pressure level after one reflection.

4. **Transmission of sound into steel from air**

   (a) Determine the power transmission coefficient and the corresponding transmission loss for sound transmitted from air into an infinite layer of steel from air. What is the result if direction of wave travel is reversed?

5. **Complex boundary impedance**

   The normalized impedance of a porous layer of thickness L at sufficiently low frequencies (wavelength λ much larger than the layer thickness) is

   $$\zeta \approx \Theta/3 + i/(H\gamma kL),$$

   where Θ is the normalized values of the total flow resistance of the layer, H, the porosity, $\gamma \approx 1.4$, the specific heat ratio for air, and $k = \omega/c = 2\pi/\lambda$.
   (a) What is the magnitude and phase angle of the complex pressure reflection coefficient in terms of the given parameters. In particular, let $R = 4$, $H = 0.95$, and $L = 4$ inches, and $L/\lambda = 0.05$.
   (b) Show that the normal incidence absorption coefficient can be expressed as

   $$\alpha \approx \frac{4\theta(kL')^2}{1 + (kL')^2(1+\theta)^2} \approx 4\theta(kL')$$

   where $\theta = \Theta/3$ and $L' = H\gamma L$.

6. **Standing wave method for impedance and absorption measurement**

   Consider a tube with a test sample at the end. The wave in the tube is a superposition of an incident wave and a reflected wave with the complex amplitudes $A \exp(ikx)$ and

$B \exp(-ikx)$. It has the maximum pressure amplitude $p_{max} = |A| + |B|$ where the incident and reflected waves are in phase and a minimum $p_{min} = |A| - |B|$. If the measured ratio p_{max}/p_{min} is denoted n, determine

(a) the absorption coefficient.

(b) If the location of the minimum pressure closest to the sample is found to be a distance d from the sample, determine from this value and n the impedance of the sample.

7. **Reflection and absorption by screen in air, steam, and water**

A plane resistive screen is stretched across the path of a plane sound wave perpendicular to the direction of propagation. The flow resistance of the screen is proportional to the kinematic viscosity of the fluid involved. The normalized flow resistance of the screen as measured in air at 70° F and 1 atmosphere is 4.

What fraction of the incident intensity is reflected from, transmitted through, and absorbed within the screen

(a) in air at 1 atm and 70 deg F,

(b) in water, and

(c) in steam at 1000°F and 1000 psi?

Air: Sound speed at 70°F : $c \approx 342$ m/sec. Density: $\rho \approx 0.0013$ g/cm^3. Kinematic shear viscosity: $\nu = \mu/\rho \approx 0.14$.

Water: $c \approx 1500$ m/sec. $\rho = 1$ g/cm^3. $\nu = \mu/\rho \approx 0.010$ CGS at 70°F .

Steam at 1000°F and 1000 psi: $c \approx 697$ m/sec. $\rho \approx 18.2$ kg/m^3. $\nu \approx 0.018$. The conditions given here for steam are rather typical for a nuclear power plant.

8. **Diffuse field absorption coefficient; angle independent impedance**

(a) Prove that the diffuse field absorption coefficient in Eq. 4.53 for a locally reacting boundary (i.e., with an angle independent normal impedance $\zeta = \theta + i\chi$) can be expressed in closed form as

$$\alpha_d = \frac{8\theta}{|\zeta|^2}(1 - A + B) \qquad \text{where}$$

$$A = (\theta/|\zeta|^2) \ln[(1+\theta)^2 + \chi^2]$$

$$B = [(\theta^2 - \chi^2)/|\zeta|^2](1/\chi) \arctan[\chi/(1+\theta)].$$

(b) Rewrite this expression in terms of the normalized admittance, $\eta \equiv \mu + i\beta \equiv 1/\zeta$.

9. **Uniform porous layer**

(a) Use the data in Fig. 4.7 and plot the normal incidence and diffuse field absorption coefficient of a 4" uniform porous layer on a rigid wall covering a frequency range from 100 to 4000 Hz. The flow resistance of the material is $0.5\rho c$ per inch.

(b) At 500 Hz, plot the absorption coefficient versus the flow resistance of the layer covering the range from 0 to 10 ρc per inch.

4.3 Sound Transmission Through a Wall

4.3.1 Limp Wall Approximation

A problem of considerable practical importance concerns the transmission of sound through a partition wall. For example, building codes contain requirements on the transmission loss of walls that separate apartments in a building and special laboratories have been established for the measurement of transmission loss.

The physics involved are, in principle, very simple. Sound incident on one side of a solid wall causes it to vibrate, and the vibrations radiate sound into the space on the other side of the wall. For a quantitative analysis, we consider a plane sound wave incident on a wall or plate of mass m per unit area placed in free field. Only the mass of the wall will be accounted for; boundary and stiffness effects will be neglected. This is a good approximation at sufficiently high frequencies, well above the resonance frequencies of a finite wall; in practice the fundamental frequency of a wall often is about 10 to 20 Hz which is well below the range of frequencies (typically 125 to 8000 Hz) involved in a transmission loss test.

The complex pressure amplitudes of the incident, reflected, and transmitted pressures at the wall are denoted p_i, p_r, and p_t. With reference to Eqs. 4.45 and 4.46, the incident, relfected, and transmitted waves are

$$p_i(x) = |p_i| e^{ik_x x},$$
$$p_r(x) = |p_r| e^{-ik_x x},$$
$$p_t(x) = |p_t| e^{ik_x x}, \tag{4.71}$$

where the factor $\exp(ik_y y)$ is contained in $|p_i|$, $|p_r|$, and $|p_t|$, to save some writing. The angle of incidence is ϕ with respect to the normal to the wall so that $k_x = k \cos \phi$ ($k = \omega/c = 2\pi/\lambda$).

Again, with reference to Section 4.2.6, the velocity fields are given by

$$\rho c \, u_{ix}(x) = |p_i| e^{ik_x x}$$
$$\rho c \, u_{rx}(x) = -|p_r| e^{-ik_x x}$$
$$\rho c \, u_{tx}(x) = |p_t| e^{ik_x x}. \tag{4.72}$$

The wall is located at $x = 0$ and we neglect its thickness compared to the wavelength. The wall is assumed impervious so that its velocity will be the same as the velocity $u_{tx}(0)$ of the transmitted wave. The equation of motion (Newton's law) of the wall is $m \partial u_{tx}/\partial t = F$, where F is the force per unit area. The corresponding complex amplitude equation is $-i\omega m, u_{tx}(0) = F(\omega)$. With $F(\omega) = p_i(0) + p_r(0) - p_t$ we get

$$p_i(0) + p_r(0) - p_t(0) = -i\omega m u_{tx}(0) = (-i\omega m/\rho c) p_t(0) \cos \phi, \tag{4.73}$$

where we have used $u_{tx}(0) = (p_t(0)/\rho c) \cos \phi$. The total velocity amplitude to the left of the wall is $u_{ix}(0) + u_{rx}(0)$ which must equal $u_{tx}(0)$ and it follows then from Eq. 4.72 that

$$p_i(0) - p_r(0) = p_t(0). \tag{4.74}$$

Addition of Eqs. 4.73 and 4.74 yields

$$p_i = [1 - i(\omega m/2\rho c) \cos \phi] p_t(0) \tag{4.75}$$

and the transmission coefficient for pressure becomes

$$\tau_p(\phi) = p_t/p_i = [1 - i(\omega m/2\rho c) \cos \phi]^{-1}. \tag{4.76}$$

With the incident and transmitted intensities denoted I_i and I_t, the *power transmission coefficient* and the corresponding transmission loss are then

---

Transmission coefficient and transmission loss. Limp wall
$$\tau(\phi) = I_t/I_i = |\tau_p|^2 = [1 + (\omega m/2\rho c)^2 \cos^2 \phi]^{-1}$$
$$TL = 10\log(1/\tau(\phi) = 10\log[1 + \cos^2 \phi (\omega m/2\rho c)^2]$$

(4.77)

[$\tau(\phi)$: *Power transmission coefficient.* I_i, I_t: *Incident and transmitted power.* τ_p: *Pressure transmission coefficient. m: Mass per unit area. ϕ: Angle of incidence. TL: Transmission loss in dB.*]

In most cases of interest, $\omega m/\rho c >> 1$, so that $TL \approx 20\log(\omega m/2\rho c)$ which is often referred to as the 'mass law' for transmission loss. According to it, a doubling of mass or of frequency results in an increase of the transmission loss of $20\log(2) \approx 6$ dB.

As an example, consider a 1/4" thick glass pane with a density of 2.5 g/cm^3 so that $m = 1.6$ g/cm^2. With $\rho c = \rho c \approx 42$ CGS and at a frequency of 1000 Hz, the transmission loss becomes ≈ 41.6 dB at normal incidence.

Diffuse Field

The average transmission coefficient in a diffuse field is obtained in the same manner as for the average absorption coefficient. All we have to do is replace $\alpha(\phi)$ in Eq. 4.53 by $\tau(\phi)$ to obtain

$$\tau_d = 2 \int_0^{\pi/2} \tau(\phi) \sin\phi \cos\phi d\phi.$$

(4.78)

With $\tau = 1/[1 + \cos^2 \phi (\omega m/2\rho c)^2]$ (Eq. 4.78), it is left for a problem to carry out the integration and show that

$$\tau_d = (1/\beta^2)\ln(1 + \beta^2),$$

(4.79)

where $\beta = (\omega m/2\rho c)$. Thus, the corresponding diffuse field transmission loss becomes

$$TL_d = 10\log[\beta^2/\ln(1 + \beta^2)].$$

(4.80)

In Fig. 4.8 are shown the transmission loss curves (thin lines) for angles of incidence 0, 30, 45, 60, and 80 degrees together with the average values in a diffuse field (thick line). The parameter that determines the transmission loss, $\beta = \omega m/2\rho c$, is proportional to the product of mass and frequency. Thus, to obtain the transmission loss for another mass m than 10 kg/m^2 at a frequency f we have to use the frequency value $(m/10)f$ in the graph.

The normal incidence value of the TL is substantially higher than the diffuse field value. Formally, this can be seen from the expression (4.78) for the transmission coefficient which has its minimum value (maximum TL) at normal incidence. Physically, it is related to the fact that the wave impedance $p_i/u_{ix} = \rho c/\cos\phi$ 'in the normal direction' of the incident sound increases with the angle of incidence so that it will be better matched to the high impedance of the wall, yielding a higher transmission and lower TL.

When $\omega m >> \rho c$, which normally is the case, the normal incidence transmission loss is $TL \approx 20 \log(\omega m / 2\rho c)$ so that it increases by $20 \log(2) \approx 6$ dB for every doubling of frequency or of mass. In a diffuse field, this increase is somewhat smaller, ≈ 5 dB.

4.3.2 Effect of Bending Stiffness

A limp panel has no bending stiffness and, like a membrane without tension, is locally reacting. There is no coupling between adjacent elements and no free wave motion. The normal impedance $-i\omega m$ is independent of the angle of incidence, and like the diffuse field absorption coefficient (Eq. 4.78) for a locally reacting absorber, the diffuse field transmission coefficient can be expressed in closed form (Eq. 4.79).

The idealization of an infinite limp panel considered so far may at first sight seem unrealistic. However, as it turns out, the results obtained are quite useful for estimates of the transmission loss and are almost always used as a comparison with experimental data. An improvement can be obtained by accounting for the bending stiffness of the wall.

Actually, in the model of an infinitely extended panel, stiffness comes into play only for waves at oblique angles of incidence at which there is a periodic spatial distribution of pressure along the panel. At normal incidence the pressure is in phase at all positions on the panel and no bending occurs.

The effect of stiffness becomes most important at short wavelengths when the radius of curvature of bending becomes small (from everyday experience we know that it becomes increasingly more difficult to bend a stiff wire as the radius of curvature of bending is decreased). Therefore, unlike an ordinary mass spring oscillator, the response of an infinite panel to an incoming sound wave will be stiffness controlled at high frequencies and mass controlled at low frequencies. In the low frequency region,

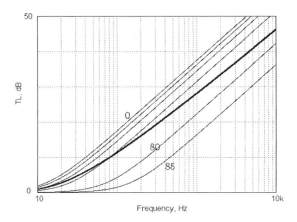

Figure 4.8: Transmission loss of a limp wall with a mass 10 kg/m$^2 \approx 2.2$ lb/ft^2. Angles of incidence: $0, 30, 45, 60, 80$, and 85 degrees. Thick line: Diffuse field average.

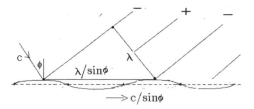

Figure 4.9: Plane wave of wavelength λ incident on a panel; angle of incidence is ϕ.

the transmission loss of a stiff panel is then expected to be essentially the same as for a limp one. For relatively thin panels, such as windows, the transition between low and high frequencies typically is about 2000 Hz; it decreases with increasing panel thickness since the bending stiffness increases faster with increasing thickness (as the third power) than does the mass.

It is not surprising then to find that the bending wave speed increases with frequency. The mass remains the same but the stiffness increases with the inverse of the radius of curvature and hence with the inverse of the wavelength. The wave speed is expected to be proportional to the square root of the ratio of the stiffness and the mass and hence to the square root of frequency. This is indeed the case as the phase velocity of the bending wave is known to be,[6]

$$v_b = \sqrt{h'v}\sqrt{\omega}, \tag{4.81}$$

where $v = \sqrt{Y/[\rho_p(1-\sigma^2)}$ is the longitudinal wave speed, Y, the Young's modulus, ρ_p, the density of the plate, $h' = h/\sqrt{12}$, h, the plate thickness, and σ, the Poisson ratio, typically ≈ 0.25.

Consider now a sound wave incident on a panel at an angle of incidence ϕ, as shown in Fig. 4.9. The intersection point between a wave front and the panel moves along the panel with a velocity

$$c_t = c/\sin\phi, \tag{4.82}$$

which is always greater than the sound speed c; it will be called the *trace velocity* c_t. When this velocity coincides with the free bending wave speed v_b *wave coincidence* or resonance is said to occur. The mass reactance of the panel is then canceled by the bending stiffness reactance and if there is no damping, the transmission loss will be zero. The *lowest* frequency at which this resonance can occur is obtained for grazing incidence of the sound, i.e., $\pi = \pi/2$, in which case the trace velocity is simply the speed of sound c. The corresponding resonance frequency, as obtained from Eq. 4.81 by putting $v_b = c$ is

$$\omega_c = 2\pi f_c = c^2/vh' \quad (h' = h/\sqrt{12}), \tag{4.83}$$

---

[6]See, for example, Uno Ingard, *Fundamentals of waves and oscillations*, Cambridge University Press, 1988.

which is called the *critical frequency*. Above this frequency there is always an angle of incidence ϕ at which coincidence $c_t = v_b$ occurs,

$$\sin \phi = \sqrt{f_c/f_r}. \tag{4.84}$$

Thus, at a given frequency, a stiff panel can be said to act like a spatial filter providing very small transmission loss at the coincidence angle.

For glass, we have $Y \approx 6 \times 10^{11}$ dyne/cm^2, $\rho_p \approx 2.5$ g/cm^3 and $\sigma = 0.25$. Then, if the thickness h is expressed in cm, the critical frequency in Hz, we get from Eq. 4.83

$$f_c = \approx 1264/h \, \text{Hz} \quad (h \text{ in cm}). \tag{4.85}$$

Actually, this expression is valid approximately also for steel, its higher value of Y being countered by a higher value for ρ (≈ 7.8 g/cm^3). Accordingly, for these materials, a 1 cm thick panel has a critical frequency ≈ 1264 Hz.

For an angle of incidence ϕ, it follows from Eq. 4.84 that bending wave resonance occurs at a frequency

$$f_r = f_c/\sin^2 \phi. \tag{4.86}$$

As for the resonance of a simple mass-spring oscillator, the effects of inertia and stiffness (in this case bending stiffness) cancel each other and the wall becomes transparent, as already indicated, and the transmission loss would be zero if there were no damping. There is always some internal damping present, however, and, as we shall see, it can be accounted for by means of a complex Young's modulus.

To account for the bending stiffness in the expression for the transmission coefficient, the impedance $-i\omega m$ for the limp panel in Eq. 4.76 has to be modified. For the linear harmonic oscillator the modification involves adding the reactance of the spring so that $-i\omega m$ is replaced by $-i\omega m + iK/\omega = -i\omega m(1 - f_0^2/f^2)$. Notice that in this case the impedance is stiffness controlled at frequencies below the resonance frequency, approaching iK/ω with decreasing frequency. For the plate, the situation is reversed, as we have indicated above. The impedance becomes stiffness controlled at high frequencies (short wavelengths, small radius of curvature) above the resonance frequency f_r and the factor f_0/f is found to be replaced by f/f_r.

To account for the bending stiffness of the plate, the impedance $-i\omega m$ in Eq. 4.76 has to be replaced by $-i\omega m[1 - (f/f_r)^2]$ (not shown in detail here) and with the expression for f_r in Eq. 4.86, the pressure transmission coefficient becomes

$$\tau_p(\phi) = \{1 - (i\omega m/2\rho c)\cos^2 \phi[1 - (\omega/\Omega_c)^2 \sin^4 \phi]\}^{-1} \tag{4.87}$$

and the transmission loss

$$TL = 10 \log(1/|\tau_p|^2), \tag{4.88}$$

where $\omega_r = 2\pi f_r$. We can express the entire frequency dependence in normalized form by replacing ωm by $(\omega/\omega_c)m\omega_c$. Then, with the expression for ω_c in Eq. 4.83 and with $m = \rho_p h$, we introduce the dimensionless parameter

$$\mu = m\omega_c/2\rho c = \rho_p h(c^2/vh')/2\rho c = \sqrt{3}(\rho_p/\rho)(c/v). \tag{4.89}$$

In terms of it and with $\Omega = \omega/\omega_c$, the pressure transmission coefficient takes the form

$$\tau_p(\phi) = [1 - i\mu\Omega\cos\phi(1 - \Omega^2\sin^4\phi)]^{-1} \qquad (4.90)$$

and the transmission loss

$$TL = 10\log|1/\tau_p|^2. \qquad (4.91)$$

The quantity $\mu = \omega_c m/2\rho c = \sqrt{3}(c/v)(\rho_p/\rho)$ depends only on the material constants v (see Eq. 4.81) and ρ_p of the plate and not explicitly on m or the thickness h.

If we wish to account for internal damping in the plate, a final modification of τ_p in Eq. 4.90 is needed. Normally, this is done by making the Young's modulus complex, i.e., Y is replaced by $Y(1 - i\epsilon)$, where ϵ is the loss factor. This means that ω_c and hence $\Omega_c = \omega/\omega_c$ becomes complex with ω_c replaced by $\omega_c\sqrt{1 - i\epsilon}$. An example

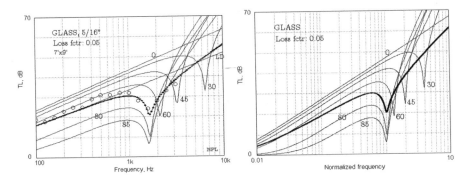

Figure 4.10: Transmission loss of a window.
Left: Glass, 5/16" thick, size: 7' × 9'. Loss factor: 0.05. Right: 'Universal' TL characteristics of glass.

of the computed TL versus frequency for the infinite panel is shown on the left in Fig. 4.10 for angles of incidence 0, 30, 45, 60, 80, and 85 together with the average transmission loss in a diffuse sound field (thick curve). It refers to a 5/16" thick 7' × 9' glass window, 7' × 9', with a loss factor of 0.05, as indicated. For comparison, refer to Fig. 4.8 for the TL of a limp panel. It should be noted that in the mass controlled low frequency region, it is essentially the same as the TL for the panel with stiffness. The normal incidence TL, corresponding to an angle of incidence of 0 degrees (the top thin line), is substantially higher than the diffuse field value over the entire frequency range.

The critical frequency, obtained from Eq. 4.83, is 1592 Hz. The resonance (coincidence) frequencies for different angles of incidence are consistent with Eq. 4.86. For example, at an angle of incidence of 45 degrees, the resonance frequency is $2f_c = 3184$ Hz. The resonance frequency decreases with increasing angle of incidence until it reaches f_c at 90 degrees. The dip in the diffuse field average transmission loss occurs somewhat above f_c.

The average TL in a diffuse sound field is obtained from Eq. 4.78 using the new value of τ in Eq. 4.90 (with a complex Ω_c). The integration has been carried out numerically. At frequencies above the critical frequency, the number of angles used in the integration has to be quite large for small values of the loss factor in order to avoid irregularities in the TL curve. If there is no loss and for $f > f_c$ there is always an angle, the *coincidence angle* given by $\sin \phi = \sqrt{f_c/f}$, at which the transmission loss is 0.

At frequencies below the critical frequency, the TL is nearly the same as for the limp panel and the loss factor has essentially no effect on the transmission loss; above this frequency, however, increased damping yields higher TL.

The experimental diffuse field data shown in the figure are in good agreement with the computed. However, if the panel size becomes smaller than the wavelength of the free bending wave of the panel, the agreement becomes less good, as anticipated. With the phase velocity of the bending wave given by Eq. 4.81 the corresponding wavelength is $\lambda = v_b/f = \sqrt{hv2\pi/12f}$. In other words, this wavelength increases with the panel thickness. Thus, for a given panel size, the deviation of the experimental from the calculated is expected to increase with increasing panel thickness. This is indeed found to be the case.

A complete analysis of transmission should include the normal modes of the panel and the coupling of these modes with the modes of the sound fields in rooms on the two sides of the panel (see Section 4.3.3). The effect of panel modes, not accounted for here, become important when the size of the panel is of the order or smaller than the wavelength of the bending wave on the panel.

Another reason for a difference between measured and calculated values involves the assumption of a diffuse sound field. In practice, the sound field in the test rooms used in the measurement of transmission loss is not completely diffuse, and the degree of diffusivity varies from one laboratory to the next and corresponding variations in the measured transmission loss are to be expected.

The set of curves on the right in Fig. 4.10 are 'universal' in the sense that TL is now shown as a function of the normalized frequency f/f_c. As already explained in connection with Eq. 4.90, the transmission loss then becomes independent of the thickness of the panel and depends only on the material. Thus, there will be one set of curves for glass, another for aluminum, etc.

Normally, in most discussions and data on TL, only the diffuse field or the normal incidence value is given. However, in many practical situations, the panel is not exposed to a wave of normal incidence or a diffuse field and neither of these TL values is representative. The difference is not trivial; the TL at an angle of incidence of 80 degrees can readily be 20 dB below the normal incidence value. For example, for traffic noise through windows in a high rise building beside a highway, the noise level inside is often found to increase with the elevation above ground despite the increased distance to the noise source.[7]

---

[7]I believe this angular dependence of the TL is probably responsible for the effect which I have noticed on several occasions sitting in the Hayden Library at M.I.T. The Memorial Drive runs along the library and the peak value of the noise from a passing car transmitted through the windows reaches a maximum value when the noise is incident at some oblique angle and not at normal incidence. This is particularly true on a rainy day when the wet pavement seems to make the tire noise rich in relatively high frequencies. This proposed explanation does not account for the possible directional characteristics of the noise, however.

4.3.3 Measurement of Transmission Loss

In the two-room method of measurement of the transmission loss, the partition to be measured is inserted in to an opening of the (heavy) wall that separates two reverberant rooms. The linear dimensions of the rooms should be at least two or three wavelengths at the lowest frequency involved in the test. Normally, the measurement is carried out in third octave bands, the lowest being at 100 Hz at which the wavelength is ≈ 11 feet. The wall between the two rooms should have a considerably larger transmission loss than the partition to be tested and often is a double concrete wall with an air space separation. The size of the opening typically is approximately $10' \times 10'$.

The rooms should be highly reverberant, so that the sound fields in the rooms can be assumed to be diffuse (there are prescribed tests to check the diffuseness in the rooms). One of the rooms contain one or more sound sources, normally loudspeakers driven by a random noise generator and power amplifiers. The spatial average rms values p_1 and p_2 of the sound pressures in the two rooms are measured. The acoustic intensity that strikes the test panel is $I_1 = C|p_1|^2$, where C is a constant, and the acoustic power that goes through the test panel is $W_1 = \tau I_1 S$, where τ is the transmission coefficient and S the area of the panel.

The transmitted power establishes a sound field in the receiving room which in steady state is such that the absorbed power in the room is equal to W_1. The spatial average of the corresponding steady state sound rms sound pressure in the receiving room is p_2, which is measured. By expressing the absorbed power in terms of p_2 and equating this power with that transmitted through the wall, the transmission loss can be expressed in terms of the sound pressure levels in the two rooms, as given by Eq. 6.14.

The two-room method is based on the assumption of diffuse fields in the source and receiver room. This cannot be fulfilled at low frequencies where only a few acoustic modes are excited in the rooms (see Chapter 6). At these frequencies the method yield large fluctuations in the measured transmission loss. For this reason, typical laboratories limit the frequency range to frequencies above 125 Hz.

Other Methods

The diffuse field transmission loss obtained in the standard two-room method test procedure yields the steady state value of the transmission loss for a diffuse sound field. The question is whether this is the relevant quantity in most cases. Sounds are generally a succession of pulses that strike the wall at some angle of incidence and the transmission is determined by the transmission coefficient for this particular angle of incidence. Furthermore, if the pulses are short compared to the reverberation time in the receiving room, it is not the reverberant level that is relevant but the direct sound transmitted through the panel. Under such conditions, the pulse transmission loss rather than the diffuse field average should be determined. Actually, such a measurement would not need a two reverberant room test facility but could be carried out anywhere with special precautions to avoid interfering reflections.

In many applications, particularly in regard to sound transmission through windows, the incident sound typically is traffic noise and the field is not at all diffuse or reverberant. Also in such a case, it appears that data on the angular dependence of the transmission loss for an incident wave is more relevant than the diffuse field value.

At least in principle, the angular dependence of the transmission loss can be determined by measuring, with an intensity probe (see Section 3.2.3), the incident and transmitted intensities as a function of the angle of incidence. The effect of reflections from walls and other objects are reduced by the directivity of the intensity probe.

An alternate way of measuring transmission loss is to apply the *time delay spectrometry, TDS*. This technique utilizes a sound source that sweeps through the frequency range at a rate that can be adjusted. The signal is received through a tracking filter which can be delayed in time with respect to the source. The bandwidth of the filter can be varied. If the delay is set to correspond to a certain travel path of the sound from the source to the receiver, only that signal is ideally measured. A signal that was emitted at an earlier time and reflected from some object has a different frequency so that when it arrives at the receiver, the filter rejects this signal.

Like the two-room method, these two alternate methods are not good at low frequencies where diffraction about the panel causes problems, particularly at large angles of incidence. The advantage with the methods is that they can be carried out anywhere without the need for a special laboratory.

4.3.4 Problems

1. **The diffuse field transmission loss**

   Carry out the integration in Eq. 4.78 to prove the expressions for the diffuse field transmission coefficient in Eq. 4.79 and the corresponding transmission loss in Eq. 4.80.

2. **Transmission into steel from air**

   (a) Determine the power transmission coefficient and the corresponding transmission loss for sound transmitted from air into an infinite layer of steel from air and from steel into air.

   (b) For transmission through a layer of finite thickness (plate) there are two interfaces. Explain qualitatively why the transmission loss in this case cannot be expected to be twice the value for one interface.

4.4 Transmission Matrices

4.4.1 The Acoustic 'Barrier'

The complex amplitude description of acoustic field variables makes possible the introduction of *transmission matrices* which provide a unified procedure in analyzing sound interaction with structures consisting of several components.

Let us consider sound transmission through a 'barrier' illustrated schematically in Fig. 4.11, be it a single or composite wall of several elements, such as air or porous layers, perforated plates, membranes, and screens.

The complex amplitudes of sound pressure and fluid velocity components in the x-direction at the front side of the barrier are $p_1(\omega)$ and $u_1(\omega)$ and the corresponding quantities on the other side are $p_2(\omega)$ and $u_2(\omega)$. It should be realized that when a sound wave is incident on the barrier, the quantities p_1 and u_1 are the sums of the contributions from the pressures and velocities in the incident and reflected waves.

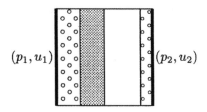

Figure 4.11: An acoustic barrier can be a combination of any number of elements. In this case four different layers are sandwiched between two panels.

If the system is linear, each of the variables at station 1 is a linear combination of the variables at station 2. In expressing this relationship, we shall use here the variable ρcu rather than u where ρc is the wave impedance of standard air; ρcu has the same dimension as p. In the following discussion, the argument (ω) will be omitted to simplify the writing somewhat; thus p will stand for $p(\omega)$, the complex amplitude of sound pressure. Thus, we express the linear relationship between the variables on the two sides of the barrier as

$$p_1 = T_{11}p_2 + T_{12}\rho cu_2$$
$$\rho cu_1 = T_{22}p_2 + T_{21}\rho cu_2 \qquad (4.92)$$

or, in matrix form,

$$\boxed{\begin{array}{l} \textit{Matrix relation for barrier (Fig. 4.11)} \\ \begin{pmatrix} p_1 \\ \rho cu_1 \end{pmatrix} = \begin{pmatrix} T_{11} & T_{12} \\ T_{21} & T_{22} \end{pmatrix} \begin{pmatrix} p_2 \\ \rho cu_2 \end{pmatrix} \end{array}} \qquad (4.93)$$

[p, u: Complex amplitudes of pressure and velocity. T_{ij}: Transmission matrix elements of barrier].

With our choice of variables, the elements are dimensionless.

The sound pressure p_1 in front of the barrier is the sum of the incident and the reflected pressures, p_i and p_r, and, likewise, the velocity is the sum of the incident and reflected wave contributions, ρcu_i and ρcu_r. The complex amplitude of the incident wave is $|p_i|\exp(ikx)$ and the reflected wave is $R_p|p_r|\exp(-kx)$, where R is the pressure reflection coefficient. The relations between pressure and velocity in the incident and reflected waves are $p_i = \rho cu_i$ and $p_r = -\rho cu_r$ (see Eq. 3.23).

If there are no reflections in the region behind the barrier, there will be a single transmitted plane, so that $p_2/\rho cu_2 = 1$, the normalized impedance at the back of the barrier being $\zeta_{=}1$. Then, with $p_1 = p_i + p_r$ and $u_1 = u_i + u_r$ and by adding the two equations in Eq. 4.92, the terms involving ρcu_i cancel each other, and we get

$$p_i = p_2(T_{11} + T_{12} + T_{22} + T_{21})/2. \qquad (4.94)$$

The corresponding transmission loss, as defined earlier, is then

$$TL = 10\,\log(|p_i/p_2|^2 = 10\,\log(|T_{11} + T_{12} + T_{22} + T_{21}|^2/4). \qquad (4.95)$$

This is a general expression for the transmission loss of the barrier for a sound wave at normal incidence. It is valid for any barrier. We shall apply it shortly to a limp wall for comparison with the result obtained earlier in Eq. 4.77.

Multiple Elements

Let us consider two barriers in series (cascade) with the matrix elements U_{ij} and V_{ij} and label the variables at the beginning and end of these elements by the indices 1, 2, and 3. Then

$$\begin{pmatrix} p_2 \\ \rho c u_2 \end{pmatrix} = \begin{pmatrix} V_{11} & V_{12} \\ V_{21} & V_{22} \end{pmatrix} \begin{pmatrix} p_3 \\ \rho c u_3 \end{pmatrix} \tag{4.96}$$

and

$$\begin{pmatrix} p_1 \\ \rho c u_1 \end{pmatrix} = \begin{pmatrix} U_{11} & U_{12} \\ U_{21} & U_{22} \end{pmatrix} \begin{pmatrix} p_2 \\ \rho c u_2 \end{pmatrix}. \tag{4.97}$$

Combining the two yields

$$\begin{pmatrix} p_1 \\ \rho c u_1 \end{pmatrix} = \begin{pmatrix} T_{11} & T_{12} \\ T_{21} & T_{22} \end{pmatrix} \begin{pmatrix} p_3 \\ \rho c u_3 \end{pmatrix}, \tag{4.98}$$

where

$$\begin{pmatrix} T_{11} & T_{12} \\ T_{21} & T_{22} \end{pmatrix} = \begin{pmatrix} U_{11} & U_{12} \\ U_{21} & U_{22} \end{pmatrix} \begin{pmatrix} V_{11} & V_{12} \\ V_{21} & V_{22} \end{pmatrix}. \tag{4.99}$$

The matrix elements of the total matrix T for the combination of the two elements is then obtained by multiplying the matrices U and V, i.e., $T_{11} = U_{11}V_{11} + U_{12}V_{21}$, etc.

In this manner, the total transmission matrix for any number of elements in cascade can be calculated. Numerically, the matrix multiplication is conveniently done by means of a computer routine.

4.4.2 Acoustic Impedance

The input impedance of the barrier is the simplest of all quantities to determine. It follows directly by dividing the two relations in Eq. 4.92 and we obtain, for the normalized input impedance,

$$\zeta_i = \frac{p_1}{\rho c u_1} = \frac{T_{11}\zeta_2 + T_{12}}{T_{21}\zeta_2 + T_{22}} \tag{4.100}$$

$\zeta_2 = p_2/\rho c u_2$.

Of particular interest is the case when the barrier is backed by a rigid wall in which case $\zeta_2 = \infty$ and

$$\zeta_i = T_{11}/T_{21}. \tag{4.101}$$

4.4.3 Reflection Coefficient

Other quantities can be expressed in a similar manner and we consider now the pressure reflection coefficient R_p. In terms of it, we have $p_r = R_p p_i$. Dividing the equations in Eq. 4.92 with each other yields $1 + R_p)/(1 - R_p) = [T_{11}p_2 + T_{12}\rho c u_1]/[T_{22}p_2 + T_{21}\rho c u_2]$ or

$$R_p = \frac{T_{11}\zeta_2 + T_{12} - T_{21}\zeta_2 - T_{22}}{(T_{11}\zeta_2 + T_{12} + T_{21}\zeta_2 + T_{22}}. \tag{4.102}$$

If there is free field on the backside of the barrier, we have $p_2 = \rho c u_2$, i.e., $\zeta_2 = p_2/\rho c u_2 = 1$.

4.4.4 Absorption Coefficient

The absorption coefficient can be expressed in terms of the impedance and the reflection coefficient as we have done earlier

$$R = (\zeta_i - 1)/(\zeta_i + 1)$$
$$\alpha = 1 - |R|^2 = 4\theta_i/((1 + \theta_i)^2 + \chi_i^2), \tag{4.103}$$

where $\zeta_i = \theta_i + i\chi_i$.

If the absorption coefficient is meant to express the power absorbed *within* the barrier (not counting the power in the transmitted wave), the expression in Eq. 4.103 is valid only if the barrier is backed by a rigid wall. Otherwise the power carried by the transmitted wave has to be subtracted. This correction is left for Problem 3.

4.4.5 Examples of Matrices

Limp Panel

As a first element, we consider the limp wall, for which we have already determined the transmission loss without the use of a transmission matrix (see Eq. 4.77). The mass per unit area of the wall is m and the frequency of the incident wave is ω. If the complex pressure amplitudes on the front and the back of the wall are p_1 and p_2, the driving force on the wall per unit area is $p_1 - p_2$ and it follows from Newton's law that $p_1 - p_2 = (-i\omega)m u$ (remember that $\partial/\partial t \to -i\omega$). The velocity of the wall is the same as the velocity of the air both on the front and on the back side of the panel, i.e., $u = u_2$. The equation of motion can then be written

$$p_1 = p_2 + (-i\omega m)u_2 \equiv T_{11}p_2 + T_{12}\rho c u_2$$
$$\rho c u_1 = \rho c u_2 \equiv T_{22}p_2 + T_{21}\rho c u_2. \tag{4.104}$$

In other words, the transmission matrix elements of the limp wall are $T_{11} = 1$, $T_{12} = -i\omega m/\rho c$, $T_{22} = 0$, $T_{21} = 1$. The corresponding matrix is

$$\mathbf{T} = \begin{pmatrix} 1 & -i\omega m/\rho c \\ 0 & 1 \end{pmatrix}. \tag{4.105}$$

Limp, Resistive Screen

Since an acoustical analysis deals with first order perturbations, the absolute velocity amplitude u_1 in front of the screen is equal to the velocity amplitude u_2 on the other side.[8]

The added feature in this problem, as compared to the limp plate, is that the screen is pervious so that the velocity of the screen is not the same as the velocity of the air at the screen. The sound pressure amplitudes on the front and back sides of the screen are p_1 and p_2, and velocity amplitude of the screen is u'. The mass per unit area of the screen is m and we assume that any stiffness reactance of the screen can be neglected (frequency higher than the resonance frequency of the screen element). Furthermore, we assume that the screen is not in contact with any other structure, such as a flexible porous layer (i.e., it has air on both sides). Under these conditions it follows from the definition of the interaction impedance $z \equiv \rho c \zeta$ that

$$p_1 - p_2 = z(u_2 - u') \tag{4.106}$$

$$-i\omega m u' = z(u_2 - u'). \tag{4.107}$$

For a purely resistive screen with a flow resistance r, we have $z = r = \rho c \theta$. Usually, this assumption is satisfactory.

It follows from Eqs. 4.106 and 4.107

$$u' = u_2 z/(z - i\omega m) \tag{4.108}$$

and

$$p_1 = p_2 + \zeta' \rho c u_2, \tag{4.109}$$

where

$$\zeta' = \zeta/[1 + i\zeta\rho c/\omega m] \tag{4.110}$$

is the *equivalent* screen impedance in which the acoustically induced motion of the screen is accounted for. With $u_1 = u_2$ the linear relation between p_1, u_1 and p_2, u_2 then can be expressed as

$$\begin{pmatrix} p_1 \\ \rho c u_1 \end{pmatrix} = \begin{pmatrix} 1 & \zeta' \\ 0 & 1 \end{pmatrix} \begin{pmatrix} p_2 \\ \rho c u_2 \end{pmatrix}. \tag{4.111}$$

Air Column. Loss-Free Tube

We consider next an air layer of length L. The general expression for a plane wave pressure field in the layer

$$p(x, \omega) = Ae^{ikx} + Be^{-ikx}, \tag{4.112}$$

where $k = \omega/c$ and A and B are complex constants.

---

[8] Rigorously, it is the mass flux that is continuous, but the difference in the density on the two sides of the screen in the absence of a mean flow is of first order, and, from conservation of mass flux, it follows that the difference in the velocities will be of second order.

The corresponding velocity field is obtained from the equation of motion $-i\omega\rho u = -\partial p/\partial x$,

$$u(x,\omega) = (1/\rho c)(Ae^{ikx} - Be^{-ikx}). \tag{4.113}$$

We now wish to relate the values of the field variable pair at the beginning and the end of the duct, p_1, u_1 and p_2, u_2, respectively To do this, we express A and B in terms of p_2 and u_2, and by placing $x = 0$ at the end of the duct (and $x = -L$ at the beginning), we get

$$A + B = p_2$$
$$A - B = \rho c u_2 \tag{4.114}$$

so that $A = (p_2 + \rho c u_2)/2$ and $B = (p_2 - \rho c u_2)/2$.

Using these values in Eqs. 4.112 and 4.113, we get

$$p_1 = \cos(kL)p_2 - i\sin(kL)\rho c u_2$$
$$\rho c u_1 = -i\sin(kL)p_2 + \cos(kL)\rho c u_2 \tag{4.115}$$

and the corresponding transmission matrix

$$\boldsymbol{T} = \begin{pmatrix} \cos(kL) & -i\sin(kL) \\ -i\sin(kL) & \cos(kL) \end{pmatrix}, \tag{4.116}$$

where $k = \omega/c = 2\pi/\lambda$ and L is the layer thickness.

It there is a mean flow in the pipe with a velocity U, the wave speeds in the positive and negative x-directions will be $c + U$ and $c - U$ and the corresponding propagation constants are then $k_+ = \omega/(c + U) = k/(1 + M)$ and $k_- = k/(1 - M)$, where $M = U/c$ is the flow Mach number. It is left for one of the problems to show that the transmission matrix in Eq. 4.116 will be modified to

$$\boldsymbol{T} = e^{i\Phi} \begin{pmatrix} \cos(k'L') & -i\sin(k'L') \\ -i\sin(k'L') & \cos(k'L') \end{pmatrix}, \tag{4.117}$$

where $\Phi = -kLM/(1 - M^2)$, $k' = k/(1 - M^2)$, and $k = \omega/c$ (see Problem 5).

4.4.6 Choice of Variables and the Matrix Determinant

The use of matrices in the present context is analogous to the treatment of linear networks in electrical engineering. In most cases, we shall deal with 2×2 matrices, which correspond to four-pole networks with two input terminals and two output terminals (Fig. 4.12). We shall deal only with passive systems, i.e., systems in which there are no sources of current or voltage within the network so that the values of the output variables depend only on the values of the input variables.

For a linear electrical four-pole network we then have the following relations

$$V_1 = A_{11}V_2 + A_{12}I_2$$
$$I_1 = A_{21}V_2 + A_{22}I_2, \tag{4.118}$$

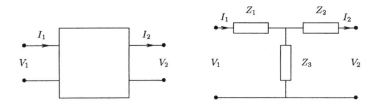

Figure 4.12: Electric four-pole and its equivalent T-network.

where (V_1, I_1) and (V_2, I_2) are the input and output values of voltage and current and A_{ij} the elements of the (transmission) matrix of the network.

In order for the network to be passive, we must have $I_1 = 0$ if $V_1 = 0$ so that $V_2/I_2 = -A_{12}/A_{11} = -A_{22}/A_{21}$ or

$$A_{11}A_{22} - A_{12}A_{21} = 1. \tag{4.119}$$

In other words, the four matrix elements are not independent but must be such that the determinant of the matrix is unity.

This condition imposed on the matrix elements can also be seen if we recall that the most general four-pole can be represented in terms of a 'T-network,' shown in Fig. 4.12, with *three* independent impedances Z_1, Z_2, and Z_3. If we express the relations between V_1, I_1 and V_2, I_2 in this network and express the four matrix elements T_{ij} in terms of the impedances, we again find the relation in Eq. 4.119.

In an acoustical circuit, the variables which correspond to voltage and current can be sound pressure p and the volume flow rate q, i.e., the product of velocity and cross-sectional area of the acoustical element involved. With this choice, q, like the electrical current, will be continuous across a discontinuity in a cross-sectional area. The determinant of an acoustical 'circuit' matrix then will be unity.

However, if *velocity u* rather than volume flow rate is chosen as a variable, the determinant of the matrix will *not* be unity but rather A_2/A_1, where A_1 and A_2 are the input and exit areas of the acoustical circuit. Despite this lack of elegance, we shall use p and u (rather than q) as the acoustical variables. In in cases where there are no changes in cross-sectional area, the determinant will be unity even with this choice.

Actually, it is convenient to use the velocity variable $\rho c u$, as we have done, where ρc is the wave impedance at of the fluid involved under normal conditions

$$p_1 = T_{11}p_2 + T_{12}\rho c u_2$$
$$\rho c u_1 = T_{21}p_2 + T_{22}\rho c u_2. \tag{4.120}$$

With this choice, the matrix elements T_{ij} become dimensionless and we shall use this choice unless stated otherwise.

4.4.7 Problems

1. **Matrices at oblique angle of incidence**

The transmission matrices given in Section 4.4.5 all referred to normal incidence. Wherever applicable, generalize these matrices when the angle of incidence is ϕ.

2. **Transmission loss of double wall**

A wall consists of two 1/8 inch glass plates separated by 4 inches. Treat the plates as limp.

(a) What is the total normal incidence transmission matrix of this double wall?

(b) Calculate the transmission loss and sketch the frequency dependence. For the mass density of glass, use $\rho_p = 2.5$ g/cm^3.

3. **Absorption within a barrier**

Derive an expression (in terms of the transmission matrix elements) for the absorption coefficient of a barrier, accounting only for the absorption within the barrier, i.e., not accounting for the power transmitted through the barrier.

4. **Absorption coefficient of a double sheet absorber**

An absorber consists of two rigid resistive sheets separated by 2 inches and backed by a 4 inch air layer in front of a rigid wall. The normalized flow resistances of the sheets are $\theta_1 = 1$ and $\theta_2 = 2$, the latter being closest to the wall (i.e., 4 inches from the wall).

(a) What is the combined transmission matrix of the two sheets and the air layer?

(b) Calculate the normal incidence absorption coefficient.

(c) If the placement of the two sheets are interchanged, will that influence the absorption coefficient?

5. **Transmission matrix for an air column**

Following the outline in the text, prove Eq. 4.117.

Chapter 5

The Wave Equation

5.1 Fluid Equations

In the introductory discussion in Chapter 3 we simply used the impulse-momentum relation to illustrate the basic idea involved in the dynamics of wave motion. To go further, the differential equations of fluid motion are more appropriate and we proceed accordingly.

The thermodynamic state of a fluid is described by three variables, such as pressure, density, and temperature and the motion by the three components of velocity. Thus, there is a total of six variables which have to be determined as functions of space and time to solve a problem of fluid motion. Therefore, six equations are needed. They are conservation of mass (one equation), conservation of momentum (three equations, one for each component), conservation of energy (one equation), and one equation of state for the fluid.

In describing the motion of a fluid, we shall use what is known as the *Eulerian description*. The velocity and the thermodynamic state (such as pressure) at a fixed position of observation are then recorded as functions of time. Different fluid particles pass the observer as time goes on. (In the Lagrangian description, the time dependence is expressed in a coordinate frame that moves and stays with the fluid element under consideration.)

5.1.1 Conservation Laws

The conservation of mass in the Eulerian description simply states that the net mass influx into a control volume, fixed with respect to the laboratory coordinate frame, must be balanced by the time rate of change of the mass within the volume. We consider first one-dimensional motion in the x-direction and let the velocity and density at x and time t be $u(x, t)$ and $\rho(x, t)$.

The *mass flux* $j(x, t) = \rho(x, t)u(x, t)$ is the mass passing through unit area per unit time at x. Similarly, the efflux at $x + \Delta x$ is obtained by replacing x by $x + \Delta x$ in j. Thus, the net mass influx to the control volume is $j(x) - j(x + \Delta x) = -9\partial j/\partial x)\,\Delta x$ in the limit as $\Delta x \to 0$.

Conservation of mass requires that the influx of mass must be balanced by the time rate of change of the mass $\rho \Delta x$ in the control volume. and we obtain

$$\partial \rho / \partial t = -\partial j / \partial x \quad \text{or} \quad \partial \rho / \partial t + \rho \partial u / \partial x = 0. \tag{5.1}$$

In the last step, the term $u \partial \rho / \partial x$ has been neglected in $\partial j / \partial x$ since it is much smaller (by a factor u/c) than $\rho \partial u / \partial x$. This follows, for example, if we express $\partial \rho / \partial x$ as $\rho \kappa \partial p / \partial x$, where $\kappa = 1/\rho c^2$ is the compressibility, discussed in Chapter 3. In a plane wave, $u = p/\rho c$ so that $\partial u / \partial x = (1/\rho c) \partial p / \partial x$. The ratio of the neglected term $u \partial \rho / \partial x$ and $\partial \rho / \partial x$ is then seen to be of the order of u/c. The neglected term is of *second order* in the field variables (product of two first order perturbations) and the omission results in the *linearized* version of the equation.

To obtain the corresponding equation for *conservation of momentum*, we proceed in an analogous manner, replacing the mass flux by the momentum flux. The momentum density in the fluid is ρu and the influx into the control volume at x is $G = (\rho u)u$. The corresponding efflux at the other side of the box is $G(x + \Delta x) = G(x) + (\partial G / \partial x) \Delta x$, making the net influx equal to $-(\partial G / \partial x) \Delta x$.

There is a contribution also from the thermal motion which is expressed by the pressure in the fluid (recall that the pressure is of the order of ρc^2 which should be compared with the convective momentum flux ρu^2). This results in a rate of momentum influx $p(x)$ at x and an efflux $p(x + \Delta x) = p(x) + (\partial p / \partial x) \Delta x$ at $x + \Delta x$, making the net influx contribution from pressure $(-\partial p / \partial x) \Delta x$.

The total influx is now $(-\partial p / \partial x - \partial G / \partial x) \Delta x$ and this must equal the time rate of change of the momentum $(\partial \rho u / \partial t) \Delta x$ contained in the box, i.e.,

$$\rho \partial u / \partial t = -\partial \rho u^2 / \partial x - \partial p / \partial x. \tag{5.2}$$

From an argument analogous to that used in the linearization of the conservation of mass equation, we find that the term $\partial \rho u^2 / \partial x \equiv \partial G / \partial x$ can be neglected in comparison with $\partial p / \partial x$ (in Chapter 10 on nonlinear aspects of acoustics it is retained) and that $\partial \rho u / \partial t$ can be replaced by $\rho \partial u / \partial t$.

Thus, the linearized form of the momentum equation is

$$\rho \partial u / \partial t + \partial p / \partial x = 0. \tag{5.3}$$

It is left as a problem to show that the omitted (nonlinear) terms are smaller than the linear by a factor of the order of u/c which normally is much less than one.

The momentum equation (5.3) contains the variables u and p and the mass equation (5.1) the variables ρ and u. The latter can also be expressed in terms of p and u since $\partial \rho / \partial t = (1/c^2) \partial p / \partial t$ (recall that $c^2 = dP/d\rho = \gamma P/\rho$).

Then, in terms of the compressibility $\kappa = 1/\rho c^2$, the linearized form of Eq. 5.1 can be written

$$\kappa \partial p / \partial t + \partial u / \partial x = 0. \tag{5.4}$$

The fluid equations 5.1 and 5.3 can readily be generalized to three dimensions. In the mass equation, the term $\partial u / \partial x$ has to be replaced by $\partial u_x / \partial x + \partial u_y / \partial y + \partial u_z / \partial z$ which can also be expressed as div $\boldsymbol{u}$.

The momentum equations for the three components of the velocity can be condensed into a vector equation in which grad $p = \hat{x}\partial p/\partial x + \hat{y}\partial p/\partial y + \hat{z}\partial p/\partial z$, where $\hat{x}, \hat{y}, \hat{z}$ are the unit vectors in the x, y, z directions. Thus, the linearized fluid equations take the form

$$\boxed{\begin{array}{l} \textit{Acoustic equations} \\ \kappa\,\partial p/\partial t = -\text{div } \boldsymbol{u} \\ \rho\,\partial \boldsymbol{u}/\partial t = -\text{grad } p \end{array}} \tag{5.5}$$

where we have introduced the compressibility $\kappa = 1/\rho c^2$.

For harmonic time dependence, the corresponding equations for the complex amplitudes $p(\omega)$ and $u(\omega)$ are, with $\partial/\partial t \rightarrow -i\omega$,

$$i\omega\kappa\, p = \text{div } \boldsymbol{u} \tag{5.6}$$

$$i\omega\rho\, \boldsymbol{u} = \text{grad } p. \tag{5.7}$$

5.1.2 The Wave Equation

From Eqs. 5.1 and 5.3, we can eliminate u by differentiating the first with respect to t and the second with respect to x to obtain a single equation for p,

$$\partial^2 p/\partial x^2 - (1/c^2)\partial^2 p/\partial t^2 = 0. \tag{5.8}$$

In three dimensions, we differentiate the mass equation (5.5) with respect to t and take the divergence of the momentum equation (5.5). Then, with div grad $p = \nabla^2 p$, it follows that

$$\boxed{\begin{array}{l} \textit{Acoustic wave equation} \\ \nabla^2 p - (1/c^2)\partial^2 p/\partial t^2 = 0 \end{array}} \tag{5.9}$$

which replaces Eq. 5.8. For harmonic time dependence this equation reduces to

$$\nabla^2 p(t) + (\omega/c)^2 p(t) = 0. \tag{5.10}$$

Plane Waves

The general solution to the one-dimensional wave equation is a linear combination of waves traveling in the positive and negative x-direction, respectively, and can be expressed as

$$p(x, t) = p_+(t - x/c) + p_-(t + x/c), \tag{5.11}$$

where p_+ and p_- are two independent functions. The validity of the solution is checked by direct insertion of this expression into Eq. 5.8.

For harmonic time dependence,

$$p(x, t) = A\cos(\omega t - kx - \phi_1) + B\cos(\omega t + kx - \phi_2), \tag{5.12}$$

where A, B, ϕ_1, and ϕ_2 are constants.

It follows from Eqs. 5.6 and 5.7 by eliminating $\boldsymbol{u}$ that the wave equation is valid also for the complex amplitude $p(\omega)$,

$$\nabla^2 p(\omega) + (\omega/c)^2 p(\omega) = 0. \tag{5.13}$$

The general one-dimensional solution is then

> Complex pressure amplitude. Plane wave solution
> $$p(x, \omega) = A\, e^{ikx} + B\, e^{-ikx}$$
                                                                    (5.14)

representing the sum of waves traveling in the positive and negative x-directions. The constants A and B are now complex, defining the magnitude and phase angles of the two waves. They are determined by the known complex amplitudes at two positions (boundary conditions).

The corresponding velocity field follows from the momentum equation (5.7) and is given by

> Complex velocity amplitude. Plane wave solution
> $$\rho c\, u(x, \omega) = A\, e^{ikx} - B\, e^{-ikx}$$
                                                                    (5.15)

[ρc: Wave impedance. A, B: Complex constants. ω: Angular frequency. $k = \omega/c$.]

The sum of several traveling waves of the same frequency, direction, and wave speed can always be represented as a single traveling wave. The complex amplitude of this wave is then the sum of the complex amplitudes of the individual waves.

Spherical Waves

So far we have been dealing with plane waves, such as encountered in a duct with rigid walls with a plane piston as a sound source. The next simplest wave is the spherically symmetrical. Such a wave, like the plane wave, depends only on one spatial coordinate, in this case the radius.

The prototype source of the spherically symmetrical wave is a pulsating sphere which takes the place of the plane piston for plane waves. At large distances from the source, a spherical wave front can be approximated locally as plane, and the relation between pressure and velocity is then expected to be the same as for the plane wave, i.e., $p = \rho cu$, so that the intensity becomes $I = (1/2)p^2/\rho c$. The total emitted power from the source is then $\Pi = 4\pi r^2 I$, and it follows that the intensity decreases with distance as $1/r^2$ and the pressure as $1/r$. As will be shown below, this r-dependence of the pressure turns out to be valid for *all* values of r.

To proceed, we need to express the wave equation in terms of the radial coordinate r. To do that, we recall that if A is a vector, the physical meaning of div A is the 'yield' of A per unit volume, where the yield is the integral of the (outward) normal component of A over the surface surrounding the volume. Our volume element in this case is $S(r)\, dr$ where $S(r)$ is the spherical surface $S(r) = 4\pi r^2$. The outflow of A from this volume element is $(SA_r)_{r+dr} - (SA_r)_r = \partial(SA_r)/\partial r\, dr$ and dividing it by the volume $S\, dr$ we get the divergence, div $A = (1/S)(\partial S/\partial A_r) = (1/r^2)\, \partial(r^2 A_r)/\partial r$.

In this case, with $A_r = \partial p/\partial r$ and $\nabla^2 p = $ div (grad p), we obtain the wave equation

> Wave equation; spherically symmetric pressure field
> $$\frac{1}{r^2}\frac{\partial}{\partial r}\left[r^2\frac{\partial p}{\partial r}\right] - \frac{1}{c^2}\frac{\partial^2 p}{\partial t^2} = 0$$
                                                                    (5.16)

By direct insertion into the equation, we find that the solution for the pressure in an outgoing wave can be written in the form

$$p(r, t) = \frac{a}{r} p(a, t - t'), \tag{5.17}$$

where $p(a, t)$ is the pressure at $r = a$ and $t' = (r - a)/c$ is the delay time of wave travel from a to r.

For harmonic time dependence, with $p(a, t) = |p(a)| \cos(\omega t)$, we get

$$p(r, t) = |p(a)|\frac{a}{r} \cos[\omega t - k(r - a)]. \tag{5.18}$$

The velocity field follows from $\rho \partial u_r / \partial t = -\partial p/\partial r$. In the case of harmonic time dependence such that $p(a, t) = |p(a)| \cos(\omega t)$ the velocity field becomes

$$u_r(r, t) = \frac{a}{r} \frac{|p(a)|}{\rho c} \left[\cos[\omega t - k(r - a)] + \frac{1}{kr} \sin[(\omega t - k(r - a)] \right]. \tag{5.19}$$

The first term represents the *far field* and dominates at distances many wavelengths from the source, i.e., $kr \gg 1$, where $k = 2\pi/\lambda$. It is in phase with the pressure field and is simply $p(r, t)/\rho c$. The second is the *near field* which dominates for $kr \ll 1$. With $\sin[\omega t - k(r - a)]$ written as $\cos[\omega t - k(r - a) - \pi/2]$, we see that this velocity lags behind the pressure by the phase angle $\pi/2$.

The complex amplitudes of pressure and velocity that correspond to Eqs. 5.18 and 5.19 are

$$\boxed{\begin{array}{l} \textit{Complex pressure and velocity amplitudes; spherical wave} \\ p(r, \omega) = (A/r)\, e^{ikr} \\ \rho c\, u_r(r, \omega) = (A/r)\, e^{ikr} (1 + i/kr) \end{array}} \tag{5.20}$$

$[k = \omega/c. \ A = (p(a, \omega)a \exp(-ika)]$.

The complex velocity amplitude in this case is obtained from momentum equation $-i\omega\rho\, u_r(r, \omega) = -\partial p/\partial r$.

The constant A is now complex and incorporates a phase factor $\exp(-ika)$. In Section 5.1.2 we return to this problem in an analysis of the sound generated by a pulsating sphere in which case the velocity at the surface of the sphere rather than the pressure is given.

To obtain the complex amplitudes for an incoming rather than outgoing wave, we merely replace ikr by $-ikr$ in Eq. 5.20.

5.1.3 Problems

1. **Linearization**

   Show that the omitted terms in the linearization of the momentum equation (5.3) are of the order of u/c, where u is the particle velocity.

2. **Sound radiation; pulsating sphere**

   The surface of a sphere of mean radius $a = 5$ cm oscillates in radial harmonic motion with the frequency 1000 Hz and with a uniform velocity amplitude 0.1 cm. Neglecting sound absorption in the air, determine the distance at which the sound pressure amplitude will be equal to the threshold of human hearing (0.0002 dyne/cm^2, rms).

3. **Sound field in a spherical enclosure**

A pulsating sphere of mean radius a is surrounded by a concentric spherical enclosure of radius b and with totally reflecting walls. The radial velocity of the pulsating sphere is $u_r = |u|\cos(\omega t)$.

(a) Determine the pressure and velocity fields in the enclosure.

(b) What is the impedance at the source?

5.2   Pulsating Sphere as a Sound Source

The spherically symmetrical wave, like the plane wave, is one-dimensional in the sense that it depends only on one space coordinate, the radius r. Next to the plane wave, it is the simplest wave form and next to the plane piston in a tube, the pulsating sphere in free field is the simplest source, being the prototype generator of a spherical wave.

There is a significant difference between the plane piston and the pulsating sphere as sound sources, however. For the plane piston in a tube, compressibility of the fluid was a necessary requirement for motion of the piston.

For a spherical source, this is not necessary, as will be shown shortly. Even if the surrounding fluid is incompressible, the reaction force on the sphere from the fluid will not be infinite, as was the case for the piston, and a finite driving force per unit area of the sphere can indeed produce an oscillatory velocity field in the fluid.

This can be shown as follows. Let the radius of the sphere be a, the radial velocity of the surface of the sphere, $u(a, t)$, and the density of the surrounding fluid, ρ. With ρ being constant because of the incompressibility, conservation of mass, $\rho u(r, t)(4\pi r^2) = \rho u(a, t)(4\pi a^2)$, requires the velocity at radius r to be inversely proportional to r^2,

$$u(r, t) = (a/r)^2 \, u(a, t). \tag{5.21}$$

The total kinetic energy of the fluid is then

$$KE = (\rho/2) \int_a^\infty u^2(r, t)(4\pi r^2)dr = (\rho a)(4\pi a^2) \, u^2(a, t)/2. \tag{5.22}$$

Thus, the kinetic energy per unit area of the sphere is $\rho a \, u^2/2$ corresponding to a mass load of ρa per unit area of the sphere.

With such a mass load on the sphere and a radial velocity of the surface being $u(a, t) = |u|\cos(\omega t)$, the pressure at the surface will be

$$p(a, t) = \rho a \frac{du(a, t)}{dt} = -\rho a \omega \, |u|\sin(\omega t) \quad \text{(incompressible flow)}. \tag{5.23}$$

If the fluid is compressible, the motion generated by the pulsating sphere no longer is limited to the velocity which varies as $1/r^2$ (near field), but it is expected to contain a spherical sound wave in which the velocity varies as $1/r$ (far field). The reason is that sufficiently far from the sphere at a radial position $r \gg a$, the wave front can be approximated locally as a plane wave. The acoustic intensity in the wave is then $I(r, t) = p^2(r, t)/\rho c$ and the power $(4\pi r^2)I(r, t)$. Conservation of acoustic energy

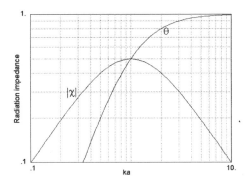

Figure 5.1: The radiation resistance θ and the magnitude of the mass reactance $|\chi$ of the normalized radiation impedance of a pulsating sphere of radius a versus $ka = 2\pi a/\lambda$.

then requires the intensity to decrease as $1/r^2$ and, consequently, the sound pressure as $1/r$.

To calculate the total sound field for all values of r, we shall use the complex amplitude approach. The analysis without complex variables is discussed in Chapter 11.

The complex pressure amplitude is expressed as

$$p(r, \omega) = \frac{A}{r} e^{ikr}, \tag{5.24}$$

representing an outgoing spherical wave, where $k = \omega/c$. The (complex) amplitude $A(\omega)$ is to be determined from the boundary condition on velocity at the surface of the sphere.

The corresponding velocity field is obtained from $-i\rho\omega u_r = -\partial p/\partial r$,

$$u_r = \frac{A}{\rho c} \frac{e^{ikr}}{r}(1 + i/kr). \tag{5.25}$$

The *radiation impedance* of the sphere is then $z = p(a)/u_r(a)$ or

$$\boxed{\begin{array}{c} \textit{Radiation impedance. Pulsating sphere} \\ \zeta \equiv \theta + i\chi = z/\rho c = [1 + i/(ka)]^{-1} = \frac{(ka)^2}{1+(ka)^2} - i\,\frac{ka}{1+(ka)^2} \end{array}} \tag{5.26}$$

[a: Sphere radius. $k = \omega/c$].

The frequency dependence of the real and imaginary parts θ and χ are shown in Fig. 5.1 versus ka.

The negative value of the reactance (signified by the factor $-i$) shows that the reactance is mass like and for small values of ka (acoustically compact sphere), the corresponding mass load, the *induced mass* per unit area is $\rho cka = \omega\rho a$, i.e., the mass reactance of an air layer of thickness a, as already noticed in connection with Eq. 5.23.

As ka increases, the radiation resistance approaches 1, and the reactance goes to 0, and the radiation impedance is the same as for a plane plane piston generating a plane wave.

At low frequencies, $ka \ll 1$, the radiation resistance is small, $\approx (ka)^2$, which means that the pulsating sphere is an inefficient radiator at low frequencies. The mass reactance has a maximum magnitude, 0.5, for $ka = 1$, the wavelength being π times the diameter of the sphere. The normalized radiation resistance is then also 0.5.

Finally, we use Eq. 5.25 to express A in terms of the radial velocity U of the sphere. Thus, with $r = a$, we get $A = \rho c U \exp(-ika)a/[1 + i/(ka)]$ and if this is used in Eq. 5.24, we find, using $p(a) = \rho c \zeta_r U$,

$$p(r, \omega) = \rho c U \frac{a}{r} e^{ik(r-a)} \frac{ka}{ka + i}. \tag{5.27}$$

The time dependence of the pressure is then obtained as $\Re\{p(r/\omega) \exp(-i\omega t)\}$. The radiated time average power in the case of harmonic motion can be expressed as

$$\Pi = (4\pi a^2)(1/2)\Re\{p(a)u(a)^*\} = (1/2)(4\pi a^2)|U|^2\theta\rho c). \tag{5.28}$$

The factor 1/2 should be removed if $|U|$ is the rms value.

5.2.1 The Point Source. Monopole

The radius of the pulsating sphere can be chosen as small as we wish but if the velocity amplitude is kept constant as the radius of the sphere goes to zero the radiated sound pressure will also go to zero. According to Eq. 5.27, the sound pressure amplitude as $ka \to 0$ will be

$$p(r, \omega) \approx -i(\rho c\, U\, ka^2)/r\, e^{ikr}) = \frac{-i\omega\rho U\, 4\pi a^2}{4\pi} e^{-ikr)}. \tag{5.29}$$

We now let the $U \to \infty$ as $a \to 0$ in such a way that the total mass flow amplitude $\rho U 4\pi a^2$ of the sphere remains constant and we refer to such a source as a *point source* or an acoustic *monopole*. Because of the factor $-i\omega$, the sound pressure is proportional to the acceleration of the total *mass flow*

$$q = (-i\omega)(\rho U\, 4\pi a^2). \tag{5.30}$$

It is the *acoustic source strength* of the point source. The mass flow rate itself will be called the *flow strength* of a source

$$q_f = \rho U\, 4\pi a^2. \tag{5.31}$$

In terms of the acoustic source strength, the sound pressure field is simply

$$p(r, \omega) = \frac{q}{4\pi r} e^{ikr}, \tag{5.32}$$

where $q(\omega) = \rho(-i\omega U)4\pi a^2$ and $(-i\omega U)$ is the radial acceleration of the surface of the sphere.

For an arbitrary time dependence of acoustic point source, the sound pressure becomes

$$p(r, t) = \frac{q(t - r/c)}{4\pi r}, \tag{5.33}$$

where $q = \partial q_f/\partial t \equiv \dot{q}_f$.

Acoustically Compact Source

The field from an arbitrarily shaped sound source can be expressed as the superposition of point sources. If the size of the source is much smaller than the wavelength, it is called an *acoustically compact source*. In the *far field*, it is a good approximation to treat the source as a point source with a source strength equal to the time derivative of the total flow strength of the source. The surface velocity of the source need not be uniform and the flow strength is obtained as the integral of $\rho u(\omega)$ over the surface of the source. The sound in the near field, close to the source, may be quite complicated and different from that of the point source but the spherically symmetrical component and the far field is determined solely by the source strength.

As an example, consider a loudspeaker mounted in one wall of a closed cabinet (box). Only the loudspeaker contributes to the flow strength since the velocity of the cabinet walls can be assumed to be zero. At wavelengths large compared to the cabinet dimensions, the source can be regarded as acoustically compact.

If the cabinet is removed and the loudspeaker is suspended in free field, the flow strength will be zero since it is positive on one side of the loudspeaker and negative on the other. Thus, with the source strength being zero, there will be no spherically symmetrical component of the sound field. As a result, the radiation efficiency at low frequencies will be reduced. An important function of a loudspeaker cabinet is to improve the low frequency radiation efficiency.

Like the Fourier decomposition of a periodic signal in Chapter 2, the *angular* dependence of the sound field of any harmonic source can be decomposed into a series of *spherical harmonics* of which the leading term is the spherically symmetrical field (no angular dependence). It can be shown that this field (if it is present) dominates for large values of r and at low frequencies.

For a broad band noise source, it should be borne in mind that in the high frequency end of the spectrum, the wavelength might not be large compared to the source dimensions so that in this regime, the source no longer can be considered acoustically compact. In that case, a more detailed analysis has to be carried out as illustrated by the line source example presented later in this section.

5.2.2 Problems

1. **Radiated power, once again**

   (a) Show that the radiated power from a pulsating sphere in Eq. 5.28 can be obtained also by integrating the intensity in the far field.

   (b) Determine the acoustic power radiated from a point source in terms of its acoustic source strength q.

2. **Threshold values of intensity and power**

   As already indicated, the rms value of the sound pressure at the hearing threshold is $p_r = 2 \times 10^{-5}$ N/m^2.

   (a) What is the corresponding value of the reference intensity I_r in a plane wave and the reference power W_r power in watts transmitted through an area of 1 m^2 at a pressure of 1 atm and a temperature of 70°F?

   (b) How does the intensity vary with static pressure and temperature?

(c) Qualitatively, how does the sound pressure vary with altitude in the atmosphere in a spherical sound wave generated at the ground levele Use conservation of acoustic power.

(d) What is the acoustic power level $PWL = 10 \log(W/W_r)$ of a source emitting 1 watt of acoustic power?

3. **Spherical fields. Power level**

Typical values for the total acoustic power emitted by a jet engine, a pneumatic hammer, and an average speaker are 10 kilowatts, 1 watt, and 20 microwatts, respectively.

(a) What are the corresponding power levels?

(b) What are the sound pressure levels at 1 m from the sources, regarded as point sources?

(c) With the opening of the mouth having an area of 3 cm^2, estimate the sound pressure level at the surface of a sphere with the same surface area.

5.3 Source and Force Distributions

The pulsating sphere and the corresponding point source can be regarded as a source of mass flow q_f and the acoustic source strength, as we have defined it in the previous section, is $q(t) = \dot{q}_f(t)$, and the corresponding complex amplitude is $q(\omega) = -i\omega q_f(\omega)$. This source does not transfer any momentum to the surrounding fluid (i.e., no net force is produced).

We consider now an acoustic source distribution $Q = \dot{Q}_f$ and force distribution F_x, both per unit volume. The conservation laws of mass and momentum, Eqs. 5.5 and 5.5, will be modified to

$$(1/c^2)(\partial p/\partial t) = -\mathrm{div}\, \boldsymbol{u} + Q_f \qquad (5.34)$$

$$\rho(\partial \boldsymbol{u}/\partial t) = -\mathrm{grad}\, p + \boldsymbol{F}. \qquad (5.35)$$

The corresponding wave equation for p, obtained by eliminating the velocity (by differentiating the mass equation with respect to t and taking the divergence of the second equation) is

$$\nabla^2 p - (1/c^2)(\partial^2 p/\partial t^2) = -Q + \mathrm{div}\, \boldsymbol{F}. \qquad (5.36)$$

Here $Q = \partial Q_f/\partial t$ is the acoustic source strength per unit volume. The corresponding equations for the complex amplitude are obtained by using $\partial/\partial t \to -i\omega$.

Let us consider first the field produced by Q. The pressure field from a single point source has already been obtained (see Eq. 5.33).

Then, by considering the field contribution from a volume element dV' as being that of a point source of strength QdV', it follows that the total field is

$$p(\boldsymbol{r}, t) = \int \frac{Q\boldsymbol{r}, \boldsymbol{r}', t')}{4\pi |\boldsymbol{r} - \boldsymbol{r}'|}\, dV'. \qquad (5.37)$$

Quantities $\boldsymbol{r}$ and $\boldsymbol{r}'$ stand for the coordinates of the field and source locations, respectively, and $|\boldsymbol{r} - \boldsymbol{r}'|$ is the distance between these locations. The 'retarded' time is $t' = t - |\boldsymbol{r} - \boldsymbol{r}'|/c$ which indicates that the sound pressure arriving at the point of observation r at time t was emitted at t' from the source point, r', $|r - r'|/c$ being the travel time.

To obtain the corresponding expression for the field from a force distribution F per unit volume, we proceed as follows. Introduce the vector A such that $p = \text{div } A$. Using this in the wave equation (5.36), we obtain the equation[1]

$$\nabla^2 A - (1/c^2)(\partial^2 A/\partial t^2) = F. \tag{5.38}$$

For each component of A, the equation has the same form as for the mass distribution in Eq. 5.36, except for a difference in signs of Q and div F, and the solution is analogous to Eq. 5.37. For the total vector A, the solution can be written (accounting for the sign difference just mentioned)

$$A(r, t) = -\int \frac{F(r, r', t')}{4\pi |r - r'|} \, dV'. \tag{5.39}$$

After having calculated A, the corresponding pressure field is obtained from $p(r, t) = \text{div } A$.

5.3.1 Point Force (Dipole)

For a point force f_x at the origin, we get

$$A_x = -\frac{f_x(t - r/c)}{4\pi r} \tag{5.40}$$

and

$$\boxed{\begin{array}{c} \textit{Point force (dipole) field} \\ p(r, t) = \frac{\partial A_x}{\partial x} = (1/4\pi r)[\dot{f}_x(t')/c + f_x(t')/r] \cos\phi \end{array}} \tag{5.41}$$

$[t' = r - r/c.$ $\dot{f}_x \equiv \partial f_x/\partial t.$ $\partial f_x/\partial x = (-1/c)(\partial f_x/\partial t)\cos\phi.$ $\partial r/\partial x = x/\sqrt{x^2 + y^2 + z^2} = \cos\phi].$

The first term is the far-field sound pressure which is proportional to $\dot{f}_x$, and the second term is the near field pressure, which is proportional to f_x. We shall return to this equation in the discussion of the sound field from an oscillating sphere. The point source of sound is often referred to as a *monopole* and the point force as a *dipole* source of sound. The monopole field has no angular dependence but the dipole field is proportional to $\cos\phi$ where ϕ is the angle between the direction of the force and the direction to the field point. The sound pressure has a maxima along the direction of the force, $\phi = 0, \pi$, and zero at the right angle thereto, i.e., at $\theta = \pm\pi/2$. (Compare the electric field distribution about an oscillating electric charge in which this field distribution is reversed.)

For harmonic time dependence, the complex pressure amplitude can be written

$$\boxed{\begin{array}{c} \textit{Point force. Complex pressure amplitude} \\ p(r, \omega) = [\exp(ikr)/4\pi r][-ik\, f(\omega) + f(\omega)/r]\cos\phi \end{array}}, \tag{5.42}$$

---

[1] If the divergences of two vectors are the same, the vectors are the same except for a possible curl of a vector. In this case this difference is not relevant.

where the factor first term is the far field pressure, proportional to the time rate of change of the force, and the second, the near field.

The field produced by two monopole sources of equal strength but with opposite signs clear has no net acoustic source strength and the resulting field has no monopole contribution. If the separation of the sources is much smaller than a wavelength, the dominant field will be equivalent to that of a point force. To prove this we put one monopole at $x' = -a_x/2$ with the complex amplitude of the acoustic source strength $q = -|q|$ and the other at $x' = a_x/2$ with the source strength $|q|$. The field point is at a distance r from the origin and an angle ϕ with respect to the x-axis. If $r >> a_x$, the distance r_1 and r_2 from the two sources to the field point are $r_1 \approx r + (a_x/2)\cos\phi$ and $r_2 = r - (a_x/2)\cos\phi$. Thus, the complex amplitude of the sum of the contributions from the two sources are

$$p_1 \approx \frac{|q|\exp(ikr)}{4\pi r}[-e^{i\,(ka_x/2)\cos\phi} + e^{-i\,(ka_x/2)\cos\phi}] \approx (-i\omega|q|a_x)\frac{e^{ikr}}{4\pi rc}\cos\phi. \tag{5.43}$$

This should be compared with the field in Eq. 5.42 from a point force f_x in the x-direction

$$p_x = (-i\omega|f_x|)\frac{e^{ikr}}{4\pi rc}\cos\phi. \tag{5.44}$$

It follows that the two monopoles are equivalent to a point force with an amplitude $|f_x| = |q|a_x$, often called the *dipole moment*.

If the complex pressure amplitudes of the monopole and dipole fields are denoted p_0 and p_1, it follows from Eq. 5.43 that

$$p_1 = (-ika_x\cos\phi)p_0. \tag{5.45}$$

Thus, if a monopole produces a sound pressure amplitude p_0 at a distance r from the source, the combination of it with an equal but opposite monopole at a separation a_x yields a maximum amplitude $|p_x| \approx p_0 ka_x\cos\phi$, where $k = 2\pi/\lambda$. With $a_x << \lambda$ this means a substantial decrease in sound pressure.

This simple example can be used to illustrate the advantage of providing a loudspeaker with a cabinet. With the cabinet, only one side of the loudspeaker radiates and it is equivalent to a monopole at low frequencies (long wavelengths). If the cabinet is removed both sides of the speaker radiate but one pushes as the other pulls. The source is then equivalent to a dipole and a reduction in the radiated sound pressure at low frequencies results. Thus, the cabinet can be thought of as a dipole-to-monopole converter at low frequencies.

5.3.2 The Oscillating Compact Sphere

An oscillating sphere does not transfer any net mass to the surrounding fluid and consequently has no flow strength, no monopole strength. It does transfer momentum, however. If the sphere is acoustically compact, the sound field produced therefore should be of the same as by the point force in Eq. 5.41. The near field in this equation is given by the second term. The corresponding velocity field is obtained from the momentum equation $\rho \dot{u}_r = -\partial p / \partial r$. The radial component of the velocity in the near field is then given by

$$\rho \frac{\partial u_r}{\partial t} = \frac{2}{4\pi r^3} f \, \cos \phi. \qquad (5.46)$$

For a solid impervious compact sphere sphere oscillating in the x-direction, the radial component of the velocity at the surface of the sphere is $u_r = u_x \cos \phi$ and if this velocity is used in Eq. 5.46 it follows that the equivalent force on the surrounding fluid produced by the sphere is

$$f = \frac{3}{2} \frac{4\pi a^3}{3} \rho \frac{\partial u_x}{\partial t}. \qquad (5.47)$$

The physical meaning of this relation is as follows. The sound pressure field reacts back on the sphere with a force equal to that required to accelerate an air mass which is 3/2 times the mass $m = (4\pi a^3 / 3)\rho$ of the air displaced by the sphere. The 'buoyancy' force caused by the air in this acceleration accounts for the force contribution $m(\partial u_x / \partial t)$; the remaining contribution corresponds to the 'induced mass' $m/2$ due to the flow outside the sphere forced to oscillate back and forth from the front to the back of the sphere.[2]

We have assumed that the sphere is small enough to allow us to use the near field in Eq. 5.47 which means that $f/r >> \dot{f}/c$ or $\dot{f}/f << c/a$. For harmonic time dependence this means that $a << \lambda$. In that case the expression for f in Eq. 5.47 can be used in Eq. 5.41 for the calculation of the complete sound pressure field from an oscillating sphere for all values of r.

5.3.3 Realization of Source and Force Distributions

In the discussion so far in the section, we have introduced an acoustic source distribution $Q = \dot{Q}_f$ and a force distribution F (with a corresponding point source and point force) without paying any attention to how such distributions can be realized in practice.

In the conservation of mass equation, the quantity Q_f is entered as a source of mass per unit volume and has to be interpreted, strictly speaking, as a mass creation. In a one-component fluid, such as air, there is no such creation. Only in a multi-component fluid is such an interpretation possible. For example, a weakly ionized gas consists of three components, the neutrals, the electrons, and the ions. Through collisions there can be a recombination of electrons and ions to form neutrals. Then, in the acoustic equations for the neutral component alone, there will indeed by a term in the mass conservation equation that accounts for this 'creation.'

As shown in Chapter 7, a fluctuating heat source in a gas, such as combustion, is acoustically equivalent to a flow strength per unit volume $Q_f = [(\gamma - 1)/c^2]H$, where γ is the specific heat ratio, c the sound speed, and H the rate of heat generated

---

[2]In the case of a pulsating sphere, we found that the induced mass was ρa per unit area of the sphere.

per unit volume of the gas. This is a common example in which a monopole distribution can be realized.

The obvious volume force in a fluid is the force of gravity ρg. In order for it to be time dependent and generate sound, the time dependence has to come from ρ. If electric charges are involved, as in the ionized gases (plasmas), there are electromagnetic volume force distributions by far more significant than gravitational.

Actually, a strong nonuniform electric field can produce a force and generate sound even in a neutral gas. More common, however, is sound generation by force distributions resulting from the interaction of fluid flow with solid objects, as discussed in Chapter 7. A typical example is the Aeolian tone from a cylinder.

5.3.4 Quadrupole and Higher Multipoles

The dipole source, consisting of two closely spaced monopole sources of opposite signs, was shown in the previous section to be acoustically equivalent to a point force. Whereas the monopole field yields the same intensity in all directions, omnidirectional, the field from the dipole was found to have two radiation lobes with the intensity having the maxima in the ±directions of the force and minima (zero) normal thereto.

Similarly, a source consisting of two closely spaced point forces of opposite sign has no dipole strength. Nevertheless, sound will be produced but it will have neither a monopole nor a dipole contribution to the field. Such a source is called a *quadrupole*.

If the dipoles, assumed to be aligned along the x-axis, are displaced with respect to each other in the x-direction by b_x, the quadrupole thus obtained is called a *longitudinal* quadrupole and if the displacement is in the y-direction, it is a *lateral quadrupole*. By analogy with the derivation of Eq. 5.43 and the corresponding relation (5.45), the complex amplitude for the longitudinal quadrupole can be expressed in terms of the dipole field as follows,

$$p_{xx} = (-ik_xb_x)p_x = (-k^2a_xb_x)p_0\cos^2\phi_x, \qquad (5.48)$$

where, as before, $k = \omega/c$.

With the dipoles displaced with respect to each other in the y-direction a distance b_y, the resulting complex pressure amplitude becomes

$$p_{xy} = (-ik_xb_y)p_x = (-k^2a_xb_xb_y)p_0\cos\phi_x\cos\phi_y, \qquad (5.49)$$

where ϕ_x is the angular coordinate of the field point with respect to the x-axis and $\phi_y = \pi/2 - \phi_x$ is the angle with respect to the y-axis. The quantities qa_xa_x and qa_xa_y are called *quadrupole moments*.

The radiation pattern for the longitudinal quadrupole has the same general form as that of the dipole, although the beams are narrower. The lateral quadrupole has a cloverleaf pattern with zeroes in the x-and y-directions.

The fields from higher order multipoles are constructed in a similar manner, and it follows that the far field amplitude of the mth order multipole will contain an amplitude factor $(kd)^m p_0$, where we have used a characteristic length to signify the relative displacements of the multipoles. At low frequencies, $kd \ll 1$, the far field is dominated by the multipole of lowest order.

As an illustration, we comment on the performance of a loudspeaker assembly at low frequencies. Thus, consider four speakers mounted in one of the walls of a sealed cabinet. By operating pairs of speakers in phase or 180 degrees out of phase, the far field produced at low frequencies can be degraded from a monopole to a dipole or a quadrupole field by choosing the phases appropriately (how?) with a corresponding reduction in the radiation efficiency.

5.3.5 Circular Piston in an Infinite Baffle

The sound radiation from a uniform oscillating piston in an infinite rigid wall is a classical problem and a summary of the analysis is given here. For details, we refer to Appendix A.

Far Field

The piston has a radius a and a velocity amplitude U, as indicated in Fig. 5.2. The sound field produced by the piston in the right hemisphere is the same as that produced by the piston pair on the right. Due to symmetry, the particle velocity normal to the plane of the pistons will be zero for $r > a$, the same as for the infinite baffle. The piston pair represents a monopole distribution with an acoustic source strength $2Q = (-i\omega)(2\rho U)$ per unit area and the sound field is obtained by integrating the corresponding monopole field contribution over the piston area, as shown in Appendix A. The resulting sound pressure distribution in the far field is found to be

$$p = 2Q \, \frac{e^{ikr}}{r} a^2 \frac{J_1(ka \sin \theta)}{ka \sin \theta} \quad \text{(far field)}, \tag{5.50}$$

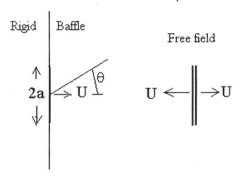

Figure 5.2: Left: Circular piston radiator in an infinite acoustically hard baffle. Right: Equivalent piston pair in free field.

where J_1 is the Bessel function of first order and $Q = (-i\omega)(\rho U)$. The corresponding normalized intensity distribution, directivity pattern, is given by

$$\boxed{\begin{array}{c} \textit{Intensity distribution. Circular piston source in a wall} \\ I(\theta)/I(0) = |2J_1(ka \sin\theta)/(ka \sin\theta)|^2 \end{array}} \qquad (5.51)$$

[a: Piston radius. θ: Angle from axis. $k = \omega/c = 2\pi/\lambda$. λ: Wavelength].

$I(\theta/I(0) = 1$ along the axis $\theta = 0$ and zero where $J(ka \sin\theta) = 0$, i.e., for $ka \sin\theta = 3.83, 7.02, 10.15$, etc. The first angle of zero intensity is given by

$$\sin\theta_1 = 3.83/ka \approx 0.61\lambda/a. \qquad (5.52)$$

The maximum intensity in the main lobe is $I(0) = [(ka)^2/8](a/r)^2 \, U_0^2\rho c)$, proportional to the square of both the frequency and the area of the piston.

The maximum in the secondary lobe between the angles θ_1 and θ_2 is only 0.02 of that in the main lobe, and the other maximum are insignificant. Thus, the bulk of the radiated power is contained in the main lobe.

We leave it for Problem 6 to calculate the total radiated power and the corresponding radiation resistance.

Radiation Impedance

By integrating the sound pressure in the near field over the piston area, the radiation impedance can be calculated as shown in Appendix A and the normalized value is

$$\zeta_r \equiv \theta_r + i\kappa_r = 1 - \frac{2J_1(2ka)}{2ka} - i\frac{2S_1(2ka)}{2ka}, \qquad (5.53)$$

where S_1 is the Struve function. In the low-frequency regime, $ka << 1$ ($\lambda >> a$), where the source is acoustically compact, the radiation impedance reduces to

$$\zeta_r \approx \frac{(ka)^2}{2} - i\frac{8}{3\pi} ka. \qquad (5.54)$$

The reactance corresponds to a mass end correction $\delta = (8/3\pi) a$ and a mass load $\delta\rho$ per unit area of the piston.

The total power radiated by the piston is $\Pi = (\pi a^2)\rho c\theta_r U^2/2$ (if U is the rms value, the factor 1/2 should be omitted). In the low-frequency regime it is proportional to the square of both the area and the frequency of the piston.

5.3.6   Problems

1.   **Sound fields from harmonic source and force distributions**

With reference to Eqs. 5.37 and 5.39, what are the expressions for the complex amplitude of the radiated sound pressure field from
(a) a harmonic acoustic source distribution with the complex amplitude $Q(r', \omega)$, and
(b) a harmonic force distribution with the complex amplitude $F(r', \omega)$?

2. **Field from a compact oscillating sphere**

   A sphere of radius a oscillates in the x-direction with the velocity $u = |u|\cos(\omega t)$. With reference to Eqs. 5.47 and 5.41, determine
   (a) the sound pressure and radial velocity fields $p(r, t)$ and $u_r(r, t)$ and
   (b) the corresponding complex amplitude fields.

3. **Sound field from the vortex shedding by a cylinder**

   A solid cylinder in a mean flow is known to generate a periodic stream of vortices (Kármán vortex street). As a result there will be a periodic force exerted on the cylinder, transverse to the flow. The radius of the cylinder is small compared to the wavelength. In calculating the corresponding sound field generated as a result of the vortex shedding, treat the cylinder as a force distribution with the force $|f|\cos(\omega t)$, i.e., complex amplitude $f(\omega) = |f|$, per unit length. The length of the cylinder is L. Derive an expression for the complex sound pressure amplitude in a plane perpendicular to the cylinder through the midpoint of the cylinder.

4. **Effect of a loudspeaker cabinet**

   A loudspeaker with a diameter of $d = 4$ inches is placed in a cabinet and is found to produce a sound pressure level of 70 dB at a distance of $r = 10$ feet on the axis of the speaker and at a frequency of 50 Hz.
   (a) Estimate the average velocity amplitude of the speaker.
   (b) Estimate the reduction in level at the same location when the cabinet is removed with the velocity amplitude of the speaker kept the same. In treating the speaker as a dipole, assume the separation of the monopoles involved to be the diameter of the speaker.

5. **Average intensity in a dipole field**

   Two monopoles a distance d apart make up a dipole field. What is the average sound intensity over the far field sphere surrounding the source in terms of the value for one monopole alone. Assume $d \ll \lambda$. In particular, if $d = 4$ inches and the frequency 50 Hz, what is the difference in the average sound pressure levels in the two cases?

6. **Radiated power and radiation resistance of a circular piston**

   Consult an appropriate mathematics text (for example, McLachlan *Bessel functions for engineers*, page 98) and confirm that

   $$\int_0^{\pi/2} \frac{2J_1(ka\sin\theta)}{ka\,\sin\theta}\sin\theta\,d\theta = (1 - \frac{J_1(2ka)}{ka})\frac{2}{(ka)^2}. \tag{5.55}$$

   Then, calculate the total power radiated by a piston in an infinite wall and determine the corresponding normalized radiation resistance of the piston.

5.4 Random Noise Sources

If a source of force distribution is random, the total mean square value of the radiated pressure is the sum of the mean square pressures from the elementary sources in the distribution. The mean square value of the acoustics source strength per unit volume at a point r' is denoted $Q^2(r')$, where Q now is the rms value. It follows from Eq. 5.37 that the mean square value of the sound pressure at a point of observation r is

$$p^2(r) = \frac{1}{16\pi^2}\int \frac{Q^2(r')}{|r - r'|^2}\,dr'. \tag{5.56}$$

If the source distribution is not random, the interference between the sound contributions from different elementary sources must be considered and the resulting sound field will have pronounced maxima and minima at locations of constructive and destructive interference.

5.4.1 Two Point Sources

Consider two random point sources located at $y = \pm d/2$, each with an acoustic source strength q, which is now taken to be the rms value. The mean square values of the sound pressures contributions from the two sources add so that the mean square pressure at the field point x, y in a plane containing the sources is

$$p^2 = \frac{q^2}{16\pi^2}\left[\frac{1}{x^2+(y-d/2)^2}+\frac{1}{x^2+(y+d/2)^2}\right]. \qquad (5.57)$$

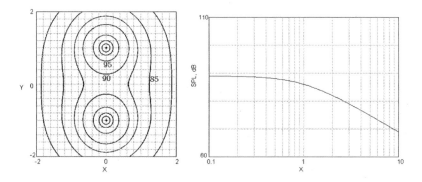

Figure 5.3: Left: SPL contours, 105, 100, 95, 90, 87,5, 85, and 82,5 dB, around two random noise point sources at $y = d/2$ and $y = -d/2$ plotted versus the normalized coordinate $X = x/(d/2)$ and $Y = y/(d/2)$. The combined power level of the two sources is 100 dB. Right: SPL versus X along the centerline of the sources at $y = 0$.

The combined acoustic power output of the sources is

$$W = \frac{2q^2}{4\pi\rho c}. \qquad (5.58)$$

This quantity can be considered known and q^2 can be expressed in terms of it. The corresponding power level is $PWL = 10\log(W/W_r)$, where the reference power is $W_r = 10^{-12}$ w as defined earlier.[3] The sound pressure level $L = 10\log[p^2/p_0^2]$, can then be written

$$SPL = PWL - 10\log(8\pi d^2/4A_r) + 10\log\left[\frac{1}{X^2+(Y-1)^2}+\frac{1}{X^2+(Y+1)^2}\right], \qquad (5.59)$$

---

[3]Recall that it can be expressed as $W_r = I_r A_r$, where $I_r = p_0^2/\rho c$ is the intensity corresponding to the reference (threshold) sound pressure (rms) $p_0 = 2\times10^{-5}$ N/m^2 and the area A_r is 1 m^2.

where $X = x/(d/2)$ and $Y = y/(d/2)$ are normalized coordinates.

In Fig. 5.3 are shown the SPL contours in the X, Y plane with two random point sources at $y = d/2$ and $y = -d/2$, i.e., at $Y = \pm 1$. In this particular example, the combined power level of the two sources is 100 dB. At distances $X > 1$, the sound pressure level approaches the value that would have been obtained if the two sources were into a single source at the origin (see Problem 1).

5.4.2   Finite Line Source

As another example, we consider a random noise source distribution along the y-axis from $y = -L$ to $y = L$, with the rms source strength $Q(y')$ per unit length so that the mean square value is $Q^2(y')$. The sound pressure field at a perpendicular distance x from the center of the line source then becomes

$$p(r)^2 = \int_{-L}^{L} \frac{Q^2(y')}{(4\pi)^2(x^2 + y'^2)}\, dy'. \tag{5.60}$$

For a uniform source distribution, Q, the integral becomes elementary and

$$p^2(x) = \frac{Q^2}{16\pi^2}\frac{2}{x}\arctan(L/x). \tag{5.61}$$

If $x >> L$, the result is $p^2(r) = (2L)Q^2/(16\pi^2 x^2)$ and if $x << L$, $p(r)^2 = (\pi/2)Q^2/(16\pi^2 x)$.

In other words, at a distance x large compared to the source size, the field is the same as that from a point source with the total mean square source strength $2LQ^2$ and the mean square pressure decreases as the square of the distance x. The corresponding sound pressure level then decreases by $10\log(x^2)$ with a decrease of ≈ 6 dB for every doubling of the distance x.

At a distance x small compared to the source size, p^2 becomes independent of the length of the source and decreases as $1/x$ with the distance x (i.e., with an x-dependence of the level given by $10\log(x)$ corresponding to a decrease by ≈ 3 dB for every doubling of the distance x).

The sound pressures in these two regions are the same at $x = 2L$, signifying the transition between the *near field* and the *far field*.

It is left for Problem 3 to show that at a field point x, y, rather than $x, 0$, as in the previous case, we obtain

$$p^2(X, Y) = \frac{Q^2}{16\pi^2 L}\frac{1}{X}\arctan(\frac{2X}{X^2 + Y^2 - 1}), \tag{5.62}$$

where $X = x/L$ and $Y = y/L$.

The total power generated is $W = 2LQ^2/(4\pi\rho c)$ and the source strength Q can be expressed in terms of W, if so desired and the sound pressure level obtained from Eq. 5.62 can be expressed in terms of the power level, as discussed in Section 5.4.1. Thus,

$$SPL = PWL - 10\log(8L^2\pi/A_r) + 10\log\left[\frac{1}{X}\arctan(\frac{2X}{X^2 + Y^2 - 1})\right]. \tag{5.63}$$

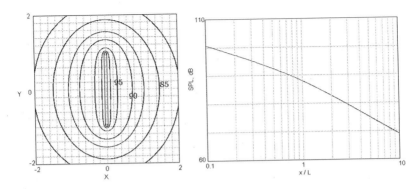

Figure 5.4: Left: Sound pressure level contours, 82.5, 85, 87.5, 90, 95, 100, and 105 dB, around a uniform line source with a power level of 100 dB and a length $2L = 2$ m. Right: The x-dependence of the sound pressure level in the mid-plane of the source with $X = x/L$ and $Y = y/L$.

In the numerical example on the left in Fig. 5.4, the line source has a power level of 100 dB and $L = 1$ m. Sound pressure level contours for 105, 100, 95, and 90 dB are shown which can be considered to describe the near field of the source. The coordinates are $X = x/L$ and $Y = y/L$ where L is half of the length of the source. As the distance from the source becomes larger than L, the contours approach the circular form characteristic of the field from a single point source at the origin.

The graph on the right in the same figure refers to the same line source and shows the sound pressure level along the X-axis. As X increases, the mean square pressure goes from a $1/X$-dependence close to the source to a $1/X^2$ dependence in the far field. The corresponding decrease of the sound pressure level (SPL) in these regions is approximately 3 dB and 6 dB per doubling of distance, respectively; the transition between these occurs approximately at $X = 1$ $(x = L)$.

5.4.3 Circular Source Distribution

As another example, consider a circular source distribution of radius R at $x = 0$ in a plane perpendicular to the x-axis. The point of observation is at x and $y = 0$, i.e., on the axis of the source. Again, the source distribution is uncorrelated with the source strength Q (rms) per unit area. The mean square sound pressure at x is then

$$p(x)^2 = \frac{1}{16\pi^2} \int_0^R Q^2(r')(2\pi r')\, dr'. \qquad (5.64)$$

For a uniform source distribution the integral is elementary and we get

$$p^2(x) = \frac{Q^2}{16\pi^2}\pi \ln(1 + R^2/x^2). \qquad (5.65)$$

Far from the source, with $x >> R$, this reduces to $(\pi R^2)Q^2/(16\pi^2 x^2)$, i.e., the same as for a point source.

5.4.4   Problems

1. **The field from two point sources**

   Examine the result shown in Fig. 5.3. At distances from the source, $X > 1$ ($X = x/(d/2)$), the sound pressure level is expected to approach the value obtained if the two sources were combined and placed at the origin. Determine whether the result in the figure is consistent with this expectation.

2. **Noise from an exhaust stack**

   Regard the end of a circular exhaust stack as a circular source with a uniform random noise distribution. The total power level of the noise emitted is 120 dB re 10^{-12} w. The diameter D of the stack is 4 m. What is the sound pressure level a distance D above the stack on the axis of the stack?

3. **Directivity of a random line source**

   (a) Show that Eq. 5.62 agrees with Eq. 5.61 for $y = 0$.
   (b) Discuss the angular distribution of the sound pressure level in the xy-plane. (Use $x = r \cos\phi$ and $y = r \sin\phi$.)

4. **Directivity of random circular source**

   Generalize the result in Eq. 5.65 and determine p^2 at a point (x, y rather than x, 0.)

   Discuss the angular distribution of the sound pressure level. Apply the result to the exhaust stack in Problem 2.

5.5   Superposition of Waves; Nonlinearity

In previous chapters we have been concerned with the basic physics of sound and specifically with the field from a single sound source. Thus, the determination of the field from a collection of sources does not involve anything basically new; the additional problems, such as wave interference, for example, are essentially geometrical (kinematic) and are pretty much the same as for other types of waves, not only sound waves. Formally, the field is obtained by adding a number of complex amplitudes representing the individual waves. Thus, for quantitative details of superposition, we can refer to general treatments of waves.[4]

Actually, in order to have wave interference phenomena, we do not necessarily have to have several sources. A single source will do, if we account for multiple reflections from several objects or wave transmission through several apertures in a screen, for example.

In regard to the superposition of waves referred to above, it is very important to realize that it was tacitly assumed that a *linear* addition was involved. Although this is indeed valid for waves of sufficiently small amplitudes, it is not true in general.

The reason is that a wave, such as a sound wave, changes the state of the material carrying the wave, albeit slightly, and hence affects the local wave speed.[5] The change of state is nonlinear, and this nonlinearity has important consequencies when it comes to the question of superposition of waves and their interference.

---

[4]See, for example, K.U. Ingard, *Fundamentals of Waves and Oscillations*, Cambridge University Press, Oxford. First printed 1988. It contains numerous examples of computed field distributions from a variety of source configurations.

[5]What is the corresponding situation for an electromagnetic wave in vacuum?

As a familiar illustration of nonliearity, consider the compression of an ordinary coil spring by a force applied at the end of the spring, compressing or stretching it. For a very large compression, the coil spring takes the form of a solid tube and for a large extension, it becomes a single strand. In both these limits, the 'stiffness' becomes very high. In general, the relation between the force F and the displacment ξ is nonlinear, i.e.,

$$F(\xi) = F(0) + \xi \, \partial F/\partial \xi + (\xi^2/2) \, \partial^2 F/\partial \xi^2 + \cdots . \tag{5.66}$$

In this power expansion of the force in terms of the displacement, the first term is a 'bias' force. The second term is the one used in a linear analysis and $\partial F/\partial \xi$ is the spring constant, normally denoted K. We have assumed here that F is independent of the velocity. The *compliance* of the spring is defined as $1/K$.

In a similar manner, we can express the displacement in terms of the driving force and conclude that a harmonic driving force will produce a displacement which contains in addition to the fundamental frequency of the driving force also harmonics thereof.

In the case of a fluid, let the relation between the pressure P and density ρ in a fluid be expressed by $P = P(\rho)$. A sound wave causes perturbations $\delta \equiv d\rho$ and dP in the density and pressure. Regarding the pressure as a function of density, a power series expansion of the pressure yields

$$P(\rho) = P(\rho_0) + \delta \, \partial P/\partial \rho + (\delta^2/2) \, \partial^2 P/\partial \rho^2 + \cdots . \tag{5.67}$$

With p being the first order perturbation and with $P/P_0 = (\rho/\rho_0)^\gamma$ (isentropic change of state) we get

$$P - P_0 = p + (p/P_0)p \, \gamma (\gamma - 1)/2\gamma^2 . \tag{5.68}$$

The second term is nonlinear and if the sound pressure p is harmonic with an angular frequency ω, the nonlinear terms will contain a frequency of 2ω. Normally, its amplitude will be small, however.

For example, the sound pressure level in normal speech is about 60 dB. With a reference rms pressure amplitude of $p_r = 0,00002 \, N/m^2$ the corresponding sound pressure amplitude is obtained from $20 \log_{10} p/p_r = 60$ or $p = 10^3 \, 0.00002 = 0.002$ N/m^2. With the normal atmospheric pressure being $P_0 \approx 10^5 \, N/m^2$, the nonlinear factor p/P_0 in Eq. 5.68 will be of the order of 2×10^{-7}. Thus, the corresponding nonlinear distortions will be quite small. For a jet engine, the level readily could be 140 dB, in which case this factor would be 10^4 times larger.

One nonlinear effect gives rise to a distortion of an initially plane harmonic wave through the generation of harmonic components. Qualitatively, this distortion can be understood also by realizing that the local speed of sound is affected by the sound in two ways. First, the sound speed increases with temperature and consequently will be slightly higher in the crest of the sound wave than in the trough; second, the particle velocity in the crest is in the direction of propagation (and in the opposite direction in the trough). These effects collaborate in making the local sound speed in the crest higher than the trough. Thus, as the wave progresses it will be distorted as the crest tends to catch up with the trough. We refer to Chapter 11 for further disucssion.

If two sound pressures of different frequencies ω_1 and ω_2 are involved, the non-linear term will produce harmonic pressures with the sum and different frequencies $\omega_1 + \omega_2$ and $\omega_1 - \omega_2$. Accordingly, experiments have been carried out to verify the existence of the scattered sound.[6]

5.5.1 Array of Line Sources. Strip Source

The field distribution from an array of N infinitely long line sources in a plane *perpendicular* to the lines will have the same form as that for the N point sources. The field will be independent of the y-coordinate which is parallel with the lines. Although there is a difference in the r-dependence of the intensity of a point source and a line source, the angular distribution will be the same.

A sound source in the form of an infinitely long vertical strip of width b can be considered to be an infinitely extended uniform distribution of finite horizontal line sources of length b equal to the width of the strip. The calculation of the angular distribution of the radiated sound field in the horizontal plane then involves an integration over the width b of the strip. (see Problem 5).

Phased Array. Moving Corrugated Board

As an example of a phased array of line sources, we consider the radiation from a moving corrugated board. For simplicity, it will be assumed to be of infinite extent. The board moves with a velocity U in the plane of the board in the direction perpendicular to the corrugations, as illustrated in Fig. 5.5. The example is relevant to our discussion of sound radiation from an axial fan in Chapter 7.

On the left in the figure, the board is moving with subsonic speed, $U < c$. The wavelength of the corrugation is Λ, the frequency of the generated sound will be $f = U/\Lambda$ and the wavelength $\lambda = c/f$. With $U < c$, we have $\lambda > \Lambda$ and it is not possible to fit a traveling plane sound wave to the corrugations regardless of the direction of propagation. A pressure disturbance is still produced by the board but it turns out to decay exponentially with the distance from the board (evanescent wave). The situation is much the same as for the generation of sound by a source in a duct below the cut-on frequency, as discussed in Chapter 6, and the physical reason for the decay is the interference between the sound from the crests and the valleys of the board which becomes destructive as the corresponding path difference goes to zero with increasing distance to the observation point.

The surfaces of constant phase of the resulting evanescent wave are perpendicular to the board and are illustrated by the thin lines in the figure. The surfaces of constant pressure magnitude are parallel to the board. To illustrate that the pressure decreases with the distance from the board, the corresponding lines of constant pressure are drawn with different thicknesses.

---

[6]Uno Ingard and David Pridmore Brown, *Scattering of Sound by Sound*, Journal of the Acoustical Society of America, June 1956

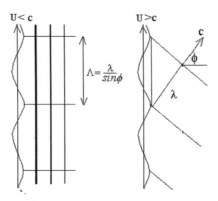

Figure 5.5: Sound radiation from a moving, corrugated board. Left: Subsonic speed, $U < c$. Horizontal lines: Surfaces of constant phase. Vertical lines (thickness indicating level), surfaces of constant sound pressure Right: Supersonic speed, $U > c$. Lines indicating surfaces of constant phase (wave fronts). The magnitude of the sound pressure amplitude is the same everywhere (plane wave).

If the velocity is supersonic, $U > c$, the situation is quite different. The wavelength of the emitted sound, $\lambda = c/f = (c/U)\Lambda$, now becomes less than the corrugation period Λ and a plane sound wave can now be matched to the boundary as it travels in a direction which makes an angle with the normal given by $\sin \phi = c/U$. As we recall from the discussion of sound radiation from the bending wave on a plate, the trace velocity of the sound wave generated is $c_t = c/\sin \phi$, and the boundary condition can be satisfied only if this velocity equals the velocity U of the board.

5.5.2 Problems

1. **Pulsating sphere**

   The radial surface velocity of a sphere of mean radius $a = 5$ cm is harmonic with a frequency 1000 Hz and with a uniform velocity amplitude 0.1 cm (rms).

   Neglecting sound absorption in the air, determine the distance at which the sound pressure amplitude will equal the threshold of human hearing (0.0002 dyne/cm^2 rms).

2. **Radiation from an array of N sources (the hard way)**

   Carry out the analysis of sound radiation from an array of N point sources with harmonic time dependence without the use of complex variables. As before, consider the far field only.

3. **Array of point sources with no net source strength**

   Reconsider the analysis of the radiation from a linear array of an even number of N point sources but rather than being in phase, there is a phase difference of π between adjacent sources. Consequently, the total source strength of the array is zero.

4. **Radiated power from a uniform line source**

   Use the far field distribution of intensity from a uniform harmonic line source and show that the total radiated power is the same as the power from a point force with the acoustic source strength equal to the total source strength of the line source.

5. **Radiation from a strip source**

   An infinitely long strip of width b has a uniform harmonic acoustic source distribution with a complex amplitude q.

   (a) Determine the angular distribution if the intensity in the far field in a plane perpendicular to the strip and the radiated acoustic power per unit length of the strip.

   (b) What would be the result for a source with a random time dependence?

6. **Radiation from circular disk**

   Derive an expression for the complex sound pressure amplitude produced by a uniform source distribution over a circular disc of radius R. Let the field point be on the axis of the disc, a distance x from the center. The complex amplitude of the acoustic source strength per unit area is q.

7. **Antenna of line sources**

   An antenna consists of an array of $N = 10$ line sources placed along a straight line with a distance between adjacent source of $d = 10\lambda$, where λ is the wavelength of the emitted wave.

   (a) How many intensity maxima are there in the radiation field around the antenna from 0 to 360 degrees?

   (b) If N is increased to 20, what is the change in the number of intensity maxima?

8. **Sound radiation from moving corrugated board**

   (a) In the sound field from a corrugated board moving at subsonic speed (Fig. 5.5), show that the pressure and axial velocity are 90 degrees out of phase. What does that mean in terms of the acoustic power radiated from the board.

   (b) For a board with a corrugation amplitude $\xi = 0.01\Lambda$, where Λ is the wavelength of the corrugation, what is the ratio of the magnitudes of the sound pressure at a distance Λ from the board obtained with the board velocities 0.5c and 2c, respectively?

Chapter 6

Room and Duct Acoustics

6.1 Diffuse Field Approximation

6.1.1 Reverberation Time

Much attention has been paid to the acoustics of concert halls and other enclosed spaces for lectures and the performing arts. Many factors, both physical and psychological, influence the judgment of the acoustic quality of rooms and many descriptors have been introduced and used in an effort to quantify various aspects of this concept. Systematic work in room acoustics began almost 100 years ago with the pioneering studies by Sabine, then a physics professor at Harvard. For further comments on Sabine, see Section 1.2.2.

We start by deriving Sabine's formula for the reverberation time in a room. The sound field is assumed to be diffuse which means that at a point in the room sound arrives from all direction with equal probability and intensity. Thus, if the point of observation is surrounded by a spherical control surface, the contribution to the acoustic energy density within the sphere from every solid angle element $d\Omega$ on the sphere will be the same and we express it as $(I_0/c)d\Omega$, where I_0 is an intensity and c the sound speed (see Chapter 3). The fact that the field is diffuse means that these contributions are all *uncorrelated* so that their energies add. Then, with a total solid angle of the spherical control surface of 4π, the total energy density becomes

$$E = 4\pi I_0/c. \tag{6.1}$$

Again, using the intensity I_0, we can express the acoustic power incident on a wall element of unit area in terms of I_0 as follows. Place a spherical control surface of unit radius with the center at the wall element and consider a wave that strikes the wall at an angle of incidence ϕ. The solid angle between ϕ and $\phi + d\phi$ is a ring of radius $\sin \phi$ on the sphere with an area $2\pi \sin \phi \, d\phi$ (see Fig. 4.4). Thus, the total intensity that strikes unit area of the wall at an angle of incidence ϕ will be $I_0 \, 2\pi \sin \phi \cos \phi \, d\phi$ since the power intercepted by a unit area of the wall will be $I_0 \cos \phi$.

Integrating over the entire control surface, we obtain the total power per unit area, the diffuse field intensity I_d,

$$I_d = \int_0^{\pi/2} I_0 \, 2\pi \sin \phi \cos \phi = \pi I_0 = Ec/4, \tag{6.2}$$

where we have used Eq. 6.1; E is the acoustic energy density introduced above. Thus, the average intensity on the wall in the diffuse field is one quarter of the intensity of a wave at normal incidence with the same energy density E.

We define the diffuse field absorption coefficient α_d such that $I_d \alpha_d$ is the absorbed power per unit area. Thus, with $I_d = Ec/4$ and with the physical area of an absorber on a wall being A, the absorbed power will be $(Ec/4)A\alpha_d$ and the quantity $A\alpha_d$ is the absorption area. The coefficient α_d is the same as the angle averaged absorption coefficient given in Chapter 4, Eq. 4.53.

These considerations imply an infinitely extended surface so that edge effects (diffraction) of the absorber can be ignored. In reality, with a finite absorptive wall element, this is no longer true, and the effective absorptive area will be larger than $A\alpha_d$. It is generally quite difficult to calculate, even for an absorber of simple shape. The actual absorption area will be denoted A_s and a corresponding absorption coefficient α_s is defined by $A_s == \alpha_s A$. The absorbed power is then $A_s(E/4c)$.

In this context of room acoustics, we shall call A_s the *absorption area* or the *Sabine (area)* and the corresponding absorption coefficient $\alpha_s = A_s/A$ will be called the *Sabine absorption coefficient*. It is this coefficient that is measured by the reverberation method, to be described below. It can exceed unity, particularly at low frequencies when diffraction effects play a significant role.

With the volume of the room denoted V, the total acoustic energy in the room is EV and the rate of decrease of it must equal the absorbed energy, i.e.,

$$\boxed{\begin{array}{c} \textit{Decay of energy density in a room} \\ V dE/dt = -A_s Ec/4) \quad \text{i.e.,} \\ E(t) = E(0)\, e^{-c(A_s/4V)t} \end{array}} \quad . \tag{6.3}$$

In other words, the energy density and hence the mean square sound pressure in the room decays exponentially, the decay constant being proportional to the total absorption area (cross section) $A_s = A\alpha_s$ and inversely proportional to the volume. If the absorption coefficient varies over the area, A_s has to be replaced by the average value.

The *reverberation time* is defined as the time in which the average sound pressure level in the room decreases 60 dB. It follows from Eq. 6.3 that

$$10 \log[E(0)/E(t)] = \frac{cA_s}{4V} t\, 10 \log(e). \tag{6.4}$$

Thus, with the left side put equal to 60 and the reverberation time denoted T_r, we get

$$\boxed{\begin{array}{c} \textit{Reverberation time} \\ T_r = 240/[10 \log(e)]\,(V/cA_s) \approx 55\, V/cA_s \end{array}} \tag{6.5}$$

[V: *Room volume.* A_s: *Absorption area (Sabine) (product of absorption coefficient and absorber area. c: Sound speed.*]

Introducing the numerical value for the sound speed $c \approx 342$ m/s, the numerical expression for the reverberation time becomes $T_r \approx 0.16\, V/A_s$, where V and A_s are expressed as m^3 and m^2, respectively.

For example, for a cubical room with a side length L, the internal surface area is $6L^2$, and the reverberation time becomes $T_r = 0.027\,L/\alpha_s$. Thus, with $L = 10$ m and a Sabine absorption coefficient of 10 percent, the reverberation time becomes 2.7 seconds.

A reverberation room in an acoustical testing facility typically may have a reverberation time of about 10 seconds at low frequencies (200 Hz, say). Among the quantities that have been proposed and used for the description of the acoustics of a room, the reverberation time is still regarded as a primary parameter. For example, there is good correlation between the intelligibility of speech in a lecture room and the reverberation time so that an optimum value can be established. Such a correlation can be determined by the fraction of randomly selected spoken words from the podium that can be understood by a listener in the audience. The optimum reverberation time depends on the room size, but typically is about one second.

To determine an optimum reverberation time for music is a more subjective matter and depends on the character of the music; typically this optimum is between 1 and 3 seconds.

The expression for the reverberation time in Eq. 6.5 accounts only for the absorption at the walls (by acoustic treatment) and formally goes to infinity when the absorption coefficient of the absorptive material on the walls goes to zero. In reality, even a rigid, impervious wall yields some absorption because of visco-thermal losses, and sound transmission through the wall is equivalent to absorption. From the discussion in Section 4.2.4, the diffuse field average absorption coefficient of a rigid wall can be shown to be

$$\alpha \approx 1.7 \times 10^{-4}\sqrt{f}, \tag{6.6}$$

where f is the frequency in Hz.

There is also sound absorption throughout the volume of the room due to visco-thermal and molecular relaxation effects. With the spatial decay of the intensity expressed as $I_0 \propto \exp(-\beta x)$, the corresponding temporal rate of decrease of the energy density in a diffuse sound field in a room, following Eq. 6.1, will be $\beta c E$. This means that the right-hand side of Eq. 6.3 has to be replaced by $-Ec(\beta V + \alpha A/4)$, where α is the visco-thermal absorption coefficient and A the total wall area.

From Eq. 6.3 it follows that the decay of the acoustic energy density due to visco-thermal losses at the walls and losses throughout the volume of a room is given by

$$E(t) = E(0)e^{-(\beta + \alpha A/4V)ct}. \tag{6.7}$$

For sufficiently large rooms and high frequencies, the absorption in the volume of the room dominates.

In regard to numerical results, we already have an expression for the frequency dependence of the absorption coefficient in Eq. 6.6. For β, we refer to the discussion in the chapter on atmospheric acoustics, in particular Fig. 10.2, where, at a temperature of 20°C, the wave attenuation is plotted in dB/km as a function frequency. For relative humidities less than 50 percent, the vibrational relaxation effects (of Oxygen) dominate, and at 50 percent, the attenuation can be written approximately as $(f/1000)^2$ dB/km. The corresponding expression for β is then

$$\beta \approx 2.29\,(f/1000)^2\,10^{-4} \quad \text{m}^{-1}, \tag{6.8}$$

where we have used $10 \log(e) \approx 4.36$.

The condition that the two absorption effects contribute equally to the decay is $\beta = \alpha A/4V$, and with the numerical values given in Eqs. 6.6 and 9.16, this corresponds to $f/1000 = 3.27(A/V)^{2/3}$, where the length unit is 1 meter. For a cubical room with a side length L, this means $f/1000 \approx 10.9(1/L)^{2/3}$. Thus, with $L = 10$ m, the volume absorption will exceed surface absorption at frequencies above 1000 Hz.

6.1.2 Measurement of Acoustic Power

One method of measuring the acoustic power of a source is to place the source in free field and integrate the acoustic intensity over a closed control surface surrounding the source. The intensity can be measured with an intensity probe described in Section 3.2.3. Free field conditions can be approximated in an anechoic room.

Another method, considered here, is to put the source in a reverberation room, discussed above. The absorption in such a room is very small and diffuse field conditions can be approximately achieved. With the source operating in steady state, the power output Π must equal the total power absorbed by the walls and by the air itself in the room. The latter absorption is usually negligible and we shall omit it here. Thus, with reference to the previous section, the absorbed power can be written $(Ec/4)A_s = [(p^2)_{av}/\rho c](c/4)A_s$, where A_s is the absorption area, defined in the previous section, and $(p^2)_{av}$ is the spatial average of the mean square sound pressure in the room. In practice, the averaging is achieved by the use of several microphones in the room. Thus, in terms of these quantities, the source power can be expressed as

$$\Pi = (Ec/4)A_s, \qquad (6.9)$$

where $E = (p^2)_{av}/\rho c$. The absorption area is determined from the measured reverberation time, Eq. 6.5.

In practice, a reference source with known power output Π_r (measured according to Eq. 6.9) is generally used to 'calibrate' the room. Then, if the average mean square sound pressure obtained with this reference source is $(p_r^2)_{av}$, the power from the actual source will be

$$\Pi = [(p^2)_{av}/(p_r^2)_{av}]\,\Pi_r. \qquad (6.10)$$

Thus, if the power level of the reference source is PWL_r the power level of the actual source is then

$$PWL = PWL_r + 10\log[(p^2)_{av}/(p_r^2)_{av}] = PWL_r + SPL - SPL_r, \qquad (6.11)$$

where SPL stands for the average sound pressure level in the room.

6.1.3 Measurement of the (Sabine) Absorption coefficient

With reference to Eq. 6.5, the absorption area (or cross section) A_s of a sample of wall treatment is obtained from the measurement of reverberation time in a room, as follows. First, the reverberation time T_r of the empty room is measured. This yields an absorption area S of the empty room. Next, one or more of the interior room surfaces is covered with the material to be tested. The reverberation time T_r'

is measured; it yields the new total absorption areas S'. The increase is attributed to the test sample and it is

$$
\boxed{
\begin{array}{c}
\textit{Measured absorption area } A_s \\
A_s = S' - S \approx 55\,(V/c)(1/T'_r - 1/T_r)
\end{array}
}
\tag{6.12}
$$

[V: Room volume. c: Sound speed. S', S: Total absorption area with and without test sample. T'_r, $T - r$: Reverberation time with and without test sample present.]

If the physical area of the test sample is A, the corresponding Sabine absorption coefficient is $\alpha_s = A_s/A$.

It is important to realize the distinction between the measured Sabine absorption coefficient α_s and the computed diffuse field absorption coefficient α_d (always less than unity). The latter implies an infinite test sample and is the quantity in Eq. 4.53 computed from the known angular dependence of the absorption coefficient for a particular material.

6.1.4 Measurement of Transmission Loss of a Wall

Following the discussion of the measurements of acoustic power and the absorption coefficient in the previous two sections, the procedure for the measurement of the diffuse field transmission loss of a partition can be readily understood.

The partition is installed in an opening in the wall between two reverberation rooms, a source room and a receiving room. The source room contains the sound source, usually one or more loudspeakers. Several microphones are used in both rooms for the measurement of the average rms sound pressure.

The transmission loss of the wall is much greater than that of the partition so that the power transmitted through the wall can be neglected. Diffuse sound fields in both rooms are assumed. With the area of the partition being A_p, the power transmitted is then $A_p \tau_s E_1 4/c$, where E_1 is the acoustic energy density in the source room and τ_s the power transmission coefficient. This power takes the place of Π in Section 6.1.2, Eq. 6.9, and gives rise to an energy density E_2 in the receiver room given by $E_2 = [4/(cA_s)]A_p\tau_s E_1(4/c)$, where, as before, A_s is the absorption area in the receiver room. It follows then that

$$
\tau_s = (E_2/E_1)(A_p/A_s).
\tag{6.13}
$$

Expressing the energy densities in terms of the rms values of the average sound pressure in the two rooms, we obtain for the transmission loss

$$
TL_s = 10\log(1/\tau_s) = SPL_1 - SPL_2 + 10\log(A_p/A_s).
\tag{6.14}
$$

As for the measured absorption coefficient, we use the subscript s to indicate that it refers to a Sabine approximation in describing the sound fields in the test rooms and the use of a finite partition. In the calculations of the transmission loss in Chapter 4, a partition of infinite extent was assumed. The transmission loss then could be expressed in a relatively simple manner in terms of the physical parameters of the partition. In this respect, there is a correspondence between the diffuse field absorption coefficient α_d and the diffuse field transmission loss TL_d, both computed quantities. This should be borne in mind when comparing experimental data with calculated.

6.1.5 Wave Modes in Rooms

We have already discussed a normal mode of a tube of length L and closed at both ends with rigid walls at $x = 0$ and $x = L_x$. It was found to be of the form

$$p_\ell(x, t) = A \cos(k_\ell x) \cos(\omega_\ell t), \qquad (6.15)$$

where $k_\ell = \ell\pi/L_x$, $\omega_\ell = ck_\ell$, and $\ell = 1, 2 \cdots$. This mode has ℓ pressure nodes, each at a distance from the rigid walls of an integer number of quarter wavelengths. We also considered briefly modes in two and three dimensions in Section 3.4.8.

In three dimensions and rectangular coordinates, a complex amplitude of the sound pressure is assumed to be of the form $p = X(x)Y(y)Z(z)$ ('separation of variables'). When inserted into the wave equation $\nabla^2 p + (\omega/c)^2 p = 0$ it leads to $X''/X + Y''/Y + Z''/Z + (\omega/c)^2 = 0$, where the primes indicate differentiation with respect to the argument of the function involved. The first term is a function of x only, the second of y only, and the third of z only, and in order for this equation to be satisfied for all values of the variables, each of the terms must be a constant such that the sum of the constants will cancel $(\omega/c)^2$. Thus, with the constants denoted $-k_x^2$, $-k_y^2$, and $-k_z^2$, the equation for X becomes $X'' + k_x^2 X = 0$, with analogous equations for Y and Z. Thus, each variable satisfies a one-dimensional harmonic oscillator type equation for which we already know that $X \propto \cos(k_x x - \phi_x)$ with similar expressions for Y and Z. The corresponding velocity field is obtained from the momentum equation, i.e., $-i\omega\rho u_x = -\partial p/\partial x$, etc. If $u_x = 0$ at $x = 0$ (acoustically hard wall), we have $\phi_x = 0$ and if $u_x = 0$ also at $x = L_x$, we must have $\sin(k_x L_x) = 0$, i.e., $k_x = \ell\pi$, where ℓ is an integer.

Then, in a rectangular room with hard walls at $x = 0$ and L_x, $y = 0$ and L_y, and $z = 0$ and $y = L_z$, the expression for the complex pressure amplitude will be of the form

$$p_{\ell,m,n} = A \cos(k_x x) \cos(k_y y) \cos(k_z z), \qquad (6.16)$$

where $k_x = \ell\pi/L_x$, $k_y = m\pi/L_y$, and $k_z = n\pi/L_z$. A mode with uniform pressure in the y- and z-directions, corresponding to $m = n = 0$, has the same form as Eq. 6.15 and is denoted $p_{\ell,0,0}$ and has ℓ nodal planes. In general, there will be nodal planes also in the y- and x-directions.

The frequency of free oscillations in the room follows from the wave equation $\partial^2 p/\partial x^2 + \partial^2 p/\partial y^2 + \partial^2 p/\partial z^2 - (1/c^2)\partial^2 p/\partial t^2 = 0$, which yields

$$k_x^2 + k_y^2 + k_z^2 - \omega_{\ell,m,n}^2/c^2 = 0 \qquad (6.17)$$

and the normal mode (angular) frequencies

$$\boxed{\begin{array}{l} \textit{Normal mode frequencies } f_{\ell,m,n} \\ \omega_{\ell,m,n} = 2\pi f_{\ell,m,n} = c\sqrt{k_x^2 + k_y^2 + k_z^2} \end{array}} \qquad (6.18)$$

$[k_x = m\pi/L_x.\ \ k_y = n\pi/L_y.\ \ k_z = n\pi/L_z 3$ *(see Eq. 6.16). L_x, L_y, L_z: Room dimensions. ℓ, m, n: Positive integers. c: Sound speed.]*

The forced motion of a room by the human voice or a musical instrument, for example, can be determined by analogy with the analysis of the closed tube response in Chapter 3. For a given source strength and in the idealized case of a loss-free room, the sound pressure in the room theoretically goes to infinity whenever the driving frequency coincides with a mode frequency. At low frequencies, with the wavelengths of the order of the room dimensions, the resonance frequencies are relatively far apart, and as the frequency is varied, the room response will be quite irregular with large variations in the sound pressure. As the frequency increases, the mode number increases rapidly, as shown in Eq. 3.92, and the response becomes more regular as the response curves of different modes will overlap.

For a square room, the frequencies $\omega_{q,0,0}$, $\omega_{0,q,0}$, and $\omega_{0,0,q}$ are all the same, and when the room is driven at this frequency, there will be three modes which will be excited at resonance. This results in a large irregularity in the frequency response of the room and this is to be avoided in order to have good room acoustics. Different modes of this kind, having the same resonance frequency, are called *degenerate* and should be avoided for good acoustics.

6.1.6 Problems

1. **Reverberation time**

   A rectangular room with the dimensions 15 m, 15 m, and 20 m, has a reverberation time of 4 seconds at a frequency of 300 Hz. It is desired to lower this time to 1 second. How large an area of wall treatment is needed (neglect diffraction effects) to obtain this reduction in the reverberation time if the absorption coefficient of the material is 0.7?

2. **Measurement of acoustic source power**

   A rectangular reverberation room with the dimensions 15 m, 20 m, and 20 m has a reverberation time of 8 seconds. A source in the room produces an average sound pressure level of 100 dB in the room. What is the power output of the source in watts?

3. **Measurement of transmission loss**

   The reverberation room in Problem 2 is used as the receiving room in a transmission loss laboratory. The source room and the receiving room are separated by a heavy (double) wall in which a test sample of a panel, 4 m× 4 m, is inserted into an opening in the wall provided for this purpose. In a certain frequency band, the average sound pressure level in the source room is 120 dB and in the receiver room 60 dB. What then is the transmission loss of the panel?

4. **Mode frequencies in a room**

   List the first ten modal frequencies of a rectangular room with the dimensions 10 m, 12 m, and 12 m. Which modes are degenerate, if any?

6.2 Waves in Ducts with Hard Walls

As before, an acoustically hard wall is one at which the normal velocity is zero. A rectangular duct can be regarded as a degenerate form of the room considered in the previous section with one side normally much larger than the others. Suppose that this side runs along the x-axis. The part of the wave function that involves this coordinate in the wave function for the room, $\cos(k_x x)$ can be considered to be the sum of one wave $\exp(i k_x x)$ traveling in the $+x$-direction and one wave $\exp(-k_x x)$ in the negative x-direction forming a standing wave. In the limiting case of an infinitely long duct or a duct in which there is no reflection, there will be no wave in the negative direction and the x-dependence of the wave function will be described by $\exp(i k_x x)$.

6.2.1 Wave Modes. Cut-off Frequency and Evanescence

Rectangular Duct

The complete wave function for the duct then will be composed of two standing wave components, $\cos(k_y y)$ and $\cos(k_z z)$, and a traveling wave component $\exp(i k_x x)$ so that the total complex amplitude of the pressure becomes

$$p(x, \omega) = A \cos(k_y y) \cos(k_z z) e^{i k_x x}. \tag{6.19}$$

The real wave function $p(x, t)$ is obtained by multiplying by $\exp(-i\omega t)$ and taking the real part of the function thus obtained.

The coefficients k_y and k_z will be the same as in the wave function for the room in Eq. 6.16. We are dealing here with the forced motion of the wave and the driving frequency is ω. The wave equation imposes the same relation as before, given in Eq. 6.17, but this time ω is given and we are seeking k_x. Thus, with $\omega_{\ell m n}$ in Eq. 6.17 replaced by ω and solving for k_x, we obtain

$$k_x = \sqrt{(\omega/c)^2 - k_y^2 - k_z^2}. \tag{6.20}$$

A uniform pressure across the duct corresponds to $k_y = k_z = 0$ and $k_x = \omega/c$, i.e., $m = n = 0$. It is the plane wave or the *fundamental mode* in the duct and it is labeled p_{00}. For other values of m, n there will be nodal planes in the wave parallel with the duct walls. For example, the wave with $k_y = m\pi/L_y$ and $k_z = n\pi/L_z$, has m nodal planes normal to the y-direction and n, normal to the z direction. The mode, denoted p_{mn}, is called a *higher order mode* and it follows from Eq. 6.20 that the propagation constant k_x for this mode can be written

$$\boxed{\begin{array}{c} \textit{Propagation constant for higher order mode} \\ k_x = (\omega/c)\sqrt{(1 - (\omega_{mn}/\omega)^2} = (\omega/c)\sqrt{1 - (\lambda/\lambda_{mn})^2} \end{array}} \tag{6.21}$$

$[\omega = 2\pi f$: *Angular frequency. c: Sound speed, free field.* $\omega_{mn} \equiv 2\pi f_{mn} = c\sqrt{(m\pi/L_y)^2 + (n\pi/L_z)^2}$: *Cut-off (or cut-on) frequency* (f_{mn}) *for* (mn)-*mode.* $\lambda_{mn} = c/f_{mn}$: *Corresponding wavelength. m, n: Positive integers. m* $= 0$, *n* $= 0$: *Plane wave (fundamental mode)*].

The real wave function for the traveling wave will be $\cos(\omega t - k_x x)$. The amplitude of the wave remains constant if the phase $\Psi \equiv \omega t - k_x x$ is constant. Thus, in order to observe an unchanging pressure in the traveling wave, we have to move in the x-direction with the velocity

$$c_p = \omega/k_x. \tag{6.22}$$

It is called the *phase* velocity (i.e., the velocity with respect to which the phase remains constant).

At frequencies above the cut-off frequency, k_x is less than ω/c, and it follows that the phase velocity is greater than the free field velocity c. For visualization, let us consider a plane wave which travels at an angle ϕ with respect to the x-axis, as shown in Fig. 6.1.

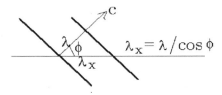

Figure 6.1: Plane wave traveling at an angle ϕ with respect to the x-axis. Two wave fronts (phase surfaces) are shown. The separation of the surface in the x-direction is $\lambda/\cos\phi$ which is proportional to the phase velocity in the x-direction.

A 'wave front' is merely a surface of constant phase, i.e., the instantaneous pressure is the same over the surface. Two wave fronts separated by a wavelength are shown. They represent a traveling plane wave which moves forward in a direction normal to the wave fronts with the speed of sound c. The important point to notice in this context is that the separation of the two wave fronts in the x-direction is $\lambda_x = \lambda/\cos\phi$ and that the intersection point of a wave front with the x-axis moves with the velocity[1] $c/\cos\phi$ in the x-direction, both greater than the free field values λ and c. If we regard the propagation constant k as a vector, the magnitude being $k = \omega/c$, the component in the x-direction is $k_x = k\cos\phi = 2\pi/\lambda_x$ and the speed of the wave fronts in the x-direction can be expressed as $c_p = \omega/k_x = c/\cos\phi$, which is the phase velocity in Eq. 6.22.

With

$$\cos\phi = k_x/k = \sqrt{1 - (m\lambda/2L_y)^2} = \sqrt{1 - (f_m/f)^2} \qquad (6.23)$$

it follows that

$$\sin\phi = m\lambda/2L_y = f_m/f. \qquad (6.24)$$

The sum of two plane waves, one traveling in the positive and the other in the negative ϕ-direction, is $(\exp(ik_y y) + \exp(-ik_y y)\exp(ik_x x) \propto \cos(k_y y)\exp(ik_x x))$, i.e., the same as the wave function in Eq. 6.19 (with $n = 0$). This means that this higher order mode, $p_{m,0}(x,\omega) = A\cos(k_y y)\exp(ik_x x)$, can be interpreted as the wave field produced by a plane wave traveling at an angle ϕ with respect to the duct axis, reflected back and forth between the walls of the duct. The general wave, p_{mn}, can be interpreted in a similar manner but is harder to visualize geometrically.

---

[1] Sometimes called the 'trace velocity.'

The phase velocity is a purely geometric quantity; the wave energy is not transported at this speed but rather with the *group velocity* which is

$$c_g = c \cos \phi, \tag{6.25}$$

the component of the actual sound speed in the x-direction. We note that the group velocity can be written $c_g = d\omega/dk_x$ and that $c_g c_p = c^2$.

Cut-off Frequency. Evanescence

Going back to Eq. 6.21, we note that the propagation constant k_x for the higher order mode $p_{m,n}$ will be real only if the frequency exceeds the *cut-off frequency* $f_{m,n}$. At this frequency the angle ϕ of the obliquely traveling plane wave that produces the mode will be 90 degrees, i.e., transverse to the x-axis. At a lower frequency, k_x becomes imaginary, $k_x = i\sqrt{(f_{m,n}/f)^2 - 1} \equiv ik_i$, and the wave amplitude decreases exponentially with x

$$p(x, \omega) = Ae^{-k_i x} \cos(k_y y) \cos(k_z z). \tag{6.26}$$

Such a wave is called *evanescent*; it will be demonstrated and discussed further in Section 6.2.2.

Circular Duct

In the circular duct, the transverse coordinates which correspond to y, z are the radius r and the azimuthal angle ϕ (Fig. 6.2). The wave equation is separable, as before, and the general solution is a combination of products of a function of r, ϕ, and x. For an infinitely long duct, the x-dependence of the complex pressure amplitude will be the same as for the rectangular duct, $\exp(ik_x x)$. In the transverse directions, there will be standing waves analogous to $\cos(k_y y)$ and $\cos(k_z z)$ for the rectangular duct; the wave in the ϕ direction will be $\cos(k_\phi \phi)$ and in the r-direction, $J_m(k_r r)$, which is a Bessel function of order m. The radial 'propagation constant' k_r will be determined by the boundary condition that the radial velocity be zero at the duct wall at $r = a$. The 'boundary condition' for the azimuthal wave function is the requirement that the function will come back to its original value when the angle is increased 2π. This means that $k_\phi = m$, where m is an integer. Actually, the wave function in the ϕ-direction could also be $\exp(\pm i\phi)$ in which case the wave corresponds to a wave 'spinning' in the positive or negative ϕ-direction. The radial velocity u_r is proportional to the derivative $J_m'(kr)$ of $J_m(k_r r)$ and the possible values of k_r, which we denote $k_{m,n}$, are determined from the boundary condition $u_r = 0$ at $r = a$, i.e.,

$$J_m'(k_{m,n}a) = 0. \tag{6.27}$$

For $m = 0$ the field is uniform in the circumferential direction the first solution to this equation is $k_{0,0}a = 0$, which corresponds to the plane wave (fundamental mode). The next solution is $k_{0,1}a = 3.8318$ which represents a pressure mode with no nodes in the circumferential direction and one node in the radial direction.

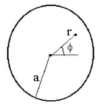

Figure 6.2: Circular tube coordinates.

The mode (1,0) which has one nodal plane in the circumferential direction and no nodes in the r-direction corresponds to the solution $k_{1,0}a = 1.8413$. In general, the mode (m, n)th pressure mode has m nodal planes in the circumferential direction and n nodes (nodal cylinders) in the radial direction.

The mode (0,1) with no nodes in the circumferential direction but one nodal cylinder has its first cut-off at the wavelength $\lambda_{01} = 2\pi a/3.83 \approx 1.64a = 0.83\,d$. Going across the duct along a diameter, we start from a maximum of the magnitude of the pressure amplitude at the wall, then go through a pressure node, then another maximum at the center of the duct and then another node to finally return to a pressure maximum at the opposite wall. The corresponding mode in the rectangular case has the cut-off wavelength equal to $1.0d_1$ which should be compared with $0.83d$ for the circular duct. In this context, we note that the width of a square cross section with the same area as the circular duct has a side $0.886d$.

A few other values of $k_{mn}a$ are

$$
\begin{array}{lll}
k_{00}a = 0.0000 & k_{01}a = 3.8318 & k_{02}a = 7.0155 \\
k_{10}a = 1.8413 & k_{11}a = 5.3313 & k_{12}a = 8.5263 \\
k_{20}a = 3.0543 & k_{21}a = 6.7060 & k_{22}a = 9.9695
\end{array}
\qquad (6.28)
$$

As for the rectangular duct, the propagation constant k_x follows from the wave equation and the analogue to Eq. 6.20 is

$$
k_x = \sqrt{(\omega/c)^2 - k_{m,n}^2} = (\omega/c)\sqrt{1 - (\omega_{m,n}/\omega)^2} = (\omega/c)^2\sqrt{1 - (\lambda/\lambda_{m,n})^2}, \quad (6.29)
$$

where $\lambda_{m,n} = 2\pi/k_{m,n}$. For example, $\lambda_{1,0} = a2\pi/1.8413 = 3.412\,a = 1.706\,d$, where $d = 2a$ is the diameter of the tube. This mode has a pressure nodal plane through the center of the duct with the pressure on one side being 180 degrees out of phase with the pressure on the other side. The velocity amplitude is of course a maximum in this plane and the mode can be regarded as the 'sloshing' of air back and forth across the duct. This should be compared to the corresponding mode in a rectangular duct with $\lambda_{1,0} = 2/d_1$. Note that in the rectangular case the cut-off wavelength is twice the width of the duct. In the circular duct it is somewhat smaller, ≈ 1.7 times the diameter, which is expected since the circular tube is narrower on the average.

Figure 6.3: Annular duct with acoustically hard walls.

Annular Duct

In some applications, as in axial compressors and in aircraft bypass engines, noise is transmitted through annular ducts.

The annular duct considered here is bounded by acoustically hard concentric circular cylinders with radii a and b (Fig. 6.3). The sound field is periodic with respect to the angular coordinate ϕ, with the angular dependence of the complex pressure amplitude being expressed by the factor $\exp(im\phi)$ or by $\cos(m\phi)$ or $\sin(m\phi)$ or a combination thereof, where m is a positive integer. The radial dependence can be represented by a combination of Bessel functions $J_m(k_r r)$ or $Y_m(k_r r)$ [corresponding to $\cos(k_r x)$ and $\sin(k_r x)$-functions] or by the Hankel functions $H_m^{(1)}(k_r r) = J_m + i Y_m$ and $H_m^{(2)}(k_r r) = J_m - i Y_m$ of the first and second kind [corresponding to $\exp(ik_x x)$ and $\exp(-ik_x x)$ in rectangular coordinates].

A wave mode $p_{m,n}$ is a linear combination of these functions such that the boundary conditions of zero radial velocity is fulfilled at $r = a$ and $r = b$. There will be several possible values for k_r which are consistent with these conditions and we denote them $k_r = \beta_{m,n}/a$.

Thus, with $\sigma = a/b$ and $k_r b = \beta_{m,n}$ and $\sigma = b/a$, the sound pressure field can then be expressed as $p(\omega) = \sum p_{m,n}$, where

$$p_{m,n} = [J_m(k_r r) + R_{m,n} Y_m(k_r r)] e^{ik_x x} e^{im\phi}$$
$$k_x^2 \equiv k_{m,n}^2 = (\omega/c)^2 - (\beta_{m,n}/b)^2. \tag{6.30}$$

The quantities $R_{m,n}$ and $\beta_{m,n}$ are determined by the boundary condition mentioned above. Since the radial velocity is proportional to $\partial p/\partial r$, these conditions can be written

$$J_m'(\beta_{m,n}) + R_{m,n} Y_m'(\beta_{m,n}) = 0$$
$$J_m'(\sigma \beta_{m,n}) + R_{m,n} Y_m'(\sigma \beta_{m,n}) = 0, \tag{6.31}$$

where $\sigma = a/b$. The prime indicates differentiation with respect to the argument.

The solutions for $\beta_{m,n}$ and $R_{m,n}$ for some different values of m, n, and σ are given in the following table.

It is interesting to compare these values of $\beta_{m,n}$ with those obtained from an approximate analysis of the problem. For a narrow annulus, for which the ratio $\sigma = a/b$ is close to unity, we expect the wave field in the annulus to be approximately the same as that between parallel walls separated a distance $d = b - a$. If we put $y = r - a$, the walls are located at $y = 0$ and $y = b - a$. The average radius of the annulus is $r_a = (b + a)/2$ and the corresponding average circumference is $2\pi r_a$. With $z = r_a \phi$

we have in essence replaced the annular duct with a rectangular duct in which the wave field is periodic in the z-direction with the period $2\pi r_a$.

Annular duct

Solutions $\sigma_{m,n}$ and $R_{m,n}$ from Eq. 1.2.9.

			$\beta_{m,n}$				$R_{m,n}$	
m	n	$\sigma = 0$	0.25	0.5	0.75	0.25	0.50	0.75
1	0	1.841	1.644	1.354	1.146	−0.129	−0.286	−0.367
	1	5.331	5.004	6.564	12.66	−0.327	2.578	1.390
	2	8.526	8.808	12.706	25.18	0.275	1.537	1.177
	3	11.71	12.85	18.94	37.73	2.160	1.324	1.115
2	0	3.054	3.009	2.681	2.292	−0.029	−0.221	−0.391
	1	6.706	6.357	7.062	12.82	−0.341	−0.350	−0.361
	2	9.970	9.623	12.949	25.258	−0.351	−0.222	−0.635
	3	13.17	13.37	19.10	37.78	−0.201	−0.431	−0.741
4	0	5.318	5.316	5.175	4.578	−0.001	−0.080	−0.363
	1	9.282	9.240	8.836	13.44	−0.038	−0.418	0.881
	2	12.68	12.44	13.89	25.58	−0.235	0.224	−0.014
	3	15.96	15.50	19.74	38.99	−0.479	0.582	−0.276
8	0	9.648	9.647	9.638	9.109	0.000	−0.004	−0.245
	1	14.12	14.12	13.8	15.71	0.000	−0.265	−4.318
	2	17.77	17.77	17.34	26.81	−0.001	−0.405	−1.684
	3	21.23	21.21	22.14	38.83	−0.015	1.119	2.446

The complex pressure amplitude in the duct then will be of the form

$$p_{m,n} \propto \cos(k_y y)\cos(k_z z), \tag{6.32}$$

where $k_y = n\pi/(b-a) = (1/b)b\pi/(1-\sigma)$ and $k_z = m2\pi/2\pi r_a = (1/b)2m/(1+\sigma)$, where we have accounted for the fact that m and n in the labeling of the Bessel functions refer to the angular and radial coordinates. Accordingly, $k_x = \sqrt{(\omega/c)^2 - k_{m,n}^2}$, where

$$k_{m,n} = k_y^2 + k_z^2 = (1/b)^2 \left[\frac{4m^2}{(1+\sigma)^2} + \frac{n\pi}{(1-\sigma)^2}\right]. \tag{6.33}$$

Comparison with Eq. 6.30 shows that the quantity within the bracket is the approximate value of $\beta_{m,n}^2$

$$\beta_{m,n}^2 \approx \frac{4m^2}{(1+\sigma)^2} + \frac{(n\pi)^2}{(1-\sigma)^2}. \tag{6.34}$$

For $n = 0$, which corresponds to a pressure field with no nodal circles in the pressure amplitude within the annulus, we have $\beta_{m0} \approx 2m/(1+\sigma)$. With $\sigma = 0.75$, this expression yields $\beta_{1,0} \approx 1.143$ and $\beta_{8,0} \approx 9.143$, which are within one percent of the values in the table. Similarly, with $\sigma = 0.5$ we get $\beta_{1,0} \approx 1.333$ and $\beta_{8,0} \approx 10.67$. The first of these is within two percent and the second within ten percent of the values in the tables.

For $n > m$ and relatively large σ, the simplified expression $\beta_{m,n} \approx n\pi/(1-\sigma)$ is a good approximation. For example, with $n = 3$ and $\sigma = 0.75$ this expression gives $\beta_{m,3} \approx 37.70$ which is in very good agreement with the tabulated results for $m = 1$ and $m = 2$.

6.2.2 Simple Experiment. Discussion

As we have seen, the sound wave in a duct with rigid walls can have many different forms. The simplest is the fundamental wave mode in which the sound pressure is uniform across the duct. It is the same field as in a plane wave in free field traveling in the x-direction. We can imagine the wave in the duct as being generated by an oscillating plane piston (approximated by loudspeaker) at the beginning of the duct. This wave will travel unattenuated along the duct at *all frequencies* if we neglect the visco-thermal effects. If the piston is simulated by the two loudspeakers in Fig. 6.4, the speakers have to be driven in phase to generate the plane wave.

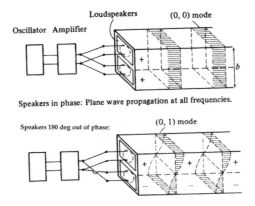

Figure 6.4: Top: If the two loudspeakers operate in phase (push-push), a plane wave will be generated. Bottom: If they are 180 degrees out of phase (push-pull), the (0,1) higher acoustic mode will propagate if the frequency exceeds the cut-on frequency $c/2D$, where c is the sound speed and D the duct width.

If the speakers are driven 180 degrees out of phase, however, so that one pushes when the other pulls, the average axial velocity amplitude in the duct will be zero and no plane wave is generated. There will still be a wave, but, unlike the plane wave, the sound pressure distribution now depends strongly on frequency. The wave components from the individual speakers travel out into the duct, one with a positive and the other with a negative sound pressure. If there were no phase shift between them, they would cancel each other. This is the case in the mid-plane of the duct where the sound pressure indeed will be zero. However, at a point in the duct not in the mid-plane between the speakers, there will be a source-to-receiver path difference so that the waves at the receiver are not completely out of phase. In fact, if the difference is half a wavelength, the waves will arrive in phase so that constructive interference results. We then get a wave that travels through the duct with a node at the mid-plane and, unlike the fundamental mode, is characterized by zero average oscillatory flow in the duct (the flows above and below the mid-plane are 180 degrees out of phase). It is referred to as a *higher order mode*.

For such a mode to propagate through the duct without attenuation, the wavelength must be short enough so that a path difference of half a wavelength can be obtained. This corresponds to a frequency above the cut-off frequency. At frequencies below the cut-off, however, the wavelength is so long that it is not possible to get a path difference of half a wavelength and constructive interference. The path length difference decreases with increasing distance from the source in the duct so that destructive interference will be more pronounced with increasing distance as we have seen from the exponential decay in Eq. 6.26.

The largest path difference is in the plane of the source, where it is the width D of the duct (from the top of one speaker to the bottom of the other). Then, if D is half a wavelength there will be constructive interference between the elementary wave from the top of one speaker and the bottom of the other. The condition for this 'cut-on' of the higher mode is $\lambda/2 = D$ and the corresponding frequency c/λ, $f_{01} = c/2D$ is the cut-on frequency of the (1,0)-mode; as we have done above, it is also called the 'cut-off' frequency, the choice depending on from what direction the frequency is approached, I suppose. At this frequency, the mode corresponds to a standing wave perpendicular to the duct axis. The label (1,0) indicates that there is one nodal plane perpendicular to the y-axis and to the z-axis. The plane wave is the (0,0) mode.

Below the cut-on frequency, the mode decays exponentially with distance from the source as mentioned above and shown in Eq. 6.26.

The arrangement in Fig. 6.4 is useful as a simple table top demonstration. To change the speakers from in-phase to out-of-phase operation simply involves switching the leads from the amplifier to the speakers, as indicated in the figure. In a particular experiment, the duct height was $D = 25$ cm corresponding to a cut-on frequency of 684 Hz. With the speakers operating out-of-phase, increasing the frequency through the cut-on value clearly produced a marked change in sound pressure emitted from the duct which could readily be observed. The duct can be said to act like a high-pass filter for the (0,1) mode.

Because of the wave decay below the cut-on frequency, the sound that radiates from the end of the duct is feeble. It is due to what is left of the evanescent wave when it reaches the end. It is also possible that a weak plane wave component may be present because of an unavoidable difference in the speakers so that their amplitudes are not exactly the same; the average velocity over the total source surface then is not exactly zero. In any event, if one of the speakers is turned off, a substantial increase in sound pressure is observed (because of the plane wave which is now generated by the remaining speaker).

Thus, in this demonstration, one speaker cancels the sound from the other, so that two speakers produces less sound than one. It demonstrates what is commonly referred to as active noise control in which sound is used to cancel sound. The term 'anti-sound' has been used to designate the contribution from the source that cancels out the primary sound.

If the frequency is increased above the cut-on frequency, a marked increase in the sound level from the end of the tube is observed because of the propagating (1,0)-mode.

The wave field in the duct can be thought of as a superposition of plane waves being reflected from the duct walls. At the cut-on frequency these waves are normal to the axis of the duct, but at a higher frequency the angle ϕ with the axis is given by $\sin\phi = (\lambda/2)/D$. There will still be a nodal mid-plane in and the mode is still referred to as a (1,0)-mode. The phase velocity of this mode will be the speed of the intersection point of a wave front with the boundary (or the axis) and this speed is $c/\sin\phi$, i.e., greater than the free field sound speed and the (0,0)-mode in the duct.

Thus, if a plane wave and a higher mode are both present in a duct, the resulting wave field will vary with position because of the difference in wave speeds.

Similar arguments show that if the wavelength is smaller than $D/2n$, where n is an integer, a mode, the (n,0)-mode, with n nodal planes and a cut-on frequency $f_{n,0} = nf_{1,0}$, can propagate and the wave field in the duct can be regarded as a superposition of plane waves which are reflected back and forth between the boundaries and traveling in a direction which makes an angle ϕ with the duct axis, where $\sin\phi = \lambda/(2nD)$. The phase velocity of a higher mode is always greater than the sound speed in free field and, like the angle ϕ, it is frequency dependent.

6.2.3 Sound Radiation into a Duct from a Piston

Piston in an End Wall

After having introduced higher modes through the experiment illustrated in Fig. 6.4 we consider now the radiation from a single piston source in the wall at the beginning of a duct. For details of the mathematical analysis we refer to Section A.3 and present here only a summary of the results.

As before, we let the x-axis be along the duct and place the source in a acoustically hard baffle wall at $x = 0$. By a piston source we shall mean a source with a specified distribution of the axial velocity across the duct. Harmonic time dependence will be assumed unless stated otherwise.

In the special case of a piston source with uniform velocity amplitude covering the entire duct area, only a plane wave will be generated. However, if the piston covers only a portion of the duct area or if the velocity distribution is non-uniform additional modes will be generated. At frequencies below the lowest cut-off frequency of the higher modes, these modes will decay exponentially with distance from the source, as discussed in the previous section, and sufficiently far from the source, the plane wave becomes dominant.

The coupling between the source and a particular mode depends on the degree of 'overlap' of the axial velocity distributions of the source and the mode. Quantitatively, the overlap is expressed in terms of the amplitude coefficient of the mode in a series expansion of the source distribution in terms of the duct modes. For example, in an acoustically hard duct, only the plane wave mode will have an average value of the velocity amplitude across the duct, and this average value must be the same as that of the source. If the average source velocity amplitude is zero, the amplitude of the plane wave will also be zero as was the case in the bottom example in Fig. 6.4.

A piston with an area A_p and a uniform velocity amplitude u_p will have an average velocity amplitude $u_0 = (A_p/A)u_p$, where A is the duct area, and this must equal the velocity in the plane wave mode. Thus, the complex pressure amplitude of the plane wave will be

$$p_0 = \rho c (A_p/A) u_p e^{ikx}, \tag{6.35}$$

where $k = \omega/c$.

The acoustic power carried by the plane wave component is $A|p_0 u_0|/2$. The corresponding radiation resistance r of the piston must be such that the same power is generated by the piston. This power is $A_p r |u_p|^2/2$ and we get

$$r = \rho c (A_p/A). \tag{6.36}$$

At frequencies below the first cut-off frequency of the higher modes, this is the only contribution to the radiation resistance. The higher modes are evanescent and contribute only a mass reactive component to the radiation impedance of the piston, as discussed below.

As the frequency increases, one higher mode after the other will be 'cut-on' and carry energy and thus contribute to the radiation resistance. Actually, if we neglect visco-thermal and other losses, linear acoustic theory indicates that if the amplitude of the piston velocity is independent of frequency, this resistance contribution goes to infinity as the frequency approaches a cut-off frequency. The direction of the fluid velocity oscillations in the corresponding mode is then nearly perpendicular to the axis of the duct, and in order to get an axial velocity component to match that of the source, a very high sound pressure amplitude will be required. This translates into a high radiation resistance as well as reactance in the vicinity of cut-off.

Beyond the cut-off frequency the higher mode involved will be cut-on to carry energy and contribute to the radiation resistance of the piston. The resistance starts out at infinity at cut-off and then decreases monotonically with frequency. This behavior is repeated for each mode. Normally, the amplitudes of the higher modes decrease with the mode number so that the fluctuation of the total resistance when more than three modes are cut on becomes small. In this high frequency limit, the modes in the duct combine to form a beam of radiation with a cross section equal to that of the piston and the specific resistance of the piston approaches ρc.

A higher mode also contributes a mass load on the piston. This can be seen from the formal solution, as demonstrated in Section A.3, but can be understood qualitatively also from kinetic energy considerations as follows.

If $|u'| = \sqrt{u_x'^2 + u_y'^2 + u_z'^2}$ is the magnitude of the velocity in a higher mode below cut-off, the kinetic energy is $(\rho/2) \int |u'|^2 dV$, where V is the volume of the tube. Since the wave field decays exponentially with distance, the integral will be finite and if we express the corresponding kinetic energy as $(1/2)M u_p^2$, we have defined the equivalent mass load M on the piston, where u_p is the velocity amplitude of the piston. The velocity u' is proportional to u_p and M will be independent of u_p.

For harmonic time dependence, the corresponding force amplitude on the piston is $-i\omega M u_p$, and the normalized reactance is

$$\chi = -i\omega M / A_p \rho c \equiv -ik\delta, \qquad (2.1.3)$$

where $k = \omega/c$. The quantity $\delta = M/A_p\rho$ can be thought of as the length of an air column which has the same mass reactance per unit area as the radiation reactance of the piston. A similar mass load also occurs at the end of an open duct or pipe and in that case δ is usually referred to as an 'end correction,' and this designation will be used here also.

The mass load contributed by a higher mode increases monotonically with frequency and theoretically goes to infinity at the cut-off frequency (for an infinitely long duct) and then decreases monotonically to zero when the sound field has become a beam. For a finite duct, the calculation of the radiation impedance must include the reflected waves from the end of the duct, for both propagating and evanescent waves.

The normalized radiation impedance of the piston at frequencies below the first cut-off frequency can be written

$$\zeta = p(0, \omega)/\rho c u_p = (A_p/A) - ik\delta, \qquad (6.37)$$

where, as before, $k = \omega/c$.

The calculation of δ requires knowledge of the higher order mode field, and such a calculation is done in Section A.3 at wavelengths much larger than the cross-sectional dimensions of the duct so the low-frequency limit value of δ can be used. Then δ depends only on the dimensions of the piston and the duct but not on frequency. In a more detailed analysis the frequency dependence of δ must be accounted for in accordance with the discussion above.

In Fig. 6.5 are shown the results of the calculations in Section A.3 of the low-frequency value of the end correction δ for square and circular pistons and it is normalized with respect to $\sqrt{A_p}$, where A_p is the piston area. We note that for a circular piston in a circular duct, or a square piston in a square duct, δ goes to zero as the piston area approaches the duct area, as expected. In this limit, as we have seen, only the plane wave mode is generated and there is only a resistive contribution to the impedance. In the other limit, when the piston area goes to zero, $\delta/\sqrt{A_p} \approx 0.48$. For a circular piston with a radius r_p this means that $\delta \approx 0.85 r_p$. This is consistent with the exact value $(8/3\pi) r_p$ for a circular piston in an infinite acoustically hard baffle in free field, discussed in Section 5.3.5.

In this context it is of interest to compare this end correction with that of a pulsating sphere of radius a. If the velocity amplitude of the surface of the sphere is u_p and we treat the fluid as incompressible, the radial velocity at a distance r will be $(a/r)^2 u_p$ and the kinetic energy $(\rho/2) \int_a^\infty u^2 4\pi r^2 \, dr = (\rho u_p^2/2)(4\pi a^2 a)$. The equivalent mass load per unit area of the sphere is then ρa which can be interpreted as that of an air layer of thickness $\delta = a$.

Piston in a Side Wall

See Appendix A.

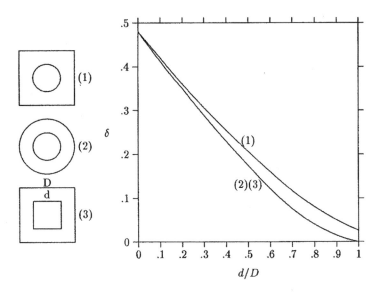

Figure 6.5: The mass end correction δ piston sources in tubes, plotted here as $\delta/\sqrt{A_p}$ vs d/D, where A_p is the piston area, d the piston width, and D, the duct width. The curves (2) and (3) are too close to be separable in the figure.

6.3 Lined Ducts

Sound propagation in lined ducts has become an important engineering problem and has been the subject of specialized texts and we shall limit the discussion here only to a review of the essentials.

The noise source involved is often the fan that drives air through the duct or it can be the ambient noise at one end of the duct. One way to achieve noise reduction is to line the duct with sound absorptive material.

We start with an approximate analysis of attenuation based on energy considerations. Thus, if the acoustic power through the duct at a location x is $\Pi(x)$ and the power absorbed per unit length by the boundary is assumed to be $\alpha\Pi(x)$, energy conservation requires that

$$d\Pi/dx = -\alpha\Pi(x), \tag{6.38}$$

which yields an exponential decay of the power, $\Pi(x) = \Pi(0)\exp(-\alpha x)$.

The assumption that the absorption per unit length is proportional to the power through the duct is justified if a couple of assumptions are made regarding the sound pressure field. If the sound pressure is assumed uniform across the duct and the wave impedance in the duct is assumed to be the same as in free field, the power in the duct will be $\Pi = Ap^2/\rho c$, where p is the rms value. Furthermore, if all the walls in the duct are treated with a locally reacting liner (see Chapter 4) with a normalized conductance (real part of the admittance) μ, the power absorbed per unit area of the duct wall is $p^2\mu/\rho c$ (see Eq. 4.44). If the treated perimeter of the duct wall is S, the corresponding absorption per unit length of the duct becomes $Sp^2\mu/\rho c$ which can be written $(S\mu/A)\Pi$. In other words, the coefficient in Eq. 6.38 is $\alpha = (S\mu/A)$. The attenuation in dB in a length x of the duct is then

$$10 \log[p^2(0)/p^2(x)] = 10 \log(e)(S\mu/A)\,x \approx 4.36(S\mu/A)\,x \quad \text{dB.} \tag{6.39}$$

On these assumptions, this expression for the attenuation is valid for any shape of the duct. For a circular duct of diameter D and with the entire perimeter treated, we have $S = \pi D$, $A = \pi D^2/4$ and $S/A = 4/D$, so that the attenuation becomes $\approx 17.4\mu(x/D)$ dB.

This is all very well and simple were it not for the assumptions made. Although they are normally valid at sufficiently low frequencies, the pressure distribution across the duct tends to become nonuniform with increasing frequency. Actually, for a higher order mode, the sound pressure at the boundary goes to zero with increasing frequency in much the same way as in free field reflection of a plane wave from a plane boundary. As grazing incidence is approached (angle of incidence going to $\pi/2$), the pressure reflection coefficient goes to -1, as can be seen from Eq. 4.50. This makes the sum of the incidence and reflected pressure at the boundary equal to zero.

This qualitative explanation is not applicable to the fundamental mode, however, since it cannot be described as the superposition of waves being reflected back and forth between the walls of the duct. The role of the boundary is now expressed in terms of its effect on the average compressibility of the air in the duct. At low frequencies, the pressure is almost uniform across the duct and the pumping in and out of the boundary results in an average complex compressibility, much like a spring in parallel with a dashpot damper. For a duct of width D_1, lined on one side where the normalized admittance is η, it is shown in Example 45 in Chapter 11 that the average complex compressibility in the channel is

$$\tilde{\kappa} = \kappa(1 + i\eta/kD_1). \tag{6.40}$$

It goes to the normal isentropic (real) value $\kappa = 1/\rho c^2$ as the frequency increases, $kD_1 >> 1$, and the corresponding attenuation goes to zero. The condition $kD_1 >> 1$ can be written $D_1/\lambda >> 1$ where λ is the (free field) wavelength. In other words, as the wavelength becomes much smaller than the width of the channel the presence of the liner will be 'felt' less and less by the wave in the channel and eventually approaches free field conditions. In this manner, also the attenuation of the fundamental mode approaches zero with increasing frequency. The interaction with the boundary can also be thought of as a relaxation effect with the complex compressibility expressed as $\tilde{\kappa} = \kappa(1 + i\eta/\omega\tau)$, where the relaxation time $\tau = D_1/c$ is the time of wave travel across the duct.

At the other end of the spectrum, with the frequency going to zero, the pressure typically becomes approximately uniform but the conductance μ typically goes to zero. The attenuation, being proportional to the product of μ and the mean square pressure at the boundary, then goes to zero; in other words, there will be essentially no absorption by the boundary in both the low- and high-frequency limits and the frequency dependence of the attenuation is expected to be bell shaped.

6.3.1 Attenuation Spectra

The qualitative discussion of attenuation above is indeed consistent with the results of a detailed mathematical analysis as demonstrated by the computed attenuation curves in Figs. 6.6 and 6.7.

Fig. 6.6 refers to a rectangular duct with one side much longer than the other and with one of the long sides lined with a uniform porous layer; Fig. 6.7 shows the attenuation in a fully lined circular duct.

These results were obtained by solving the wave equation subject to the boundary conditions at the walls of the duct. To illustrate the procedure, we consider the duct on the top left in Fig. 6.8. It refers to a rectangular duct with one side lined with a locally reacting liner. The results obtained can be used for the fundamental acoustic modes in any of the acoustically equivalent duct configurations shown in the figure.

The x-axis is placed along the length of the duct. The plane of the surface of the liner is at $y = D_1$ and the opposite unlined wall is at $y = 0$. The two other unlined walls are at $z = 0$ and $z = D_2$ at which the normal velocity is zero (acoustically hard walls). The wave field will be of the same form as in Eq. 6.19. With the walls placed at $y = 0$ and $z = 0$ the wave functions expressing the y- and z-dependence of the field are expressed by $\cos(k_y y)$ and $\cos(k_z z)$. The corresponding velocity fields, being proportional to the gradients of pressure, are then $\propto \sin(k_y y)$ and $\propto \sin(k_z z)$ and automatically satisfy the boundary conditions of being zero at the acoustically hard walls $y = 0$ and $z = 0$.

The factors k_y and k_z are determined by the boundary conditions at the walls at $y = D_1$ and $z = D_2$. With the wall at $z = D_2$ being acoustically hard, we must have $\sin(k_z D_2) = 0$ which means $k_z = n\pi/D_2$, where n is an integer. At the lined wall at $y = D_1$ the admittance is given and if its normalized value is denoted η, the boundary condition requires that $u_y/p = 1/(\rho c \eta)$. From the momentum equation $(-i\omega\rho)u_y = -\partial p/\partial y$, it follows that $u_y = |p|(ik_y/\omega\rho)\sin(k_y D_1)$, where we have used $p = |p|\cos(k_y y)$, and the boundary condition then imposes the following condition on k_y,

$$(k_y D_1)\tan(k_y D_1) = -ikD_1\eta, \tag{6.41}$$

where $k = \omega/c$. The admittance is complex, and this equation for k_y generally has to be solved numerically. At low frequencies, such that $kD_1 \ll 1$, the left-hand side can be approximated by $(k_y D_1)^2$, in which case

$$(k_y D_1)^2 \approx -ikD_1\eta. \tag{6.42}$$

It follows from the wave equation that $k_x^2 + k_y^2 + k_z^2 = (\omega/c)^2$, i.e.,

$$k_x = \sqrt{(\omega/c)^2 - k_y^2 - k_z^2}. \tag{6.43}$$

With $k_z = n\pi/D_2$, the mode that has a uniform pressure in the z-direction corresponds to $n = 0$ which yields

$$k_x = \sqrt{(\omega/c)^2 - k_y^2}. \tag{6.44}$$

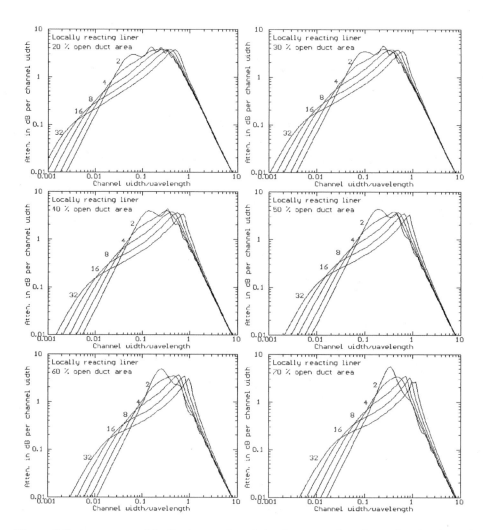

Figure 6.6: Attenuation of the fundamental mode (in dB per unit length equal to the channel width D_1) in a rectangular duct with one side lined with a locally reacting rigid porous layer with a total normalized flow resistance Θ (2, 4,..32). Liner thickness: d. Fraction open area $D_1/(D_1+d) = 20$ to 70%. The results can be used with two opposite walls lined with identical liners if the channel width (distance between liners) is $2D_1$.

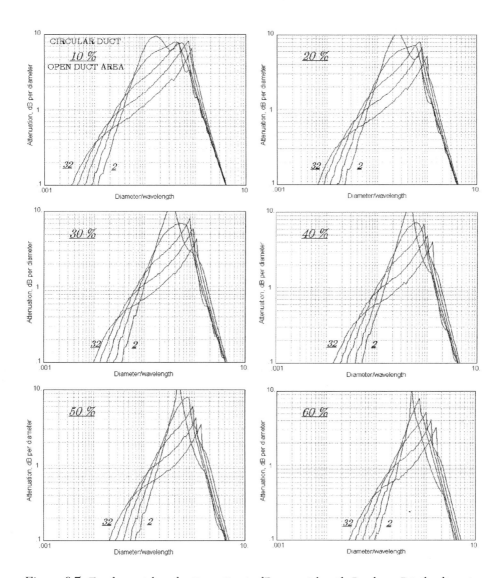

Figure 6.7: Fundamental mode attenuation in dB per unit length D, where D is the diameter of the air channel of a *circular* duct lined with a locally reacting porous layer. Frequency variable is D/λ, where λ is the wavelength. Total normalized flow resistance of the liner: 2, 4, 8, 16, and 32, as indicated. Open area fractions of duct: 10, 20, 30, 40, 50, and 60 percent, as shown, corresponding to liner thicknesses d such that d/D is 1.08, 0,62, 0.41, 0,29, 0.21, and 0.15, respectively.

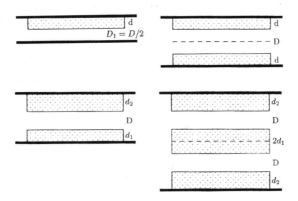

Figure 6.8: Equivalent duct configuration which yields the same attenuation for the fundamental acoustic mode.

With $k_x = k_r + i k_i$, the x-dependence of the complex pressure amplitude is given by

$$p(x, \omega) = p(0, \omega) e^{-k_i x} e^{i k_r x}. \tag{6.45}$$

If the time dependence of the pressure at $x = 0$ is $|p| \cos(\omega t)$, we get

$$p(x, t) = |p| e^{-k_i x} \cos(\omega t - k_r x). \tag{6.46}$$

In other words, the imaginary part of the propagation constant determines the exponential decay of the sound pressure so that the attenuation in dB in a distance x is

$$20 \log(|p(0)|/|p(x)|) = 20 \log(e) k_i x \approx 8.72 \, k_i x. \tag{6.47}$$

The real part determines the phase velocity $c_p = \omega/k_r$. In a coordinate frame moving with this velocity, the phase $(\omega t - k_r x)$ remains constant.

In the low-frequency approximation, with k_y given by Eq. 6.42, we get

$$k_x \approx \sqrt{(\omega/c)^2 + i \eta k / D_1} = (\omega/c) \sqrt{1 + i \eta / k D_1}, \tag{6.48}$$

where, as before, $k = (\omega/c)$. For a porous liner, the frequency dependence of η is such that at sufficiently low frequencies $\eta/k D_1 << 1$. In that case, expanding the radical to first order in η, the corresponding approximation is

$$k_x \approx (\omega/c) + (1/2) i \eta / D_1. \tag{6.49}$$

It is left as a problem to show that the attenuation obtained in this approximation is the same as that in Eq. 6.39.

Wave Impedance of the Duct

One of the approximations made in the derivation of Eq. 6.39 is that the wave impedance of the duct is the same as in free field, i.e., ρc. We are now in a position to determine the actual value of the wave impedance p/u_x, where p and u_x are the complex amplitudes of pressure and axial velocity in a traveling wave. With

$$u_x = (1/i\omega\rho)\partial p/\partial x = pk_x/\omega\rho \qquad (6.50)$$

the normalized value for the wave impedance and admittance becomes

$$\zeta_w = \frac{1}{\eta_w} = \frac{1}{\rho c}\frac{p}{u_x} = k/k_x, \qquad (6.51)$$

where $k = \omega/c$. With reference to the discussion of Eq. 6.48, we note that for a porous liner $k_x \approx k$ at low frequencies which is consistent with the assumption contained in Eq. 6.39.

Since the imaginary part of k_x is positive, it follows that the reactive part of the wave impedance is negative, i.e., mass-like. Physically, this is due to a transverse component of the fluid velocity in the duct; for a given axial velocity amplitude, the kinetic energy per unit length when expressed as $\rho_e u_x^2/2$ requires that the equivalent inertial mass density ρ_e will be larger than ρ when the kinetic energy contribution from the transverse velocity is accounted for.

6.3.2 Problem

1. **Low-frequency approximation of attenuation**

   Show that the low-frequency attenuation resulting from Eq. 6.49 is consistent with the result in Eq. 6.39

Chapter 7

Flow-induced Sound and Instabilities

7.1 Introduction

The term aero-acoustics has come to mean the part of acoustics which deals with problems encountered in aerodynamics, for example, sound generation by flow and by devices such as fans and compressors. Sometimes the term 'aero-thermo-acoustics' is used to indicate that also acoustical problems related to heat release and other thermal effects are included.

The second part in the title of this chapter, instabilities, indicates that we are going to treat problems which result from the intrinsic instability of fluid flow. It means that oscillations (and sound) can be produced without any external oscillatory driving force, but merely as a result of the flow breaking into spontaneous oscillations. The sound or vibrations produced as a result can feed back on the flow and promote or stimulate the instability so that the amplitude will grow exponentially with time until some damping mechanism eventually limits the amplitude. Such vibrations often cause mechanical (acoustic fatigue) failure.

In regard to sound generation, in general, we have considered so far mainly vibrating boundaries or piston sources producing density fluctuations and corresponding pressure fluctuations. In the case of a pulsating sphere, for example, the acoustic flow source strength, the rate of mass transfer to the surrounding fluid, was denoted q_f and the corresponding acoustic source by $q = \dot{q}_f$ (or $q(\omega) = -i\omega q_f(\omega)$ for harmonic time dependence). The oscillating acoustically compact sphere was found to be equivalent to a point force acting on the fluid. Corresponding continuous distributions were denoted Q and F per unit volume as 'drivers' in the acoustic wave equation (5.36). It was pointed out that a heat transfer H per unit volume is equivalent to a mass flow source $Q_f = (\gamma - 1)H/c^2$ as far as sound generation is concerned. Through this equivalence, sound generation by sparks, combustion, lightning, etc., can be analyzed mathematically. The acoustic source Q and the force F densities were found to be monopole and dipole densities, respectively.

An acoustic mass flow source can be obtained directly, without a pulsating sphere, by modulation of flow from an external source. In a siren, puffs of air are produced when flow is forced through the holes in a disk which are periodically blocked by a rotating disk with a similar hole pattern. In speech, the sound is produced by the flow from the lungs through the vocal chords and the vocal tract; in this case the modulation of the flow is caused by moving boundaries.

Even in the absence of moving boundaries, sound can be produced by flow. The reason is that fluid flow is generally unstable which results in the time dependence of the velocity required for sound generation. The frequency of oscillation is then not imposed by a moving boundary as in the siren or vocal tract but is a characteristic of the flow itself; the frequency is then typically proportional to the flow velocity.

Then, if a flow generated sound source is of the monopole type, with the sound pressure being proportional to $\dot{q}_f$, it will be proportional to ωU where U is the flow velocity in a frequency band centered at ω. Then, with ω being proportional to U, it follows that the sound pressure will be proportional to U^2 and the radiated power to U^4. Similarly, we find in analogous manner that if the source is of the dipole type, the power will be proportional to U^6 and if it is of the quadrupole type, to U^8. As we shall see, the latter applies to a region of uniform turbulence.

If the flow interacts with acoustic or mechanical resonators, the velocity dependence of the acoustic power output given in the previous paragraph can be drastically modified as indicated earlier.

In preparation for our study of these problems, we present some notes on the instability of a vortex sheet and the interaction of fluid flow with a solid boundary including the characteristics of drag and wake formation behind a blunt body. On this basis, a classification of flow-induced instabilities is proposed and examples are discussed. For example, both mechanically and acoustically stimulated vortex streets behind cylinders are considered with application to heat exchangers, for example. Similarly, orifice and pipe tones are analyzed and experimental data are presented and discussed. An example of the often encountered whistle produced by flow through perforated plates in industrial dryers is given.

Valve and seal instabilities in fluid machinery are potential sources of mechanical failure and they are given due attention. The same applies to the problems of flow excited side-branch resonators and conduits in ducts. A unique feature is the mode-coupling between resonator and duct modes and between different resonator modes. Another aspect of the problem deals with the characteristics of slanted side branch tubes and the effect of flow direction on the excitation of pipe modes.

7.2 Fluid-Solid Body Interaction

7.2.1 Boundary Layers and Drag

The potential inviscid flow about a sphere is completely symmetrical on the upstream and downstream sides, as indicated in Fig. 7.1, and the same applies to the pressure distribution over the surface of the sphere. As a result, there is no net change in the momentum flux in the fluid resulting from the interaction and no drag force on the sphere. Although the idealization of an inviscid fluid leads to flow fields in relatively good agreement with experiments in regions away from boundaries, the prediction of zero drag force is unrealistic.

The assumption of inviscid fluids leads to other unsatisfactory results. For example, the computed flow velocity frequently becomes infinite, as is the case at sharp corners of a solid object or at the sharp edge of an orifice. Inclusion of viscosity eliminates these mathematical deficiencies, however. In macroscopic description of fluid motion, the tangential flow velocity at a rigid boundary will be zero but it increases to the 'free stream' value U_0 (which in most cases can be taken to be the value obtained for an inviscid fluid) in a (thin) *viscous boundary layer*. It can be thought of also as a layer of vorticity with a vorticity $\partial U / \partial y$, where y is the coordinate perpendicular to the boundary.

The force on an object in laminar viscous flow with a constant unperturbed velocity U_0, depends on U_0, the area A of the object that obstructs the flow, and the coefficient of shear viscosity of the fluid. The viscous stress (force per unit area) in shear flow is given by $\mu \partial u_x / \partial y \approx \mu U_0 / D$, where the velocity gradient has been expressed as U_0 / D, where D is a characteristic dimension of the object. The 'dynamic' stress (momentum flux) in the fluid is of the order of ρU_0^2. The ratio of this and the viscous stress is of the order of

$$R = \rho U_0 D / \mu = D U_0 / \nu, \qquad (7.1)$$

which is called the *Reynolds number*. The Quantity $\nu = \mu / \rho$ is the coefficient of *kinematic viscosity*. It is, loosely speaking, and in resonator terminology, a kind of Q-value of the fluid with the inverse being a damping factor.

The calculation of the drag force on an abject in fluid flow is complex and normally has to be done numerically. From dimensional considerations it follows, however, that the dependence of the force on the variables involved will be of the form

$$f = C(R, M)(\rho U_0^2 / 2) A, \qquad (7.2)$$

where U_0 is the (free stream) fluid velocity and A a characteristic area of the object (usually the projection normal to the flow). C is then a dimensionless constant called the *drag coefficient*, which generally is a function of all variables involved; the combination of the variables must be dimensionless in order for C to be dimensionless. Thus, the variable which contains the viscosity is the Reynolds number $R = L U_0 \rho / \mu$, where L is the characteristic length of the object. For a compressible fluid, also the compressibility enters and one way of introducing it in dimensionless form is through the Mach number $M = U_0 / c$, where c is the sound speed. For small Mach numbers, it is often a good approximation to treat the flow as incompressible. The quantity $\rho U_0^2 / 2$ is the dynamic pressure of the flow and $P_0 + \rho U_0^2 / 2$, the stagnation pressure (P_0 is the static pressure in the free stream).

Eq. 7.2 can be thought of as a scaling law, and with the Reynolds number kept constant, experimental results on a model in air can be used for the prediction of the interaction force on a body in water or some other fluid or in air at some other condition (pressure and temperature) than in the test.

Sphere. Stokes Law

To calculate the viscous drag force on a sphere in an incompressible fluid, we have to solve the Navier Stokes equation, which, for incompressible flow ($\vec{\nabla} \cdot \vec{U} = 0$), becomes

$$\rho \frac{D\vec{U}}{Dt} = -\text{grad } P + \mu \nabla^2 \vec{U} \qquad (7.3)$$

subject to the condition that the velocity be zero on the surface of the sphere and equal to the free stream velocity at infinity. The quantity μ is the coefficient of viscosity. The solution is complicated and beyond the present scope. Instead, we shall make an estimate of the force as follows.

The last term in Eq. 7.3 is the viscous force per unit volume. The radius of the sphere is the characteristic length and the order of magnitude of this force is $\propto \mu U/a^2$. The characteristic volume is the volume of the sphere $\propto a^3$, and the total drag force on the sphere will be $f \propto \mu a U$. For sufficiently small velocities corresponding to Reynolds number less than 1, the constant of proportionality can be shown to be 6π resulting in

$$\boxed{\begin{array}{c} \textit{Viscous drag force on a sphere} \\ f = 6\pi\mu a U \end{array}} \qquad (7.4)$$

[mu: Coefficient of shear viscosity. U: Free stream velocity. a: Sphere radius].

It is often called the *Stokes force* on the sphere. It may be familiar to some readers from elementary physics laboratory in the Millikan oil drop experiment for the determination of the electron charge.

It may seem surprising at first that the drag force is proportional to the *radius* of the sphere rather than to the surface *area*, since the viscous force acts along the surface. It should be realized, however, that the force is not only proportional to the area but also to the velocity gradient. If we normalize the velocity with respect to the free stream velocity U_0 and put $U' = U/U_0$ and $r' = r/a$, the gradient is $dU/dr = (U_0/a)dU'/dr'$. Multiplication by the surface area thus yields a drag force proportional to the radius.

In terms of the Reynolds number $R = DU/\nu$, based on the diameter $D = 2a$ of the sphere, where $\nu = \mu/\rho$ is the kinematic coefficient of viscosity, the corresponding drag coefficient (Eq. 7.2) is

$$C(R) = 24/R \qquad \text{(sphere, viscous drag, } R < 1\text{),} \qquad (7.5)$$

where $R = U_0 D/\nu$, based on the diameter $D = 2a$ of the sphere.

It is an experimental fact that above a certain Reynolds number, $R \approx 100$, the flow starts to separate from the boundary of the sphere, as indicated schematically in Fig. 7.1 and starts out as a (circular) vortex sheet behind the sphere. This vortex sheet is unstable, as will be demonstrated shortly, and the flow becomes turbulent, eventually filling up the wake behind the body, as indicated schematically in the figure.

Flow separation in (b) in Fig. 7.1 results in an increase in the drag force f on the sphere, and the velocity dependence of f goes from the Stokes law $f \propto U$ to approximately $f \propto U^2$ for large values of R. The corresponding dependence of the

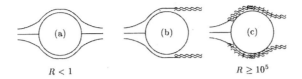

$R < 1$ $\qquad\qquad\qquad R \geq 10^5$

Figure 7.1: Flow interaction with a sphere. (a). Laminar flow. (b). Laminar boundary layer and flow separation. (c). Turbulent boundary layer.

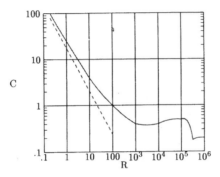

Figure 7.2: Drag coefficient C of a sphere versus the Reynolds number R. For comparison is shown (dashed line) the coefficient which corresponds to viscous drag, $C = 24/R$.

drag coefficient on the Reynolds number goes from $C(R) \propto 1/R$ to $C(R) \approx$ constant as shown in Fig. 7.2.

For values of R between 2×10^5 to 5×10^5, an interesting phenomenon occurs, sometimes called the *drag crisis*. The drag coefficient drops sharply from about 0.4 to about 0.15, and this drop occurs over such a small range of velocity that the actual drag force decreases with increasing velocity, which can be a source of instability. This effect has been found to be a result of a transition in the boundary layer on the upstream side of the sphere from laminar to turbulent, as indicated in (c) in Fig. 7.1. This, in turn, moves the flow separation point in the downstream direction, and the cross-sectional area of the wake and the drag force behind the sphere are both reduced. A further increase of R beyond this range restores the drag coefficient to an approximately constant value of 0.2 with the corresponding drag force being proportional to U^2.

The flow can be tripped by an irregularity in the surface of the sphere to make the boundary turbulent at lower Reynolds number than that given above for a smooth surface. This results in a reduced drag. A familiar application of this principle is the golf ball provided with dimples for the purpose of drag reduction.

7.2.2 Model of a Porous Material; Lattice of Spheres

The acoustic interaction with a sphere in the presence of a mean flow leads to two resistive components, one due to the viscous drag on the sphere and the other due to the acoustic modulation of the turbulent wake. If the velocity amplitude of the incident wave is u, the viscous contribution to the acoustic drag force amplitude is $6\pi\mu a u$. The acoustic modulation of the turbulent drag force is $C\rho\pi a^2[(U_0 + u)^2 - U_0^2] \approx C2\rho U_0 u\pi a^2$, where C is the drag coefficient.

Thus, if the cubical cell of a lattice of spheres has the volume L^3, there will be $N = 1/L^3$ spheres per unit volume and if the spheres are sufficiently far apart so that the expression for the drag force on a sphere in the lattice can be taken to be the same as for a single sphere in free field, the flow resistance per unit length in this model of a 'porous material' becomes

$$r = N(6\pi\mu a + C2\rho U_0 \pi a^2) = 2\rho U_0 \pi (a^2/L^3)(C + 3/R), \qquad (7.6)$$

where $R = U_0 a/\nu$ is the Reynolds number. The first term, the drag coefficient C, is of the order of unity in the turbulent regime.

Fig. 7.2 shows that the turbulent regime is not fully developed unless R exceeds 1000, approximately, and this means that when turbulent wakes are present, the acoustic flow resistance is dominated by the sound-flow interaction, represented by the first term in Eq. 7.6.

The induced mass of the oscillatory flow about a sphere is known to equal half of the mass of the fluid displaced by the sphere. Thus, the induced mass per unit volume of the lattice of spheres becomes $\rho_i = N(2\pi a^3/3)\rho$. The equivalent mass density is $\rho_e = \rho + \rho_i = (1 + G)\rho$, where the induced mass factor is $G = \rho_i/\rho$. The structure factor (see Chapter 4) $\Gamma = \rho_e/\rho = 1 + G$ becomes

$$\Gamma = 1 + \frac{2\pi}{3}(a/L)^3. \qquad (7.7)$$

This model of a porous material can be used to calculate approximately the propagation constant for sound in rain, fog, or suspension of particles. As an industrial application, the measurement of the attenuation has been used as a means of monitoring the density of coal powder in furnaces.

7.3 Flow Noise

7.3.1 Sound from Flow-Solid Body Interaction

The second sound generator in Eq. 5.9 refers to a time dependent force on the fluid. For a concentrated force $f(t)$, we found that the sound pressure a distance r from the source could be expressed as

$$p(r, t) = (1/4\pi rc)\,\dot{f}(t - r/c)\,\cos\phi, \qquad (7.8)$$

where ϕ is the angular position of the observation point as measured from the direction of the force.

Of particular interest here are flow-induced force fluctuations. At sufficiently high flow speeds, the *drag coefficient*, typically is of the order of unity. For air, the kinematic viscosity is $\nu \approx 0.15$ CGS at 1 atm and 70°F , and for water $\nu \approx 0.01$ CGS.

If the flow is turbulent, the velocity U is not constant. The time dependence is normally a result of irregularities in the flow ('eddies'). A model of a 'frozen' pattern of flow irregularities being convected by the mean flow is often approximately valid. The spatial variation of the irregularities can be represented by a spectral distribution of eddies.

If the velocity fluctuation caused by an eddy is U', the variation of the interaction force on the body will be $f_d = C\rho/2[(U + U')^2 - U^2] \approx C\rho U'U$. If the eddy size is Λ, the frequency of the fluctuation will be $\nu = U/\Lambda$ and $\dot{f}$ in Eq. 7.8 will be $\dot{f} \propto 2\pi\rho U'U^2/\Lambda$. Normally, the velocity fluctuation is proportional to the mean flow velocity, $U' = \beta U$, and, with $C \approx 1$, the expression for the radiated sound pressure in Eq. 7.8 will be of the form

$$p(r, t) \propto \frac{C A \beta}{4\pi r c \Lambda} (\rho U^3/2) = \frac{C A \beta}{4\pi r \Lambda} M^3 \frac{\gamma P}{2}, \qquad (7.9)$$

where C, the drag coefficient, is of the order of unity, $M = U/c$ is the Mach number of the mean flow, $\beta = U'/U$, Λ, the eddy size, and A, the projected area of the body. In the last step in this equation, we have expressed the speed of sound as $c = \sqrt{\gamma P/\rho}$, i.e., $\rho c^2 = \gamma P$, where P is the static pressure and γ the specific heat ratio (≈ 1.4 for air). In this manner, the sound pressure is expressed in terms of the static pressure and the Mach number of the mean flow. With P_0 being the ambient pressure, the quantity $P_0 + \rho U^2/2$ is the stagnation pressure of the flow. With $M = U/c$, we note that the radiated sound pressure is proportional to the *third power* of the mean flow velocity. The corresponding radiated acoustic power is then proportional to the *sixth power* of the velocity.

The Aeolian Tone

Even if the incident flow is uniform and steady, the interaction of this flow with a solid body can lead to a time dependent force on the object and a related emission of sound. Such is the case in the interaction of flow with a cylinder. The wake behind the cylinder turns out to be oscillatory over a wide range of flow velocities (Kármán vortex street) and this in turn results in sound emission, the Aeolian tone with a corresponding fluctuating force on the cylinder and sound emission. This phenomenon is discussed in some detail in Chapter 10.

Sound Generation by a Fan

It is the relative motion of the fluid and the solid body which is relevant as far as the interaction force is concerned and in a device such as a fan this relative motion is dominated by the speed of the fan blades. In this case, fluid motion is induced so that both the solid body and the fluid are in motion. This is such an important example of sound generation by flow-solid body interaction that the entire book could have been devoted to this subject. A brief account is presented in Chapter 8.

7.3.2 Noise from Turbulence

The generator of sound in a region of turbulence with no flow injection of mass flow or momentum (no external force) can have no net monopole or dipole source strength.

This does not mean that local pressure fluctuations are absent throughout the turbulent flow; it merely means that the *average* over the entire region will be zero. A local pressure fluctuation over a region of size a ('eddy' size) in the turbulent flow will be of the order of $\rho U'^2$, where, as before, U' is the velocity perturbation. Actually, in this case, we shall assume that there is no mean flow, so U' is the actual local velocity.[1] We recall that a pulsating sphere with a pressure $p(a)$ at the surface produced a pressure field $p(r) = (a/r)p(a)\exp(ikr)$ and, similarly, a local pressure fluctuation in the flow can be represented by a local monopole which will generate an elementary sound wave contribution with a pressure of the form $p_0 \propto (a/r)\rho U'^2$ (see Eq. 5.17).

We consider here a control volume of the flow large enough to include a large number of eddies with a characteristic size a assumed small compared to the wavelength of the radiated sound. If a fluctuation in density or momentum is positive at one location, there will be a corresponding negative fluctuation in the vicinity, consistent with the observation that there is no net mass flow into the control region. The time dependence is random but if we consider a small frequency band at the frequency ω and a corresponding period T, the period will be of the order of a/U', where U' now stands for the velocity fluctuation in that frequency band. Were it not for the spatial separation of these adjacent pressure pulsations and the corresponding difference in time of wave travel to the point of observation, the pressure waves would cancel each other. The travel time difference is of the order of a/c and if we consider the harmonic component of the fluctuation at the frequency ω, the sum of the pressure contributions will be that of a dipole with the pressure $(ka\cos\phi)p_0$ (see Eq. 5.43) where p_0 is the pressure from the single pressure fluctuation (monopole) and $ka = \omega/c$, which is proportional to U'.

Since the dipole is equivalent to a force distribution as discussed in Section 5.3.1 and since there is no net force on the region considered, the net dipole moment must be zero. Thus, for each elementary dipole there will be one equal and opposite with a separation of the order of a, the eddy size. The combination of the two yields a quadrupole, as shown in Section 5.3.4, with a pressure field obtained from the dipole field by multiplying by the factor $\propto ka\cos\phi$. Following the arguments in that section, the far field pressure from the quadrupole pressure field will then be $\propto (U'/c)^2 (a/r)\rho U'^2$.

We no longer can proceed in the same manner as for the monopoles and dipoles to require the quadrupoles to occur in pairs, since we have no stipulation similar to those of zero mass and momentum transfer to the turbulent region. Thus, we conclude that the lowest order equivalent multipole of the turbulent region will be a quadrupole.

The intensity from an elementary quadrupole volume element in the flow becomes

$$I_4 \propto (p_4^2/\rho c) = (1/\rho c)\, M'^4\, (a/r)^2\, (\rho U'^2)^2. \tag{7.10}$$

---

[1] If the turbulent region has a mean velocity, our discussion refers to a frame of reference moving with the same velocity.

After multiplication by $4\pi r^2$, we obtain the corresponding expression for the radiated power

$$W_4 \propto M'^5 \left(\rho U^3/2\right) a^2) \propto U'^8. \tag{7.11}$$

The second factor $\rho U'^3 a^2/2 = (\rho U'^2 a^3/2)(U'/a)$ can be interpreted as the average rate of building the kinetic energy of an eddy, $\propto \rho U'^2 a^3/2$, a/U' being the period. We can also interpret this term as the power carried by the kinetic energy flux $\rho U'^3/3$ through an area occupied by one eddy.

The first factor in Eq. 7.11, M'^5 can then be interpreted as the fraction of the kinetic energy of the eddy that is converted into sound ('efficiency' of conversion). The acoustic power is proportional to the *eighth* power of the velocity fluctuation. Assuming that the different eddies in the turbulent region are uncorrelated, their individual power contributions to the sound field add and the total power obtained by an integration over the turbulent region will be proportional to U'^8. This velocity dependence was first obtained by Lighthill from a detailed analysis based on the fluid equations in which the Reynolds stress was included in the momentum equation; it is a seminal contribution to this field.[2]

An interesting analogy to this 8th power law involves the electromagnetic (heat) radiation from a black body. It is known from Planck's radiation law that the total power of electromagnetic radiation from the thermal molecular motion is proportional to T^4, where T is the absolute temperature. The temperature, in turn, is a measure of the average molecular kinetic energy of the thermal motion, expressed by $m\langle u^2 \rangle = kT/2$ per degree of freedom, where k is the Boltzmann constant. Consequently T^4 can be expressed as being proportional to u^8, where u is the rms value of the molecular velocity. In this sense, the black body radiation, like sound from turbulence, is also an 8th power law.

7.3.3 Jet Noise

Consider a jet being discharged from a nozzle at speed U. It carries a kinetic energy $A\rho U^3/2$ per second, where A is the area of the nozzle. The velocity fluctuations U' in the jet will be proportional to the mean velocity U, U'/U typically being of the order of a few percent. Thus, on the basis of the radiation from a single eddy (quadrupole) in Eq. 7.11, the acoustic power radiated by the jet, as a first approximation, is expected to be of the form

$$\Pi = C\,M^5\,(A\rho U^3/2) \propto U^8 \quad \text{(subsonic)}, \tag{7.12}$$

where C is a constant. For a circular subsonic air jet at $70°\text{F}$, the constant has been found experimentally to be $C \approx 10^{-4}$.

Jet Noise Spectrum

Measurements of the frequency spectrum of the noise from circular subsonic jets indicate that the average spectrum density (over all directions of radiation) can be

---

[2]M. J. Lighthill, Proc. Roy. Soc. (London) A222, 1 (1954).

expressed as

$$E(f)/E(f_0) = F(f/f_0), \qquad (7.13)$$

where $E(f_0)$ is the peak value of the spectrum density and f_0 a characteristic frequency given by

$$f_0 \approx 0.15U/D. \qquad (7.14)$$

As shown in Eq. 7.16, $E(f_0) \approx \Pi/3.1$, where Π is the total acoustic power from the jet.

On the basis of measured jet noise spectra[3] we propose, with $\xi = f/f_0$, the following empirical (smoothened) spectrum function[4]

$$\boxed{\begin{array}{c} \textit{Empirical noise spectrum of subsonic jet (Fig. 7.3)} \\ (E(f)/E(f_0) \equiv F(\xi) = \dfrac{9^3 \xi^2}{(5+4\xi^{1.5})^3} \end{array}} \qquad (7.15)$$

[f: Frequency. f_0: See Eq. 7.14. U: Flow speed. d: Nozzle diameter. $E(f)$: Spectrum density. $E(f_0)$: See Eqs. 7.13 and 7.16. Total power and $E(f_0)$: See Eq. 7.16].

The total radiated power is obtained from

$$\Pi = \int_0^\infty E(f)df \equiv E_0 f_0 \int_0^\infty F(\xi)d\xi \approx 3.1\, E_0 f_0. \qquad (7.16)$$

The last step is a result of a numerical integration based on the spectrum function in Eqs. 7.13 and 9.8. With W being proportional to U^8, as given in Eq. 7.12 and with $f_0 \propto U$, it follows that the maximum spectrum density depends on the velocity as $E(f_0) \propto U^7$. Furthermore, the spectral function 9.8 shows that at low frequencies, $f << f_0$, the Mach number dependence of the spectrum density is $\approx M^5$ and for $f >> f_0$, it is $\approx M^{9.5}$. At the peak frequency, the dependence is M^7 and for the overall power it is $\propto M^8$.

In other words, the 8th power law applies only to the *total* power and generally not to the power in a finite frequency band. For example, in an experimental study of the spectra of fricative speech sounds, a Mach number dependence of $\approx M^5$ has been reported. In this experiment, however, the peak frequency f_0 was very high, and the analyzing equipment covered only the portion of the spectrum below f_0. Under these conditions, the experimental result is consistent with our predicted Mach number dependence.

Jet spectrum density functions are shown in Fig. 7.3 for a subsonic air jet for Mach numbers $M = 1$, 0.9, and 0.8. The frequency variable is normalized with respect to frequency at $f_0 = 0.15c/D$ at the peak of the $M = 1$ spectrum and its spectrum density peak value $E_0 = W/(3.1\,f_0)$ (see Eq. 7.16) is used for the normalization of the spectrum densities.

---

[3]H.E. von Gierke, *Handbook of Noise Control*, edited by C.M. Harris (McGraw-Hill Book Company, Inc., New York, 1957) Chapter 33, p 35.

[4]Uno Ingard, *Attenuation and regeneration of sound in jet diffusers*, JASA, Vol. 31, pp 1202-1212, 1959.

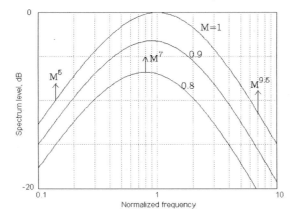

Figure 7.3: The power spectrum function $E(f)/E_0$ of the noise from a subsonic air jet vs the normalized frequency f/f_0 for Mach numbers 1, 0.9, and 0.8 of the jet. (c: sound speed, D: nozzle diameter.) $f_0 = 0.15c/D$ is the frequency of the maximum at the spectrum peak for $M = 0$ and $E_0 = \Pi/f_0$ is the corresponding maximum spectrum density and Π the total power output for $M = 1$. (Eq. 9.8)

The shapes of the octave band and one-third octave band spectra will be altered to some extent because the band width increases with frequency. For details, we refer to Problem 5.

As the Mach number decreases, the spectra are displaced toward lower frequencies as the peak value decreases, as shown. As a result, the curves come closer together at low frequencies and further apart at high. The corresponding Mach number dependencies, M^5 and $M^{9.5}$, in these regions are indicated together with the M^7 dependence of the peak spectrum density. For example, a decrease from $M = 1$ to $M = 0.8$ leads to reductions in the spectrum density levels in these regions by 4.8, 9.2, and 6.8 dB, respectively, as the overall power is reduced by 7.8 dB, corresponding to its M^8-dependence.

Because of the strong velocity dependence of the emitted noise from a jet, a reduction of the exit velocity by means of a diffuser would be an effective means of noise reduction if flow separation in the diffuser could be avoided. For a given mass flow, this approach reduces the thrust of the jet and cannot be used in propulsion.

The fact that the frequency of the peak value of the spectrum depends on the diameter of the jet has been used in order to shift the jet noise from low to high frequencies. The motivation for such a shift is that low frequencies attenuate less than the highs in propagation through the atmosphere, as discussed in Chapter 9. Multitube nozzles have been used on jet engines to accomplish this frequency shift. Actually, for ground-based jet engines, such a reduction can be achieved by means of a lossy diffuser such as a perforated cone which disperses the flow to reduce the average exit velocity. Such basket diffusers have been used in ground run-up tests of aircraft jet engines.

Comment. The 'acoustic efficiency' of a subsonic jet expressed by the factor $\approx$ $10^{-4}M^5$, i.e., the fraction of the kinetic energy of the jet converted into sound, cannot be expected to be valid for arbitrarily large values of the Mach number. After all, the acoustic power cannot exceed the mechanical power (which is proportional to M^3) of a jet and there is an obvious upper limit to M in the M^8 law given approximately by $10^{-4}M^5 = 1$, i.e., $M \approx 6.3$. Actually, when M exceeds 1, experiments show that the Mach number dependence of the efficiency gradually approaches a constant, typically of the order of a tenth of a percent, representing the fraction of the mechanical energy converted into sound. The problem is more complex than that, however, since in the supersonic regime, intense pure tones can be generated as a result of instabilities in the jet, such as shock cell oscillations.

7.3.4 Problems

1. **Sound from a sphere in turbulent flow**

   At sufficiently high Reynolds numbers, typically $R > 10^5$, ($R = Ua/\nu$, U, for velocity, a, radius, and ν, kinematic viscosity ≈ 0.15 for air), it is a good approximation to put the drag force on a sphere equal to $A\rho U^2/2$, where $A = \pi a^2$ and a, the radius of the sphere. In a frequency band centered around 50 Hz, the measured rms value of the velocity fluctuation is 0.1 percent of the mean flow velocity. The Mach number of the mean flow is 0.5. The radius of the sphere is 1 cm.
   (a) What is the Reynolds number (based on the diameter) of the flow?
   (b) What is the rms sound pressure in the same frequency band centered at 50 Hz and the corresponding sound pressure level at a distance of 1 m from the center of the sphere and an angle of 60 degrees from the flow direction?
   (c) What is the radiated acoustic power in watts?

2. **Low-frequency sound radiation from an aeroplane in bumpy flight**

   The fluctuations in the lift force on an aeroplane in turbulent flow might have an amplitude equal to the entire weight of the plane and the period of the corresponding oscillations typically might be of the order of 1 s.
   (a) Treat the plane as a point force and calculate the acoustic power generated at a frequency of 1 Hz by the fluctuating lift force.
   (b) Show that the integrated near field pressure over ground equals the weight of the plane.

3. **Sound radiation from a subsonic jet**

   Air is discharged from a nozzle at a Mach number $M = 1$. The diameter of the nozzle is $D = 0.5$ m. The air density is $\rho \approx 1.3$ kg/m^3 and the sound speed $c = 340$ m/s.
   (a) Determine the total acoustic power emitted by the jet.
   (b) What is the corresponding average sound pressure level about the jet at a distance of 50 m?
   (c) What is the maximum value of the power spectrum level?
   (d) What is the change in the level of the emitted sound at a frequency $f \gg f_0$ for a 10 percent change of the thrust of the engine?

4. **Jet noise reduction trough spectrum shift**

   Consider the jet in Problem 3. If the nozzle is replaced by ten parallel tubes creating ten parallel jets with the same total area and thrust, what would be the reduction of the sound pressure level at a frequency 100 Hz?

5. **Octave and 1/3 octave band jet noise spectra**

(a) Let the center frequencies (on a log scale) of the octave band jet noise spectrum be $f_n = f_0 \, 2^n$, where $f_0 = 0.15 U/d$ is the frequency at the peak value of the spectrum density and n an integer, positive, negative, or zero. The nth band covers the frequency range from $f_n(1/\sqrt{2}$ to $f_n\sqrt{2})$. Calculate the octave band power levels for $n = -5$ to $n = 5$ with reference to the total power level and check if the following values are correct, -38.4, -29.5, -20.6, -12.7, -5.0, -1.6, -0.43, -1.9, -5.2, -9.2, -13.5 dB.

(b) For the 1/3 octave bands the center frequencies are given by $f_n = f \, 2^{n/3}$ and the band covers the frequency range from $f_n \, 2^{-1/6}$ to $f_n \, 2^{1/6}$. Calculate the third octave band levels with respect to the total power level for $n = -10$ to $n = 10$ and check if the following values are correct, -29.1, -26.2, -23.4, -20.6, -17.9, -15.4, -13.1, -11, -9.1, -7.6, -6.4, -5.6, -5.2, -5.2, -5.4, -5.9, -6.7, -7.7, -8.8, -10, -11.3.

7.4 'Spontaneous' Instabilities

7.4.1 Single Shear Layer

A single shear layer is unstable as a result of the interaction with itself, so to speak, and this instability progresses into turbulence. The sheet can be thought of as a continuous uniform distribution of line vortices. If they are all aligned, they remain in the sheet but if a fluctuation brings one line vortex out of the plane of the sheet, its flow field will affect the position of the other line vortices so that initially, in the linear regime of perturbation, the displacement out of the sheet grows exponentially with time. As a result, the sheet breaks up and sound will be emitted in the process. The time dependence of the motion of the sheet can be followed numerically on a computer. One might refer to this instability as a spontaneous creation of vortex sound.

7.4.2 Parallel Shear Layers. Kármán Vortex Street

Similarly, two parallel shear layers are also unstable but now, through the interaction between the sheets, individual isolated vortices are developed to form a periodic stable zig-zag pattern, generally known as the Kármán vortex street. Actually, this street refers to the wake formed behind a cylinder in a uniform flow field. Initially, this wake is in the form of parallel vortex sheets which develop into the vortex street as shown in Fig. 7.4. Unlike the single vortex sheet, the double sheet contains a characteristic length, the separation d between the sheets (width of the wake), and there will be a corresponding characteristic frequency of the order of U/d, where U is the velocity of the incident flow.

Figure 7.4: Kármán vortex street (from Milton van Dyke, *An Album of Fluid Motion*, The Parabolic Press, Stanford, 1982. (courtesy of Professor Van Dyke).

The number of vortices formed per second is found to be

$$\boxed{\begin{array}{c} \textit{Kármán vortex; shedding frequency (Fig. 7.4)} \\ f_U = S\,\frac{U}{d} \end{array}} \qquad (7.17)$$

[*U: Free stream flow speed. d: Cylinder diameter (or width of wake for any blunt body). S: Strouhal number (≈ 0.2 for cylinder)*].

The Kármán vortex sheet occurs only at Reynolds numbers in the approximate range between 300 and 10^5 in which the Strouhal number is found experimentally to be ≈ 0.2. Below 300, the flow is essentially laminar, and above 10^5, the vortices are not well correlated along the length of the cylinder and the wake quickly develops into a fully turbulent flow. However, the broad band turbulent spectrum has a peak close to f_U. The noise spectrum of a subsonic turbulent jet with a diameter d and velocity U turns out also to have its maximum approximately at $0.15\,U/d$, as already discussed in Section 7.3.3, Eq. 7.14.

The drift velocity of the vortices in the wake can be shown to be $\approx 0.7U$ so that the spacing between successive vortices will be $\approx 0.7U/f_U \approx 3.5d$.

The frequency f_U in Eq. 7.17 can be used for a blunt body in general, not necessarily a cylinder, if d is taken to be the width of the wake. For a cylinder, the width is about the same as the diameter of the cylinder, but for a flat plate, it is wider than the plate. For an airfoil, the width of the wake is about the same as the thickness of the trailing edge, and the vortex shedding frequency can be quite high. With a thickness of the trailing edge of 0.5 cm and a flow Mach number of 0.8 the frequency is ≈ 10.9 kHz.

The Interaction Force

The force on the cylinder contains both an axial and a transverse component. The axial is dominated by a time independent part corresponding to a drag coefficient which can be shown to be $C \approx 1.5d$ per unit length of the cylinder.

The transverse force is oscillatory. As a vortex is shed on one side of the cylinder, a counter circulation is induced to conserve angular momentum, as shown schematically in Fig. 7.5. It is well known that when a mean flow is superimposed on a circulation, there will be a transverse force on the object (compare the lift on a wing) which in the figure will be directed downwards (the superposition of velocities results in higher velocity and lower pressure below the cylinder than above). The reaction force on the air will be upwards so that the sound generated by the vortex shedding will be positive above the cylinder at the instant shown in the figure. Since the vortex shedding is

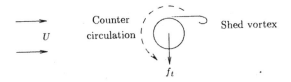

Figure 7.5: Counter circulation about a cylinder resulting from vortex shedding and the corresponding periodic transverse force.

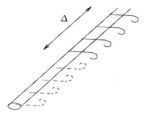

Figure 7.6: Concerning the degree of correlation of vortex shedding along a cylinder.

periodic, moving from one side of the cylinder to the other, sound will be generated at the same frequency.

The transverse force amplitude per unit length has been estimated to be $f_t \approx 6d(\rho U^2/2)$. Experimental values have been found to be considerably lower $\approx 0.5d(\rho U^2/2)$. One reason for the discrepancy is presumably a relatively poor correlation between the vortex shedding along the length of the cylinder, as illustrated schematically in Fig. 7.6. Since the vortices shed out of phase tend to cancel each other's contribution to the transverse force on the cylinder, the resulting sound emission is normally quite weak. Typically, the correlation may extend over 4 to 5 diameters of the cylinder, as shown schematically in Fig. 7.6, where Δ indicates qualitatively the correlation length. This length decreases with with increasing degree of turbulence in the incident flow.

If the cylinder has a transverse resonance which is the same (or nearly so) as the forcing frequency, the oscillation amplitude can be quite large, not only because of the resonance per se, but because the motion of the cylinder apparently has a tendency to increase the correlation length of the vortices along the cylinder. We shall return to this question in the next section.

The vortex tone can be reduced and even eliminated by substantially reducing the correlation length by making the surface of the cylinder irregular. This can be done, for example, by spiraling a wire around the cylinder. The 'singing' of Pitot tubes in a wind tunnel is often eliminated in this manner. Similarly, the periodic wake behind (tall) chimneys can be eliminated by an helix of protruding bricks around the chimney. Another method (usually more complicated to implement) is to have a fin applied on the downstream side of the cylinder that separates the vortex sheets and prevents coupling between them.

7.4.3 Flow Damping

The interaction of flow not only can lead to an instability and excitation of vibrations of a mechanical oscillator but it can also produce damping. We refer to Section 10.3.2 for an example, demonstration, and analysis of this effect.

7.5 'Stimulated' Flow Instabilities, a Classification

The sound emitted by the vortex shedding of a cylinder, often called the Aeolian tone, is generally quite weak. However, if this instability is stimulated through feedback from a mechanical or acoustic resonator, the intensity can be considerably enhanced. This is but one example of many in which feedback occurs. The instabilities which will be considered in the remainder of this chapter are the flow-induced instabilities referred to in Fig. 7.7. In these examples, there is at least one of the interactions shown between characteristic modes of motion of flow, structures, and sound and they have been termed 'flutter,' 'flute,' and 'valve' instabilities.

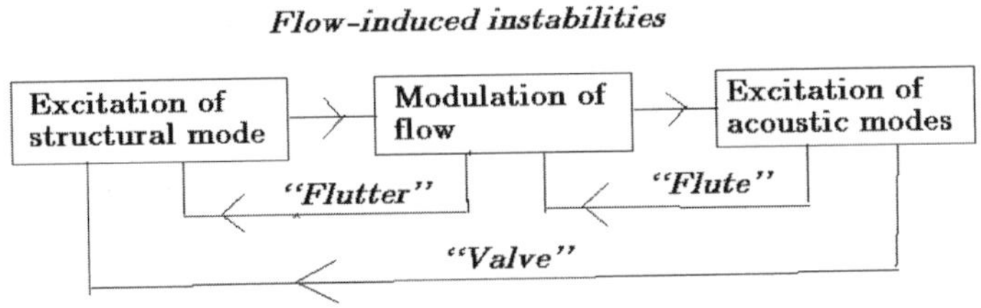

Flow-induced instabilities

Figure 7.7: A classification of flow induced instabilities.

To these should be added *heat driven* instabilities. They are frequently encountered in combustion chambers where acoustic modes of the chamber can stimulate the rate of heat release. Another example is the old Rijke tube in which the heat transfer from a heated screen is modulated by acoustic pipe modes. Instabilities in weakly ionized gases is still another example in which the heat transfer from the electrons to the neutral gas component is modulated by acoustic modes in the gas. Another class of instabilities are driven by *friction*, as in the violin, the squeal of tires, hinges, brakes, and the chalk on the black board. They are in a sense similar to the flute instabilities where the resonances now are mechanical rather than acoustical and the friction force, like the flow, can provide either positive or negative damping.

7.6 Flutter; Mechanically Stimulated Flow Interaction

7.6.1 Kármán Vortex Street

In what we have referred to as flutter instability, fluid flow interacts with an elastic body in such a way that a displacement or deformation of the body alters the fluid-flow interaction to promote or stimulate the displacement and make it grow. We restrict the discussion to instabilities in which both the fluid and the structures individually have characteristic modes and frequencies of motion.

For example, in the periodic shedding of vortices by a (rigid) cylinder, the characteristic frequency or mode of spontaneous oscillations of the flow itself arises from the instability of two parallel vortex sheets in the wake of the cylinder and the frequency is known to be $f_U \approx 0.2U/d$, as already indicated (Eq. 7.17). If the cylinder has a transverse vibrational resonance with a frequency equal to (or close to) that of the vortex shedding, the cylinder will be excited at resonance by the flow. The motion of the cylinder then in turn is likely to affect the vortex shedding and the amplitude of the cylinder and the emitted sound can then be considerably increased. The stimulation by the vibration to a great extent apparently can cause an increase in the correlation of the vortex shedding along the cylinder so that the elementary reaction forces on the cylinder will be in phase (see Fig. 7.6).

7.6.2 Instability of a Cylinder in Nonuniform Flow

A cylinder in a nonuniform flow, as indicated schematically in the Fig. 7.8, can be driven in oscillatory motion by the flow without the presence of a Kármán vortex street, as follows.

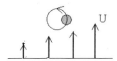

Figure 7.8: Oscillations of a cylinder driven by nonuniform flow.

The cylinder is anchored at both ends, at $z = 0$ and $z = L$; typically, it could be a tube in a heat exchanger. It is located in a nonuniform flow with the velocity increasing from left to right, as shown. Due to the turbulence in the flow, there will always be some forced lateral oscillations of the cylinder. When it is displaced to the right, the drag force on the tube increases and the tube is forced into a swirling mode of motion. As it returns on the left side along the circular path, it encounters a drag force in the opposite direction but it is smaller than the force on the right. When the tube again moves into the flow on the right the process is repeated. Consequently, there will be a net energy transfer to the cylinder in one cycle. As a result, the amplitude of the swirling motion increases as will the energy transfer per cycle. This leads to an instability which we can analyze as follows.

The radius of the swirling motion (a superposition of a vertical and horizontal harmonic motion) by $r(z, t)$, where z is the coordinate along the cylinder. The z-dependence is the same as that of the fundamental oscillatory mode of the cylinder, $r(z, t) = |r(t)| \sin(kz)$, where $k = \omega/v$ and v the bending wave speed and $|r(t)|$ is the amplitude at $z = L/2$. The excursion in the x- and y-directions are $\xi(z, t) = r(z, t) \cos \omega t$ and $\eta(z, t) = r(z.t) \sin(\omega t)$. If the relative growth of r is small in one period of swirling, the velocity in the y-direction is

$$\dot{\eta}(z, t) \approx r(z, t)\omega \cos(\omega t). \tag{7.18}$$

The steady flow velocity increases with x as shown schematically in the figure but is assumed independent of z. We place $x = 0$ at the equilibrium position of the cylinder. The flow velocity at the displaced position ξ of the cylinder is then

$$U(\xi) \approx U(0) + \frac{\partial U}{\partial x}\xi + \cdots . \tag{7.19}$$

The drag force per unit length of the cylinder is then of the form

$$F_y(\xi) = C\rho D\, U(\xi)^2 \approx C\rho D\left[U(0)^2 + 2U(0)\frac{\partial U}{\partial x}\xi + (\frac{\partial U}{\partial x}\xi)^2\right], \tag{7.20}$$

where C is a drag coefficient and D the cylinder diameter.

The power transferred to the cylinder per unit length is $F_y\dot{\eta}$. When this is integrated over one period, only the second term in Eq. 7.20 contributes and we get the average rate of transfer of energy in one period. We also have to integrate over the length of the cylinder which contributes the factor $L/2$. The energy of the cylinder which equals the maximum kinetic energy, integrated over its length, is $E = (M/2)\omega^2|r(t)|^2/2$, where M is the total mass of the cylinder. It follows then from Eqs. 7.18 and 7.20 that

$$dE/dt = \alpha E$$
$$\alpha = (C\rho DL)[U(0)\frac{\partial U}{\partial x}]\frac{2}{M\omega}$$
$$E = E(0)e^{\alpha t}. \tag{7.21}$$

In the absence of damping, the oscillation grows exponentially with time. There is always some energy $E(0)$ initially because of the lateral oscillations caused by the turbulence in the flow. If damping is present and in the absence of flow the average energy loss per cycle would be βE, say. In that case, the exponential growth would not occur unless $\alpha > \beta$.

We note that the cylinder gets more unstable the lower the resonance frequency ω. As was discussed in Chapter 4, the wave speed of a bending wave is proportional to $\sqrt{\omega}$. The fundamental resonance frequency is determined by $kL = (\omega/v_b)L = \pi$. With v_b being proportional to $\sqrt{\omega}$ it follows that $\omega \propto 1/L^2$. This means (see Eq. 7.21) that for a given material, the growth rate coefficient α will be proportional to L^3. Thus, one obvious way to stabilization is to support a long cylinder at several positions along its length to reduce L.

7.7 Flute Instabilities; Acoustically Stimulated Vortex Shedding

7.7.1 Cylinder in a Flow Duct. Heat Exchangers

Acoustic stimulation of vortex shedding such as in the vortex street behind a cylinder can enhance the sound emission considerably in much the same way as in an optical laser. It is the mechanism involved in many musical wind instruments. The vortex sheet is then stimulated by the sound in an acoustic resonator excited by the flow, as indicated by the feed back loop in Fig. 7.7. The instability occurs when the vortex

Figure 7.9: Acoustically stimulated vortex shedding from a cylinder in a duct with the pressure and velocity distribution in the fundamental acoustic transverse mode in the duct.

frequency and the acoustic resonance frequency are the same (or nearly so). A particular example will be discussed shortly which involves the periodic vortex shedding from a cylinder stimulated by the acoustic modes in a duct. There are many other examples of this type of instability, which will be considered also.

The characteristic feature of this instability is that, unlike the flutter instability, it is affected by *temperature*. This arises because of the $\sqrt{T}$-dependence of the sound speed and hence the acoustic resonance frequencies which must equal the vortex frequency for the instability to occur.[5] The knowledge of the temperature dependence of an instability is a useful aid in diagnosing the nature of the instability. It indicates that an acoustic resonator and feedback from it are likely to be involved in the process. As for the mechanically stimulated vortex, a likely reason for the increase in the force and sound emission is the increase in the correlation length of the vortices along the cylinder.

As an example, consider a cylinder of diameter d mounted across a rectangular duct perpendicular to the axis, as indicated in Fig. 7.9. The width of the duct is D. If the flow velocity in the duct is U, the frequency of vortex shedding at the cylinder will be $f_U \approx 0.2\,U/d$, as explained earlier. If this frequency coincides with one of the acoustic (transverse) resonances of the duct, $f_a = nc/2D$, where c is the sound speed and n an integer, resonant self-sustained stimulated vortex shedding can occur. In many respects, this is analogous to the stimulated emission of light in a laser with the duct walls representing the mirrors in the optical cavity.

Equating the vortex frequency and the frequency of the nth transverse mode in the duct (the first mode occurring when the duct width is half a wavelength), i.e., $0.2U/d = nc/2D$, yields the condition

$$M = \frac{U}{c} = 2.5n\frac{d}{D} \qquad (n = 1, 2, 3...).\tag{7.22}$$

It is assumed that the cylinder is not placed in a velocity node of the sound wave. If the cylinder is located at the center of the duct, only modes for *odd* values of n will be excited.

---

[5]Since the vortex frequency is practically independent of the Reynolds number, the temperature dependence of the kinematic viscosity can be neglected.

Example, Heat Exchangers

A heat exchanger often consists of an array of parallel pipes perpendicular to the axis of a rectangular duct carrying flow. Due to the coupling between a transverse acoustic duct mode (usually the fundamental) and the vortex shedding from the pipes, a very high sound pressure amplitude can result if the vortex shedding frequency is close to the acoustic mode frequency. The amplitudes can be so large that the oscillatory stresses produced in pipes and duct walls can lead to acoustic fatigue failure.[6]

The simplest means of eliminating the instability in the heat exchanger is to introduce a partition wall in the center of the duct which cuts the transverse dimension in half and doubles the acoustic mode frequency. This is usually sufficient to bring the acoustic frequency away from the vortex frequency for the flow velocities encountered.

Feedback oscillations of the heat exchanger type are frequently encountered in many different contexts. One example involved an exhaust stack of a jet engine test cell. A number of parallel rods had been installed between two opposite walls in the rectangular exhaust stack to reduce wall vibrations. As it turned out, at a certain operating power of the jet engine, in that case 60 percent of full power, the frequency of the vortex shedding frequency of the rods coincided with the first transverse mode of the exhaust stack and the resulting oscillation produced a tone which could be heard several miles away. This environmental noise problem caused the test facility to be shut down.

An interesting aspect of this example was that the amplitude of the sound varied periodically with time at a period of about a second or two. In some manner, the instability shut itself off at a certain amplitude and then started again. Nonlinear damping or an amplitude dependent flow resistance in the duct could have been the reason. In the latter case, there would be a reduction of the mean flow and the vortex frequency in the duct with increasing amplitude thus removing this frequency from the coincidence with the acoustic mode frequency and shutting off the instability. The pressure drop then would decrease and the mean flow velocity would increase again to its original value to reestablish the vortex frequency and the instability. To eliminate the tone in this example, the rods were cut out and the walls stiffened by outside reinforcement.

The same type of feedback mechanism applies also in a circular duct with one or more radial rods. The feedback now results from the excitation of the circumferential acoustic modes in the duct, the first occurring at a frequency $\approx 1.7c/D$, where c is the sound speed and D the duct diameter. At one time it was feared that the guide vanes in a fan duct could give rise to such tones.

In regard to heat exchangers there is good reason to suggest that the instability in Section 7.6.2 is a likely cause of observed pipe failures. It appears that pipes close to a flow inlet are vulnerable. In this region, the flow is apt to be nonuniform which is the required condition for the instability in Section 7.6.2 to occur.

---

[6]Acoustic fatigue failure is similar to the well-known effect of breaking a metal wire by bending it back and forth a (large) number of times. If the stress exceeds a critical value, failure results after a certain number of cycles which depends on the stress.

7.7.2 Pipe and Orifice Tones

The flow through a circular orifice separates at the entrance to the orifice and forms a vortex sheet with circular symmetry. Like the two parallel shear layers behind a blunt body, this sheet also can form periodic vortices which now takes the form of rings with a characteristic frequency proportional to the flow velocity, $f_U = SU_0/D'$, where D' is some characteristic length, a combination of the orifice diameter D and the length L_0 of the orifice, and S is a constant. By analogy with the parallel vortex sheet instability, D' should be a measure of the separation of interacting vortex sheets, and, as a first approximation at least for sufficiently short lengths of the orifice, we shall assume here that $D' = D$. The value of the constant S, according to our experiments, is approximately 0.5.

If L_0 is considerably larger than D, the influence of L_0 cannot be ignored, however, since experiments indicate that the orifice whistle does not seem to occur when L_0 is greater than $\approx 4D$. An explanation might be that the vena contracta[7] then falls well inside the orifice and as the flow expands, it will strike the wall of the orifice and possibly ruin the coherence of the sheet oscillations.

Even for small values of L_0, it has an effect on the instability, albeit indirectly, since with L_0 less than $\approx D/4$, flow-induced instability does not seem to occur. A likely reason is that the acoustic frequency and the flow velocity are then so high that the acoustic radiation resistance and the flow-induced resistance (see Chapter 10) prevent the instability from developing.

The orifice shown in Fig. 7.10 is located in a pipe and if the frequency of any of the modes of this system is sufficiently close to the vortex frequency, acoustically stimulated self-sustained oscillations can occur. Only axial modes will be considered here. The lowest frequency is approximately that of an open-ended pipe with a wavelength approximately twice the length of the pipe. The high frequency end of the spectrum starts with the first mode of the orifice itself with a wavelength approximately twice the acoustic thickness $L_0' \approx L_0 + \delta$ of the orifice plate, where L_0 is the physical orifice length, $\delta \approx (1 - \sigma)0.85D$, the two-sided end correction, and σ, the ratio of the orifice area and the pipe area. In this high-frequency regime, the orifice modes are essentially decoupled from the pipe modes but at lower frequencies.

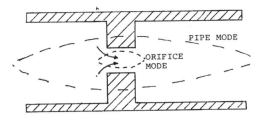

Figure 7.10: Flow excitation of orifice and pipe tones.

---

[7]The area of the separated flow (jet) in the orifice contracts to a minimum, the vena contracta, and then expands again.

The presence of the orifice will affect the frequencies of the pipe modes. Instead of a pipe, any other resonator will produce pipe tones such as a Helmholtz resonator with flow through it, as will be discussed in an example below. The simplest example, familiar to all, is the mouth whistle.[8]

If the characteristic vortex frequency is denoted $f_U = SU_0/D$, the flow velocity at which an acoustic mode of frequency f_a can be stimulated by the flow into a self-sustained oscillation is given by $f_U \approx f_a = c/\lambda_a$, i.e., $M_0 = U_0/c \approx D/(S\lambda_a)$. The lowest pipe mode has a wavelength $\lambda_a \approx 2L_p$, where L_p is the pipe length, and with $D/L_p << 1$, the flow velocity, frequency, and sound intensity, will be correspondingly small. However, in some applications involving perforated plates containing a large number of orifices, the pipe tones can be quite intense.

As the flow speed increases, higher order modes of the pipe system will be excited until the pure orifice mode is reached. With the acoustic length of the orifice being L_0', as given above, the frequency of the lowest mode is $f_a = (c/2L')(1 - M_0^2)$, and the overtones frequencies are $f_n = nf_1$, where the factor $1 - M_0^2$ is due to the wave speeds in the upstream and downstream directions being $c - U_0$ and $c + U_0$, respectively (see transmission matrix of a pipe with flow in Section 4.4). The critical flow Mach number for the lowest orifice tone follows from $f_U = f_a$, i.e., from the equation $M_0 \approx (D/2L_0'S)(1 - M_0^2)$. For example, with $S \approx 0.5$ and $D = L_0$, we get $M_0 \approx 0.44$.

Extensive measurements have shown that the Mach numbers and the corresponding frequencies indeed cluster around the predicted values. As a rule of thumb, the excitation of intense orifice tones usually can be expected to occur in the Mach number range between 0.25 and 0.5.

The conditions for the excitation of a higher orifice mode is obtained in an analogous manner. The vortex shedding, although periodic, is not harmonic, and overtones of the f_U exist and can be involved in the stimulation of acoustic modes.

Acoustic Whistle Efficiency

In experiments with an orifice with a diameter $D = 2r_0 = 0.5"$ and a thickness $L_0 = 0.5"$, the sound pressure level in free space at a distance $r = 100$ cm from the orifice was found to have a maximum value of 115 dB, obtained when the pressure drop across the orifice was ≈ 0.13 atm, corresponding to a Mach number of ≈ 0.44 in the orifice. (An orifice with $D \approx L_0$ seems to give the highest intensity.)

To determine the corresponding *acoustic efficiency* of the orifice, defined as the ratio of the radiated acoustic power and the flow losses, we express the latter as $W_f \approx A\rho U^3/2 = A(\gamma P)^2 M_0^3/2\rho c$, where we have treated the flow as incompressible and where $P = \rho c^2/\gamma$ is the static pressure, $M_0 = U_0/c$, and $A = \pi r_0^2$.

On the assumption of an omni-direction source, the acoustic power radiated into free field half-space can be written $W_a = 2\pi r^2 p^2/\rho c$, where p is the rms value of the sound pressure at the distance r from the source. The acoustic efficient is then $\eta_a = W_f/W_a = 4(r/r_0)^2(p/P)^2/(\gamma^2 M_0^3)$. The observed sound pressure level of

---

[8]Speech production is different. Here the time varying acoustic modes of the vocal tract are excited by a periodic pulsation of air through the glottis.

115 dB corresponds to $p/P \approx 10^{-4}$. Then, with $r = 100$ cm, $r_0 = 0.64$ cm, $\gamma = 1.4$, and $M_0 = 0.44$, we get $\eta_a \approx 6 \cdot 10^{-3}$.

If the orifice is placed in a duct as in Fig. 7.10, simulating a valve, for example, we can estimate the sound pressure level in the pipe.

Elimination of Orifice/Pipe Tones

The results given above express only necessary conditions for the occurrence of the orifice tones. Other factors, such as the uniformity of the shear layer at the entrance and flow-induced sound absorption (see Chapter 10) at the exit end are also important. At Mach numbers above 0.5, the latter becomes so large as to prevent resonances from occurring. This can readily be demonstrated by exciting an open-ended pipe by random noise from a source outside and measuring the response by a microphone placed at the center of the pipe. The flow through the pipe can be obtained by connecting the pipe to a plenum chamber which is connected to a pump. With the microphone placed at the center of the duct and with no flow through the duct, the spectrum obtained clearly show the odd number duct modes resonances as narrow spikes. As the flow speed is increased, the resonances are broadened and at a Mach number of ≈ 0.5, they are essentially gone. (No organ music would be possible with a Mach number above 0.5 in the pipes!)

For a conical orifice, as obtained by countersinking a circular orifice, no whistling occurs if the apex angle of the countersink is larger than 60 degrees, regardless of the direction of the flow. If the vortex sheet in the separated flow at the inlet of the orifice is broken up by making the edge of the orifice irregular, the chance of whistling is markedly reduced and a simple means of eliminating the whistle is to place a wire mesh screen across the entrance. The effect is similar to that of the helical wire used to prevent the periodic vortex behind a cylinder from occurring.

Example

In many applications, such as in various forms of industrial dryers, flow through orifice plates is often used and this can give rise to problems associated with whistling. One example is illustrated schematically in Fig. 7.11. It involves a film dryer in which the film is transported below a set of plenum chambers which supply air of different temperatures to the film through perforated plates.

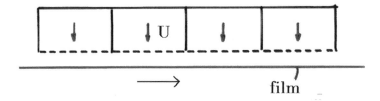

Figure 7.11: Example of flow generated tone in a film dryer facility.

This dryer turned out to generate an intense tone at the resonance frequency of the plenum-orifice combination, as explained in Section 7.7.2. Not only was the tone an environmental noise problem but it also affected the film drying process. The acoustic oscillations turned out to modulate the drying rate so that the film came out with striations at a spacing that corresponded to the whistle frequency.

Similar dryers are used in many other applications in processing facilities, for example, in the textile industry for drying fibers. One way to eliminate such a tone is to place a wire mesh screen on the upstream side of the orifice plate.

7.7.3 Flow Excitation of a Resonator in Free Field

Like most problems in acoustics, the flow excitation of a resonator has a long history going back to Helmholtz (1868) including attempts to understand the whistle mechanism. Since then, numerous papers have been written on the subject and there probably will be more to come. Our own studies of the problem[9] included Schlieren photography with stroboscopic and high intensity flash illumination of the flow around the mouth of a tube resonator in free field identifying periodic vortex generation. The experiments were limited to only one tube, however, 2 cm in diameter and 30 cm long. The air stream was uniform over an area of 3 cm × 3 cm with speeds up to 3500 cm/sec. The angle of attack of the flow could be varied over the range 0 to 50 degrees and it was found that for angles less than 15 degrees, no oscillations occurred. A wire mesh screen was used in the resonator to provide damping which could be changed by varying its location in the resonator. Actually, the screen was a package of three screens, each with an open area of 29 percent and a diameter of the strands of 0.1 mm. Only the fundamental mode was considered. With the screens placed at the rigid end wall of the resonator, essentially no damping was obtained, and with the screen at the opening, maximum damping. In this manner, the Q-value of the resonator could be varied over the range from 10 to 43. The frequency of the flow-induced oscillations was close to the acoustic frequency of the fundamental mode of the resonator, $f_a \approx c/4L'$, where $L' = L + 0.32d$ is the acoustic length of the pipe. With $L = 30$ and $d = 2$ cm, this frequency is ≈ 277 Hz. The flow velocity U at which the mode was excited extended over a range about a mean value of about 1150 cm/sec.

The simple kinematic model we use for the excitation mechanism is that a perturbation of the shear layer which starts at the leading edge of the orifice is convected on the vortex sheet with a speed $U' = \beta U \approx 0.5U$. (An intuitive mechanical model of a shear layer is a board moving on roller bearings on a plane boundary with the fluid velocity U. The relative velocity of the contact point of a roller with the boundary is zero and the center of the roller bearing, corresponding to the average speed of the vortex sheet, moves forward with a velocity $U/2$.) As the disturbance reaches the downstream edge of the orifice, an acoustic signal is fed back to the upstream edge to stimulate the vortex sheet. Thus, the roundtrip time of this fluid oscillator will be $\approx d/U' + d/c \approx d/U'$. Assuming that the self-sustained oscillation occurs when this time equals the period of the acoustic mode, the corresponding flow velocity will be

[9]Uno Ingard and Lee W. Dean III, *Excitation of acoustic resonators by flow*, Trans. of the Second Symposium on Naval Hydrodynamics, 1958, pp 137-150.

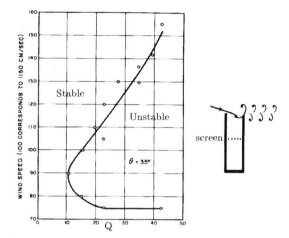

Figure 7.12: Stability contour in a velocity Q-value space for flow excitation of a tube resonator in free field. The velocity scale is linear and the value 100 in this example corresponds to a flow velocity of 1150 cm/sec. The flow was inclined 35 degrees, as indicated. The curve divides the space into a stable (left) and unstable region (right).

$U \approx cd/(4\beta L') = f_a d/\beta$. With $\beta \approx 0.5$, $d = 2$, $L' = 30.64$ cm, and $f_a \approx 277$ Hz, as given above, this velocity becomes $U \approx 1109$ cm/sec, in good agreement with our experiments.

Although the highest emitted sound occurs close to this predicted flow velocity, oscillations are maintained over a range of velocities, the range depending on the damping of the resonator. Fig. 7.12 shows a stability diagram of the oscillator, showing the critical flow velocity plotted as a function of the Q-value of the resonator. The region of instability is on the right side of the curve and the region where no oscillations occur is on the left. It should be noted that the velocity scale is linear with the value 100 corresponding to a flow velocity of 1150 cm/sec.

It is clear from the diagram that below a Q-value of 10, no oscillations occur, and that the flow velocity range of instability increases with an increasing Q-value. The diagram refers to an angle of attack of the flow of 35 degrees. Starting at the lower bound of the critical velocity curve and moving toward the upper bound at a constant Q-value, the amplitude of oscillation starts from zero, reaches a maximum and then goes back to zero.

As indicated in the diagram, a sufficiently high damping and a corresponding low Q-value will prevent the resonator from being excited by flow. This is familiar from using a soft drink bottle as a whistle by blowing over its mouth. Assuming that the bottle is half full, say, it is fairly easy to make it whistle. However, after shaking the bottle so that a foam is formed, it is generally not possible to excite the bottle because of the sound absorption provided by the foam.

In our experiments, the sound pressure was measured not only outside the resonator but also inside, at the end wall. Even in the stable region, weak sound pres-

sures at the resonance frequency were detected inside the resonator, responding to
the turbulence in the incident flow and signifying a linear response of the resonator
to an oscillating driving force.

7.7.4 Flow Excitation of a Side-Branch Resonator in a Duct

Instead of a cavity resonator in free field, we now consider a side-branch cavity resonator in a duct. The (kinematic) model of flow excitation used in the previous section
will be used also here. Thus, a shear layer is started at the upstream edge of the resonator opening. If U is the free stream velocity, a flow perturbation of the shear layer
is carried by the shear layer at a speed $U' = \beta U$ across the opening and interacts with
the downstream edge which feeds back to the upstream edge (not unlike an edge tone
oscillator). This defines a characteristic roundtrip time and frequency for the shear
layer. If the feedback is assumed to be carried by the speed of sound, the roundtrip
time will be $D/U' + D/c$ and the corresponding frequency $f_U = (\beta U/D)/(1+\beta M)$,
where $M = U/c$. As in the previous section, the coefficient β is approximately 0.5.

The frequency of the nth mode of the resonator is $f_n = (2n - 1)c/4L'$, where
the acoustic length of the resonator is $L' = L + \delta$ and $\delta \approx 0.43D$ (one-sided end
correction). The flow frequency f_U has also overtones and a condition for instability
or coupling between the m:th fluid mode and the n:th acoustic is $mf_U = f_n$ where m
and n are 1,2,3....

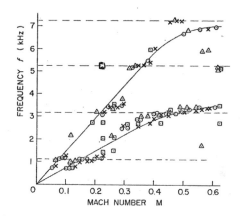

Figure 7.13: Data points are the measured frequencies of flow excited acoustic modes in a
side-branch cavity in a duct. The dashed lines are the acoustic resonance frequencies and the
solid lines the Mach number dependence of the shear flow frequencies for $m = 1$ and $m = 2$.
The predicted instability frequencies are represented by the intersection of the dashed and
solid lines but as for the flow excitation of the resonator in free field (see Fig. 7.12), there is a
range of flow velocities in which tones are produced.

The data points in Fig. 7.13 show measured frequencies of flow excited tones of
a side-branch resonator in a duct as a function of the Mach number in the duct.
Both the duct and the resonator had a square cross section of width $w = 0.75''$ and

the length of the resonator tube was $L_0 = 3"$. With an end correction $\delta = 0.3w$, the corresponding acoustic length of the cavity was $L_0' \approx 3.23"$. The resonance frequencies thus obtained are 1040, 3120, 5200, 7280 Hz, etc. These frequencies are shown as dashed lines in the figure. Duct length: 84". Area: 3/4" by 3/4". Resonator placed 11" from the beginning (flow entrance) of the duct.[10]

The Mach number dependence of the frequencies of the shear layer of the first two modes, corresponding to $m = 1$ and $m = 2$, as given in the discussion above, are drawn as solid lines. As for the resonator in free field, flow excitation occurs over a range of Mach numbers centered around the values predicted by equating the characteristic frequencies of the flow and of the sound, represented by the intersection of the dashed and solid lines. Although most of the data points are consistent with this view, there are others that fall outside. This deviation will be discussed shortly.

Over most of the range of Mach numbers, more than one frequency is usually excited. For example, at a Mach number of $M = 0.2$, the measured acoustic spectrum contained two pronounced peaks close to the predicted first and second resonances. At this Mach number, the levels of the peaks were about the same. The relative strength of the tones depends on the Mach number, however, and for $M = 0.22$, the second peak became approximately 20 dB above the first, and a weak third peak at the third mode was present. This shift of amplitude toward higher modes with a Mach number was consistent at all locations of the resonator along the duct.

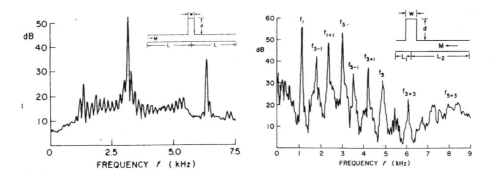

Figure 7.14: Flow excitation of a side-branch resonator in a duct showing coupling of acoustic modes. Duct length: 84". Area: 3/4" by 3/4". Resonator placed 11" from (flow) exit end of pipe. The reference level on the dB scale is not specified (0 typically corresponds to 70 dB re 0.0002 dyne/cm^2 at 12" from duct opening). Left: Coupling between resonator mode and the axial mode of the duct (Mach number ≈ 0.3). Right: Coupling of resonator modes (Mach number ≈ 0.5.)

---

[10]Ingard and Singhal, unpublished. Experiments carried out in the M.I.T. Gas Turbine Laboratory in the 1970s.

Mode Coupling

The phenomenon is more complex than what has been implied above, however, since unexpected frequencies often occur in the spectrum, particularly at higher Mach numbers. Thus, with the resonator placed at the center of the duct, the spectrum shown on the left in Fig. 7.14 contains 'satellite' frequencies around the major peak in the spectrum. The difference between these frequencies turned out to be the fundamental frequency of the duct. With a length of 84", and a Mach number of 0.3, this frequency is $\approx$ 73 Hz. In other words, the frequencies of the satellites are $f_a \pm s\,73$, where s is an integer. These frequencies occur as a result of coupling between a resonator mode and the axial modes in the main duct.

An even more spectacular example of mode coupling is shown in the graph on the right in Fig. 7.14 at a Mach number of 0.5. In this case, interaction between the modes in the resonator is involved. In addition to the first three resonances, corresponding to $n = 1$, 3, and 5, and denoted f_1, f_3, and f_5, there are combination frequencies $f_3 - f_1$, $f_1 + f_1$, $f_5 - f_1$, $f_3 + f_1$, etc. This coupling effect is not the same if the resonator is placed at the flow entrance rather than at the flow exit of the duct, in this case 11" from the end. (The reference level on the dB scale in the figure is not specified. The value 0 dB typically corresponds to 70 dB re 0.0002 dyne/cm^2 at 12" from the end of the duct in our experiments.)

Sound Pressure of Cavity Tones

The sound pressure level outside the pipe, a distance of 12" from the pipe entrance, was found to reach a maximum value of 125 dB at 12" in front of the pipe opening at a Mach number of 0.5 in the duct. However, this refers to merely one resonator-duct configuration. As far as I know, no data are available from which the acoustic power

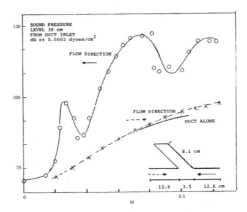

Figure 7.15: Flow excitation of a slanted side-branch resonator in a duct. Sound pressure level, at a distance of 30 cm in front of the duct opening, versus the Mach number in the duct. Upper curve: Flow from right to left. Lower curve (crosses): Flow from left to right. Slant angle: 45 degrees. Length of resonator: 8.1 cm. Length of duct: 28.7 cm. Cross section: 1.9 cm $\times$ 2.5 cm. Duct cross section: 1.9 cm $\times$ 1.9 cm.

generated by the cavity tone can be reliably predicted. In the case considered here, the empirical formula $SPL = 169 - 10\log(A_d/A) + 30\log(M)$ (A_d, duct area, A, resonator area) for the prediction of the sound pressure level in the pipe has been used with a corresponding expression for the acoustic power level.

Slanted Resonator in a Duct. Effect of Flow Direction

Normally, there is no difference between the upstream and downstream edges of the orifice in a side branch resonator in a duct and the direction of the flow then does not influence the excitation of the resonator. However, if the resonator or any other side-branch, such as an air bleed vent, is slanted with respect to the duct axis as shown in Fig. 7.15, the direction of flow is quite significant.

With the flow going from right to left (i.e., against the pointed downstream edge of the opening) the resonator is excited strongly creating a maximum sound pressure level of about 125 dB at a Mach number of 0.37 in the duct and a distance of 30 cm from the duct entrance. In the other direction, i.e., with the flow striking the blunt downstream edge, no tone is produced. The excitation mechanism, as indicated earlier, is reminiscent of the jet edge oscillator.

7.8 Valve Instabilities

Characteristic of a valve type instability is that a structural deformation gives rise to a fluctuation in the fluid velocity proportional to the *displacement* amplitude of the structure and the fluid fluctuation excites an acoustic mode which in turn deforms the structure. This completes the feedback loop, as indicated in Fig. 7.7.

7.8.1 Axial Valve Instability

Fig. 7.16 is a sketch of a control valve and its regulating characteristic, i.e., the mass flow rate $Q_f(P_1, P_2, y)$ versus the lift y of the valve. This rate is a function of the upstream and downstream pressures P_1 and P_2 and of y. The figure also shows a downstream pipe which, in the case of a power plant, might lead to a turbine.

A perturbation η in the lift produces a fluctuation $q_f = (\partial Q_f/\partial y)\eta \equiv \beta\eta$, where β is the slope of the regulating curve. The corresponding velocity perturbation at the entrance of the pipe is then

$$u = (1/\rho A)\beta\eta, \qquad (7.23)$$

where A is the area of the pipe and ρ the density. Perturbations in P_1 and P_2 also contribute to q_f, but only η leads to an instability.

The velocity perturbation u in Eq. 7.23 gives rise to a pressure fluctuation p at the entrance of the pipe which produces a force pA on the valve, where A is the valve area. The power transmitted to the valve by this force is $pA\dot{\eta}(t)$ and the average value over a period T is (harmonic motion assumed)

$$\Pi = (1/T) \int_0^T p\,\dot{\eta}\,dt. \qquad (7.24)$$

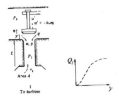

Figure 7.16: Concerning flow-induced axial oscillations of a valve.

Exactly at the resonance of the pipe the response of the pipe is purely *resistive* and the pressure will be proportional to u, i.e.,, $p = Ru$. The acoustical length of the pipe (physical length plus end correction) is then an odd number of quarter wavelengths. Since $u \propto \eta$ (Eq. 7.23), the power (Eq. 7.24) transferred to the valve then becomes proportional to $\int \eta\dot{\eta}\,dt$ which is zero. Consequently, the system is stable, no power is fed back to the valve as a result of its motion.

At a frequency below the resonance frequency of the pipe, the load is a *mass reactance* (at very low frequencies, it is simply the reactance of the air in the pipe moving like a piston) and reaches a maximum value just below the frequency at which the acoustical length of the pipe is an odd number of quarter wavelengths. In that case power to the valve is $\propto \int (\ddot{\eta}\,\eta\,dt \propto \int (\dot{\eta})^2\,dt$ which is positive so that the system becomes potentially unstable.

Finally, above the resonance of the same mode, the load is stiffness like and $p = -K\eta$, where K is an equivalent spring constant (recall that η refers to the *upwards* displacement of the valve plug opposite the displacement of the air column in the pipe, hence the minus sign) and the feedback power is $\propto \int \dot{\eta}(-\eta)\,dt$ which is negative. The air column in the pipe then acts like an additional damper.

To summarize, the power transferred to the valve during one period is

$$
\boxed{
\begin{array}{ll}
\textit{Average feedback power. Stability criterion of axial valve} & \\
\Pi = \begin{cases} \propto \int \eta\dot{\eta}\,dt = 0 & \text{resistance controlled, neutrally stable} \\ \propto \int \dot{\eta}^2\,dt > 0 & \text{mass controlled, potentially unstable} \\ \propto \int(-\eta^2\,dt < 0 & \text{stiffness controlled, stable} \end{cases} &
\end{array}
}
\qquad (7.25)
$$

In what follows, harmonic time dependence is assumed and all perturbations are regarded as complex amplitudes. The velocity of the piston is u', the perturbation in the mass flow $q_f = \beta\eta$, as given above, the velocity of the air at the entrance to the pipe, u. Conservation of mass requires $\beta\eta = \rho A(u + u')$. With $u' = -i\omega\eta$ follows

$$i(\beta/\rho A\omega) = 1 + u/u'. \qquad (7.26)$$

The input impedances (per unit area) of the pipe and the valve are denoted z and z', respectively, so that

$$p = zu = z'u'. \qquad (7.27)$$

From these equations it follows

$$z' + z - i(\beta/\rho A\omega)z = 0. \qquad (7.28)$$

To solve this equation for the complex frequency $\omega = \omega + r + i\omega_i$, the frequency dependence of z and z' must be known. However, even without a solution, the essential aspect of the problem can be observed by expressing the impedances in their real and reactive parts. Thus, with $z = r + ix$ and $z' = r' + ix'$, the equation becomes

$$r' + r + r_f + i[x' + x - r(\beta/\rho A)] = 0, \qquad (7.29)$$

where the flow-induced resistance, $r_f = (\beta/\rho A)x$, is a result of the coupling between the flow and the sound. This resistance will be negative if x is negative (i.e., the pipe impedance has a mass reactive part (remember our choice of time factor $\exp(-i\omega t)$), which confirms the simple calculations of the power fed back to the value piston by the sound field in the pipe as given in Eq. 7.25). The equation also shows that the coupling gives rise to an increase in the mass reactance of the system through the term $r(\beta/\rho A)$. In other words, if Eq. 7.29 can be regarded as an equation of motion of the valve piston, modified by the aero-acoustic coupling, this coupling takes the form of an added resistance r_f and an added reactance $x_f = -r(\beta/\rho A)$. The fact that the resistance r_f can be negative shows that the oscillator can be unstable, leading to an exponential growth of amplitude. The strength of the aero-acoustic coupling is proportional to the slope $\beta = \partial Q/\partial y$ of the valve regulating characteristic. In addition, the coupling resistance r_f is proportional to the pipe reactance x, which reaches a high negative maximum at a frequency just below the resonance of the pipe. Thus, unless the sum $r + r'$ of the resistances of the pipe and the valve exceeds r_f, the system will be unstable.

Eq. 7.29 is analogous to Eq. 6.17 for free oscillations of a damped harmonic oscillator in Section 2.3.3 and it should be solved for the complex frequency. In the case of the damped harmonic oscillator the solution in Eq. 6.17 yielded a negative imaginary part of the frequency which resulted in a decaying motion. Now, we expect that a positive imaginary part and a growing (unstable) motion is also possible.

To investigate the solution for ω, we need the explicit frequency dependence of the impedances z' and z in Eq. 7.29. We express the valve impedance z' simply as that of a harmonic oscillator of mass m_v, spring constant k_v, and resistance coefficient r', all per unit area of the valve plug, taken to be the same as the area of the pipe. Thus,

$$z' = r' - i\omega m_v + ik_v/\omega. \qquad (7.30)$$

The impedance $z = \rho c \zeta$ is the input impedance of the air column in a pipe with an acoustic length L' (i.e., the sum of the physical length and an end correction). The Mach number of the flow in the pipe is M. With reference to Eq. 4.117, the transmission matrix of the pipe is

$$\mathbf{T} = e^{i\Phi} \begin{pmatrix} \cos(k'L') & -i\sin(k'L') \\ -i\sin(k'L') & \cos(k'L') \end{pmatrix}, \qquad (7.31)$$

where $\Phi = -kLM/(1 - M^2)$, $k' = k/(1 - M^2)$, and $k = \omega/c$.

The normalized flow-induced acoustic resistance at the end of the pipe is $\theta_2 \approx M$, as discussed in Chapter 10. We need not consider the mass reactive component of the impedance since it is accounted for in the acoustic length L'.

The input impedance at the entrance of the pipe is then obtained from Eq. 4.100,

$$\zeta_1 = \theta_1 + i\chi_1 = \frac{T_{11}\zeta_2 + T_{12}}{T_{21}\zeta_2 + T_{22}}, \tag{7.32}$$

where the termination impedance $\zeta_2 = \theta_2 \approx M$ is the flow-induced acoustic resistance.

With these expressions for z' and z in Eq. 7.28, the solution to the frequency equation (7.29) for the complex frequency $\omega = \omega_r + i\omega_i$ has to be obtained numerically, in general. The complex frequency contains both the actual frequency of oscillation and the damping. The time dependence of the amplitude is then expressed as $p(t) = p(0)\exp(\omega_i t)\cos(\omega_r t)$ so that a positive imaginary part of the frequency, ω_i, corresponds to exponential growth of the amplitude and hence instability.

Stability Diagram

If $\omega_i = 0$, the stability is marginal (neutral), representing the borderline between stability and instability. Thus, if we put $\omega_i = 0$, i.e., $\omega = \omega_r$, in the equations above, we obtain relations between the parameters of the system expressing marginal stability.

Thus, with $\omega = \omega_r$, the real and imaginary parts of the input impedance to the pipe in Eq. 7.32 become

$$r/\rho c \equiv \theta_1 = \theta_2[1 + \tan^2(k'L')]/[1 + \theta_2^2 \tan^2(k'L')]$$
$$x/\rho c \equiv \chi_1 = -(1 - \theta_2^2)\tan(k'L')/[1 + \theta_2^2 \tan^2(k'L')], \tag{7.33}$$

where $\theta_2 \approx M$ is the flow-induced termination resistance of the pipe, with M being the flow Mach number, $k' = (1 - M^2)\omega/c$. L' is the acoustic length of the pipe (see Eq. 7.30).

The maximum negative value of the reactance is obtained when $\tan(k'L') = 1/\theta_2$ and is $-(1 - \theta_2^2)/2\theta_2$. Since θ_2 usually is of the order of 0.1, the corresponding value of $k'L'$ is somewhat smaller than $(2n - 1)\pi/2$, where n is 1, 2, ...

The frequency equation (Eq. 7.28) can be written as

$$(r' + r + r_f) + i[-\omega m_v + x + (k_v - k_f)/\omega] = 0 \qquad (\omega_i = 0), \tag{7.34}$$

where $r_f = (\beta/\rho A)(x/\omega)$ is an equivalent feedback resistance and $k_f = (\beta/\rho A)r$ are equivalent feedback spring constant. The equation can be regarded as the frequency equation for an oscillator with an impedance $z' + z$ augmented by the feedback resistance r_f and a reactance $-k_f/\omega$ which are proportional to the coupling constant $\beta = \partial Q_f/\partial y$.

Both the real and imaginary parts of Eq. 7.33 must be zero, i.e.,

$$r' + r + r_f = 0 \tag{7.35}$$
$$-\omega m_v + x + (k_v - k_f)/\omega = 0, \tag{7.36}$$

where the frequency dependence of r, x, r_f, and k_f have been given above.

Neutral stability requires that the feedback resistance be negative so as to cancel the resistances r' and r presented by the valve and the fluid column in the pipe. This condition is expressed by Eq. 7.35. A negative value of r_f requires that x be negative and this, in turn, means that the pipe reactance must be mass-like which is consistent with our earlier observation related to the acoustic feedback power to the valve. A negative value of x adds to the inertial mass reactance of the valve and since the spring constant is reduced by k_f, the frequency of instability oscillations will be smaller than the frequency of the free uncoupled valve oscillator.

By combining Eqs. 7.35 and 7.36, we can calculate the frequency of oscillation in terms of the system parameters. The equations determine not only the frequency, however, but impose also a relation between the system parameters which must be fulfilled in order that the system be neutrally stable. This relation is obtained by introducing the calculated frequency in either of the two equations.

In discussing Eqs. 7.35 and 7.36 it is convenient to introduce dimensionless system parameters. They can be chosen in many different ways and among the options can be mentioned the following. One quantity, to be called the *instability parameter*, is $\gamma = \beta/(\rho A \omega_0) = \beta'/(A\omega_0)$, where $\beta' = \beta/\rho = \partial V/\partial y$, $V = Q/\rho$ the volume flow rate through the valve, and $\omega_0 = \sqrt{k_v/m_v}$ the angular frequency of the free (uncoupled) oscillation of the valve.

The dimensionless quantity used for the pipe length will be $k_0 L' = \omega_0 L'/c = 2\pi L'/\lambda_0$. The mass m_v of the valve will enter in combination with the mass of the fluid in the pipe as $m_v/(\rho L')$ and the resistances r and r' will be expressed in terms of $\theta = r/\rho c$ and $\theta' = r'/\rho c$.

Eqs. 7.35 and 7.36 yield the frequency of oscillation and determine the value of the instability parameter γ required to make the feedback resistance overcome the friction in the system and make it neutrally stable. If we plot this value of γ as a function of $k_0 L$ for given values of θ_2 and θ', we obtain a stability diagram or contour. If the operating point $(\gamma, k_o L)$ of the system lies on the contour, the system is neutrally stable, but if the point lies above (below) the contour, the system is unstable (stable). The entire contour will not be computed here; only some general characteristics will be given by an analysis of the special cases $k_0 L' \ll 1$, $k_0 L' \approx (2n - 1)\pi/2$, and $k_0 L' \approx n\pi$, where $n = 1, 2, \ldots$.

First, consider $k_0 L' \ll 1$. The expressions for the pipe resistance and reactance in Eq. 7.33 reduce to

$$r/\rho c \approx \theta_2$$
$$x/\rho c \approx -(1 - \theta_2^2)k'L' = -kL, \tag{7.37}$$

where, in the last step, we have used $k' = k/(1 - M^2)$ and $\theta_2 \approx M$. With these values used in Eq. 7.35 and with $\gamma = \beta/(\rho A \omega_0)$, this equation can be expressed as

$$\gamma = (1/k_0 L')(\theta_2 + \theta') \qquad k_0 L \ll 1, \tag{7.38}$$

which is the low-frequency approximation of the stability contour.

The corresponding frequency of oscillation is obtained from Eq. 7.36. Thus, with $\omega_0^2 = K/Am$, we get

$$(\omega/\omega_0)^2 = \frac{1 - \gamma(\rho c/\omega_0 m_v)\theta_2}{1 + (\rho L/m_v)}. \tag{7.39}$$

In the majority of cases, $\rho c << \omega_0 m_v$ and $\rho L << m_v$, and it follows then that the frequency of oscillation will be only slightly lower than frequency of the uncoupled oscillations of the valve.

To determine the instability contour and the frequency of oscillation over the entire range of the frequency parameter $k_0 L'$ requires a numerical solution of the frequency equation and this is left as a project. However, the essence of the problem can be illustrated without such a solution if we assume that the mechanical resistance r' in the valve oscillations can be neglected in comparison with the acoustic input resistance r of the pipe. Then, assuming that the frequency of oscillation of the coupled system is equal to the frequency of free oscillations of the valve, the entire stability contour can be obtained, as follows.

With reference to Eq. 7.33, the reactance can be expressed in terms of the resistance as $x/r = (1 - \theta_2^2)\tan(k'L')/[\theta_2(1 + \tan^2(k'L')]$ and if $r' = 0$ in Eq. 7.35 and with $r_f = (\beta/A\rho)(x/\omega)$, it follows from Eq. 7.35 that with $\omega \approx \omega_0$ and $k'L' \approx k_0'L'$, the stability contour becomes

$$\boxed{\begin{array}{l} \textit{Approximate stability contour (Fig. 7.17)} \\ \gamma = \dfrac{\theta_2(1+\tan^2(k_0'L'))}{(1-\theta_2^2)\tan(k_0'L')} \end{array}} \qquad (7.40)$$

$[\gamma = \beta/(\rho A\omega_0$ *Instability parameter (see Eq. 7.38).* $k_0' = (1-M^2)\omega_0/c.$ ω_0: *Angular frequency of free valve oscillations.* $\theta_2 \approx M$: *Flow-induced termination resistance of pipe.* L': *Acoustic length of pipe (see Eq. 7.30)].*

This is valid when $\tan(k_0'L') > 0$ corresponding to mass-like reactance of the fluid column. For $\tan(k_0'L') < 0$, the system is stable. Thus, in the stability contour in

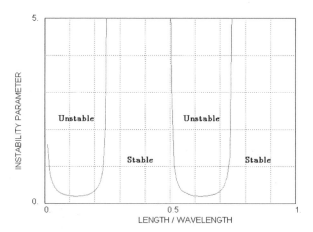

Figure 7.17: Approximate stability contour (diagram) for axial control valve in which the mechanical damping of the valve is assumed to be zero and the frequency of oscillation equals the free valve frequency ω_0. The instability parameter is $\gamma = \beta/(\rho A\omega_0)$. The length used in the abscissa is the acoustical length L' of the pipe and the wavelength is $\lambda_0 = 2\pi c/\omega_0$.

Fig. 7.17, the system is unstable above the contour and stable below, as indicated.

In this simplified version of the analysis, only the flow-induced losses in the pipe have been included and expressed in terms of the termination resistance $\theta_2 \approx M$ (M is the flow Mach number). The minimum value of γ, as obtained from Eq. 7.40, is then $2\theta_2/(1 - \theta_2^2)$ which occurs for $\tan(k_0'L') = 1$, i.e., at $L'/\lambda_0 = 1/8 + n0.5$ for the n:th minimum.

The unstable region, in essence, corresponds to the pipe length for which the input reactance is mass-like and the feedback resistance negative. The sound pressure and valve vibration amplitude can then be considerable, and this effect has been found to lead to valve failures in the processing industries and in power plants. In nuclear power plants, in which the pipe length can be quite large, the axial vibration frequency generally is low, typically in the range 40 to 100 Hz.

7.8.2   Lateral Valve Instability

A transverse rather than axial acoustic mode of the pipe can also be involved in an instability which is the case for lateral oscillations of a valve. Unlike the axial valve instability, there is now no net feedback force produced by the acoustic mode but rather a feedback *torque* on the valve which can lead to an instability in the lateral displacement of the valve (bending motion of the valve stem). Whereas an acoustically induced axial instability typically occurs at a relatively low frequency (typically of the order of 50 to 100 Hz in a nuclear power plant, depending on the pipe length), the lateral instability normally has a considerably higher frequency, of the order of 1000 Hz (depending on the pipe diameter). In the following discussion, we shall neglect the effect of a reflected wave from the end of the pipe.

In a circular pipe, the first lateral mode resonance occurs at a frequency $\approx c/1.7D$, where c is the sound speed and D the pipe diameter. With reference to Fig. 7.18, such a mode can be excited by an oscillating nonuniform flow entering the pipe which can be a result of a lateral oscillation of the valve. The sound pressure p thus produced acts on the valve and can produce a torque on the valve to promote the

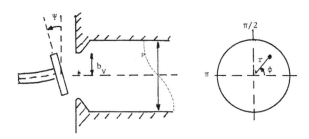

Figure 7.18: The flow is assumed to go from left to right. A bending of the valve stem, as shown, increases the flow into the pipe on the right side thus exciting a transverse mode which with positive pressure on the right-hand side, thus promoting the initial diplacement (potentially unstable).

angular displacement of the valve and cause instability, as indicated in Fig. 7.18. In order for a transverse acoustic mode to be generated, the velocity at the entrance of the duct must depend on the angle ϕ. A lateral displacement of the value, even though it may not affect the mean velocity of produce an axial acoustic wave in the duct, it generally will excite a higher mode with an axial velocity distribution of the form $u_x = A_{mn} J_m(k_{mn}r) \cos(m\phi)$. For small angles of the valve displacement, the amplitude A_{mn} is proportional to the angle of displacement ψ, and we set $A_{mn} = u_{mn}\psi$, where u_{mn} is a characteristic of the valve, an 'asymmetry' parameter which can be measured in a static test. For a valve plug in the form of a spherical ball with a matching spherical valve seat, the asymmetry parameter will be zero and the valve will be stable; for a valve in general, however, it will not be zero and the valve becomes potentially unstable. It should be noted, however, that in the configuration shown in Fig. 7.18, a reversal of the flow direction would make the valve stable for acoustically induced lateral oscillations with the pressure in the acoustic mode now being negative on the upper side in the figure, thus producing a counter torque on the valve.

The axial propagation constant of the mn:th mode is $k_x = \sqrt{k^2 - k_{mn}^2}$ and the pressure perturbation corresponding to the velocity perturbation, as obtained from $-i\omega\rho u_x = -\partial p/\partial x$, will be

$$p_{mn} = \rho c \frac{u_{mn}}{\sqrt{1 - (k_{mn}^2/k^2)}}\psi = \rho c \frac{u_{mn}}{\sqrt{1 - f_{mn}^2/f^2}}\psi, \qquad (7.41)$$

where f_{mn} is the cut-off frequency of the mn:th mode and f, the frequency of oscillation.

The reaction torque τ_{mn} on the valve from such a pressure mode can now be calculated by integrating over the valve surface and we express the corresponding torque amplitude as $\tau_{mn} = p_{mn} A_v r_{mn} \equiv z_{mn}\psi$, where A_v is the valve area and r_{mn} an average lever arm for the mn-mode. The quantity z_{mn} is then

$$z_{mn} = \rho c A_v u_{mn} r_{mn} / \sqrt{1 - f_{mn}^2/f^2}. \qquad (7.42)$$

Friction in the valve results in a torque which we express as $-R\partial\psi/\partial t$ and the bending stiffness in a restoring torque $K\psi$. Then, if the moment of inertia is I, the complex amplitude equation of motion leads to the frequency equation $-I\omega^2 - i\omega R + K = z_{mn}$, which, with $z_{mn} \equiv z_r + iz_i$, becomes

$$-I\omega^2 - i(\omega R + z_i) + (K - z_r) = 0. \qquad (7.43)$$

This oscillator equation includes the feedback from the higher mode and it follows that the equivalent resistance contains the feedback torque through the quantity z_i. If this quantity is negative, the equivalent resistance can be negative and instability will result.

It is clear from Eq. 7.42 that if $f > f_{mn}$, $z_i = 0$ and a higher mode will not affect the damping. With $f < f_{mn}$, however, z_i will be negative, so that the total damping in the system will be reduced and can be zero or negative at a frequency sufficiently close to f_{mn}. The valve is then unstable and will be driven by the aero-acoustic coupling in lateral oscillations.

Stability Diagram

The frequency equation (7.43) is to be solved for the complex frequency $\omega = \omega_r + i\omega_i$. To establish a stability contour, we put $\omega_i = 0$ and for this condition of marginal stability the equation reduces to

$$-I\omega_r^2 + K - z_r = 0$$
$$\omega_r R + z_i = 0. \tag{7.44}$$

In terms of the quantities $\omega_v = \sqrt{K/I}$, $\Omega = \omega/\omega_v$, and $Q = I\omega_v/R$, these equations become

$$1 - \Omega^2 = z_r/K$$
$$\Omega/Q = z_i/K. \tag{7.45}$$

Below the cut-off frequency, $z_r = 0$ and $z_i = -C/\sqrt{f_{mn}^2/f^2 - 1}$, where $C = \rho c A_v r_{mn} u_{mn}$, and the equations reduce to

$$1 - \Omega^2 = 0$$
$$\Omega/Q = \eta/\sqrt{f_{mn}^2/f^2 - 1}$$
$$\eta = \rho c A_v r_0 u_{mn}/K. \tag{7.46}$$

With $\Omega = 1$, as obtained from the first equation, the frequency of oscillation is the (known) frequency f_v of free lateral oscillations of the valve. For each lateral acoustic mode in the pipe, with a cut-on frequency f_{mn}, the stability boundary curve in a space with coordinates $Q\eta$ and f_v/f_{mn} takes the form

$$\boxed{\begin{array}{c} \textit{Stability contour for lateral valve oscillations (Fig. 7.19)} \\ Q\eta = \sqrt{(f_{mn}/f_v)^2 - 1} \qquad (f_v < f_{mn}) \end{array}} \tag{7.47}$$

[η: 'Instability' parameter (Eq. 7.46). $Q = I\omega_v/R$: See I: Moment of inertia of valve. R: Resistance coefficient (see Eqs. 7.43 and 7.44). $f_v = \omega_v/2\pi$: Frequency of free lateral valve oscillations (Eq. 7.44) f_{mn}: Cut-on frequency of the (m,n) acoustic mode in the duct].

The cut-on frequency for the lowest mode is $\approx c/1.7D$, where D is the pipe diameter.

The stability diagram (contour) for the mn:th mode is shown in Fig. 7.19. The instability parameter on the ordinate axis is $Q\eta$, where $Q = \omega_v I/R$ is the 'Q-value' of the free oscillations of the valve and $\eta = \rho c A_v u_{mn} r_{mn}/K$ contains the non-uniformity parameter u_{mn} and the 'lever arm' r_{mn}. The region of instability increases with Q and the non-uniformity coefficient of the valve. The valve is stable to the right of the contour.

Although the discussion above referred to the idealized case of a flat valve plug, the results apply also to other valve configuration. For example, if the valve plug is provided with a cone with a sufficiently small apex angle which protrudes into the pipe, the torque produced on the valve can be in the opposite direction to that obtained for a valve plug without a cone but the effect will be the same with the

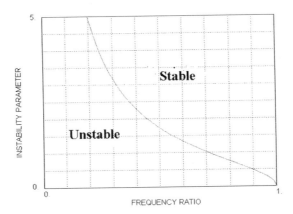

Figure 7.19: Stability diagram for flow driven lateral oscillations of a valve. Instability parameter: $Q\eta$, defined in the text. Frequency ratio: f_v/f_{mn} where f_v is the frequency of free lateral oscillations of the valve and f_{mn} is the cut-on frequency of the mn-mode involved (Eq. 7.47).

acoustic mode promoting the intial displacement when the flow is in the 'normal' direction, as explained in the caption to Fig. 7.20. Again, the valve is potentially unstable for acoustically induced lateral oscillations with the flow going in the 'normal' direction but stable when the flow is reversed. Many a problem involving lateral valve oscillations has been eliminated by simply reversing the orientation of the valve.

The lateral instability resulted from the bending of the valve stem. Under such conditions, the acoustic modes in the pipe involved predominantly involve modes with one nodal diameter, i.e., $m = 1$. It is possible, however, that the valve plug itself can be excited into vibrations and modes with higher values of m then have to be considered. The analysis of these is completely analogous to what has been said for the $m = 1$ mode.

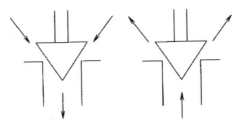

Figure 7.20: Left: A displacement to the valve plug increases the volume flow on the right side and a transverse acoustic mode in the pipe is generated with a positive pressure on the right-hand side. This results in a force on the valve plug to the left and a corresponding torque promoting the original displacement (potentially unstable). Right: With the flow reversed the same acoustic-mode is excited but 180 degrees out of phase since the increased outflow on the right-hand side corresponds to a decrease of the pressure in the acoustic mode (stable).

It is possible that a lateral valve instability can be of the 'flutter' type in which the coupling between the flow and the valve involves vorticity or flow-induced pressure distributions not related to acoustic modes (see Fig. 7.7). For valves used at supercritical pressure ratios, shock waves add to the causes of valve instabilities. One mechanism which is proposed here is that a lateral displacement of a valve plug with a cone attachment extending into the pipe will influence the angular dependence of the axial location of a shock and this can lead to a feedback torque which can promote the displacement and cause instability.

7.8.3 Labyrinth Seal Instability

There are many other valve-type instabilities, i.e., instabilities where both aeroacoustic and structural vibrations are involved in the feed back loop (see Fig. 7.7). One example has to do with an annular labyrinth seal between two stages in a jet engine compressor, as shown schematically in Fig. 7.21.

The seal barriers are mounted on an outer cylinder and will provide (flow) ports between these barriers and an inner concentric cylinder. The center line of the cylinders are indicated by the dash-dot line. The flow goes from right to left through the annular ports.

Qualitatively, the mechanism is the same as for the axial valve instability discussed in the previous subsection. The basic difference is that our discussion of the valve involved only one structural degree of freedom, that of the valve plug. Now the flow goes through an annular gap which generally varies with the angular position and the structural vibrations involve a cylinder with its many modes of vibration.

The gap size can be modulated by vibrations of both cylinders involved in the system but we shall consider only the contributions from the inner cylinder shown in the figure. The mechanism of the instability is as follows. A structural mode of the cylinder produces a variation of the port width with the same angular dependence as the redial displacement of the cylinder mode. This variation produces a modulation of the air flow through the ports and there will be a modulated flow entering the annular region between the port barriers. The angular dependence of it is the same as the

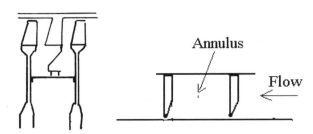

Figure 7.21: Labyrinth seal. The flow goes through two annular gaps between the two seal barriers and a cylinder. The gaps are modulated by the vibrations of the cylinder which leads to flow modulation and coupling to the acoustic modes in the annulus and the structural modes.

angular dependence of the radial displacement of the cylinder. This flow excites a circumferential acoustic mode in the annulus with the same angular dependence as the structural mode of the cylinder. The acoustic mode interacts with the cylinder (i.e., deforms it) and if the phase relations are right, this feedback will promote the vibrations of the cylinder and create an instability. Conditions for instability are much the same as for the valve instability. If the acoustic mode frequency of the annulus is somewhat below the frequency of the free vibration of the corresponding mode of the cylinder, the acoustic feed back is equivalent to a negative damping, and an instability will occur unless the combined damping of the acoustic and structural modes is larger than the equivalent negative damping. The mathematical analysis of the problem is analogous to that for the valve.

There is an interesting aspect of this problem which involves the difference in the dispersion relation for ordinary sound and for bending waves on a cylinder. We are dealing with circumferential bending waves on the cylinder which are described in the same way as the waves on a thin plate. For the sound wave traveling in the annulus between the seal barriers, the wave speed is independent of frequency but for a bending wave on a thin plate, it is proportional to the square root of frequency. As a result, the modal frequency of the sound wave will be proportional to the modal number of the wave in the annulus (i.e., the number of wavelength around the annulus), the first mode occurring when the circumference of the annulus is one wavelength. For the bending wave, on the other hand, the modal vibrational frequency will be proportional to the *square* of the mode number. The modal frequencies are marked by the open and filled circles in Fig. 7.22, where f_a and f_s refer to the acoustic and structural modes. The structural mode frequencies are essentially independent of temperature. The acoustic frequencies f_a, however, do depend on temperature and move up and down with the temperature. As for the valve instability, exact coincidence of the acoustic and structural frequencies f_a and f_s does not lead to an instability. Rather, the instability region can be shown to be limited to a narrow band

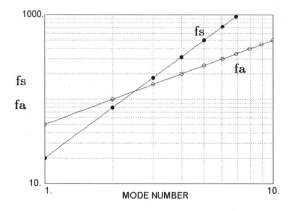

Figure 7.22: Concerning seal instability. Acoustic and structural mode frequencies, fa and fs versus the mode number n.

below the structural frequencies and in order for instability to occur, the acoustic frequency for a given mode has to fall in this region. At the condition shown in the figure, the third mode, $n = 3$, satisfies this condition. If the temperature is increased, the fourth mode will be unstable, and if the temperature is decreased, the second mode will eventually become unstable.

7.9 Heat Driven Instabilities

An acoustically stimulated heat source in a gas can lead to instabilities which come about as a result of the influence of fluid velocity and/or pressure on the rate of heat release. Mechanically maintained oscillations are also possible in which thermal expansion deforms a body in such a manner that the heat transfer to the body is altered thus providing feed back. An example is the rocking motion of a tea kettle.

In regard to sound generation, a fluctuating heat release in a gas is equivalent to a mass flow source. This can be seen as follows. When an acoustic flow source Q_f (mass rate transfer per unit volume) is present under isentropic conditions, the mass conservation equation is $\frac{\partial \rho}{\partial t} + \rho \text{div} \, \boldsymbol{u} = Q_f$, as discussed in Eq. 5.34.

Under nonisentropic conditions, we let the pressure be a function of ρ and S, where S is the entropy. Thus, with $P = P(S, \rho)$, we get $dP = [\partial P/\partial \rho]_S \, d\rho + [\partial P/\partial S]_\rho \, dS$. With dP being the acoustic perturbation p of the pressure, the first term becomes $c^2 d\rho$, where c is the isentropic sound speed. Normally, this is the only term which is used. To evaluate the second term, we use the equation of state expressing the relation between P, ρ, and S, the derivation of which is reviewed as follows.

The first law of thermodynamics $dH = c_v dT + P dV = c_v dT - (P/\rho^2) d\rho = c_v dT - (rT/\rho) d\rho$, where dH is the heat transfer per unit mass, c_v the specific heat and r the gas constant, both per unit mass. From $P = r\rho T$ follows $dP/P = dT/T + d\rho/\rho$. Then, with $dH = TdS$, where S is the entropy per unit mass, it follows from these equations that $dS = c_v dP/P - (r + c_v) d\rho/\rho$, or $dP/P = \gamma d\rho/\rho + S/c_v$, where we have used $r = c_p - c_v$ and $\gamma = c_p/c_v$.

The relation between first order perturbations p, δ, and σ in pressure, density, and entropy is then

$$p = (\gamma P_0/\rho_0)\delta + (P_0/c_v)\sigma = c^2\delta + (P_0)/c_v)\sigma, \qquad (7.48)$$

where c is the isentropic sound speed. The first order equation for mass conservation $\partial \delta/\partial t + \rho \text{div} \, \boldsymbol{u} = 0$ then will be modified to read

$$(1/c^2)\partial p/\partial t + \rho \text{div} \, \boldsymbol{u} = (P_0/c^2 c_v)\partial \sigma/\partial t. \qquad (7.49)$$

Finally, if the rate of heat transfer per unit mass is H and with $dH = TdS$, the right-hand side becomes $(P_0/Tc^2 c_v)\partial H/\partial t = (\gamma - 1)/c^2)\rho \partial H/\partial t$, where we have used $P_0 = r\rho T$, $r = c_p - c_v$, and $\gamma = c_p/c_v$.

In the presence of a mass source term with a flow rate Q_f per unit volume, the right-hand side is $\partial Q_f/\partial t$. Thus, a heat source is acoustically equivalent to a mass flow source $[(\gamma - 1)/c^2]\rho H$, where H is the heat transfer per unit mass. Had we defined the heat transfer per unit volume, $H' = \rho H$, the equivalence would have been $Q_f \leftrightarrow (\gamma - 1)/c^2]H'$.

7.9.1 The Rijke Tube

The Rijke tube is an ancient example of a heat maintained oscillation and is still occasionally used as a simple lecture demonstration. It is simply a vertical steel tube, typically 1 to 2 m long with a diameter of about 10 cm. The tube contains a wire mesh screen in the lower half of the tube.

The screen is heated by a Bunsen burner (or an electric current), and after removal of the burner (to avoid vortex formation), a tone gradually builds to a high amplitude at the frequency of the fundamental acoustic mode of the air column in the tube. As the screen cools, the amplitude of the tone decreases. It is observed that a tone is produced only if the screen is located in the lower half of the tube. A qualitative explanation of the phenomenon goes as follows.

The heating of the air in the tube by the screen creates a convection of air up through the tube at a velocity U. Consider now the effect of a superimposed acoustic (axial) mode which can be considered to be initiated by an unavoidable fluctuation. In the fundamental acoustic mode, the oscillatory flow velocity u, goes in and out of the ends of the tube in counter motion, the velocity at the center of the tube being zero. During the half cycle when the flow is inwards (outwards) the sound pressure p in the tube increases (decreases) with time.

At a time when the flow is inwards (i.e., sound pressure increasing with time), the oscillatory flow velocity u is in the *same* direction as the convection velocity U in the *lower* half of the tube. Then, with the screen located in this half, the rate of heat transfer to the gas from the screen will be increased through the cooling action of the air motion in the sound field. This tends to increase the sound pressure in the tube (i.e., promote the growth already under way and an instability normally will occur).

In the upper half of the tube, on the other hand, the inflow in the acoustic mode opposes the convection flow, and with the screen located in this half, the rate of heat transfer to the gas is reduced through the action of the acoustic mode. This tends to decrease the sound pressure in the tube (i.e., oppose the existing rate of increase) and the acoustic feed back in this case leads to an additional damping of the acoustic mode and no instability.

7.9.2 Combustion Instabilities

The rate of heat release H in combustion depends, among other things, on the pressure P, $H = H(P)$. A combustion chamber (in a gas turbine, for example) can be regarded as an acoustic resonator with several acoustic modes. The sound pressure p in such a mode will produce a fluctuation $q = \delta H = (\partial H/\partial P)\delta p$ in the heat release and if $\partial H/\partial P > 0$, this corresponds to positive feedback and a corresponding negative acoustic damping of the chamber mode. Then, if the actual damping of the mode is not sufficient to counteract the negative damping, the system will be unstable and the acoustic oscillations will grow until nonlinear effects will limit the sound pressure in the mode. However, the pressure induced structural vibrations may then be large enough to cause fatigue failure.

The modal damping can be increased by adding acoustic absorbers to the interior of the chamber, such as porous material or resonators. In the analysis of their effects, the new modes and their damping in the modified chamber have to be determined. As a first approximation in such an analysis, the absorbers are designed for maximum damping the frequency of the original mode and if the bandwidth is large enough (by providing resistive screens in resonators, for example) they will perform well also at the modified modal frequency. Since the problem is to prevent the onset of the instability, the damping should be optimized at relatively low sound pressure levels (i.e., without reliance on nonlinear damping effects which can be considerable in resonator dampers). Actually, if resonator dampers are used, the performance can be made relatively independent of the sound pressure level if porous screens are used in the necks of the resonators.

Chapter 8

Sound Generation by Fans

In Chapter 7, flow interaction with a stationary solid body was considered. As far as sound generation is concerned, however, it is the relative motion of the body and the fluid that matters. In devices such as fans, propellers, pumps, and compressors, it is primarily the solid object (blade) that is moving but in the process it induces motion of the fluid also. This interaction often results in significant sound generation which has become an important engineering (environmental) problem and a field of its own with an extensive literature (refer to Appendix A6 for supplementary notes).

As far as sound generation is concerned, there is essentially no difference between a propeller and a fan except for the translational motion of a propeller and related acoustical effects, such as Doppler shift. A fan often operates in a duct and this introduces the problem of coupling with the acoustic modes in the duct but the flow-induced force distributions on the blades on fans and propellers are essentially the same.

The extensive literature on fans includes empirical relations between the sound power and operating parameters, such as flow rate and pressure change across the fan. We shall consider here merely some of the physics involved in sound generation which forms the basis for analytical studies (refer to Appendix A6 for supplementary notes).

8.1 Axial Fan in Free Field

The sound generated by a fan contains a periodic as well as a random part. The former is related to the impulses produced by the blades on the surrounding air and the latter is caused by turbulence. Only the periodic part will be discussed here. The turbulent part arises from the flow in the wakes of the blades and from turbulence in the incident flow.

The interaction of a blade of a fan with the surrounding fluid is much the same as that of an airplane wing and results in a lift and a drag force on the blade and a corresponding momentum transfer to the fluid. The tangential component of the interaction force, the drag, results in a torque on the fan.

The axial force component produces an average pressure change across the fan, a *driving pressure* or *internal pressure*, and a corresponding flow in the external flow

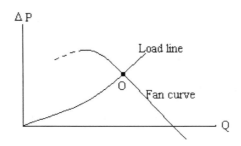

Figure 8.1: The fan curve, i.e., the pressure change ΔP across the fan versus the volume flow rate Q, and the load line of the external circuit. The operating point of the fan is O.

loop. By analogy with an electrical circuit, the driving pressure corresponds to the electromotive force in a generator or battery and the flow corresponds to the electrical current in the circuit attached to the generator.

The flow that results from the driving pressure ΔP depends on the load on the fan, i.e., the resistance of the external flow path. This resistance (ratio of ΔP and the average flow velocity) is generally velocity dependent. In the electrical analogy, the effect of the external circuit on the electrical current is often expressed in terms of a circuit resistance (or impedance in an AC circuit) and an equivalent internal resistance (or impedance) of the source is also introduced. There are analogous quantities for the fan.

The relation between the driving pressure ΔP (average pressure change across the fan) and the volume flow rate Q (reduced to standard conditions of temperature and pressure)[1] through the fan is generally expressed in terms of a 'fan curve,' as shown schematically in Fig. 8.1, which accounts for the 'internal' characteristics of the fan. The corresponding description of the external flow loop is contained in the 'load line,' which is also shown in the figure. The slope of this line yields the resistance of the loop. The intersection of the fan curve and the load line yields the operating point O of the fan. There will be one fan curve for each rotational speed of the fan.

The driving pressure ΔP of the fan generally increases monotonically with decreasing flow rate Q up to a maximum value, but a further decrease in Q often leads to a decrease in the pressure. This yields a region of a positive slope of the fan curve and a potentially unstable operation of the fan.

The fan curve extrapolated into regions of negative driving pressure can be interpreted as describing the performance of the fan when driven as a turbine. The closest corresponding electrical analogue would be the charging of a battery or running an electrical generator as a motor.

The fan curve shows the relation between the *average* values of the flow rate through the fan and the pressure change across it. From the standpoint of sound radiation, these average quantities per se are of little interest. Rather, the relevant

---

[1] Not to be confused with Q_f the mass flow rate or Q, the acoustic monopole strength per unit volume used in Chapter 5.

characteristic has to do with the local variations or ripples in the pressure and the flow which move with the fan and produce time dependent perturbations of the fluid as the blades move by a stationary observer.

In our analysis of this problem we simulate the fan by an equivalent acoustic force distribution in a (pillbox-like) control volume enclosing the fan. Such a model can be developed on several levels of sophistication as will be discussed below.

8.1.1   Sound Generation; Qualitative Observations

A stationary observer in a plane just outside the fan experiences a momentum transfer or force from each blade as it passes by. Since there is a net force provided by the fan as a whole, there is a corresponding mean value of the local force. The time dependent part is the oscillation about the mean value caused by the ripples mentioned above and this time dependence is responsible for sound radiation. In a coordinate system attached to the moving blades, the mean value, of course, is independent of the angle but the remaining part is a periodic function of the angle with the period $2\pi/B$ if there are B identical blades in the fan. In the observers stationary frame of reference, there will be a corresponding periodic function of time with the period equal to T/B where T is the period of revolution of the fan.[2] An observer at an adjacent angular position records the same periodic function except for time delay. In our first model, we shall replace the distributed force on a blade by an average point force located at a radius which we denote a'. In this acoustic simulation, the sound is produced by a continuous distribution of point forces (dipoles) over a circle of radius a'. The magnitude of the force is the same at all angular positions but the phase angle varies.

The calculation of the field from this model is analogous to that for a linear array of sources. Because of the periodic spatial variation of the force distribution (with a period equal to the angular blade separation) there will be no net acoustic dipole source of the fan. For each Fourier component of this periodic function, a positive contribution to the force is always matched by a corresponding negative one. The distance from a positive source point to the axis of the fan will be the same as that from a negative source point; the corresponding radiated sound contributions then arrive at the axis with equal strength but with opposite signs and cancel each other through destructive interference. Thus, under these conditions, the sound pressure along the entire exis of the fan will be zero both upstream and downstream of the fan.

To obtain a sound pressure different from zero requires a path difference between the positive and negative parts of the force distribution region in the plane of the fan. Such a path difference exists at locations away from the axis. The path difference depends on the angular position of the field point and for each Fourier component of the angular force distribution there will be a corresponding directivity pattern for the radiated intensity resulting from wave interference in much the same way as for the point source or line source arrays. The discussion of sound radiation from a moving corrugated board or a one-dimensional cascade of blades is particularly relevant (see Fig. 8.2). If the speed U of the board or cascade is subsonic, we found that the pressure perturbation in the surrounding fluid decreased exponentially with distance

---

[2]If the blades are not identical or their spacing nonuniform, the fundamental period will be T.

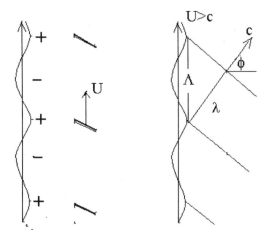

Figure 8.2: Regarding the radiation of sound from a moving corrugated board and an analogous linear cascade of blades. For $U < c$ only a local periodic disturbance ('sloshing') of the air is produced corresponding to an evanescent sound field which will occur even in an incompressible fluid. The difference in path lengths from adjacent positive and negative source regions to the field point is less than half the acoustic wavelength and constructive interference cannot occur. Only if $U/c > 1$ is such an inteference possible.

from the board and the surfaces of constant pressure amplitude (wave fronts) are parallel with the board. Such an evanescent field is similar to that of a higher order acoustic mode in a duct, discussed in Chapter 6, particularly in Section 6.2.2. The reason for the decay is the (partial) destructive interference between the positive and negative sound pressures emitted from adjacent positive and negative source regions (crests and valleys). If the velocity of the board is U and the spatial period of the corrugations Λ, the frequency of the generated sound is $f = U/\Lambda$ and the wavelength $\lambda = c/f = (c/U)\Lambda$. For subsonic motion, $U/c < 1$, the wavelength λ of the emitted sound is larger than the spatial period Λ of the board. The maximum path difference of travel for positive and negative signals to a field point is $\Lambda/2$. The path difference required for signals to arrive in phase (constructive interference) is $\lambda/2 = (1/M)\Lambda/2$. Thus, for subsonic motion $M < 1$ this cannot be achieved and the interference is partially destructive. As the distance to the field point increases, the path difference decreases, and the destructive interference becomes more and more complete and the resulting sound pressure closer to zero. The correspondence spatial decay of the sound pressure turns out to be exponential.

In this evanescent field (near field), it can be shown that the sound pressure and velocity perturbations are 90 degrees out of phase and, in steady state, there is no acoustic power emitted into the far field (see Problem 1). The near field merely corresponds to a periodic sloshing of the air back and forth between the alternating regions of crests and valleys. The kinetic energy in this motion is established during the transient period of starting the motion during which energy is transferred to the fluid to build up this motion. Once steady state has been established, however, there will be no net flow of energy from the board to the fluid, neglecting viscous drag.

Only if the speed of the board or blade cascade becomes supersonic will the wavelength λ become smaller than Λ which makes possible constructive interference and an energy carrying sound wave, traveling at an angle θ with respect to the normal to the board (see Fig. 8.2).

From this discussion of the board and the linear blade cascade, it is tempting to conclude that a fan with subsonic tip speed would not radiate sound into the far field. Unlike the board or the one-dimensional cascade of blades, a fan in free field, however, *does* radiate sound even if the rotational tip speed of the fan is subsonic. As a reason one might suggest (loosely) that even if the tip speed is subsonic, the pressure field which moves with the propeller and extends beyond the blades will be supersonic at a radial position sufficiently far out from the tip thus causing radiation. This qualitative picture is appealing because if this outer supersonically moving pressure field is eliminated by putting a duct (shroud) around the fan, the radiation disappears and we will be left with a decaying field in the duct similar to that produced by the board.

In this discussion, it has been tacitly assumed that the inflow is uniform; a nonuniform flow results in a qualitatively different interaction, as will be shown next.

Effect of Nonuniform Flow

An inhomogeneous inflow flow with a circumferential variation at the entrance to the fan has several important consequences, even if the flow is stationary. First, the force on a blade as it plows through this flow will be time dependent. If the flow is also time dependent (caused by turbulence or pulsations, for example) additional force variations result. Second, the pressure disturbance produced by the interaction of the fan with the flow generally moves in the circumferential direction at a speed different from the speed of the fan; the speed can be supersonic even if the tip speed is subsonic. This is obvious when the circumferential period of the inhomogeneity is the same as that of the blades. All the blades then interact with an inhomogeneity at the same time and the corresponding pressure pulses about the blades are all in phase. This is equivalent to saying that the speed of the corresponding excess pressure perturbation travels with infinite speed in the circumferential direction. Strong acoustic radiation is then to be expected and there will be a net sound pressure even on the axis of the fan (as before, the field from the *uniform* part of the flow still will be zero on the axis, of course).

The specific example in Fig. 8.3 is instructive to illustrate the notion of a spinning pressure interaction field. Thus, let the inflow be stationary but periodic in the circumferential direction with three cycles, which, for example, could be produced by the wakes from three struts in front of the fan. Let these regions be denoted I1, I2, and I3, located at the angular positions 0, 120, and 240 degrees. The fan has two blades, B1 and B2, and we shall designate an interaction between blade B1 and inhomogeneity I1 by (11) with similar designations (23), etc., for other interactions. As B1 interacts with I1 at the angular location 0, the blade B2 is at 180 degrees and therefore 60 degrees from I3. Thus, after the (11)-interaction has produced a

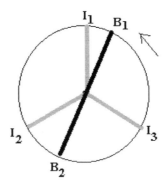

Figure 8.3: The interaction of a two-bladed fan with a nonuniform flow with three periods in the circumferential direction.

pressure pulse at I1 (zero degrees), the next interaction, (23), will take place at I3 when the fan has turned 60 degrees to bring B2 to interact with I3. The pressure pulse then has moved from I1 to I3, 120 degrees in the negative direction.

After another 60 degrees of rotation of the fan, B1 interacts with I2 so that the pulse has moved from I3 to I2, another 120 degrees in the negative direction. In other words, a displacement of the fan by 60 degrees in the positive direction produces a rotation of the pressure field by 120 degrees in the negative direction and its angular velocity will be 2Ω, twice that of the fan. The sequence of interactions is (11), (23), (12), (21), (13), (22), etc.

We also detect a pressure field component that rotates in the positive direction. The sequence of interactions in this process is (11), (22), (13), (21), (12), (23), etc. In this case, a displacement of the fan by 300 degrees produces a displacement of the pressure pattern by 120 degrees in the positive direction. Thus, this field will rotate with an angular velocity which is $(2/5)\Omega$ in the positive direction.

This result can be generalized to the interaction between the mth harmonic of the blade passage frequency and the qth spatial harmonic component of the flow nonuniformity. Thus, the angular velocities of the resulting spinning pressure fields in this case become

$$\boxed{\begin{array}{c} \textit{Angular velocities of spinning pressure fields in Fig. 8.3} \\ \Omega_{\pm} = \Omega/(1 \pm q/mB) \end{array}} \qquad (8.1)$$

[Ω: *Angular velocity of fan rotation. q: Order of the Fourier component of the flow inhomogeneity. B: Number of blades. m: Multiple of the blade passage frequency*].

This result emerges automatically from a general mathematical analysis of the interaction without the need for the type of counting used in the description above. In our example, with $q = 3$ and $mB = 2$, we get $\Omega_- = -2\Omega$ and $\Omega_+ = (2/5)\Omega$, as found above. The plus and minus signs correspond to rotations in the positive and negative directions, respectively. It should be noted that if the harmonic order of the flow nonuniformity is larger than mB, the fast moving pressure field will rotate in the negative direction; if it is smaller, both waves travel in the positive direction with a speed less than that of the fan.

The nonuniformity of the inflow can contribute significantly to and even dominate the radiation from a fan or propeller, particularly at low speeds. This is true even to a higher degree for a fan in a duct. Since lowering the speed of rotation of the pressure field reduces the efficiency of sound radiation, as we shall see shortly, it is desirable to make q (number of struts) as large as possible. The same applies to the number of guide vanes in rotor-stator interaction, as will be discussed later.

8.1.2   Point Dipole Simulation

The simplest acoustic modeling of a fan is to let each blade be represented by a point force (dipole). For a fan with B identical and uniformly spaced blades, the forces are distributed uniformly on a circle of radius a' and moving with a constant angular velocity Ω of the fan, as indicated schematically in Fig. 8.4. A straight-forward extension of this model is to let the radial positions, the angular spacings, and the strengths of the forces to be different but it will not be considered here. We shall comment, however, on an extension of the model in which the point forces are replaced by line forces in which the spanwise (radial) load distribution on the blades is accounted for.

Each point force in our model is a result of the interaction of a blade with the surrounding fluid and is determined by the fluid velocity relative to the blade. The flow incident on the fan is assumed to be uniform and axial so that there is no time dependence of the interaction force in the frame of reference S' moving with the fan (see Fig. 8.4). The axial component of the force is denoted X and the tangential or drag component T. As indicated in the figure, the coordinates in the frame of reference attached to and moving with the fan blades are denoted r' and ϕ' and the plane of the fan is placed in the yz-plane at $x = 0$. The spatial period of the force distribution in S' is $2\pi/B$.

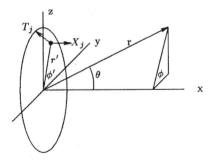

Figure 8.4: Coordinates used in the analysis of sound radiation from point dipoles moving along a circle. Each source is specified acoustically in terms of an axial and a tangential force amplitude, X_j and T_j.

The description of the distribution in the stationary frame of reference S, which is used in the calculation of the sound field, is obtained by the coordinate transformation $\phi' = \phi - \Omega t$, where Ω is the angular velocity of the fan. Thus, with the complex amplitude of the mth harmonic component in the Fourier expansion of the force function in S' being proportional to $\exp(imB[\phi'])$, it is $\exp(imB\phi)\exp(-imB\Omega t)$ in S. This represents the complex amplitude distribution of the force over a circle of radius a' with the frequency of the mth harmonic component being $mB\Omega$ and with a phase angle $mB\phi$.

To obtain the radiated sound field we have to integrate over this distribution, the integral being expressed by Bessel functions.

The far field sound pressure thus obtained from the axial and tangential force distributions is expressed as a Fourier series, i.e., a sum of harmonics of the fundamental (blade passage) frequency $B\Omega$ and the mth harmonic of the total field is found to be

$$\boxed{\begin{array}{c} \textit{Point force simulation of fan (Fig. 8.4)} \\ p_m(r,t)/F = mB\frac{\eta M \cos\theta + T/X}{2(r/a')}P_m(\theta)\sin[mB(\phi - \Omega t + kr + \pi/2)] \\ P_m(\theta) = J_{mB}(mB\eta M \sin\theta) \end{array}} \qquad (8.2)$$

$[F = BX/(\pi a^2) \approx \Delta P$: Average axial force per unit area of the fan. ΔP: Static pressure change across fan. p_n: Sound pressure of the nth harmonic of the blade passage frequency. $\eta = a'/a$. a': Radius of force circle. a: Radius of fan. P_m: Directivity pattern of radiation field of the mth harmonic. J_{mB}: Bessel function of order mB. X, T: Axial and tangential force components on a blade. B: Number of blades. $M = a\Omega/c$: Tip Mach number of the fan. θ: Polar angle (Fig. 8.4). a: Radius of the point force circle. Ω: Angular velocity of the fan].

The average axial force F per unit area is approximately equal to the static pressure change ΔP across the fan (the 'driving pressure' in Fig. 8.1 and we have intentionally brought Eq. 8.2 into this form so as to normalize the sound pressure with respect to ΔP.

The Sound Field from the Axial Force Distribution

In the sound field contribution from the axial force component X, the X-field, the magnitude of the pressure will be the same on the two sides of the fan but the signs are different, as expressed by the factor $\cos\theta$ in Eq. 8.2. This is typical of the radiation field from an axial dipole; as it pushes on one side and pulls on the other. Each of the axial forces (dipoles) in the distribution yields zero pressure in the transverse direction, i.e., at $\theta = \pi/2$, and maximum (magnitude) in the axial directions $\theta = 0$ and $\theta = \pi$. Although this is true for each individual source, the *combined* effect of *all* the axial dipoles results in zero amplitude on the axis (at $\theta = 0$ and $\theta = \pi$) because of destructive interference of the sound from positive and negative regions of the source distribution. These regions occur since the force density distribution varies periodically with the angular position ϕ' for each component in the Fourier decomposition of the field. The path lengths from the positive and negative regions to a point on the axis of the fan are exactly the same and the corresponding sound pressures arrive out of phase on the axis and cancel each other for all values of m.

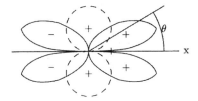

Figure 8.5: Qualitative radiation pattern of a fan in the xz-plane as viewed from a point on the negative y-axis in Fig. 8.4 with the contributions from the axial (solid curve) and the tangential (dashed curve) force distributions. Note that the phase changes by 180 degrees from the front to the back of the propeller for the axial force but not for the tangential.

Thus, the X-field is zero at $\theta = 0$ and π. It will be zero also at $\pi/2$ because of the directivity of an individual dipole. The angular distribution of the sound field from the axial force distribution is illustrated schematically by the solid lines in Fig. 8.5.

The Sound Field from the Tangential Force Distribution

The directionality of the sound field from the tangential force distribution, the T-field, is quite different. Each force produces maximum sound pressure in the direction $\pi/2$ and zero pressure in the axial direction; also the total pressure on the axis will be zero because of the same destructive interference which occurred for the axial force distribution. However, there will be no complete destructive interference at $\theta = \pi/2$ since the path lengths from the positive and negative portions of the source distribution to a point in the plane of the fan (or to any point not on the axis) will be different. Unlike the X-field, the T-field on the two sides of the fan will be in phase as illustrated by the dashed lines in Fig. 8.5.

The Total Sound Field

Fig. 8.5 can be thought of as an instantaneous picture of the X- and T-fields in the plane containing the axis of the fan. In the particular instant shown in the figure, the sound pressure in the T-field is positive both at the top and at the bottom. At another instant the pressure may be both negative or have opposite signs. However, whatever the signs may be, the sound pressure in the X-field on the downstream side will always have the same sign as that in the T-field. The magnitude of the total sound pressure in the downstream direction then will be the sum of the magnitudes of the two contributions but in the upstream direction, it will be the difference. As a result, the total sound pressure will not be the same on the two sides of the fan; the contributions from the X- and the T-field are in phase downstream and out of phase upstream. In fact, with reference to Eq. 8.2, the two fields cancel each other at an angle given by

$$\cos\phi = -\frac{T}{\eta M X},$$

\hfill (8.3)

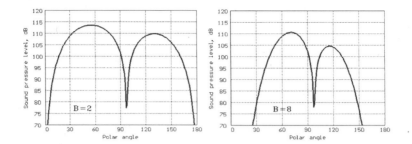

Figure 8.6: Sound pressure level versus the polar angle θ of the first harmonic component of the sound from a fan in free field as simulated by moving point dipoles.
Number of blades: Left: $B = 2$. Right: $B = 8$. Tip Mach number: $M = 0.8$. Average thrust per unit area of fan: $F = 1000$ bar. Tangential/axial force ratio: $T/X = 0.1$. Distance: $r/a = 10$. Span-wise location of sources: $a'/a = 1$. Harmonic component of blade passage frequency: $m = 1$.

where M is the tip Mach number of the fan. This angle lies in the upstream hemisphere (see Fig. 8.6).

The Interference (Bessel Function) Factor

The Bessel function in the X- and the T-fields formally shows that these fields are both zero on the axis ($\theta = 0$ and π) through destructive interference, as discussed above. The Bessel function $J_m(z)$ will be zero for values $z = \alpha_{mn}$ other than zero and the corresponding angle is obtained from $mB\eta M \sin\theta = \alpha_{mn}$, where $\eta = a'/a$. However, this has no real solution for θ for a subsonic source speed. This is readily seen if we recall that the first maximum of the Bessel function $J_{mB}(z)$ occurs for $z \approx mB$ and the first zero (after $z = 0$) at a somewhat higher value. The first maximum then corresponds to an angle given by $\sin\theta = 1/(\eta M)$ which for subsonic speed of the sources, i.e., $\eta M < 1$, has no real solution for θ; the function increases monotonically as θ goes to $\pi/2$ (recall that $\eta = 1$ corresponds to the tip of the blade). For supersonic motion, however, with ηM sufficiently large, solutions for θ other than $\theta = 0$ exist.

There is another useful way of expressing the angular and Mach number dependence of the sound field. Using the first term in the power series expansion of $J_m(z)$, we have

$$J_m(z) \approx \frac{1}{m!}\left(\frac{z}{2}\right)^m. \tag{8.4}$$

Next we use Stirling's formula to express the factorial as

$$m! \approx \sqrt{2\pi m}\,(m)^m e^{-m} \tag{8.5}$$

to obtain an approximate expression for the sound pressure amplitude in Eq. 8.2,

$$p_m/F \approx \frac{\sqrt{mB}[M\cos\theta) + T/(\eta X)]}{2\sqrt{2\pi}}\left(\frac{1}{2}eM\eta\sin\theta\right)^{mB}. \tag{8.6}$$

This illustrates explicitly the important characteristic of the field that the strength of the Mach number dependence of the radiated sound increases with the number of blades and the order of the harmonic component. If the tangential force is neglected, the sound pressure amplitude in this approximation increases as

$$p_m \propto M^{mB+1}. \tag{8.7}$$

Furthermore, the factor $[(e\eta M/2) \sin\theta]^{mB} \approx [1.4(\eta M) \sin\theta]^{mB}$ in Eq. 8.6 increases rapidly toward a sharp maximum at $\pi/2$ for small Mach numbers. Notice the strong dependence of the pressure on the Mach number for large values of mB. For $m = 1$ and $B = 2$, $p_1 \propto M^3$ and for $B = 8$, $p_1 \propto M^9$.

The interference between sources discussed above relies on the assumption that the relationship between the phases of the various field components is maintained during wave propagation. This may not be the case in practice because of turbulence and other inhomogeneities in the fluid. The sound pressure on the axis of the fan will then be different than the predicted value of zero.

8.1.3 Numerical Results

As a first example (Fig. 8.6), we have computed the angular dependence (directivity) of the total sound pressure amplitude from Eq. 8.2. The point forces are assumed identical and uniformly spaced on a circle with a radius equal to the radius of the fan, i.e., $\eta = a'/a = 1$; their speed will then be the tip speed of the fan.

The ratio of the tangential and axial force is chosen to be $T/X = 0.1$ and the average thrust per unit area of the fan is $F = 1000$ N/m^2, i.e., about one percent of the atmospheric pressure. The distance to the observation point from the fan is $r = 10a$ which can be considered to be in the far field of the radiation for which our calculations are valid. For these parameter values, the calculated angular dependence of the sound pressure level (re 20 μbar) is shown in Fig. 8.6 for fans with 2 and 8 blades with a tip Mach number of 0.8.

On the axis, corresponding to the angles 0 and 180 degrees, the sound pressure is zero so that the corresponding levels are $-\infty$. The pressure amplitude is zero also at the angle given by Eq. 8.3; in this case, with $\eta = 1$, it is $\theta = \cos^{-1}(-0.1/0.8) \approx 97.2$ degrees. (The data in the graph are computed with a resolution of one degree and the minimum does not reach its true depth.) The two maxima in the directivity pattern occur at approximately 50 and 170 degrees for 2 blades and 70 and 115 for 8 blades.

The dependence of the sound pressure on the number of blades depends on whether the total thrust or the thrust per blade is kept constant as the number of blades is increased. If the amplitude contributions from the blades were all in phase, the total sound pressure amplitude obviously would increase proportionally to the number of blades. In reality, there will be phase differences and a corresponding competition between this additive effect and the effect of interference. The outcome of this competition depends on the values of the parameters involved.

The result in Fig. 8.6 refers to the fundamental harmonic component. However, higher harmonics can be important. An example of the calculated spectrum is shown in Fig. 8.7. Since we are dealing with point forces, the Fourier components

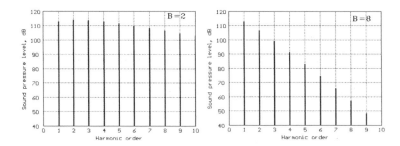

Figure 8.7: Sound pressure level spectrum. Fan simulated by point dipoles in circular motion. Number of blades: Left: $B = 2$. Right: $B =8$. Tip Mach number: $M = 0.8$. Average thrust per unit area of fan: $F = 1000$ bar. Transverse/axial force ratio: $T/X = 0.1$. Span-wise location of sources: $\eta = a'/a = 1$. Polar angle: *theta* = 70 degrees. Distance: $r/a = 10$.

of the source spectrum have all the same strength, and the frequency dependence of the sound pressure level is due solely to the frequency dependence of the coupling to the surrounding fluid and to interference. The shape of the spectrum depends significantly on the polar angle. This angle is 70 degrees in the figure, and at that location the level does not vary much with the harmonic order for the 2-bladed fan; actually, there is an initial slight increase in level above the level of the fundamental. The situation is quite different for 8 blades, and as the number of blades is increased, the higher harmonics become less significant compared to the fundamental.

Radiated Power

The sound pressure amplitude of the nth harmonic in Eq. 8.2 can be written

$$p_m = F \frac{D(\theta)}{2(r/a)}$$
$$D(\theta) = mB(M \cos \theta + T/(\eta X)) J_{mB}(mB\eta M \sin \theta). \qquad (8.8)$$

The corresponding acoustic intensity is $I_m = p_m^2/\rho c$, where the amplitude is assumed to be expressed as an rms value to avoid a factor of 1/2. The power radiated in the downstream hemisphere is then

$$W_m = \int_0^{\pi/2} 2\pi r^2 I_m \sin \theta \, d\theta = \frac{1}{2} \frac{|F|^2}{\rho c} (\pi a^2) \int_0^{\pi/2} D^2(\theta) \sin \theta \, d\theta. \qquad (8.9)$$

The corresponding power radiated in the upstream direction is obtained by integrating from $\pi/2$ to π.

Radiation 'Efficiency'

The angular dependence of the sound pressure in Fig. 8.6 refers to point forces placed at the tip of the propeller, i.e., with $\eta = a'/a = 1$. As the sources are moved inward, their speed is decreased, of course, and the radiated pressure is reduced accordingly.

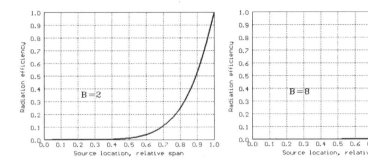

Figure 8.8: Point dipoles in circular motion; re fan acoustics. The radial position of the sources is a' and the radius of the fan is a. The radiation efficient is defined as $W(a')/W(a)$, where W is the radiated power.
Number of blades: Left: $B = 2$. Right, $B = 8$. Tip Mach number: $M = 0.8$. Tangential/axial force ratio: $T/X = 0.1$. Harmonic component: $n = 1$.

As a measure of the radiation efficiency of a source at a given radial position for a fixed tip Mach number, we use $W(a')/W(a)$, the ratio of the radiated powers corresponding to the source locations a' and a (at the tip). This quantity is shown as a function of a'/a in Fig. 8.8. The sound reduction obtained by moving the source inward (unloading the tip of a blade) is seen to depend very strongly on the number of blades, a fact that has already been expressed in a different way in the discussion of Eq. 8.6. In other words, as the number of blades of a propeller is increased, the acoustically 'active' part becomes more and more dominated by the regions close to the tip.

8.1.4   Simulation with Span-wise Distributions of Dipoles

To improve the acoustic simulation of a fan, we replace the point dipole forces in the previous section by dipole line sources which are assumed identical and uniformly spaced. As before, X will be the axial force on one of the blades and, with $\eta = r'/a$, the force distribution in the span-wise (radial direction) of the blade is described by a distribution function $\beta_x(\eta)$ such that $(X/a)\beta_x(\eta)$ is the force per unit length, where a is the span of the blade (in this case the radius of the fan). In an analogous manner, we introduce a force distribution function $\beta_t(\eta)$ for the tangential component of the force so that the force per unit length is $(T/a)\beta_t(\eta)$.

The field from the point force simulation in Eq. 8.2 is now modified to

$$
\begin{array}{c}
\textit{Line force simulation of a fan} \\
p = \sum_{m=1}^{\infty} p_m \quad \text{where} \\
p_m(r, \theta, \phi, t)/F = mB\,(a/2r)P_m(\theta)\sin[nk_r r + nB(\phi + \pi/2) - nB\Omega t] \\
P_m(\theta) = [I_x M \cos\theta + I_t(T/X)]P_m(\theta)\sin[mk_r r + mB(\phi + \pi/2) - mB\Omega t] \\
I_x = \int_0^1 \beta_x(\eta)J_{mB}(mB\eta M \sin\theta)\,d\eta \\
I_t = \int_0^1 (1/\eta)\beta_t(\eta)J_{mB}(mB\eta M \sin\theta)\,d\eta
\end{array}
$$

$$(8.10)$$

As before, a is the radius of the fan, $F = BX/\pi a^2$, the average thrust per unit area, and T/X the tangential-to-axial force ratio, assumed independent of the span-wise position coordinate $\eta = r'/a$. Ω is the angular velocity of the shaft, M the tip Mach number, and $mk_r r = (\omega_m r/c)$, where $\omega_m = mB\Omega$ is the mth harmonic of the blade passage frequency.

As a first numerical example, based on Eq. 8.10, the angular dependence of the sound pressure amplitude P_m is shown in Fig. 8.9. The amplitude is expressed in terms of the sound pressure level with respect to the standard reference 20 μbar. Results for $B = 2$ and 16 blades are shown at a distance from the fan of 10 fan radii, i.e., $r/a = 10$. In each case, three different span-wise force distribution functions have been used, assumed to be the same for both the axial and the tangential force. The three distribution functions are:

$$\text{(a)} \quad \beta(\eta) = 1, \quad \text{(uniform load)}$$
$$\text{(b)} \quad \beta(\eta) = 2\eta, \quad \text{(linear increase toward tip)}$$
$$\text{(c)} \quad \beta(\eta) = (\pi/2)\sin(\eta), \quad \text{(mid-span maximum)} \quad (8.11)$$

Each distribution is normalized, $\int \beta\eta = 1$, so that the force on each blade is the same for all distributions. In this case it has been chosen so that the average thrust per unit area of the fan $F = BX/(\pi a^2)$ is 1000 N/m², about 1 percent of the atmospheric pressure. In this example, the ratio of the tangential and axial loads has been assumed to be the same, 0.1, at all radial positions of the blade. The tip Mach number is 0.8.

For distribution (b), the load at the tip is the largest, and, as expected, it yields the highest level. In distribution (c), the load has been moved inboards to have its maximum midspan, and, again as expected, it yields the lowest level. Distribution (a) represents a uniform load on the blade and it leads to a sound pressure level between the previous two. For two blades, the maximum sound pressure level difference

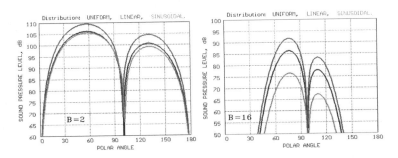

Figure 8.9: The angular distribution of the sound pressure level (in dB re 20 μbar) produced by a fan simulated acoustically by swirling dipole line sources. Number of blades: Left: $B = 2$. Right: $B = 16$. Tip mach number: $M = 0.8$. Average thrust per unit area of fan: $F = 1000$ N/m². Harmonic component: $m = 1$. Distance to field point: $r = 10a$, where a is the fan radius. Tangential/axial force ratio: T/X: 0.1. Distributions: Upper curve: Linear increase toward the tip. Middle curve: Uniform. Lower curve: Sinusoidal, maximum at midspan (see Eq. 9.10).

between these distributions is about 4 dB, for 8 blades, $\approx$ 10 dB, and for 16 blades, $\approx$ 15 dB. In comparing the maximum values of the sound pressure levels for different number of blades, it should be kept in mind that the total pressure change across the fan is the same in all cases rather than the force per blade. From a practical standpoint of noise reduction, the result shows that moving the load away from the tip becomes a more effective strategy as the number of blade increases.

With the sound pressure expressed in terms of the average static pressure change ΔP across the fan, as we have done, the sound pressure level depends only on the product nB and the angle. Increasing this product moves the angular distribution of the sound pressure toward a polar angular region in the vicinity of 90 degrees and it also becomes narrower in the process.

Sound Pressure Level Spectrum

The results in Fig. 8.10 refer to the fundamental harmonic component, $m = 1$, in the Fourier decomposition of the sound pressure. As indicated, higher harmonics, corresponding to $m > 1$, can contribute substantially to the overall sound field, however, particularly for fans with few blades and at polar angles close to 90 degrees. The marked difference in level between the higher harmonics at the two angular positions in the figure is noteworthy.

The Pressure Signature

The superposition of the harmonic components of the pressure in Eq. 8.10 can make the time dependence of the overall pressure far from the simple harmonic function of the fundamental. The significance of the harmonics depends on the number of blades, the Mach number, and the polar angle. We have already seen in the spectrum in Fig. 8.10 that at an angle of 30 degrees the harmonics play a much less significant

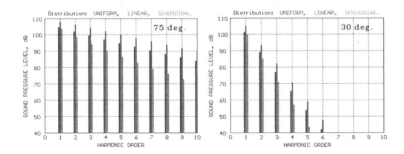

Figure 8.10: The sound pressure level (re 20 μbar) spectrum produced by a fan simulated acoustically by swirling dipole line sources. Polar angle: Left: $\theta = 75$ deg. Right: $\theta = 30$ deg. Number of blades: 2. Tip mach number: 0.8. Average thrust per unit area of fan: $F = 1000$ N/m^2. Distance to field point: $r = 10a$, where a is the fan radius. Tangential/axial force ratio: $T/X = 0.1$. Spectrum lines, from left to right: span-wise load distributions (a), (b), and (c) in Eq. 8.11.

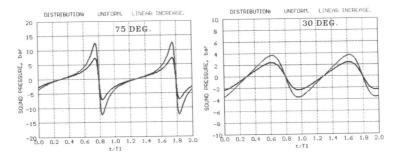

Figure 8.11: The time dependence of the sound pressure at polar angles 75 and 30 degrees (corresponding sound pressure level spectra, see Fig. 8.10). The time coordinate is normalized with respect to the blade passage period T_1. Span-wise load distribution: Lower curves: Uniform, upper curves: Linear increase toward tip. (see Eq. 8.11). Number of blades: 2. Tangential/axial force ratio: $T/X = 0.1$. Tip mach number: $M = 0.8$. Average thrust per unit area of fan: $F = 1000$ bar.

role than at 75 degrees. This is reflected also in the time dependence of the sound pressure, as shown in Fig. 8.11.

This result was obtained by adding 40 terms in the Fourier series in Eq. 8.10. The values of the system parameters are the same as in Fig. 8.10.

The signature of the pressure at 30 degrees is seen to be much closer to a harmonic curve than at 75 degrees, as expected from the spectrum in Fig. 8.10. The pressure signature at 75 degrees is much sharper and gives the appearance of a periodic shock wave (in comparing amplitudes at 75 and 30 degrees, note the difference in scales). A reduction in Mach number has essentially the same effect on the signature as a reduction in the polar angle.

Total Radiated Power and its Spectrum

From the sound pressure distribution in Eq. 8.10, the corresponding acoustic powers radiated in both the upstream and downstream directions are obtained in the same way as for the point source simulation in the previous section. An example of the results obtained is given in Fig. 8.12, where the dependence on the tip Mach number is shown. It is qualitatively quite similar to that for the point dipoles, and we find again that the power is dominated by the contribution in the downstream direction. In fact, it is a good approximation to consider it to be the total power. It should be noted that the Mach number dependence of the power becomes stronger as the number of blades is increased.

The result for the total radiated acoustic power in Fig. 8.12 referred to the fundamental component $m = 1$ (blade passage frequency). To get an idea of the relative significance of higher harmonics, we show in Fig. 8.13 the power levels of the first 10 harmonic components at a Mach number of 0.8 and for 2 and 8 blades.

Again, an increase of the number of blades produces a stronger reduction of level with harmonic order.

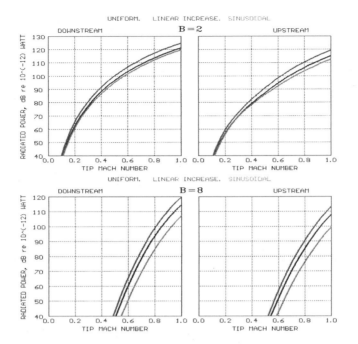

Figure 8.12: The tip Mach number dependence of the acoustic powers in the downstream and upstream hemispheres radiated by a fan which is simulated acoustically by swirling dipole line sources.

Number of blades: Upper figure, $B = 2$, lower, $B = 8$. Harmonic component: $m = 1$. Average thrust per unit area of fan: $F = 1000$ bar. Tangential/axial force ratio: T/X: 0.1. Span-wise load distributions: Upper curves: Linear increase toward tip, middle: Uniform, lower: Sinusoidal, midspan maximum (see Eq. 8.11).

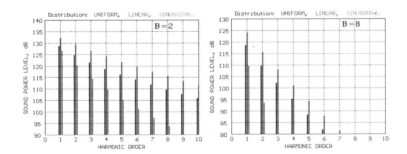

Figure 8.13: The power level spectrum of the total radiated acoustic power by the fan in Fig. 8.12 at a tip Mach number of 0.8. The three spectral lines shown for each harmonic, starting from the left, correspond to the three span-wise force distributions (a),(b), and (c) in Eq. 8.11, i.e., uniform, linear increase toward tip, and sinusoidal with midspan maximum.

Number of blades: $B = 2$ and $B = 8$, as shown. Average thrust per unit area of fan: $F = 1000$ bar. Transverse/axial force ratio: $T/X = 0.1$.

8.1.5   Effect of Nonuniform Inflow

As we have seen, the interaction of the fan with a nonuniform inflow produces two spinning pressure fields with the angular velocities $\Omega_- = mB\Omega/(mB - q)$ and $\Omega_+ = mB\Omega/(mB + q)$, where q is the order of the spatial harmonic component of the (stationary) nonuniform flow velocity and m is the harmonic component of the blade passage frequency $B\Omega$. The radiation efficiency of the pressure field spinning with the higher angular velocity Ω_- is greater than the field with Ω_+ and usually dominates the radiation field. If the flow nonuniformity matches the harmonic variation of the force distribution, i.e., with $q = mB$, angular velocity becomes infinite and the fan will in effect radiate as a piston.

We shall not carry out the detailed analysis here but merely point out that it shows that the order of the Bessel function factor now is changed from J_{mB} to J_{mB-q}. The lowering of the order of the Bessel function can increase the magnitude of the sound pressure by several orders of magnitude and the angular distribution is also changed. For example, with $q = mB$ the sound pressure on the axis no longer will be zero, but a maximum.

Polar Amplitude Variation

The role of nonuniform flow is demonstrated in Fig. 8.14 where the angular sound pressure distributions for both uniform and nonuniform flow are shown for fans with 2 and 16 blades. In each case we consider the fundamental of the blade passage frequency, $m = 1$, and the harmonic q of the flow nonuniformity is 2 and 16, respectively, so that in each case $q = mB$. The significance of the flow inhomogeneity increases with increasing blade number, as can be seen in this case. For 2 blades, the magnitude of the relative force fluctuation on a blade is ≈ 5 percent to make the maximum sound pressure amplitude resulting from the inhomogeneity about the same as for the uniform flow. For 16 blades, only about 0.1 percent is required.

Circumferential Amplitude Variation

At a given frequency of the sound field, $mB\Omega$, let us consider the sum of the pressure fields that result from two harmonics of the flow irregularity, q_1 and q_2. These fields rotate with different speeds, as given by Eq. 8.1, and are generally not in phase at a fixed position ϕ.

In particular, we consider here the case when $q_1 = 0$. The corresponding field is due to the interaction with the uniform flow and the second to a nonuniform flow field with a harmonic component q_2.

The sum of the two fields yield a circumferential amplitude variation which is independent of time and represents a 'map' of the corresponding variation of the flow into the fan. In the figure to the left, we considered the second harmonic component, $q_2 = 2$, of the circumferential variation of the flow and we get two periods of variation in the sound pressure level, as shown; in the figure to the right, we have used $q = 4$, and there are 4 periods. The amplitude variation generally is not as great as in the special case shown in the figure, and if one of the sound pressure components of the sound field dominates, the amplitude variation will be insignificant.

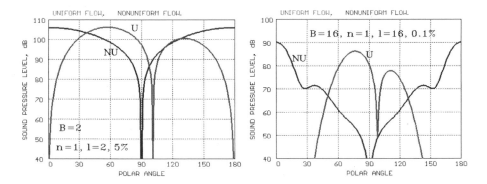

Figure 8.14: Effect of inhomogeneous flow on sound radiation from a fan simulated acoustically by swirling dipole line sources with uniform span-wise force distribution. The nonuniform flow contribution to the sound field (NU) has a maximum on the axis, at 0 and 180 degrees. Number of blades: Left: 2, Right: 16. Tip Mach number: 0.8. Average static pressure change across the fan: $\Delta P = 1000$ bar. Tangential/axial force ratio: $T/X = 0.1$. Relative force fluctuation due to flow nonuniformity: 5 percent (2 blades), 0.1 percent (16 blades), and harmonic orders of Fourier expansion of the flow, 2 and 16, respectively. Harmonic order of blade passage frequency, $m = 1$. Distance: $r/a = 10$.

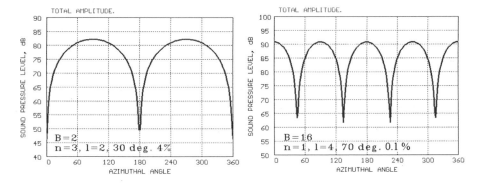

Figure 8.15: The circumferential variation of the sum of the sound pressure contributions from blade interaction with the average (uniform) flow component ($q = 0$) and one circumferential harmonic component in nonuniform inflow; fan simulated acoustically by swirling dipole line sources with uniform span-wise force distribution.
Number of blades: $B = 2, 16$. Tip Mach number: 0.8. Average static pressure change across the fan: $\Delta P = 1000$ bar. Tangential/axial force ratio: $T/X = 0.1$. Relative force fluctuation due to flow inhomogeneity: 4 and 0.1 percent. Harmonic order: Figure left: sound, $m = 3$; flow, $q = 2$. Right: $m = 1$, $q = 4$. Polar angles: Left: $\theta = 30$, right: 70 deg. Distance: $r/a = 10$.

8.2 Fan in a Duct

We consider here a fan in an annular duct; the inner radius of the annulus is the radius of the hub of the fan and the outer radius is the tip radius of the fan. The essentials of sound radiation by a fan in such a duct can be captured by 'unwrapping' the annular region and treat it as a duct between two parallel walls and the fan as a linear cascade of blades. This is a good approximation, at least geometrically, if the hub-to-tip ratio is sufficiently large, say above 0.5, and we shall use it here. The sound field returns on itself in the annulus after an angle change of 2π which must hold also in the unwrapped version. This means, that if the average radius of the annulus is a, the sound field must be such that the sound pressures at $z = 0$ and $z = 2\pi a$ or any multiple thereof are the same (periodic boundary condition).

The separation of the walls is d, the width of the annulus. The coordinate x is along the duct, y is perpendicular to the walls which are at $y = 0$ and $y = d$, and z is the direction of motion of the cascade of fan blades. The plane of the fan is at $x = 0$ (the fan is here assumed thin compared to the wavelength).

The plane of the fan is regarded as an acoustic source plane with a dipole distribution of sources in the same manner as for the fan in free field. On one side of the plane there is only an axial (inflow) component of the flow but on the other there is also a z-component (swirl). The Mach numbers of these components are M_x and M_z; typically, $M_z \approx 0.5 M_x$. The pressure distribution in the y-direction (radial) will be of the form $\cos(n\pi y)$, $n = 0$ being the fundamental radial mode. The z-dependence of the field is expressed by a traveling (swirling) wave of the form $\exp(i k_z z)$. The wavelength of the m:th harmonic of the blade passage frequency is $\lambda_z = 2\pi a/mB$ and $k_z = 2\pi/\lambda_z = mB/a$.

8.2.1 Modal Cut-off Condition and Exponential Decay

The dispersion relation for the spinning mode in the combined axial and swirling flow generated by the fan in the unwrapped annular duct discussed above is derived in Example 47 in Ch.11 and the result is

$$K_{x\pm} = -\frac{M_x(1 - M_z K_z)}{1 - M_x^2} \pm \frac{1}{1 - M_x^2}\sqrt{(1 - M_z K_z)^2 - (K_y^2 + K_z^2)(1 - M_x^2)}, \quad (8.12)$$

where $K_x = k_x/k$, $K_y = k_y/k$, $K_z = k_z/k$, and $k = \omega/c$. M_x and M_z are the flow Mach numbers in the axial and transverse directions. With $k_z = mB/a$ (see the end of the last section) and $k = \omega/c = mB\Omega/c$, the normalized value of k_z is $K_z = (mB/a)/(mB\Omega/c) = 1/M$, where M is Mach number of the blade. The width of the annulus is d (the distance between the walls in the unwrapped version of the annulus) and $k_y = n\pi/d$ corresponding to the pressure wave function $\cos(k_y)$ and the velocity wave function $\propto \sin(k_y y)$. The normalized value of k_y can also be expressed in terms of M as $K_y = n\pi a/mdBM$.

M_z is different from zero only on the downstream side of the fan. The plus and minus signs in the second term correspond to a wave in the positive and negative x-direction, respectively. The cut-off value of the (m,n)-mode corresponds to the value of M which makes the imaginary part of K_x equal to zero. With reference to Example 47 (see Ch. 11), the cut-off value of the blade Mach number for the (m,n)-mode is

$$M_c = M_z \pm \sqrt{[1 + (n\pi a/mBd)^2](1 - M_x^2)}. \tag{8.13}$$

The plus sign is used if the mode spins in the same direction as the swirling flow move in the same direction (the normal condition[3]), otherwise the minus sign applies. It should be recalled that a is the mean radius of the fan and d is the width of the annulus. The integer n is the number of pressure nodes of the wave function in the span-wise direction and m the harmonic order of the blade passage frequency. For the lowest order, $n = 0$, the critical Mach number M_c will be independent of the harmonic m of the blade passage frequency. If, in addition, $M_x = 0$, a mode will propagate if the Mach number of the blades relative to the swirling flow exceeds unity, as expected. This result is modified by the axial flow speed which reduces the critical Mach number. As an example, with $M_z = M_x = 0.5$, we get $M_c = 1.37$ for downstream radiation. For the field radiated in the upstream direction, however, where $M_z = 0$, the critical fan Mach number is only $M_c = 0.87$.

For subsonic speeds of the blades, the fields on the upstream and downstream sides of the fan will decay exponentially with distance from the fan, the downstream decay rate being greater than the upstream since the swirl on the downstream side reduced the relative speed of the blades. The decay rate is found to increase with the number of blades and with the harmonic order of the blade passage frequency.

8.2.2   Effect of a Nonuniform Flow

Our discussion of the effect of a nonuniform flow on sound radiation from a fan in free field in Section 8.1.1 indicates that even a fan with subsonic tip speed can produce a spinning pressure field with supersonic speed with an angular velocity larger than that of the fan. Then, even a modest nonuniformity can result in a large increase in the radiated sound, particularly for a fan with many blades.

In uniform flow, a fan in a duct cannot produce a plane wave, only higher modes and the sound field then will contain a propagating mode only if the frequency exceeds the cut-off frequency of the mode. As we shall see, cut-off conditions corresponds approximately to a sonic tip speed of the fan. Below this speed, the wave field will be evanescent like the field from the moving corrugated board in Section 5.5.1.

In nonuniform flow, as we have seen in Section 8.1.1, Fig. 8.3, the interaction pressure between the fan and the flow can move at a higher speed than the fan; a mode that is normally cut off can then be cut on. Even a plane wave can be produced. This means that the uniformity of the flow in a duct can have a much more important effect on the sound field than in free field.

With reference to Eq. 8.1 for the speed of rotation of the pressure field resulting from nonuniform flow, we have to replace k_z in the dispersion relation (8.12) by $k_z/(1 \pm q/mB)$ and the expression for the critical value of the blade Mach number

---

[3]If rotor-stator interaction is involved, the waves from the stator will produce waves spinning both with and against the swirl.

in Eq. A.56 has to be replaced by

$$M_c = M_z \delta \pm \sqrt{[\delta^2 + (n\pi r_a / dmB)^2](1 - M_x^2)}, \tag{8.14}$$

where $\delta = 1 \pm (q/mB)$.

In the particular case when the harmonic order of the flow nonuniformity is $q = mB$, the minus mode yields $\delta = 0$ and, if $m = 0$, a propagating mode (the plane wave) is generated for all Mach numbers, i.e., $M_c = 0$, in both the upstream and downstream directions. It should be noted that not only the critical Mach number but also the decay constant for the (m,n):th mode depends on q/mB.

8.2.3   Rotor-Stator Interaction

In a fan duct of a by-pass aircraft engine there is a set of guide vanes downstream of the fan. The purpose of these is to eliminate the swirl of the flow caused by the fan to improve the efficiency. In addition to having a swirl, the flow also contains the wakes from the blades of the fan which are convected by the swirling flow. As these wakes strike the guide vanes (stator) there will be a fluctuating force component on the stator and hence sound generation. Thus, although the flow incident on the fan may be uniform it will always be swirling and nonuniform at the stator.

The calculation of the sound field produced by the stator is analogous to that for the fan with nonuniform inflow. There is an important difference, however. For the rotor we considered a stationary nonuniform flow interacting with moving fan blades. For the stator the inflow is rotating and the guide vanes stationary. Normally the wakes from the rotor are identical and uniformly spaced so that the fundamental angular period of the incident flow will be $2\pi/B$ and the corresponding fundamental angular frequency of the time dependent interaction $B\Omega$. For both the rotor and the stator, the sound emission is due to the relative motion between flow and hardware and the characteristics of the sound in the two cases are similar.

As was the case for the rotor in inhomogeneous flow, the swirl with its wakes interacting with the stator gives rise to rotating pressure fields with angular velocities which differ from the angular velocity of the fan. Thus, with the stator having V vanes, the interaction between the mth harmonic of the wake flow and the ℓth harmonic in the Fourier expansion of the periodic obstruction of the guide vanes gives rise to rotating pressure waves with the angular velocities

$$\Omega_\pm = mB\Omega/(mB \pm \ell V). \tag{8.15}$$

The dominant contribution to the sound field corresponds to the minus sign which yields an angular velocity greater than Ω if $|mB - \ell V| < mB$. Furthermore, if $\ell V > mB$, the wave spins in the negative direction (i.e., against the swirling flow). For example, with $B = 16$ and $V = 40$ and $\ell = 1$, this is the case for $m = 1$ and $m = 2$. For $m = 3$, the wave will spin in the same direction as the flow.

From the standpoint of noise reduction, it is desireable to make the rotational speeds $\Omega_\pm$ as small as possible to prevent cut-on of higher modes. To achieve this, the guide vane number V should be as large as possible.

There are some differences to consider in comparing rotor and stator sound generation, however. For the rotor, an incident uniform flow will produce sound but for the stator it will not in the present model. Another difference has to do with the relative roles of the axial and tangential force components. For the rotor, the axial force component is normally considerably larger than the tangential. For the stator, however, the tangential component is expected to be of the same order of magnitude as for the fan but the axial component should be considerably smaller than for the fan. Furthermore, whereas the swirling flow in the duct was on the downstream side of the fan, it is on the upstream side of the stator. Thus, the expressions for the wave impedances on the two sides of the stator have to be chosen accordingly. For the stator, an additional parameter needs to be considered; namely, the angle between the wakes and the guide vanes. It turns out to have an important effect on the sound field.

Effect of Refraction and Reflection

The model we have used of the fan-stator combination contains a swirling flow in the region between the rotor and the stator and the transition at the rotor and the stator is modeled as a shear layer with a discontinuity in the tangential velocity component. So far, we have not accounted for the reflection of sound that takes place at such a layer. Furthermore, the radiation from the end of the fan duct into free field requires consideration but will not be considered here.

8.3   Centrifugal Fan

Our discussion of sound generation by a centrifugal fan will be limited to some observations related to what we call the 'whirling pipe' model, illustrated schematically in Fig. 8.16. Here the impeller consists simply of a single tube whirling about a fixed horizontal axis with an angular (shaft) velocity Ω. It is surrounded by a casing (shroud), as indicated. The length of the whirling pipe is ℓ so that the tip speed becomes $\ell\Omega$.

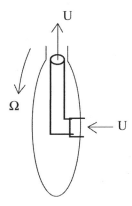

Figure 8.16: The whirling tube; a conceptual model of a centrifugal fan.

The flow exits through an opening in the shroud so that when the tube passes this opening there will be a pulse of air emerging. In the frame of reference moving with the tube there is a centrifugal force density $\rho\Omega^2 r$, where ρ is the density and r the distance from the axis of rotation. With the velocity of the fluid relative to the pipe denoted U and the pressure p, the equation of motion then becomes[4]

$$\partial U/\partial t + \partial U^2/\partial r = -(1/\rho)\partial p/\partial r + \Omega^2 r. \qquad (8.16)$$

At sufficiently low frequencies we can treat the fluid as incompressible and neglect the inertia term. This corresponds to acoustic wavelengths considerably larger than the length of the tube. Then, integration of the equation from the inlet to the exit, where the pressures are denoted P_1 and P_2, we get

$$P_2 - P_1 = \rho\Omega^2\ell^2/2, \qquad (8.17)$$

where ℓ is the length of the tube.

The flow into the inlet can be considered to be laminar. Then, if the ambient pressure is P_0, we have $P_1 = P_0 - \rho U^2/2$. At the discharge, on the other hand, there is a pressure loss which we denote $\beta\rho U^2$ so that $P_2 = P_0 + \beta\rho U^2/2$, where β depends on the angular position of the pipe and is a function of time, $\beta \equiv \beta(t)$. When the pipe lines up with the discharge opening in the shroud of the fan, the discharge can be considered to be a turbulent jet and $\beta \approx 1$.

It follows then from Eq. 8.17 that $U^2[1 + \beta(t)]/2 = \Omega^2\ell^2/2$ or

$$U(t) = \Omega\ell/\sqrt{1 + \beta(t)}, \qquad (8.18)$$

which expresses the time dependence of the flow velocity. It will have a maximum value when the tube lines up with the discharge opening at $t = 0$. Then, with $\beta(0) \approx 1$, we get the maximum value $U \approx (\Omega\ell)/\sqrt{2}$. The time dependence is periodic with the fundamental angular frequency Ω but it will contain overtones which can be determined once $\beta(t)$ is given.

The time dependence of the discharge velocity, which equals the inlet velocity, gives rise to sound. In the low frequency approximation the corresponding sound sources can be modeled as an acoustic monopole, one at the discharge and one, out of phase, at the inlet, with the far field sound pressure contributions from each being proportional to the time derivative of the mass flow rate.

If there are N symmetrically spaced tubes rather than one, the emitted sound will have a fundamental frequency $N\Omega$ with overtones determined by $\beta(t)$.

At wavelengths not large compared to ℓ, we have to account for the compressibility of the fluid within the fan and the analysis has to be modified accordingly; for example, acoustic resonances of the tube have to be considered. However, the essential features of the mechanism of sound radiation still applies.

---

[4]There is no need to include gravity because its effect on the velocity is cancelled by the variation in the external pressure difference with position in the vertical plane of rotation.

8.3.1 Problems

1. **Sound radiation from a moving corrugated board**

   Study Example 46 in Chapter 11, and consider a board in which the amplitude ξ of the corrugation is one percent of the wavelength Λ of the corrugation. The board moves with a velocity U. Calculate the magnitude $|p|$ of the sound pressure at a distance from the board equal to Λ if (a) $U = 2c$ and (b) $U = 0.5c$.

2. **Sound pressure amplitude; point force simulation of a fan**

   From the expression for the total sound pressure field in Eq. 8.2, check the approximate expression for the sound pressure amplitude in Eq. 8.6

3. **Angular SPL distribution; point force simulation of fan**

   With reference to Eq. 8.2 and Fig. 8.6 and with the data used in this figure, what is difference between the fundamental components of SPL at a polar angle of (a) 30 degrees and (b) 90 degrees in going from a fan with 2 blades to one with 8 blades at the same tip Mach number and thrust per unit area? If the comparison is made between the 4th harmonic of the SPL from the 2 blade fan and the fundamental of the 8 blade fan (the same frequency in both cases), what then is the result?

4. **Mach number dependence of acoustic power; line force simulation of a fan**

   With reference to Fig. 8.12 and the fan data in this figure, what is the change in the total radiated acoustic power from (a) a 2 blade fan and (b) from an 8 blade fan resulting from a reduction in the tip speed Mach number from 0.8 to 0.6. Assume that the thrust per unit area varies as the square of the tip Mach number.

5. **Effect of nonuniform flow on sound radiation; line force model of a fan**

   Fig. 8.14, left, shows the angular distribution of the radiated sound from a 2 blade fan in uniform as well as nonuniform flow, the latter producing a fluctuation in the force on a blade equal to 5 percent of the force from the uniform flow.
   (a) What is the combined total sound pressure level from these contributions at 0 degrees and 90 degrees?
   (b) At what angles are the individual SPL values equal?
   (c) What can you say about the combined total SPL at these angles?

Chapter 9

Atmospheric Acoustics

9.1 Historical Notes

Many of the problems in acoustics of current interest were formulated and studied a long time ago, and atmospheric acoustics is a prime example. Systematic investigations of the 'acoustic transparency' of the atmosphere can be traced back with certainty to the beginning of the 18th century. A report by Derham (1708) was an authoritative source for many years, but some of Derham's results and conclusions, particularly in regard to the influence of fog and rain on sound transmission, were challenged about 50 years later by Desor and were conclusively shown to be incorrect by John Tyndall, the prominent English scientist, who in 1874 directed an extensive experimental study of sound propagation. Unlike Derham, Tyndall found that fog, rain, hail, or snow did not cause any noticeable attenuation of sound, at least in the frequency range covered by the signaling devices involved in his experiments. Mounted 235 ft above high water on a cliff overlooking the ocean in the vicinity of Dover, England, several types of foghorns and cannons were used as sound sources. In Fig. 9.1 is reproduced a drawing of a steam-driven siren used in the experiments. Manufactured in the United States and furnished by the Washington Lighthouse Board, this particular sound source is a steam-driven siren with one fixed and one rotating disk with radial slots. The disks are mounted vertically across the throat of a conical horn, 16 1/2 feet long and 5 inches in diameter at the throat, gradually opening to reach a diameter of 2 feet and 3 inches at the mouth. The horn is connected to a boiler and driven by steam at a pressure of 70 psi.

Other sound sources used in the experiments were two brass trumpets (11'2" long and 2" at the mount). The vibrating reed in the trumpet was 9" long, 2" wide, and 1/4" thick and was made of steel. The trumpet was sounded by air at 18 psi. Other sources were a locomotive whistle and three cannons, one 19-pounder, a 5 1/2 inch howitzer, and a 13 inch mortar.

Observations of the range of audibility of the sound over the ocean were made under various weather conditions. In his very lucid account of his work, 'Researches on the Acoustic Transparency of the Atmosphere, in Relation to the Question of Fog-signaling,' Tyndall attempted to explain the various observations in terms of reflections from 'flocculent acoustic clouds,' consisting of regions of inhomogeneity in humidity

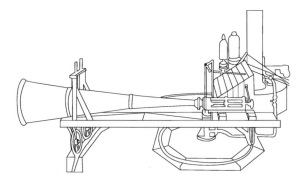

Figure 9.1: Steam-driven siren used in sound propagation studies by Tyndall (1845).

and/or temperature. Tyndall explains: "The intercepted sound is wasted by repeated reflections in the acoustic clouds, as light is wasted by repeated reflections in an ordinary cloud. And, as from the ordinary cloud, the light reflected reaches the eye, so from the perfectly invisible acoustic cloud, the reflected sound reaches the ear."

The observed temporal fluctuations in the range of transmission of the sound were ascribed to drifts of the acoustic clouds. Tyndall remarks: "An interval of 12 hours sufficed to change in a surprising degree the acoustic transparency of the air. On the 1st of July, the sound had a range of nearly thirteen miles; on the 2nd, the range did not exceed 4 miles."

Associated with the range fluctuations, Tyndall observed 'echoes' sent back from the ocean to the source, and he described the phenomenon as follows: "From the perfectly transparent air the echoes came, at first with a strength apparently a little less than that of the direct sound, and then dying away."

"...In the case of the siren, moreover, the reinforcement of the direct sound by its own echo was distinct. About a second after the commencement of the siren-blast, the echo struck in as a new sound. This first echo, therefore, must have been flung back by a body of air not more than 600 or 700 feet in thickness. The few detached ordinary clouds visible at the time were many miles away and could clearly have had nothing to do with the effect."

"...On again testing the duration of the echoes, it was found to be from 14 to 15 seconds.... It is worth remarking that this was our day of longest echoes, and it was also our day of greatest acoustic transparency, this association suggesting that the direction of the echo is a measure of the atmospheric depths from which it comes."

Tyndall's report contains not only a description of the systematic studies of sound propagation over the ocean, but also many accounts of everyday observations on sound transmission effects. As an example, we quote the following: "...On reaching the serpentine this morning, a peal of bells, which then began to ring, seems so close at hand that it required some reflection to convince me that they were ringing to the north of Hyde Park. The sound fluctuating wonderfully in power. Prior to the striking of eleven by the great bell of Westminster, a nearer bell struck with loud clanger. The first five strokes of the Westminster bell were afterward heard, one of them being extremely loud; but the last six strokes were inaudible. An assistant was stationed to attend the 12 o'clock bells. The clock which had struck so loudly at 11 was unheard at 12, while of the Westminster bell eight strokes out of twelve were inaudible. To such astonishing changes is the atmosphere liable."

Several other observations of this type are described in the report, and in regard to the influence of fog, we note the following: "...To these demonstrative observations one or more subsequent ones may be added. On several of the moist and warm days, at the beginning of 1974, I stood at noon beside the railing of St. James Park, near Buckingham Palace, three-quarters of a mile from the clock tower, which was clearly visible. Not a single stroke of 'Big Ben' was heard. On January 19th, fog and drizzling rain obscure the tower; still from the same position I not only heard the strokes of the great bell, but also the chimes of the quarter bells."

On the basis of such observations, which were consistent with the main body of results from the studies of propagation over the ocean, Tyndall concluded that fog and rain per se do not markedly cause any attenuation of sound. On the contrary, there appears always to be an increase in the transparency of the atmosphere when fog and rain are present. Tyndall explained this as a result of an increase in the uniformity of the air when fog and rain are present. We now know that there is another and probably more important effect to account for these observations as will be discussed in the next section.

A conclusion similar to Tyndall's was reached at about the same time by Joseph Henry, under whose supervision sound propagation studies were carried out in the United States, as described in the 'Report of the United States Lighthouse Board of 1874.' Henry took exception to Tyndall's theory of 'flocculent clouds,' however, and their effect on the penetration depth of fog signals. Rather, Henry proposed an explanation of the variation in range, not as a result of scattering from such clouds, but in terms of the refraction of sound resulting from vertical wind gradients. The paradoxical result observed by both Tyndall and Henry, that foghorn signals often carry farther over the ocean *against* the wind than *with* the wind, Henry ascribed to a reversal of wind direction at a certain height. Then, for sound transmission against the wind in the lower regions, the sound is first refracted upwards, but as it reaches the upper reversed layer, sound is turned back to carry far over the ocean. For transmission in the opposite direction, the sound reaching the upper layer is bent upwards and is not returned to produce a deep penetration; the rays staying in the lower layer are bent downwards, and their range is limited by the height of the lower layer. This controversial question no doubt helped to stimulate further interest in the field, both in this country and abroad.

In Section 9.3.3 we propose an explanation of what we have called 'Tyndall's paradox' in terms of the 'molecular' absorption of sound, not known until about 60 to 70 years after Tyndall's experiments.

Even Lord Rayleigh got involved in the foghorn signaling project. As Scientific Advisor to Trinity House, he turned his attention more to problems related to the sound source than to the propagation and raised many questions regarding the power efficiency of sirens and their directivity pattern. He pointed out, for example, that by using a vertical array of sources, the sound could be concentrated in a horizontal plane over the ocean rather than wasted in other directions. Rayleigh analyzed this problem theoretically and demonstrated his results by means of model experiments. This work probably is the first systematic study of 'phased (antenna) arrays' and is described in a paper 'On the Production and Distribution of Sound.' In this paper he also brought up the question of the role of nonlinearity as a limiting factor in the power output.

The question of nonlinearity was considered also by King in his account of one of the most extensive studies of sound propagation that has been made. The sound detection in previous studies had been mainly subjective as they involved a determination of the audibility of the sound. It also made use of primitive detectors such as the excitation of membranes by the sound and the motion of sand particles on the membranes, and Tyndall often employed sensitive flames in his laboratory experiments with sound. King, on the other hand, had a transducer (phonometer) which enabled him to make quantitative objective measurements of the sound pressure. His studies involved sound propagation over both land and sea. A powerful 40 hp siren was used as a sound source, and King, like Rayleigh, was interested in the acoustic efficiency as well as nonlinear effects of the source. The observation included such phenomena as silent zones, and King attempted to explain these effects in terms of refraction caused by wind stratification in much the same way as Henry had done.

Among the early experiments on sound propagation should be mentioned the work of Baron, which was carried out in 1938 but not reported until 1954 (for understandable reasons). The practical purpose of this study was the determination of the range of various warning signals in the city of Paris, but it included also sound transmission over ground at different frequencies and source elevations. As in previous studies, large sound pressure fluctuations were noted, sometimes as large as 50 dB.

These early studies of sound transmission were motivated mainly by the obvious practical problem of the range of audibility of (warning) signals. The more recent studies have resulted more from the environmental noise ('pollution') problem caused by jet aircraft and a variety of industrial noise sources. The basic questions to be considered are still the same although a wider range of frequencies now has to be included. In the past few decades, a number of research programs, both in this country and abroad, have been focused on the problem of sound propagation from aircraft under various conditions of testing and flight. Improved instrumentation for both acoustic and meteorological measurements has made possible more serious attempts to correlate sound transmission characteristics with the state of the atmosphere. As a result, considerable strides have been made toward an understanding of this problem although several questions still remain to be answered.

The instrumentation and data processing in this field has reached the point where sound can be used as a diagnostic tool in the study of the atmosphere (atmospheric SONAR). Thus, use of sound to determine the average temperature and wind profiles in the upper atmosphere (using very low-frequency sound) has been used for some time and acoustic monitoring of the vortices produced by large aircraft at airports is now feasible.

Localization of sound sources by sound ranging techniques with highly directive arrays of sound detectors is another area of considerable interest. This technique depends on the variation in phase of the sound field at different location. Phase and amplitude fluctuations produced by turbulence is then a problem to overcome in particular when combined with the effects of reflection from the ground.

In this chapter we shall review some of these aspects of atmospheric acoustics and attempt to bring out the essentials without extensive mathematical analysis.

9.2   The Earth's Surface Boundary Layer

9.2.1   The Stratification of the Atmosphere

In regard to the wind, the atmosphere can be divided into three layers, the *surface boundary layer*, the *transition region*, and the *free atmosphere*. In the lowest of these, extending up to about 50 to 100 meters above the ground surface, the motion of the air is turbulent and determined by local pressure gradients and the friction of the surface. In the highest layer, the free atmosphere, the motion is approximately that of an inviscid fluid under the action of the forces arising from the rotation of the earth. The resulting wind is known as the *geostrophic wind*. In the transition region, the wind is influenced both by the earths rotation and by surface friction. This layer extends up to 500 to 1000 meters and forms, together with the surface boundary layer, the so-called *planetary boundary layer*. It is known that the wind changes direction with height in the planetary boundary layer. In the surface boundary layer, the average wind direction is approximately independent of height and we shall start with sound propagation in this region. In Section 9.6, propagation from an aircraft in flight at an altitude of 10,000 m is analyzed.

9.2.2   Wind Profile

The study of the wind structure in the surface boundary layer is one of the important problems in micro-meteorology. We shall give a brief summary of some of the known properties of the wind which is of interest for sound propagation.

Both theoretical and experimental results indicate that the average wind velocity increases approximately as the logarithm of the height over ground. This has been expressed in various empirical formulas containing parameters which depend on the roughness of the ground. One such formula is

$$u(z)/u* \approx 2.5 \ln(1 + z/z_0), \tag{9.1}$$

where $u*$ is called the *friction velocity* and z_0, the *roughness distance*. For example, for a very smooth ground, such as a mud flat or ice, $z_0 \approx 0.001$ m and for thin grass, up to 5 cm in height, $z_0 \approx 5$ m. The friction velocity defined as $u* = 2/\ln(1+2/z_0)]$ m/s, has been determined for different ground conditions; it can be chosen to be the value required to produce a mean velocity of 5 m/s at a height of 2 m. Then, from Eq. 9.1 and for very smooth ground, $u* \approx 2/\ln(1 + 2000) \approx 25$ cm/sec and for 5 cm grass, $u* \approx 55$ cm/sec. The ratio $u*/u(2)$ has been found to be approximately constant for values of $u(2)$ from 0.2 to 5 m/sec.

As an example, for the very smooth surface and a mean velocity of 5 m/sec at $z = 2$ m, the formula gives $u(2) \approx 4.9$ m/s rather than 5, but this discrepancy probably indicates that the value of z_0 has been rounded off to 0.1 cm from a slightly lower value.

If, instead, the measured mean velocity at a height of 2 m had been 3 rather than 5 m/sec, the value of the friction velocity would have been $u* = (3/5)16 \approx 9.6$ cm/sec (since the friction velocity is known to be approximately proportional to the velocity at 2 m). The velocity at a height of 1 meter, according to Eq. 9.1, then becomes $u \approx 1.7$ m/sec.

Eq. 9.1 for the z-variation of the average wind velocity has been found to be in relatively good agreement with experiments when there is no vertical temperature gradient in the atmosphere. In the daytime, it turns out that the velocity gradient is less than that indicated by the equation and the opposite is true during nighttime. Accordingly, the variation of the velocity over ground is sometimes given in the form $du/dz \approx a(z + z_0)^{-\beta}$, where a is a constant. The exponent β is approximately 1 for a small or close to zero temperature gradient, larger than 1 for a negative and smaller than 1 for a positive gradient.

The average wind velocity has a diurnal variation. It reaches a maximum around noon and a minimum at midnight. At higher altitudes, this variation is displaced in time. In fact, there is some experimental evidence that indicates a reversal of the diurnal variation at heights above 50 m over land. The corresponding variation over large bodies of water is not very well known, and in coastal regions, the diurnal variation is often masked by land and sea breezes.

9.2.3 Wind Fluctuations

The spectrum of wind fluctuations covers a wide range of frequencies. It is sometimes divided into three frequency regions referred to as large-scale turbulence (a period of about an hour), intermediate-scale (a period of about 2 to 3 minutes), and small-scale (a period of a few seconds). Experiments have shown that at least two-thirds of the energy in the fluctuations correspond to frequencies in the region of 0.5 to 20 Hz. The frequency distribution of the eddy velocities is essentially Maxwellian.

Close to the ground, say, 2 meters above ground, the horizontal fluctuations are often about 50 percent stronger than the vertical. All three components of fluctuations at this level are approximately proportional to the average wind speed. This is true at least in the daytime when the temperature decreases with height (negative gradient). In layers approximately 20 to 25 meters above ground, the fluctuations are about the same in all directions. The turbulence is then almost isotropic, and the energy of the fluctuations partitioned equally on the three directions.

The roughness of the surface should have some effect on the turbulent structure, but little information is available on this point. The wind structure is strongly dependent on the temperature gradient, however. In the daytime, when the temperature decreases with height (lapse rate), there is a tendency for the air layer at the surface to move upwards and stir up the air above, thus inducing turbulence. During the night, when the temperature gradient normally is reversed (inversion), the atmosphere is more stable and the air flow has a tendency to be laminar.

9.2.4   The Temperature Field

For an atmosphere at rest under adiabatic conditions, the temperature decreases with altitude, the gradient being

$$\frac{dT}{dz} = -\frac{g}{R}\frac{\gamma - 1}{\gamma}, \tag{9.2}$$

where g is the acceleration of gravity, R, the universal gas constant, and $\gamma = C_p/C_v$, the specific heat ratio. This temperature gradient, with the magnitude denoted Γ, is often referred to as the *adiabatic lapse rate*. With $g = 9.81$ m/s^2, $R = 8.31$ joule/mole/K, $\gamma = 1.4$, and 1 mole of air being ≈ 0.029 kg, we find $\Gamma \approx 0.98$ °Cper 100 m. Heat transfer due to turbulence and convection makes the temperature distribution close to ground much more complicated and yields gradients much higher than the adiabatic lapse rate; temperature gradients as large as 1800 Γ have been measured.

The temperature has a well-defined seasonal as well as diurnal variation. As a very rough approximation, the temperature gradient during the summer can be expressed as $dT/dz = -0.5/z$ during the day and $dT/dz \approx 0.15/z$ during the night, at least in the region $z > 0.5$ m. During the winter, the result is $dT/dz \approx -0.1/z$ for daytime and $dT/dz \approx 0.15/z$ for nighttime for $z > 0.1$ m. For a small value of z, these rates are considerably higher than the lapse rate.

The transition between lapse rate and inversion depends on the height above ground and is also different in summer and winter. The evening transition is about the same in winter and summer and occurs about one and a half hours before sunset; the morning transition in the summer takes place about half an hours after sunrise, and in the winter, about one and a half hours after sunrise.

The temperature fluctuates in a similar way as does the wind and the fluctuations correlate with temperature gradients and show diurnal variations.

9.3   Sound Absorption

By sound absorption in the atmosphere we mean the conversion of acoustic energy into heat without the interaction with any boundaries. As a result of this absorption, the amplitude of a sound wave will decrease with distance of propagation, and in this chapter we shall deal with the dependence of this attenuation on frequency and atmospheric conditions.

The heat producing mechanisms are all related to intermolecular collisions and the time it takes for thermal equilibrium to be established when the state of a gas or fluid is forced to change with time as it is in a sound wave. It is expressed through a *relaxation time* to be defined quantitatively later.

The molecular motions are translations, rotations, and vibrations, and each of these motions have a characteristic relaxation time. Viscosity and heat conduction involve the translational motion and are often referred to as translational relaxation and exist in all gases. Rotational and vibrational relaxation occur only in poly-atomic gases.

The attenuation is intimately related to the ratio of the period and the relaxation times. As we shall see later, the dominant effect in sound attenuation in air under normal conditions is related to the vibrational relaxation of the oxygen and nitrogen molecules.

9.3.1 Visco-Thermal Absorption

Although it is normally not the most important (except in a monatomic gas), we start by considering the visco-thermal attenuation caused by viscosity and heat conduction. It is often referred to as the 'classical' attenuation. It was known more than 100 years ago through the works of Stokes (1845) and Kirchhoff (1868). The early investigators on sound transmission through the atmosphere used relatively low frequencies, of the order of 100 Hz and less, and the visco-thermal attenuation is then small and no particular attention was paid to it. As we shall see, this attenuation increases with the square of the frequency and needs to be accounted for at high frequencies in typical practical problems of sound transmission.

Normally, we associate losses due to viscosity as a result of shear motion and, qualitatively, one might wonder at first how shear arises in a compression (and rarefaction) of a gas as it occurs in a sound wave. One way to understand it qualitatively is to consider the compression of a square along one of its sides. Look at the resulting deformation of the interior square that is formed by connecting the midpoints of the sides. This square will be deformed into a rhombus with a corresponding angular displacements of opposite sides representing shear.

The qualitative understanding of the effect of heat conduction is more straightforward. When a volume element is compressed, the pressure in the element increases and when it is expanded, the process is reversed. Without heat conduction, the work done on the gas during the compression would be recovered during the expansion and no losses would occur on the average. With heat conduction, however, the pressure and temperature buildup during the compression is affected by heat leakage from the element so that the pressure and temperature 'relax' and will not reach the values they would have achieved in the absence of heat flow. During the expansion, there is no instantaneous return to the pressure and temperature that existed during the compression and the pressure will be lower than during the compression and a net work is done on the element in a cycle. This work is drawn from the sound wave and causes attenuation.

For a very short compression-rarefaction cycle (i.e., at the high-frequency end of the spectrum) it is tempting to say that there is no time for the heat to flow and no average absorption is to be expected over a cycle. At a second thought, however, we realize that the wavelength becomes so small and the thermal gradients so large that heat does indeed have time to flow, the better the higher the frequency, and the conditions in that range will be isothermal and reversible and there will be no net loss during a period. Similarly, in a very long cycle corresponding to a low frequency, the inflow of heat during expansion will be a copy of the outflow during the compression. In such a 'quasi-static' change of state, no net work is to be expected.

When we talk about attenuation we refer to the spatial variation in sound pressure and when we relate attenuation to energy loss per cycle, we refer to the attenuation in a distance equal to a wavelength, since that is the distance traveled during one cycle. Consequently, the attenuation per wavelength will be zero at both ends of the frequency spectrum and there will be a maximum at some finite period, the *thermal relaxation time*. Actually, for viscosity and heat conduction, this thermal relaxation time is so short, of the order of the time between collisions, $\approx 10^{-9}$ seconds for normal air, that in practice the frequency never becomes so high that the period will be of the order as the thermal relaxation time. The attenuation, therefore, will increase with frequency over the range of interest. Actually, the concepts of viscosity and heat conduction tacitly are based on the assumption that the relaxation time is zero and the results obtained, as we shall see, leads to an attenuation which increases with frequency over the entire range. The situation will be different, however, for vibrational relaxation, as we shall see.

The effect of viscosity on the attenuation enters formally in terms of the *kinematic viscosity*, $\nu = \mu/\rho$, the ratio of the coefficient of shear viscosity and the density, which can be interpreted as the ratio of the rate of 'leakage' of momentum out of a volume element and the momentum of the element.

Similarly, the effect of heat conduction enters through the quantity $K/\rho C_p$, where K is the heat conduction coefficient and C_p the heat capacity per unit mass at constant pressure. The quantity $\chi = K/\rho C_p$, which might be called the *kinematic heat conduction coefficient*, can be interpreted as the ratio of the rate of heat leakage out of a volume element and its heat capacity ρC_p. The latter represents the 'thermal inertia' and corresponds to the inertial mass ρ in the expression for the kinematic viscosity.

From kinetic theory of gases it follows that both ν and χ are of the order of ℓc, where ℓ is the mean free path and c, the sound speed. The ratio $\nu/\chi = P_r$ (sometimes called the Prandtl number) is ≈ 0.7 for normal air. Kinetic theory also shows that μ and K are very nearly independent of density which means that ν and χ are both inversely proportional to density.

If the attenuation in the sound pressure of a plane wave is expressed by the factor $\exp(-\alpha x)$, the theory shows that α can be expressed in the simple-looking form

$$\alpha \ell = (\omega \tau)^2 \quad \text{where}$$
$$\ell = (1/c)[4\nu/3 + (\gamma - 1)\chi]/2, \quad \tau = \ell/c$$
$$\nu = \mu/\rho, \quad \chi = K/\rho C_p, \tag{9.3}$$

which shows a square law dependence of the attenuation on frequency.

The quantity $\gamma = C_p/C_v$ is the ratio of the specific heats at constant pressure and constant volume and is ≈ 1.4 for air. The characteristic length ℓ is of the order of a mean free path. Under normal conditions it is very small, of the order of 10^{-5} cm. The kinematic viscosity for normal air ($20°$C and 1 atm) is $\nu \approx 0.15$ and the kinematic heat conduction coefficient, $\chi = K/\rho C_p \approx 0.15/0.7 \approx 0.21$.

Inserting these values in Eq. 9.3, we get $\ell \approx 0.82 \times 10^{-5}$ cm and, numerically, the attenuation is then

$$\text{Attenuation} \approx 0.12\,(f/1000)^2 \quad \text{db/km}, \tag{9.4}$$

where the frequency f is expressed in Hz. The decay constant in Eq. 9.3 refers to the pressure field. The decay constant for intensity is 2α, a fact sometimes forgotten in numerical analysis. Thus, with the pressure field expressed as $\propto \exp(-\alpha x)$ the attenuation in dB is $20\log[p(0)/p(x)] \approx 20\log(e)(\alpha x) \approx 8.7\alpha x$. The attenuation is small indeed; to make it 1 dB per km, the frequency has to be about 2000 Hz. At 10,000 Hz, however, the attenuation becomes substantial, about 12 dB/km.

The kinematic viscosity increases with temperature; at the temperatures 100, 0, 20, 100, 500, and 1000°C the values are 0.06, 0.13, 0.15, 0.23, 0.79, and 1.73. The ratio ν/χ of the kinematic viscosity and heat conduction, the Prandtl number decreases slightly with temperature, being ≈ 0.72 at -100°C and 0.70 at 1000°C.

In the qualitative explanation of the visco-thermal attenuation, we considered the average energy transfer over one period. This argument then applies to the attenuation per wavelength, which, from Eq. 9.3, can be expressed as

$$\alpha\lambda = 2\pi\,\omega\tau, \tag{9.5}$$

where we have used $\tau = \ell/c$.

9.3.2 'Molecular' Absorption

At a given temperature, the total internal (thermal) energy in a gas in thermal equilibrium is made up of translational, rotational, and vibrational motion of the molecules, the proportions depending on temperature. If the temperature is increased from T_1 to T_2, a redistribution of the energy amongst the various modes takes place from one equilibrium value to another as a result of intermolecular collisions. The translational energy responds to a perturbation in a very short time, of the order of the time between molecular collisions which at atmospheric pressure and 20°C is about 10^{-9} seconds. Also, the rotational energy adjusts itself quickly in a time of the same order of magnitude. The vibrational motion, on the other hand, requires a much longer time; the probability of excitation is low and many molecular collisions are required to excite the vibrational motion of the molecules (a minimum of energy of one vibrational quantum is required for a successful energy transfer in a collision).

The average rate of change of the vibrational energy is proportional to the difference $(E_e - E)$ between the final equilibrium value E_e and the instantaneous value E, and the constant of proportionality is expressed as $1/\tau_v$, where τ_v is a measure of the response time of the vibrational motion. It is called the *relaxation time*. In pure Oxygen, for example, τ_v is of the order of 0.003 seconds. For pure Nitrogen, it is about 10^{-9} seconds. Such a vast difference can be explained only by the characteristics of molecular vibrations and intermolecular forces and, in this context, will be left as an accepted fact.

Under the periodic perturbation in pressure and temperature produced by a sound wave, the rate of excitation of vibrational motion will also be periodic but there will be a phase lag in the pressure variations with respect to the variations in volume. As a result, the work done on the gas by the sound wave during the compression will not be completely regained during the expansion. The difference is converted into heat and an attenuation of the sound wave results. Actually, the explanation is qualitatively similar to that given for the attenuation produced by viscosity and heat conduction for which the frequency dependence of the attenuation constant could be expressed as $\alpha\ell = (\omega\tau)^2$, where ℓ was the mean free path and $\tau = \ell/c$ (see Eq. 9.3).

An analysis based on this mechanism leads to an attenuation $p \propto \exp(-\alpha x)$ in which the attenuation constant can be expressed as

$$\alpha_v \ell_v = \frac{(\omega \tau_v)^2}{1 + (\omega \tau_v)^2}, \tag{9.6}$$

where ℓ_v is a mean free path and $\tau_v = \ell_v/c$ (c, sound speed) is the vibrational relaxation time. If $\omega \tau_v \ll 1$, this expression has the same form as for the visco-thermal attenuation for which the relaxation time was so short that the denominator could be put equal to 1.

This expression for the attenuation per mean free path in Eq. 9.6 is applicable to any relaxation process if the mean free path is properly chosen. For the visco-thermal attenuation, the mean free path is the average distance of travel of a molecule between collisions since each collision implies a transfer of translational momentum. For the attenuation related to the molecular vibrations, it is the average travel distance ℓ_v between collisions in which vibrational motion is excited (transfer of a quantum of vibrational energy).

From Eq. 9.6 it follows that the attenuation per wavelength can be written

$$\alpha_v \lambda = 2(\alpha_v \lambda)_{max} \frac{\omega \tau_v}{1 + (\omega \tau_v)^2}, \tag{9.7}$$

where $(\alpha_v/\lambda)_{max} = \pi/\ell_v$. The maximum value is obtained for $\omega \tau_v = 1$, i.e., at the frequency $f = 1/2\pi \tau_v$. Since it is caused by the vibrational molecular energy it is proportional to the specific heat contribution from vibrational molecular motion in *thermal equilibrium*. This contribution is temperature dependent and is expressed by the function $F(T) = C_v/R$, where R is the gas constant. It is known from statistical mechanics that

$$F(T) = \frac{x^2 e^x}{(e^x - 1)^2}, \tag{9.8}$$

where $x = \Theta/T$ and Θ is the Debye temperature for the vibrational motion. It is the temperature required to make the vibrational energy a substantial part of the thermal energy. For oxygen, $\Theta = 2235$ K. At temperatures of main interest here, we can put $F(T) \approx x^2 \exp(-x)$, and at a temperature of 20°C ($T = 293$ K), we get $F \approx 0.0276$.

In a non-equilibrium situation, the vibrational energy strives toward this equilibrium value at a rate proportional to the deviation of the instantaneous vibrational energy from the equilibrium. Working out the dynamics based on this premise, the maximum attenuation in Eq. 9.7 can be shown to be

$$(\alpha_v \lambda)_{max} = \frac{\pi}{2} \frac{(\gamma - 1)^2}{\gamma} F(T), \tag{9.9}$$

where γ is the equilibrium value of the specific heat ratio C_p/C_v. Then, with $\gamma = 1.43$, we find from Eq. 9.9 $(\alpha \lambda)_{max} \approx 0.00557$ at a temperature of 20°C, which is in good agreement with experimental results for pure oxygen. Since oxygen constitutes only 20 percent of air, the corresponding maximum attenuation per wavelength in air should be only one-fifth of the value for pure oxygen, i.e.,

$$(\alpha \lambda)_{max} = 0.0011 \quad \text{Corresponds to 0.0096 dB per wavelength.} \tag{9.10}$$

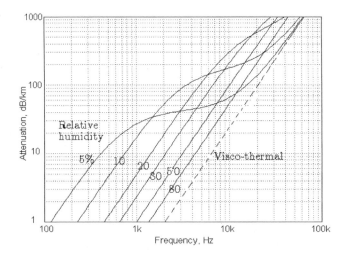

Figure 9.2: Total attenuation in air at 20°C in dB per 1000 m. Dashed curve: visco-thermal.

This is also in good agreement with experiments.

The maximum attenuation per wavelength is obtained at a frequency $f_m = 1/2\pi\tau_v$ which for pure oxygen is $f_m \approx 50$ Hz. The presence of water vapor in the air has been found to decrease τ_v and therefore increase the corresponding value of f_m. Apparently, the probability of excitation of vibrational motion of an Oxygen molecule in a collision is increased by the presence of a water molecule which then acts like an 'impedance matching' device.[1]

In terms of the water content in the air expressed by p_w/p_t, where p_w is the vapor pressure and p_t the total pressure, experiments have shown that the vibrational relaxation frequency can be expressed approximately by the empirical expression

$$f_v = 5.7 \cdot 10^8 (p_w/p_t)^2 + 50 \quad \text{Hz.} \tag{9.11}$$

The relative humidity is $\Phi = p_w/p_s$, where p_s is the saturated vapor pressure at the temperature in question which depends on temperature; the dependence of p_s/p_t on temperature is given by
Temperature, °C: −10, −5, 0, 5, 10, 20, 30, 40
p_s/p_t: 0.00264, 0.00395, 0.006, 0.0085, 0.012, 0.023, 0.042, 0.072.

Thus, at a temperature of 20°C, we have $p_s/p_t = 0.023$ and $p_w/p_t = 0.023\Phi$. Then, from Eq. 9.11 it follows that the dependence of the vibrational relaxation frequency on the relative humidity at 20°C is

$$f_v \approx 3 \cdot 10^5 \, \Phi^2 + 50 \quad \text{(at 20°C).} \tag{9.12}$$

---

[1]This is, of course, a naive view. A detailed analysis requires a quantum mechanical study of a three body collision involving two oxygen molecules and a water molecule.

In Fig. 9.2 is shown the computed total attenuation in dB/km, i.e., the sum of the visco-thermal and 'molecular' attenuation caused by Oxygen, at a temperature of 20°C. In the frequency range normally of interest in most problems, a low relative humidity results in high attenuation, contrary to what one might expect.

Effect of Temperature

A change in temperature at a constant value of the relative humidity leads to a change in the water vapor content, i.e., a change in p_w/p_t (p_w vapor pressure, p_t, total pressure) and hence in the vibrational relaxation frequency and the attenuation. This temperature dependence can result in large variations in the received sound level in propagation over large distances. At a given relative humidity, a decrease in temperature results in a dryer air in terms of the actual water content as measured by p_w/p_t and this means a reduction in the relaxation frequency and higher attenuation at relatively low frequencies. It is left for one of the problems to prove these assertions.

The temperature dependence resulting from the variations in water content ordinarily is the most important. There is also the temperature dependence of the factor $F(T)$ in Eq. 9.9 (expressing the equilibrium value of the vibrational energy contribution to the specific heat) and of the sound speed.

The temperature dependence of the classical (visco-thermal) attenuation arises from the temperature dependence of the kinematic viscosity and heat conduction coefficients and of the sound speed.

Influence of Air Pressure

The numerical results presented so far, including Fig. 9.2, refer to an air pressure of 1 atm. At higher altitudes, where the pressure is lower, the relaxation frequency will be lower since the time between collisions is longer. In fact, the relaxation frequency is proportional to the square of the particle density (why?). The vibrational relaxation contribution to the attenuation has to be changed accordingly. This will be discussed in more detail in Section 9.6 dealing with sound propagation from an aircraft in high altitude flight.

9.3.3   Proposed Explanation of Tyndall's Paradox

The paradox is that Tyndall, in his experiments on sound propagation, often found the attenuation to be higher for sound propagation in the direction of the wind than in the opposite direction. Henry proposed an explanation based on refraction in wind and a reversal of wind direction at a certain height above ground. We propose here that the paradox is merely a result of molecular relaxation attenuation and its dependence on the water content in the air. This phenomenon was discovered some 50 to 60 years after Tyndall's and Henry's experiments.

It should be remembered that in Tyndall's experiments, the sound source was located at the seashore and the acoustic signal strength was observed on the sea. Therefore, in the case of 'downwind' propagation, the wind was blowing from the land toward the sea and in upwind propagation from the sea toward land. Then, it is reasonable to assume that for downwind propagation (wind from land toward sea), the air was drier than in the upwind propagation and consequently causing higher attenuation at low frequencies in accordance with the attenuation curves in Fig. 9.2.

9.3.4 Effect of Turbulence

It is an exception rather than the rule that the atmosphere is calm and homogenous. Turbulent flow and thermal inhomogeneities produce variations of the speed of sound both in space and time and sound scattering will result. In contrast to sound absorption, scattering causes a redistribution rather than absorption of sound. Therefore, the effect of turbulence on sound propagation is expected to be most pronounced in highly directional fields. Sound will then 'diffuse' from regions of high to regions of low intensity turbulence, and the directivity will be less and less pronounced with increasing distance from the sound source. Although the total acoustic energy will be conserved (neglecting absorption), the sound pressure level as a function of distance in any one direction will be affected.

In the idealized case of a purely spherical steady state sound source in isotropic turbulence, a spherically symmetrical wave (averaging out fluctuations) will be produced. In practice, when absorption is accounted for, the back-scattered wave must be considered partially lost from the acoustic wave since when it reappears at another point in space, it has been attenuated. Actually, in an acoustic pulse wave, it is clear that at least the back-scattered wave will produce a reduction of the primary acoustic pulse.

Some Sound Transmission Experiments

Field Measurements

In a study of scatter attenuation, a harmonic wave train was transmitted over a distance of one mile. To eliminate as much as possible the effect of reflections from the ground, the sound source, a 60 watt driver with an exponential horn with a maximum diameter of 3 feet, was located at the ground level with the axis inclined with respect to the ground. Eleven microphones, 300 ft apart, were placed along a line at different heights increasing in proportion to the distance from the source so that they fell on the extension of the axis of the horn. The last microphone was located at a height of 62 feet. An analysis of such a sound field indicated that free field conditions existed along the line beyond the point were the vertical distance to the ground was approximately 3 to 4 wavelengths.

Each microphone was supplied with a battery-operated pre-amplifier located at the base of the microphone tower. The amplified signals from the microphones were transmitted over transmission lines to a field station for recording and processing.

Pulsed wave trains of pure tones were emitted by the source. The duration of each train was approximately 0.25 sec. Each receiver was gated in such a way that it was turned on only during the time the pulse passed the microphone. This gating was accomplished by means of a stepping switch that probed one microphone after another with a speed corresponding to the travel speed of the pulse.

The resulting series of pulses received from the microphones was displayed on an oscilloscope and photographed. The trace on the oscilloscope then directly represented the sound pressure as a function of distance from the source. The oscilloscope scale was made logarithmic by means of a logarithmic amplifier. The gains on the individual microphone amplifiers were made proportional to the distance of the microphones from the source. In this way, the microphone signals would all be the same in magnitude in the ideal case of an undamped spherical wave in which the sound pressure decreases inversely with the distance from the source. Consequently, if there were no sound absorption or scattering, the envelope of the oscilloscope traces simply would be a horizontal straight line. On the other hand, an exponential decay of the sound pressure of the form $p \propto \exp(\alpha r)/r$ would be represented by a straight line with a slope proportional to the attenuation constant α.

Results of this kind were obtained over a range of frequencies. In each case the pulse duration was 0.3 seconds. The decays were found not to be perfectly straight lines but fluctuating, varying somewhat from one pulse to another. This behavior is to be expected on account of the turbulence in the atmosphere.

Experimental data of this sort were collected intermittently over a period of approximately one year at various conditions of relative humidity, temperature, and wind. To study the influence of wind on the attenuation, the data were grouped into categories corresponding to wind velocities of 0−1, 1−3, 3−6, 6−10, and 10 to 15 mph at 20 ft above ground. In the wind range 0 to 1 mph we would expect the attenuation to be dominated by absorption alone. Although, on the whole, these results are consistent with the theoretical attenuation for air attenuation, the observed attenuation was somewhat smaller than the calculated at frequencies above 2000 Hz and somewhat larger at frequencies below 2000 Hz.

As the wind velocity increases, there is a tendency toward an overall increase in the attenuation. Except at frequencies below 2000 Hz, the scatter attenuation appears to be small compared with the sound absorption and only weakly dependent on frequency. Also the wind velocity dependence of the attenuation was found to be quite weak.

To attempt an explanation of this behavior, it is necessary to study the scattering process in some detail. Lighthill and others have found the total scattered energy to be proportional to the square of the frequency and to the mean square of the turbulence velocity fluctuations. However, if we consider the effect only of the *back-scattered* energy, the frequency dependence of the corresponding attenuation is indeed found to be weak. The eddies involved in this scattering are those with a size of the same order of magnitude as the wavelength under consideration.

To determine the velocity dependence of the scattering attenuation at a particular frequency, it is necessary to examine the spectrum of atmospheric turbulence. From studies of such spectra taken under various conditions, one finds that there is no unique relationship between the average wind speed and the turbulent spectrum. For example, the strength and spectral distribution of turbulence depend strongly on the thermal gradient. A comparison of turbulent spectra at different wind velocities frequently shows that although the intensity of the large eddies increase with the average flow speed, the eddies responsible for back-scattering in the frequency region of interest are only weakly dependent on the average wind speed. Thus, if back-scattering is assumed to be responsible for the attenuation of non-directional or weakly directional sound fields, the weak dependence of the attenuation on the wind velocity can be understood. However, from the results obtained, it appears that the scatter attenuation for an omni-direction field is small compared with the effect of sound absorption.

Sound beam. For a highly directional sound field, the situation can be considerably different. For a size L of the energy carrying eddies much larger than the wavelength and an rms value of the velocity fluctuation v, the attenuation in intensity is expected to be $I = I_0 \exp(-2\alpha x)$, where

$$\alpha\lambda \approx 4\pi (L/\lambda)\,(v/c)^2. \tag{9.13}$$

If we compare this scatter attenuation with the maximum value 0.0011 of the vibrations relaxation attenuation (see Eq. 9.9), we find that the scattering attenuation dominates if $v/c >> 0.01\sqrt{\lambda/L}$. Typically, with $L \approx 10$ m and with a frequency of 1000 Hz, we get the condition $v >> 1$ m/sec for scatter attenuation to be dominant. If we assume the rms value of the turbulent velocity to be about 10 percent of the mean velocity, we find that for a frequency of 1000 Hz, the scatter attenuation begins to dominate at wind speed of about 20 mph. As the frequency f increases, this critical wind velocity is reduced in proportion to $1/\sqrt{f}$.

Pulse Height Analysis of Scattered Sound

The scattering of a beam of sound from turbulence has been demonstrated[2] in a laboratory experiment in which the effect of turbulence on the amplitude distribution of acoustic pulse waves was measured. In this experiment, carried out in an anechoic chamber, a 100 kHz pulse modulated sound beam, produced by a specially designed electrostatic transducer, was transmitted through a region of turbulence produced by four centrifugal blowers directed toward a common center. In another version of the experiment, the flow from 250 pairs of opposing nozzles fed from a common manifold was employed. The pulses were received by a specially designed condenser microphone whose output was fed to a ten-channel pulse-height analyzer. Each channel could be set in such a manner that only pulses in a certain prescribed amplitude range would be accepted. The number of pulses received in each channel was counted and displayed on counters. The pulse height distribution at any point of observation thus could be read directly from the ten counters.

If no turbulence interrupts the transmission of sound, the pulses received have all the same height and only one channel of the analyzer is activated. The corresponding pulse-height distribution is then a 'line' with a width equal to the channel bandwidth. However, when turbulence is present, the detected pulses vary in height according to the turbulence fluctuations, and several channels in the analyzer will be activated thus defining a pulse height distribution. Consider first the case when the microphone is located within the main lobe of the sound beam. The expected effect of turbulence is then to reduce the average pulse amplitude, since the scattered energy is largely

---

[2]Michael D. Mintz and Uno Ingard, *Experiments on scattering of sound by turbulence*, J. Acous. Soc. Am. 32, 115(A), (1959).

removed from the sound beam. This was clearly indicated by the experimental results. The average pulse height was displaced to a value markedly lower than the pulse height obtained without turbulence. The average pulse-height reduction increases with increasing velocity. (In these experiments no attempt was made to measure the turbulent strength except in a qualitative manner.) Some of the pulses reaching the microphone actually had a higher amplitude then in the absence of turbulence. This can be thought of as an occasional focusing effect of the turbulent flow.

With the microphone positioned outside the sound beam, that is, in the geometrical shadow, the most probable pulse height was larger with turbulence than without it. Again, this illustrates that sound is scattered out of the beam into the shadow zone, and, again, the effect increases with increasing flow velocity.

It follows that the overall effect of turbulence on a sound field with high directivity is to redistribute the sound in such a manner as to make the sound pressure distribution about the source more uniform. Thus, the term 'diffusion' of sound in turbulent flow is sensible.

9.3.5 Effect of Rain, Fog, and Snow

According to Tyndall's observations, the presence of water in the form of rain, fog, or snow in the atmosphere does not significantly affect the attenuation of sound. However, as Tyndall also mentioned, this result may be due indirectly to a more uniform atmosphere which is present at least in the case of fog and gentle rain.

One direct effect of rain and snow is the friction that results from the interaction with sound. There is also an effect resulting from the acoustic modulation of the vapor pressure, but this effect is small and we consider here only the effect of viscosity.

At very low frequencies, the water droplets are expected to move along with the air, and no friction and corresponding attenuation results. At high frequencies, on the other hand, the induced motion of the droplets will be negligible and the relative motion of the air and the droplets will produce viscous losses and attenuation. If the velocity of the air is u and the velocity of the droplet v, the viscous drag force is known to be

$$f_v = 6\pi a\mu(u - v), \tag{9.14}$$

where a is the radius of the droplet and μ the coefficient of shear viscosity of the air (Stokes relation).

The mass of a droplet is $m = 4\pi a^3 \rho_w/3$ and the equation of motion is $mdv/dt = f_v$, or in terms of complex amplitudes, $-i\omega mv = f_v$. Combining this equation with Eq. 9.14, v (and $u - v$) can be expressed in terms of u. In fact, we find

$$(u - v)^2 = \frac{(f/f_r)^2}{1 + (f/f_r)^2} u^2, \tag{9.15}$$

where $f_r = 6\pi\mu a/m2\pi$.

The energy dissipation caused by one droplet is $f_v(u - v) = 6\pi a\mu(u - v)^2$ where the quantities involved are rms values. The corresponding loss per unit volume, containing n droplets, is then n times as large. The intensity in the sound wave is expressed as $I = \rho_a c u^2$, where ρ_a is the air density. Then, with the loss per unit volume being $L_v = 6\pi na\mu(u - v)^2$ and with $u^2 = I/\rho c$, we obtain $\partial I/\partial x = -\beta I$, i.e., $I = \exp(-\beta x)$, where

$$\beta = \frac{2nmf_r}{\rho_a} \frac{(f/f_r)^2}{1 + (f/f_r)^2}. \tag{9.16}$$

This decay constant refers to the acoustic intensity and is twice that for sound pressure, $\beta = 2\alpha_w$.

We leave it for one of the problems to compare this attenuation with the molecular relaxation attenuation and show that the attenuation caused by water droplets should be negligible under normal conditions, at least at sufficiently high frequencies, say, above 100 Hz.

9.3.6 Problems

1. **'Molecular' attenuation**

   From Eq. 9.7, express the attenuation constant α_v as a function of f/f_v suitable for numerical computations ($f_v = 1/\tau_v$ is the relaxation frequency) and check the results in Fig. 9.2. Use $(\alpha_v \ell_v)_{max} = 0.0011$ and the data for f_v versus humidity, Eq. 9.12.

2. **Influence of temperature on vibrational relaxation and attenuation**

   From the data given in the text can you modify Eq. 9.12 so that it applies to a temperature of 0°C. How does this change in temperature affect the attenuation curves in Fig. 9.2?

3. **The attenuation caused by rain, fog, and snow**

   Compare the molecular attenuation in air with the attenuation caused by small particles, density n per unit volume and radius a, and discuss the relative importance of the two.

4. **Tyndall's paradox**

   With reference to the discussion of Tyndall's paradox in the text, estimate the difference in penetration depth over the ocean of a 200 Hz tone emitted from the land-based sound source for wind against and with the sound. Assume that the corresponding relative humidities are 50 and 30 percent. What can you say about the difference in penetration depth for upwind and downwind propagation under these conditions?

9.4   The Effect of Ground Reflection

9.4.1 Pure Tone

The main effect of the ground is to produce a reflected sound field that interferes with the primary field from a sound source located above ground. If the sound source emits a pure tone, this interference can lead to considerable variations in sound pressure level with distance from the sound source. In order for such an interference to take place, it is necessary that the phase relationship, or coherence, between the direct and reflected sound field be maintained. As we shall see, turbulent fluctuations in the atmosphere can destroy this coherence, particularly at high frequencies. Therefore, the effect of ground depends in no small measure on the state of the atmosphere above it.

We start with the ground regarded as a rigid, totally reflective plane boundary and an omni-directional point source S, as indicated in Fig. 9.3. The reflected sound field

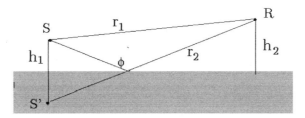

Figure 9.3: Point source S above a reflecting plane. Image source: S.' Receiver at R.

is then the same as that which would be produced by an image source S' with the same strength as S but located at the image position, as shown. The total sound field is then the sum of the direct and the reflected fields. When the path difference between the direct and reflected waves is an integer number of half (full) wavelengths, destructive (constructive) interference occurs.

If the horizontal distance between the source and receiver is x and with the notation in Fig. 9.3, the travel path of the direct sound from the source S to the receiver R is $r_1 = \sqrt{x^2 + (h_2 - h_1)^2}$ and from the image S' to R it is $r_2 = \sqrt{x^2 + (h_1 + h_2)^2}$. The difference in path lengths between the reflected and the direct sound is $r_2 - r_1$, and it decreases with increasing x. The largest difference is at $x = 0$ and equals $2h_1$ if $h_2 > h_1$ and $2h_2$ if $h_2 < h_1$.

For large values of x compared to $(h_1 + h_2)$, we have $r_2 \approx x + (h_1 + h_2)^2/2x$ and similarly $r_1 \approx x + (h_1 - h_2)^2/2x$ so that the path difference between the reflected and direct sound is $r_2 - r_1 \approx 2h_1h_2/x$. If this equals half a wavelength, there will be destructive interference between the reflected and the direct sound and a minimum in the total sound field at the distance

$$x_m = 4h_1h_2/\lambda. \tag{9.17}$$

Since the path difference decreases with x, it follows that beyond this distance x_m, there can be no interference minimum in the field. Thus, x_m represents the boundary between an interference zone with maxima and minima for $x < x_m$ and a far field zone, $x > x_m$, in which the sound pressure decreases monotonically with x in the same way as in free field. For example, with $h_1 = h_2 = 6$ ft, the distance to this last minimum at a frequency of 1121 Hz ($\lambda \approx 1$ ft) will be 144 ft. Note that the range of the interference zone increases with increasing frequency.

If the boundary is not totally reflecting but has a finite impedance, the pressure reflection coefficient for a plane wave is (see Eq. 4.102)

$$R = \frac{\zeta \cos\phi - 1}{\zeta \cos\phi + 1}, \tag{9.18}$$

where ζ is the normalized boundary impedance and ϕ the angle of incidence (see Fig. 9.3). It has been assumed that the boundary is locally reacting and the impedance ζ is the normal impedance. As an approximation, we shall use this reflection coefficient also for our spherical wave, thus ignoring the curvature of the wave front as it strikes the boundary. This is a good approximation at large distances from the source but, in a general analysis, the 'sphericity' should be accounted for. The detailed analysis of it, however, is beyond the scope of this text.

The most important consequence of a finite boundary impedance is correctly expressed by the plane wave approximation; thus, in the far field, beyond the interference zone, the dependence of the sound pressure on distance is not $1/x$ as in free field or for a totally reflecting boundary but rather $1/x^2$. To see this, we express the sound pressure $p_b = p_i + p_r$ at the boundary as

$$p_b = (1 + R)p_i = \frac{2\zeta \cos \phi}{\zeta \cos \phi + 1}\, p_i \approx 2\zeta(h_1/x)p_i \propto (1/x^2), \qquad (9.19)$$

where, for $h_1/x \ll 1$, we have used $\cos \phi \approx h_1/x$ and $p_i \propto 1/x$. The omission of the term $\zeta \cos \phi \approx \zeta(h_1/x)$ in the denominator implies that $(h_1/x)\zeta \ll 1$. In other words, the larger the value of ζ, the further out we have to go before the $1/x^2$ dependence sets in and in the limit of $\zeta = \infty$, i.e., the totally reflecting boundary, the $1/x^2$ zone will not occur for a finite distance. Physically, the $1/x^2$ dependence is related to the reflection coefficient going to -1 as the angle of incidence goes to 90 degrees and the reflected pressure tends to cancel the pressure from the incident wave. For an estimate of the distance to the $1/x^2$ for a given boundary impedance, the condition $|\zeta|(h_1/x) < 0.02$ (i.e., $x \approx 20|\zeta|h_1$) is normally satisfactory. It should be borne in mind, though, that there is an additional condition that depends on the wavelength. The onset of the $1/x^2$ region must fall beyond the interference zone. According to Eq. 9.17, the distance to the last minimum in the interference zone is $4h_1h_2/\lambda$.

This ground effect depends on the interference between the direct and the reflected sound and assumes that the atmosphere is uniform and static so that the phase difference between these sounds depends only on the difference in path lengths. In a nonuniform and turbulent atmosphere, this relationship is broken and the result obtained above will be modified. If the turbulence is strong enough, the two waves will be uncorrelated at the location of the receiver in which case the mean square sound pressures add. In that case the region of the inverse x^2 dependence of the sound pressure will be eliminated and the rms value will decrease as $1/x$ as in free field.

In practice, this behavior has been observed for sound propagation over snow covered ground. In calm whether, the inverse x^2 dependence was clearly seen with a correspondingly low sound pressure level sufficiently far from the source. On a windy day, with the wind going perpendicular to the sound path, the sound pressure level in this region *increased*. This is explained as a result of a destruction of the correlation between the direct and reflected sound by turbulence and a corresponding reduction of destructive interference.

Similarly, in the interference zone, with pronounced minima in the pressure field in a calm atmosphere, the minima will be reduced when turbulence is present. The path length difference at a minimum then will not remain at an integer number of half wavelengths but will fluctuate, thus producing corresponding fluctuations in the sound pressure level.

9.4.2 Random Noise

If instead of being a pure tone, the emitted sound pressure has a more general time dependence $p_1(t)$, the interference between the direct and the reflected waves applies to each component in the Fourier spectrum. If the time dependence is completely random, there will be no variation in the *total* sound pressure due to interference and the mean square values of the incident and reflected sounds add.

However, even if $p_1(t)$ is random, there will be an interference pattern for a band limited portion of the field and as the bandwidth goes to zero, the interference pattern for the pure tone is approached. For a detailed discussion of this problem, we refer to one of the examples in Chapter 11.

In Fig. 9.4 are shown the computed pressure distributions for both a totally reflecting boundary (infinite impedance) and for a resistive boundary with a normal impedance of 2 ρc. over a boundary with a purely resistive impedance of 2 ρc, for both. In both cases the results obtained for a pure tone as well as an octave band of random noise are shown. The source height is $h = 4\lambda$, where, in the case of the octave band, the wavelength refers to the center frequency of the band.

There is a significant difference in the result for the two cases which is largely due to the variation of the reflection coefficient with the angle of incidence of the sound. As explained in connection with Eq. 9.19, the pressure distribution in the far zone, beyond the interference zone, the sound pressure decreases as $1/x^2$ for the boundary with a finite impedance rather than $1/x$ for the infinite impedance. This means that the slope of the SPL curve versus the distance approaches $40 \log 2 \approx 12$ dB per doubling of distance rather than the 6 dB for the totally reflecting boundary. In the interference zone, the maxima decrease with distance as ≈ 6 dB per doubling of distance and the change to the 12 dB slope in the far zone is quite apparent.

The distance to the last minimum has been reduced in comparison with that for the hard boundary. The reason is that for an impedance $\zeta = 2$, the incident sound

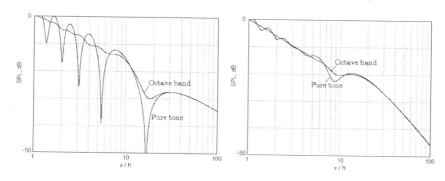

Figure 9.4: The sound pressure level versus distance from a source over a plane boundary. Left: Boundary totally reflecting, i.e., with infinite impedance. Right: Boundary impedance purely resistive with a normal impedance of 2 ρc. Source height: $h = 4\lambda$. Both pure tone and an octave band of random noise are considered. For the octave band, the wavelength refers to the center frequency of the band.

will be totally absorbed by the boundary at an angle of incidence of 60 degrees and the reflected wave that causes an interference at the last minimum is weakened considerably as is the destructive interference.

9.4.3 Problems

1. **Angle of incidence and destructive interference**

   A spherical point source emits a pure tone at a wavelength λ over a plane, totally reflecting the boundary. The height of the source, the same as of the receiver, is $h = 4\lambda$. What is the angle of incidence of the sound which causes the destructive interference at the most distant minimum from the source?

2. **Fraction of sound absorbed by a plane impedance boundary**

   A spherically symmetrical point source is located above a locally reacting plane boundary with a purely resistive normal impedance θ (normalized).

   (a) Use the plane wave approximation for the reflection coefficient and show that the fraction of the acoustic power of the source absorbed by the boundary is

   $$W/W_0 = 2\left[\frac{\ln(1+\theta)}{\theta} - \frac{1}{1+\theta}\right]. \tag{9.20}$$

   (b) If the normalized admittance of the boundary is $\eta = 1/\zeta = \mu + i\sigma$, show that the fraction of power absorbed by the boundary in the angular range from 0 to ϕ is

   $$W/W_0 = \mu\left[\ln\frac{\sigma^2 + (1+\mu)^2}{\sigma^2 + (\mu + \cos\phi)^2} - \frac{2\mu}{\sigma}\arctan\frac{\sigma(1-\cos\phi)}{\sigma^2 + (1+\mu)(1+\cos\phi)}\right]. \tag{9.21}$$

   Is this consistent with the answer in (a)?

9.5 Refraction Due to Temperature and Wind Gradients

9.5.1 Introduction

In addition to the propagational effects in the atmosphere considered so far, there is also the refraction of sound due to variations in the temperature and the wind. Of particular importance is the variation of these quantities and the related sound speed with altitude.

Consider first the effect of temperature. The local speed of sound is $c \propto \sqrt{T}$ so that the relation between the gradients of the sound speed and temperature becomes $(1/c)dc/dz = (1/T)dT/dz$, where T is the absolute temperature. To explain qualitatively what the effect of the gradient will be, let us consider a plane wave front which starts out vertical, corresponding to sound propagation in the horizontal direction. Since the wave speed is different at different vertical positions, the wave front will not remain vertical, as indicated in Fig. 9.5. If the temperature decreases with height, for example, the lower part of the wave front will travel faster than the upper and as a result the front will be tilted upwards. This is best described by using the idea of an acoustic ray, which is simply the normal to the wave front; it indicates the

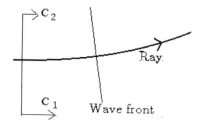

Figure 9.5: Refraction of sound. Temperature decreases with height.

direction of propagation. Thus, the ray will not remain horizontal but will be bent upwards, as shown in the figure.

In this figure the temperature decreases with altitude. If the direction of propagation is reversed, the wave will still be bent upwards as a result of *refraction*.

For a vertical wind gradient, we use the same argument to demonstrate that it produces a curved sound ray. We can again use Fig. 9.5 for illustration but let the arrows labeled c_1 and c_2 represent the different wind speeds U_1 and U_2. These velocities are the horizontal components of the wind in the plane of sound propagation. Thus, if the total horizontal wind velocity is $|U|$, only the component $|U| \cos \phi$ will influence refraction. In general U_2 is larger than U_1. The local wave speed is $c + U$, and with $U_2 > U_1$, it follows that the wave will be bent downwards. If we now reverse the direction of propagation, the resulting wave speed $c - U_2$ at the upper part of the wave front now becomes smaller than the value $c - U_1$ at the lower part and the wave will be refracted upwards. Thus, unlike refraction in a temperature stratified atmosphere, the refraction due to flow causes sound to be bent downwards in the downwind direction and upwards in the upwind direction; flow makes the atmosphere anisotropic in this respect.

9.5.2 Law of Refraction

In a moving fluid, the phase velocity of a sound wave with respect to a fixed coordinate system is the sum of the local sound speed in the fluid and the component of the fluid velocity in the direction of propagation of the wave,

$$c_p = c + U \cos \theta. \tag{9.22}$$

It is the propagation speed of a surface of constant phase (wave front). A spatial variation of the fluid velocity produces a corresponding variation of the phase velocity, and this can lead to flow-induced refraction, as indicated qualitatively in Section 9.5.1.

We consider first the idealized case of two fluids moving with different speeds U_1 and U_2 in the x-direction and separated by a plane boundary which can be thought of as a thin, mass-less membrane. The sound speeds in the two fluids are c_1 and c_2. A sound wave is incident on the boundary and the angle between the direction of propagation and the plane is θ_1 as shown in Fig. 9.6. We prefer to use this angle rather

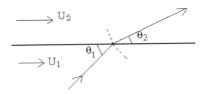

Figure 9.6: The refraction of sound at the interface between two moving fluids.

than the angle between the direction of propagation and the normal to the boundary for reasons that will become clear shortly.

In the figure, the fluid velocity above the boundary has been assumed higher than below and this causes a change in direction of propagation as indicated in the figure, where the dashed lines are the wave fronts (surfaces of constant phase) and the solid lines are the corresponding 'rays' (directions of propagation).

As a wave front moves forward with the phase velocity $c_p = c + U \cos \theta$, the intersection point between the front and the boundary moves along the boundary with the higher velocity, the trace velocity, $c_t = c_p / \cos \theta = c / \cos \theta + U$. Since the wave fronts hang together at the boundary, this velocity must be the same on both sides of the boundary, i.e.,

$$c_1 / \cos \theta_1 + U_1 = c_2 / \cos \theta_2 + U_2, \qquad (9.23)$$

which can be written

$$\cos \theta_2 = \frac{c_2 \cos \theta_1}{c_1 - (U_2 - U_1) \cos \theta_1}. \qquad (9.24)$$

The entire angular range of incidence and refraction is 0 to 180 degrees; for angles larger than 90 degrees, $\cos(\theta)$ has to be taken with its negative sign. Eq. 9.24 may be considered to be the acoustic equivalence of 'Snell's law' of refraction in optics.

Consider first the case when there is no fluid motion. The angle of refraction is then given simply by $\cos \theta_2 = (c_2/c_1) \cos \theta_1$. This will yield a real value of θ_2 only if the right-hand side is between -1 and $+1$, i.e., $-1 \le (c_2/c_1) \cos \theta_1 \le 1$. In the limiting case when it is 1 we get $\theta_1' = c_1/c_2$, the refracted wave is then parallel with the boundary. In the other limit, -1, $\cos \theta_1'' = -(c_1/c_2)$ and $\theta_1'' = 180 - \theta'$.

If θ_1 is smaller than the critical angle of incidence θ_1', total reflection occurs. Then there will be no propagating wave in the second region, only a pressure field which turns out to decay exponentially with distance from the boundary; the wave is *evanescent*. The presence of such a field is necessary in order to satisfy the boundary condition of continuity of sound pressure across the boundary.

Next consider another special case with $c_1 = c_2$, $U_1 = 0$. From Eq. 9.24 follows that $\cos \theta_2 = c \cos \theta_1 / (c - U_2 \cos \theta_1)$. It is left as a problem to show that the critical angle for total reflection is now $\cos \theta_1' = 1/(1 + M_2)$, where $M_2 = U_2/c$ is the flow Mach number. Total reflection occurs if $\theta_1 < \theta_1'$ or if $\theta_1 > 180 - \theta_1'$. This case corresponds qualitatively to a sound wave impinging on a jet.

As a final example, we choose $U_2 = 0$, which would correspond qualitatively to the sound emerging from a jet. Again it is left for a problem to show that as the emerging sound is confined to an angular region between $\theta_2' = \arccos[1/(1+M_1)]$ and $180 - \theta_2'$.

Ray Curvature

To study sound propagation in the atmosphere, we need to extend the discussion above to a continuous inhomogeneous medium. Thus, we let the sound speed $c(z)$ and, as before, the horizontal flow velocity component $U(z)$ in the plane of propagation vary continuously with the height z above ground. The conservation of the trace velocity still applies so that

$$c_t = U(z) + \frac{c(z)}{\cos\theta(z)} = \text{constant}. \tag{9.25}$$

As before, $\theta(z)$ is the angle between the direction of propagation and the x-axis. Differentiation with respect to z yields

$$\frac{dU_z}{d+}c\frac{\sin\theta}{\cos^2\theta}\frac{d\theta}{dz} + \frac{1}{\cos\theta}\frac{dc}{dz} = 0 \quad \text{or}$$

$$\frac{d\theta}{dz} = -\frac{\cos\theta}{c\sin\theta}\left(\frac{dU}{dz} + \frac{1}{\cos\theta}\frac{dc}{dz}\right). \tag{9.26}$$

For almost horizontal rays, $\cos\theta \approx 1$, and the differential equation for θ reduces to

$$\frac{d\theta}{dz} = -\frac{1}{\sin\theta}\left(\frac{1}{c}\frac{d(U\cos\phi + c)}{dz}\right). \tag{9.27}$$

The radius of curvature R of the ray can be expressed as follows. The elementary arc ds along the ray that corresponds to an angular increment $d\theta$ can be expressed as $ds = R d\theta$ and $\sin\theta$ as dz/ds so that

$$d\theta/dz = (1/[R\sin\theta]. \tag{9.28}$$

Using this expression in Eq. 9.27, we get

$$1/R \approx -(1/c)\,d(U + c)/dz, \tag{9.29}$$

where $(1/c)dc/dz = (1/2T)dT/dz$. A positive value of the radius of curvature corresponds to a ray which turns upwards. A frequently observed phenomenon is illustrated

Figure 9.7: Under conditions of temperature inversion and/or for propagation downwind, sound will be refracted downwards and can reach a location that normally would be shielded from the sound.

in Fig. 9.7. In the case of temperature inversion and/or downwind propagation in an atmosphere where the wind velocity increases with height, sound will be bent toward the ground. Then the sound from the traffic on a highway which normally would not be heard because of shielding, can be become quite noticeable (and annoying). The distance between source and receiver can be several miles. Furthermore, under certain conditions, rays traveling at different altitudes can have different radii of curvature and lead to focusing at the receiver which can drastically increase the sound pressure level.

9.5.3 Acoustic 'Shadow' Zone

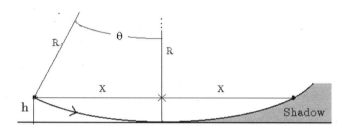

Figure 9.8: Acoustic shadow formation due to refraction above a plane boundary.

As we have seen, with a temperature lapse rate or for propagation against the wind (with the wind velocity increasing with height), the sound is refracted upwards which we illustrate schematically by means of rays (in the geometrical or high frequency approximation). Then, if we draw rays emitted in all directions from a source above ground there will be rays that do not reach the ground and others that are reflected from the ground. There is a limiting ray such that the ground surface is tangential to the ray. This ray defines a geometrical shadow zone, as indicated in Fig. 9.8, which cannot be reached by any ray.

This behavior is similar to a sound wave above a curved boundary in a uniform atmosphere. The rays are then prevented from reaching the acoustic shadow beyond the horizon defined by the ray that is tangent to the boundary. It should be realized, though, that the visualization of the sound pressure field distribution by means of rays is meaningful only at very high frequencies. Therefore, shadow zones predicted on the basis of ray acoustics merely give a qualitative indication of a region of low sound pressure. The actual sound field can be computed only from a wave theoretical analysis, as will be discussed later.

To determine the distance from a source to the shadow zone, illustrated in Fig. 9.8, we consider here the simple case of a constant gradients of wind and temperature sufficiently small so that the ray trajectories can be approximated by circles, as we have done. In the figure, the radius of curvature is R but, as drawn, the center of the circular ray is positioned too high to show in the figure. If the height of the source and the receiver are both h and the distance from the source to the shadow zone at the height h

is $2X$, it follows from simple geometry that $h = R(1 - \cos\theta) \approx R\theta^2/2 \approx R(h/X)^2/2$ so that $X = \sqrt{2hR}$. The distance to the shadow zone then becomes

$$2X \approx 2\sqrt{2hR}. \tag{9.30}$$

Normally, during daytime, the temperature decreases with height and refracts sound upwards, regardless of the direction of propagation. With the wind increasing with height, the refractions by temperature and wind are in the same direction downwind and in opposite direction downwind. In this section, dealing with the combined effect of wind and temperature, we need to specify the direction of the wind in the horizontal plane and we express the projection of the wind velocity in the plane of propagation as $U = U_h \cos\phi$, where U_h is the total horizontal wind velocity. The effects of refraction due to wind and temperature then will will cancel each other at an angle such that $(dU_h/dz)\cos\phi + dc/dz = 0$, where $(1/c)dc/dz = (1/2T)dT/dz$. Actually, for propagation in the earth's surface boundary layer, which typically is 50 to 100 m high, the effect of wind is often so strong that it overcomes the effect of temperature so that the downwards refraction results in the direction of the wind.

For propagation upwind, both wind and temperature refract the sound upwards and the distance to the shadow zone is a minimum. Typically, this distance is between 50 and 100 m for a source and receiver height of 2 m, and can easily be observed in a field; a person B on the downwind side can easily hear a voice signal from A located upwind, but A, most likely, will not be able to hear B.

On the left in Fig. 9.9 are shown examples of shadow formation typical for summer around a sound source located 10 ft ($\approx$3.3 m) above ground in the presence of a wind of 7 mph ($\approx$4.4 m/sec). The shadow distance refers to a receiver height of 10 ft above ground and it is shown at noon and at midnight. The two shadow boundaries refer to

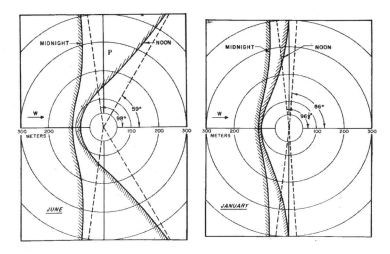

Figure 9.9: Examples of shadow zone formation. Source and receiver height: 10 ft ($\approx$3.3 m). Wind velocity: 4.4 m/sec ($\approx$7 miles/hour). Ground surface: 50 cm thin grass.

noon and midnight. The ground is covered with thin grass 50 cm high. The shortest distance to the shadow, about 84 m, is obtained at noon in the direction against the wind. At right angles to the wind, the distance is about 140 m and increases continuously with angle until the limiting value of 58 degrees is reached. Beyond this angle, no shadow can form since the effect of wind in the downwind sector more than cancels the effect of temperature. In the nighttime, with a considerably reduced effect of temperature, the shadow boundary is 'folded' back so that a larger area of the ground is exposed to the sound. The location P in the figure which is in the shadow during the day will be exposed during the night.

We have assumed that the average wind velocity is the same at night- and daytime. This may not always be the case. If there is no wind at night and a temperature inversion, there will be no shadow whatsoever.

On the right in Fig. 9.9 are shown the corresponding results in the winter. They are qualitatively the same as during the summer but the distances to the shadow boundary are greater than in the summer. Furthermore, the difference between daytime and nighttime results are not so pronounced as in the summer. In Fig. 9.10, the effect of the nature of the ground surface is illustrated. The wind velocity is again 7 mph at a height of 10 ft and the distance to the shadow boundary corresponds to this height. Curve 1 refers to a very flat ground like a mud flat or smooth concrete. The down refraction by the wind gradient is not strong enough to eliminate the upwards refraction of the temperature gradient and shadow formation occurs all around the source although the distance to the shadow zone is greater downwind than upwind, of course.

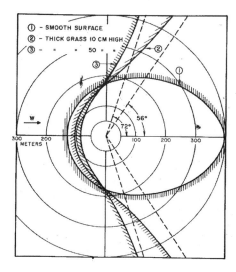

Figure 9.10: Effect of ground surface on shadow formation. Wind velocity: 7 mph at 10 ft above ground. Summer conditions.

Curve 2 refers to a ground covered with thick grass, 10 cm high. The wind gradient is now larger than for curve 1, and the downwards refraction of the wind more than compensates for the upwards refraction by the temperature gradient and this results in sound exposure in a 112 degree wide downwind sector. Upwind, the distance to the shadow boundary is decreased from about 130 to 103 m in comparison with the smooth.

Finally, curve 3 refers to a ground cover of 50 cm high thick grass and the result is again exposure in a downwind sector covering a total angle of 144 degrees. The upwind distance to the shadow is now reduced further.

It appears that a roughening of the ground surface (by vegetation) enhances rather than reduces the sound pressure in the downwind direction since it increases the wind gradient and the strength of the downwards refraction. In the upwind direction, however, a reduction occurs, albeit less striking. The effect consists only in the shortening of the distance to the shadow zone in the upwind half-plane. To evaluate the overall effect of vegetation, 'ground absorption,' discussed earlier, must also be included.

The Sound Field in the 'Shadow'

So far, in the discussion of the effect of refraction, we have considered only the geometrical aspects of the problem, which led to the idea of shadow formation and the calculation of the distance to the shadow zone. Such an analysis predicts no sound in the shadow zone. In practice, the conditions for the validity of geometrical acoustics are not fulfilled, and only a complete wave theoretical analysis from the solution of wave equation for an inhomogeneous moving medium, makes possible a determination of the sound field inside the shadow zone. It is found to be of

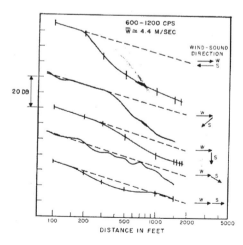

Figure 9.11: Measured sound pressure level versus distance from a source, 10 ft above ground. The wind velocity is 7 mph ($\approx$ 4.4 m/s). The different curves refer to different directions of the sound relative to the wind.

the evanescent type, decreasing exponentially with distance of penetration into the shadow.

Before discussing the results of such an analysis, we shall look at the experimental data in Fig. 9.11 showing the measured octave band sound pressure level (SPL) in the band 600 to 1200 Hz versus distance from a source. The source is located 10 ft above ground and the wind speed at that height is 7 mps wind. The five curves in this figure refer to different directions of sound propagation relative to the direction of the wind. The top curve corresponds to propagation against the wind ($\phi = 180$). Close to the source, the sound pressure level decreases approximately as in free field (i.e., by $20 \log(2) \approx 6$ dB per doubling of distance). At a distance of about 200 ft, a marked decrease in sound pressure level occurs indicating the beginning of a shadow zone. The decrease is at first rapid but levels off with distance into the shadow zone. At about 1000 ft from the source, the reduction of the SPL caused by the shadow, which will be called the *shadow attenuation*, is about 25 dB.

The next curve in the figure refers to propagation at an angle of 45 degrees from the upwind direction ($\phi = 135$ degrees). Again, there is a shadow boundary, although the decrease of the SPL is not as rapid as in the previous case. Even for propagation at a right angle to the wind, corresponding to the third curve, a shadow is realized. Since the wind does not cause refraction in this direction, the shadow is caused only by the temperature gradient.

The next curve corresponds approximately to the conditions along a limiting line in Figs. 9.9 and 9.10 along which the refraction by the wind and the temperature cancel each other, i.e., $(dU/dz)\cos\phi + dc/dz = 0$, the distance to the shadow is theoretically infinite. Although we find a small attenuation along this direction, nothing like a shadow is obtained.

The last curve shows the SPL distribution downwind, i.e., at $\phi = 0$. As in many other downwind measurements, there is an indication of sound reinforcement at a certain distance from the source. This may be caused by the directivity of the source. If the source emission is stronger in a direction above the horizontal, the sound emitted in this (rather than the horizontal) direction will be received at the observation point due to downwards fraction, thus exceeding the level that would be obtained in a uniform atmosphere.

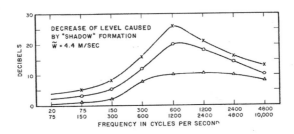

Figure 9.12: Frequency dependence of the attenuation caused by a shadow zone above ground. Source and receiver height: 10 ft. Wind speed: 7 mph. Distance from source: 1000 ft.

Similar results were obtained also at other frequencies. The observed frequency dependence of the shadow attenuation at a distance of 1000 ft from the source is shown in Fig. 9.12. The three curves correspond to sound propagation in different directions: against the wind, 45 degrees from the upstream direction, and at right angles to the wind ($\phi = 180$, 135, and 90 degrees, respectively).

The frequency dependence of the shadow attenuation has one unexpected feature. Rather than increasing monotonically with frequency, the attenuation reaches a maximum and then decreases. In this case, for sound propagation against the wind, the maximum is about 25 dB and it occurs in the 600 to 1200 Hz band. As we shall see shortly, wave theory (diffraction) alone predicts a shadow attenuation that increases monotonically with frequency. Therefore, the experimental data suggests the presence of an additional mechanism of sound transmission into the shadow zone which dominates at high frequencies. It seems reasonable to assume that this mechanism is scattering of sound by turbulent fluctuations in the atmosphere, as explained below.

The Shadow Attenuation

The wave analysis of sound propagation in a vertically stratified atmosphere shows that the wave field within the shadow zone can be expressed as a sum of an infinite set of modes that decay exponentially with the horizontal distance from the shadow boundary. The mode with the lowest decay rate will dominate sufficiently far into the shadow. This mode is found to be of the form

$$p \propto \frac{A}{\sqrt{x}} e^{-\alpha x}, \tag{9.31}$$

where x is the horizontal distance from the geometrical shadow boundary to the point of observation inside the shadow. The decay constant can be shown to be

$$\alpha = n\,(c' + U'\cos\phi)^{2/3}(\omega/c)^{1/3}, \tag{9.32}$$

where ω is the frequency, $c' = dc/dz$ is the gradient of the sound speed, and $U' = dU/dz$, the wind gradient. Both gradients are evaluated at the ground surface. The angle ϕ as before, is measured from the downwind direction. The constant n depends on the ground impedance; it is found to be 2.96 for a pressure release boundary and 1.29 for a rigid boundary.

The important thing to notice is that the pressure decreases exponentially with distance and that the attenuation rate increases as $f^{1/3}$. The shadow attenuation, as defined above, has the same frequency dependence.

In the case of no wind, the diffraction analysis has been verified in a laboratory experiment. It involved the measurement of the amplitude of acoustic wave pulses in a two-dimensional propagation chamber in which a vertical temperature gradient was maintained by means of appropriate heat sources and sinks (for atmospheric stability, the atmosphere was turned upside down, with the warm ground surface on top).

Although the sound field diffracted into the shadow zone can be calculated in a fairly rigorous manner from first principles, the scattering of sound by turbulence into the shadow is made complicated by the statistical nature of the turbulence.

In the high-frequency region, where the wavelength is considerably shorter than the characteristic eddy size, the total scattering cross section of turbulence is known to be proportional to the square of both the frequency and the rms value of the velocity fluctuation.

In the following semi-empirical discussion we shall assume that the scattering intensity into the shadow zone increases as f^2, where f is the frequency. The contribution to p^2 in the shadow zone, p being the rms value, will be of the form $p_s^2 \approx Bf^2U^2/r^2x^2$, where r is the distance from the sound source to the scattering region and x, the distance from it to the point of observation. We have assumed that the fluctuations ΔU are proportional to the mean wind speed U, the constant of proportionality being absorbed in A. The dependence of the total mean square sound pressure in the shadow zone on the wind velocity U and the frequency f, using the expression for the diffracted field in Eq. 9.31, is then of the form

$$p^2 \approx B\left(\frac{A'}{x}e^{-n(c'+CU)^{2/3}f^{1/3}x} + \frac{f^2U^2}{r^2x^2}\right), \tag{9.33}$$

where $A' = A^2/B$ and C are constant based on the assumption that the velocity gradient $U' = \partial U/\partial z$ in Eq. 9.32 is proportional to U.

The diffracted field decreases and the scattered field increases with frequency so that the total field will have a minimum and the corresponding shadow attenuation a maximum at a certain frequency, in qualitative agreement with the experimental data. Actually, for the purposes of developing a semi-empirical prediction scheme for the sound pressure in the shadow zone, the experimentally determined maximum shadow attenuation and the corresponding frequency for a given U can be used to obtain numerical values for the constants C and A'. To go much beyond such a rough estimate of the field distribution requires considerably more effort. However, the main purpose here was to explain qualitatively the observed fact that the shadow attenuation does not increase monotonically with frequency as diffraction theory predicts but reaches a maximum at a certain frequency.

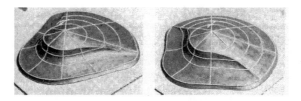

Figure 9.13: SPL distribution about a source as affected by shadow formation. Left: Conditions essentially the same as in Fig. 9.9. Right: Higher wind speed.

To visualize the SPL distribution, particularly the shadow effect, we refer to Fig. 9.13. The height of the surface above the base plane represents SPL and the distance from the source is to be interpreted as being on a logarithmic scale. The left model in the figure corresponds essentially to the conditions described in Fig. 9.9 where the refractions and temperature cancel each other at an angle of about 60 degrees. In the right model, the wind velocity is higher than on the left so that the upstream distance to the shadow zone is reduced and more of the downstream sector is fully exposed to the sound.

9.5.4 Problems

1. **Refraction due to a temperature discontinuity**

   (a) Consider a horizontal plane at which there is a discontinuity in the temperature from 40°C below to to 10°C above the plane. A sound wave is incident on the plane from below at an angle of incidence of the sound wave (as defined in Fig. 9.6). What is the angle of refraction? Can total reflection occur? If so, determine the critical angle of incidence.

   (b) Answer the same questions if the wave is incident from above.

2. **Refraction due to a flow discontinuity (shear layer)**

   Consider again Problem 1 but instead of a temperature discontinuity there is now a discontinuity in the flow velocity. Let $U_1 = 0$ and $U_2 = 0.5\,c$, where c is the sound speed.

   (a) Discuss in detail the reflection and transmission for all angles of incidence.

   (b) Show that the critical angle of incidence for total reflection is given by $\cos\theta_1 = 1/(1 + M_2)$, where $M_2 = U_2/c$.

   (c) Repeat (a) with the sound entering from above. Can total reflection occur?

3. **Sound emerging from a jet. Shadow zone**

   In Fig. 9.6, let U_1 represent the flow in a jet and let $U_2 = 0$. Sound generated within the jet is incident on the boundary. Discuss the relation between the angles of incidence and refraction over the entire angular range from 0 to 180 degrees for the incident angle. Show that the emerging sound is confined to an angular region and determine this angular range.

9.6 Propagation from a High Altitude Source

The previous discussion dealt with 'short range' propagation in the earth's surface boundary layer with altitudes less than 50 m. Although no new physics is involved, the propagation to ground from a source at an altitude of 10,000 m or more brings some new aspects to the problem which we shall discuss in this section. It is based primarily on a study of sound propagation from a propfan flying at altitudes 20,000 and 35,000 ft at a speed corresponding to a local Mach number of 0.8.

The propfan was mounted on the wing of an experimental aircraft which also carried several microphones to monitor the sound pressure from the propfan. The fan had 8 blades and a diameter of 108 inches and the blade passage frequency was 237.6 Hz. Sound pressures at a distance of 500 feet from the flight path were also measured by a 'chaser' plane. The sound pressure level at the ground in fly-over experiments were made during each flight with microphones placed under the flight path (centerline position) and also along sidelines. Each microphone was placed on a 'reflector' plate (40 inch diameter) on the ground.

The fact that the sound source is moving at high speed adds another dimension to the problem as the Doppler shift has to be considered. The experiment serves as a good illustration of the analysis of a propagation problem which contains several aspects of acoustics and propagation in a 'real' atmosphere. Measured data on the altitude dependence of static pressure, temperature, wind velocity, and humidity in this atmosphere were obtained, as discussed in the next section.

9.6.1 The 'Real' Atmosphere

Data on temperature, static pressure, humidity, and wind velocity were obtained at intervals of 30, 150, and 300 m in the altitude ranges between 15 and 2085 m, 2175 and 5015 m, and 5250 and 11850 m, respectively.[3]

Temperature and Pressure Distributions

The temperature data are summarized in Fig. 9.14. It is a fairly good approximation to expect the temperature to decrease linearly with height over entire range under consideration, i.e., up to 35,000 ft, (10,668 m). The temperature at this altitude is ≈ 218 K ($-55°$C) and at the ground level ≈ 293 ($20°$C). The average slope over this range is then $|\alpha| \approx 0.7°$C per 100 m. During the two days data were taken there was no significant difference in the temperature distribution except below an altitude of about 100 m and the variation of the temperature at different times of the day were small.

The pressure variation with height z is obtained from $dP = -\rho g\, dz$, where ρ is the density and g, the acceleration of gravity. Furthermore, $P = r\rho T$, where r is the gas constant per unit mass. With T assumed to decrease linearly with z, $T = T_0 - \alpha z$, where $\alpha = [T(H) - T(0)]/H$, it follows that the corresponding variation of pressure in this region is given by

$$P(z) = p(0)(1 - \alpha z)^{\gamma g/\alpha c^2},\tag{9.34}$$

where $\gamma = c_p - c_v \approx 1.4$ and $g = 9.81$ m/s^2. We have here introduced the sound speed c_0, where $c_0^2 = \gamma P_0/\rho_0 = \gamma r T_0$.

The function $P(z)$ is shown in Fig. 9.14 and it is in good agreement with the measured pressure distribution.

Humidity Distribution

Unlike temperature, the relative humidity distribution can vary significantly from one day to the next, as shown in Fig. 9.14. As we have seen earlier, the relaxation frequency for vibrational excitation of Oxygen and Nitrogen depends on the ratio of the vapor pressure and the atmospheric pressure and this, in turn, depends on the relative humidity, temperature, and static pressure.

---

[3]Meteorological as well as acoustical data were provided by DOT/TSC and refer to 30-31 October, 1987.

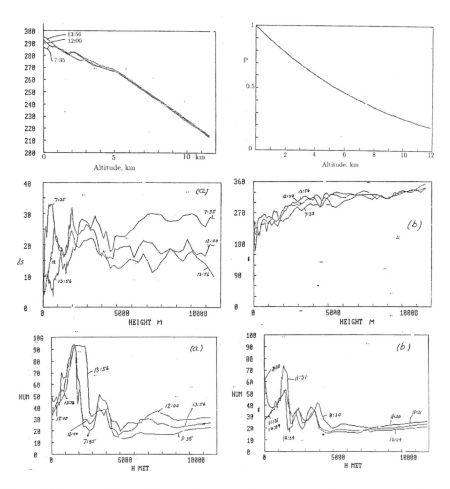

Figure 9.14: From left to right: Distributions of temperature, static pressure, wind speed, wind direction, and humidity, as indicated. In the last two figures, the humidity is shown at two different days. (Based on data from DOT/TSC, 1987).

Wind Distribution

Like the relative humidity, also the wind speed is apt to vary with time more than temperature. In Fig. 9.14 are shown the altitude dependence of the measured wind speed and direction at three different times during the day. The direction is measured from the direction of the flight. Although the wind speed is seen to vary considerably with time of the day, the direction is relatively constant. There is almost a reversal in direction going from ground level to a height of 10,000 m.

9.6.2 Refraction

In Fig. 9.15 is indicated schematically the problem under consideration. A sound source, in this case a propfan, is flying at an altitude H (20,000 or 35,000 ft) with a constant speed V (local Mach number 0.8). The blade passage frequency of the fan is f_b (237.6 Hz). We wish to calculate the sound pressure on the ground during a pass of the plane accounting for the atmospheric effects such as refraction and absorption. There are also the variation of the acoustic impedance with altitude and the Doppler shift of the emitted sound to be considered. The Doppler shifted frequencies (from the source frequency of 237.6 Hz) that reached the ground were found to be in the range 150 to 500 Hz.

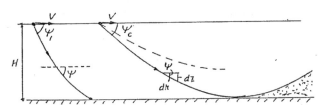

Figure 9.15: Sound emission from a sound source moving with a velocity V at an altitude H.

The temperature $T(z)$ and the wind velocity $U(z)$ are assumed to depend only on the height z over the ground level. A ray, indicating the direction of propagation of a sound wave, then remains in one and the same vertical plane from the source to the receiver on the ground. The intersection of this plane and the ground is chosen as the x-axis. Our analysis will be restricted to the case when the flight path is in the plane of sound propagation either in the positive or negative x-direction. The corresponding sound pressure levels on the ground under the flight path are often referred to as 'centerline' data.

As before, in describing low altitude sound propagation, the component of the wind velocity in the x-direction is expressed as

$$U_x = U(z) \cos \phi, \tag{9.35}$$

where ϕ is the angle between the wind direction and the x-axis. This direction varies with altitude as indicated in Fig. 9.14.

We consider a sound wave emitted from the altitude H in a direction specified by the emission angle ψ_1, as indicated in Fig. 9.15. With $c = c(z)$ being the sound speed, and c_1 the value at the altitude H of the source, the law of refraction discussed earlier in this chapter requires the trace velocity c_t to remain constant, i.e.,

$$c_t = \frac{c(z)}{\cos\psi(z)} + U_x(z) = \frac{c_1}{\cos\psi_1} + U_{1x}. \qquad (9.36)$$

Thus, the directional cosine of the ray at the altitude z is

$$\cos\psi = \frac{(c/c_1)\cos\psi(H)}{1 + \Delta M_x \cos\psi(H)}, \qquad (9.37)$$

where $\Delta M_x = (U_{1x} - U_x)/c_1$.

As indicated schematically in Fig. 9.15, a sound ray emitted at an angle below a certain critical angle $\psi_c(H)$ will not reach the ground so that the point of observation will lie in the (geometrical) acoustic shadow caused by refraction. At the emission angle ψ_c, the ray is tangential to the boundary so that $\psi(0) = 0$ or π, i.e., $\cos\psi(0) = 1$ or $\cos\psi(0) = -1$. The corresponding values for $\cos\psi(H)$, as obtained from Eq. 9.37, are

$$\cos\psi_c(H) = \frac{\pm 1}{(c(0)/c(H) \mp (U_x(H) - U_x(0))/c(0)\Delta M_x(z)}. \qquad (9.38)$$

Unlike sound propagation in the lower atmosphere, the temperature dependence of the sound speed now dominates the overall refraction and it is usually a good approximation to neglect the effect of the wind. We use this approximation in the following estimate. With $c(0)/c(H) = \sqrt{T(0)/T(H)} \approx \sqrt{293/220} \approx 1.15$, the critical angles of emission obtained from Eq. 9.38 are then ≈ 30 and ≈ 150 degrees. This means that when the source is moving toward the observer, all rays emitted at angles below the critical value of 30 degrees will not reach the observer and the same holds true when the source has passed overhead and is moving away from the observer if a ray is emitted at an angle greater than 150 (180−30) degrees. In other words, the rays that reach the ground originate within an emission angles range between 30 and 150 degrees.

The *emission angle* is not the same as the angle under which the source is seen when the sound arrives at the observer. The difference between the *viewing angle* and the emission angle will be discussed shortly.

First we determine the travel time of the sound from the source to the receiver. The elementary distance of wave travel that corresponds to an altitude interval of Δz is $\Delta r = \Delta z/\sin\psi(z)$. For emission angles between the critical values (30 and 150) the total travel distance is

$$r = \sum \Delta r = \sum_0^H \Delta z/\sin\psi(z). \qquad (9.39)$$

In regard to the present numerical analysis, atmospheric data on temperature, wind, pressure, and relative humidity were available at a total of 112 altitudes, 70 in the range from 15 to 2085 m with $\Delta z = 30$ m, thus covering the altitude range from

0 to 2100 m, 20 in the range from 2175 to 5025 m with $\Delta z = 150$ m, covering the altitude range from 2100 to 5100 m, and 22 in the range from 5250 to 11550 m with $\Delta z = 300$ m, covering the altitude range from 5100 to 11700 m. For an altitude of 35,000 ft (10,668) m, the closest altitude interval is centered at 10650 m, with the upper value of the interval being 10650+150=10800 m, exceeding the flight altitude by 132 m.

To obtain the travel time t_r of the sound wave along the refracted path from source to receiver, we merely have to replace $\Delta z/\sin\psi(z)$ in Eq. 9.39 with $\Delta z/c(z)\sin\psi(z)$ and compute the sum. We normalize the travel time and express it as

$$t_r = \sum_0^H \Delta z/c(z)\sin\psi(z) \equiv [H/c(H)]F_1(\psi(H)).\tag{9.40}$$

The horizontal distance traveled during the time t_r is

$$x_r = \sum_0^H \frac{\Delta z\cos\psi(z)}{\sin\psi(z)} \equiv HF_2(\psi(H)),\tag{9.41}$$

where $F_2 = x_r/H$ is a function of the emission angle. During the time of sound travel from the emission point, the source has advanced a distance Vt_r along the flight path, and the angle ψ_v at which the source is seen at the time of arrival of the sound is then given by

$$\tan\psi_v = (x_r - Vt_r)/H = F_2 - M(H)F_1,\tag{9.42}$$

where F_1 and F_2 are known functions of the emission angle defined in Eqs. 9.40 and 9.41. We shall call ψ_v the *viewing angle*, indicated in Fig. 9.16.

The sound pressure at the ground station during a flight test is recorded as a function of time and for simple comparison with the experimental data, it is useful to express our results in terms of time of arrival of the sound. We choose as $t = 0$ the time when the source is seen overhead at $x = 0$, i.e., when the viewing angle is 90 degrees. Then, if the source is seen at an angle less than 90 degrees, the time is negative,

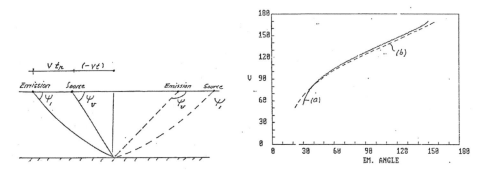

Figure 9.16: Left: Emission and viewing angles defined. Right: Viewing angle versus emission angle.

otherwise positive. From Eq. 9.42 it follows that the x-coordinate of the source at the time the signal is received on the ground, i.e., a time t_r after emission, will be $-H(F_2 - M(H)F_1)$. Then, with our choice of $t = 0$, the location of the source at the time of arrival can also be expressed as $x = Vt$, and it follows that the corresponding arrival time is $t = -(H/V)(F_2 - M(H)F_1)$. If we normalize the time with respect to $H/c(H)$, it can be written

$$\tau = F_1(\psi(H)) - F_2(\psi(H)/M(H), \tag{9.43}$$

which establishes what angle of emission corresponds to a particular time in the experimentally obtained pressure trace. This makes possible a direct comparison of the measured pressure with the calculated. Actually, a more elegant way of determining the angle of emission would have been to measure the frequency of the received signal and from the known speed of the sound source determine the emission angle from the Doppler shift (see Eq. 9.46).

The characteristic time $H/c(H)$ used in the normalization is 35.7 seconds if $H = 35000$ ft and 19.1 seconds with $H = 20,000$ ft, where $c(H) = 35,000$ ft is based on an absolute temperature of 222 K.

9.6.3 Attenuation Due to Absorption (Vibrational Relaxation)

The blade passage frequency of the propfan under consideration was 237.6 Hz and the Doppler shifted frequencies observed on the ground ranged from approximately 150 to 500 Hz. In this frequency range, the attenuation is due primarily to the vibrational relaxation of Oxygen and to some extent of Nitrogen. Visco-thermal losses and molecular rotational relaxation effects are negligible.

The frequency dependence of the attenuation of each of these effects is expressed in terms of the ratio of the frequency and the corresponding relaxation frequency. The vibrational relaxation frequencies of Oxygen and Nitrogen depend on the ratio of the water vapor pressure and the total static pressure. The vapor pressure, p_w, the product of the saturation pressure, p_s, and the relative humidity, depend strongly on the temperature through the temperature dependence of p_s. Furthermore, relaxation frequency increases with the intermolecular collision frequency and hence with the static pressure.

The altitude dependence of the relative humidity is shown in Fig. 9.14. The calculation of the corresponding vapor pressure involves the use of the temperature dependence of the saturated vapor pressure and the relaxation frequency is obtained from the ratio of the vapor pressure and the air pressure by means of an empirical relation, as was described earlier.

Examples of the computed altitude dependence of the attenuation per unit length is shown in Fig. 9.17 at two different times of the day. The ten curves in each figure refer to different frequencies, from 100 Hz to 1000 Hz in steps of 100 Hz starting from the bottom. The attenuation is concentrated to a region below approximately 7000 m with a pronounced peak at approximately 5000 m thus forming an attenuation 'barrier.' This has the important consequence that the total attenuation over the entire path of sound propagation from an aircraft at 35,000 ft will be essentially the same as

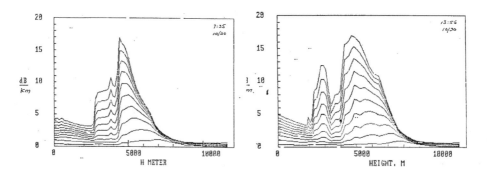

Figure 9.17: Attenuation in dB/km due to vibrational relaxation absorption in the air as a function of the altitude with frequency as a parameter from 100 Hz to 1000 Hz in steps of 100 Hz, starting at the bottom.

one at 20,000 ft. The attenuation is due mainly to the relaxation vibration of Oxygen except at low altitudes where the contribution from Nitrogen plays a role.

There is a noticeable variation of the attenuation with the time of the day. In the middle of the day (13:56) the attenuation barrier is somewhat wider than in the morning (7:35) and the integrated attenuation over the entire path of sound propagation will be somewhat larger.

Other Attenuations

Spherical divergence. The geometrical spreading of the wave gives rise to a level reduction

$$L(r) - L(r_0) = 20 \log(r/r_0), \tag{9.44}$$

where r is the distance of sound propagation along the refracted path from source to receiver. This distance depends on the emission angle, as indicated in Eq. 9.39. The reference r_0 in this case is taken to be $r_0 = 500/\sin \psi(H)$ ft since the sound pressure level at this distance was measured by the chaser plane mentioned at the beginning of this section. For an emission angle of 90 degrees, we have $r = H$. Then, with $r_0 = 500$ ft and $H = 35,000$ ft and $H = 20,000$ ft, we get $L(r) - L(r_0) = 37.9$ dB and 32 dB, respectively.

Effect of wave impedance variation. For a plane wave, the intensity $p^2/\rho c$ is conserved in the absence of absorption in the air. Since the wave impedance ρc varies with altitude, the intensity conservation requires that the relation between the sound pressure amplitudes at z and $z = 0$ is

$$p(z)/p(0) = \sqrt{Z(0)/Z(z)} = [T(0)/T(z)]^{1/4}[P(z)/P(0)]^{1/2}. \tag{9.45}$$

With the altitude variation of temperature and pressure given in Figs. 9.14, we find that the wave impedance related decrease in sound pressure level with altitude is almost linear with height with a total value of 5 dB from the ground level to an altitude of 10 km.

Total Attenuation

The total reduction in sound pressure level is the sum of the integrated attenuation due to sound absorption, the effect of spherical divergence, and the effect of the wave impedance variation with altitude. For the atmospheric conditions that existed during the flight on October 31 at 14:00, we have shown the computed total attenuation as a function of the emission angle in Fig. 9.18. The source height was 35,000 ft and the local flight Mach number 0.8. Shown separately are the contributions due to sound absorption in the air (including the effect of the altitude dependence of the wave impedance) and to spherical divergence. The effect of wind was found to be negligible so the refraction is due to temperature alone. For comparison is shown (dashed curve) the corresponding results if refraction is not accounted for.

If the refraction is omitted, the attenuation is significantly smaller at small and large emission angles and there are no critical angles of emission (30 and 150 degrees) defining the boundaries outside which the receiver is left in a shadow zone. In practice, background noise on the ground may make it difficult to determine the received signals at emission angles outside the range between 55 and 120 degrees.

The asymmetry in the total attenuation curve in Fig. 9.18 is due to the Doppler shift which increases the frequency of the received signal for emission angles less than 90 degrees and decreases it for angles larger than 90 degrees. The Doppler shifted frequency is given by

$$f(\psi_H) = \frac{f_0}{1 - M \cos \psi_H}, \tag{9.46}$$

where f_0 is the blade passage frequency, M, the Mach number of the source, and ψ_H, the emission angle.

The difference between the sound pressure levels measured at the monitoring location at 500 ft from the source and levels measured at the ground station yields, in principle, the experimentally determined total level reduction of the sound as it travels through the atmosphere to the ground. However, there may be a practical

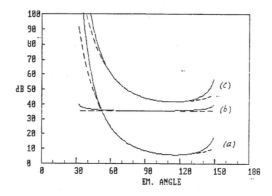

Figure 9.18: Total attenuation versus emission angle of blade passage tone from propfan. (a): Attenuation due to absorption and effect of wave impedance variation. (b): Spherical divergence. (c): Total attenuation. The dashed curve: Refraction omitted.

problem of the background noise level at the ground station which contaminates the signal at low emission angles where the attenuation is the greatest and hence the received signal strength is the lowest. It was found that on the particular day referred to here and for a source height of 35,000 ft, the data for emission angles less than 65 degrees were background noise limited and could not be used. However, at other angles it was possible to demonstrate that the calculated total attenuation was in good agreement with the measured.

9.6.4   Fluctuations

Observed fluctuations in the measured sound pressure level at the ground station of more than 5 dB is typical for pure tone sound propagation in the atmosphere in which interference between two or more sound waves transmitted over different paths to the receiver. This problem has been studied in detail for sound propagation along a horizontal path above ground in which the fluctuations occur as a result of the interference between the direct and reflected waves. Due to the difference in path lengths, destructive interference results in pronounced minima in sound pressure and if the atmosphere is not static but turbulent, the phase difference will fluctuate as does the total sound pressure. In such experiments with a source height of 5 ft, observed fluctuations of 10 dB are common even at relatively short distances, less than 300 ft.

In the present propfan sound propagation study, the receiver is placed on the ground (actually about 1 cm above), and there is no significant path length difference between the incident and reflected sound. Thus, the situation is not the same as for sound propagation over the ground surface referred to above, where fluctuations could be explained in terms of the path length difference and the interference between a direct and a reflected wave.

In seeking an explanation for the observed fluctuations, it is crucial to realize that the propfan cannot be treated as a single point source; rather, it is an extended source with a diameter of $D = 108$ inches, approximately twice the wavelength of the blade passage tone. Actually, the fan is approximately equivalent to a distribution of point sources (dipoles and monopoles) of different signs placed along a circle of diameter D (the source distribution has a spatial periodicity with the number of periods being the same as the number of blades) and interference of the individual sound waves emitted from these separate sources will give rise to fluctuations in a turbulent atmosphere.

As shown in Chapter 8, the directional characteristics of the sound from the propfan is determined by the interference of the sound from these elementary sources. For example, at a point on the axis of the fan, the acoustic path length to all the sources is the same and the sound waves from the positive and negative sources cancel each other, making the total sound pressure zero on the axis. In other directions, the path lengths are not the same, and the interference no longer will be destructive making the resulting sound pressure different from zero.

·In a nonuniform atmosphere, with a nonuniform acoustic index of refraction, the *acoustic* path lengths (i.e., the path lengths expressed in terms of the wavelength) from the observer to the various elementary sources and the corresponding directional characteristics will be affected. For example, the sound pressure on the axis no longer will be zero, in general. With the atmospheric characteristics being time dependent fluctuations in the sound pressure will occur. For an observer on the ground, the sound waves from the elementary sources on the fan have traveled over a long distance (35,000 ft) along different (adjacent) paths. Even it the separation between these paths is relatively small (a few feet on the average), the difference in the average sound speed along these paths required to produce a phase shift of 180 degrees at the blade passage frequency is exceedingly small, of the order of 0.01 percent. Consequently, such phase shifts are to be expected to occur with corresponding large fluctuations in the received sound pressure. A typical example of the time dependence of the total sound pressure recorded at the ground station is shown in Fig. 9.19.

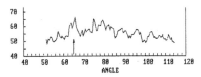

Figure 9.19: Typical record of total SPL versus emission angle in fly-over experiment with propfan at 35,000 ft flying at Mach number $M = 0.8$.

Furthermore, since the source is in motion, temporal fluctuations in the received sound pressure will occur even if the atmosphere characteristics were time independent. All that is required is a small spatial nonuniformity in the direction of the flight. If the characteristic length of such a nonuniformity is L, the characteristic fluctuation time would be L/V. As an example, a fluctuation period of 0.1 to 0.5 seconds and a flight Mach number of 0.8 (≈ 296 m/sec) would correspond to a characteristic scale of $L \approx 30 - 150$ m, which is reasonable.

At a relatively small distance from the fan, the atmospherically induced phase shifts of the elementary sound waves will be negligible and this result in an approximately stationary sound pressure distribution with little or no fluctuation. The sound waves are generally not in phase at the point of observation; the phase difference determines the directivity pattern.

For long range propagation, with significant phase fluctuations, the probability of a resulting increase in the received sound pressure should be approximately the same as a decrease. Although a decrease yields a larger excursion on a dB scale, it is important to realize that with several elementary sources involved (8 blades), the peak level in a fluctuation can be considerably higher than the level observed in the near field. This should be kept in mind in the analysis of experimental data in a comparison of the data at 35000 ft and at 500 ft.

Another important point to consider is that the scattering of sound from turbulence tends to redistribute the sound, transferring acoustic energy from regions of high to regions of low sound pressure. Thus, deep and pronounced minima in the near field directivity pattern (such as in the forward direction) will be less distinct in the far field radiation pattern.

Chapter 10

Mean-flow Effects and Nonlinear Acoustics

10.1 Review of Fluid Equations

With the density, pressure, and fluid velocity denoted ρ, P, and U_i (component form), the equations for mass and momentum conservation are

$$\partial\rho/\partial t + \partial\rho U_j/\partial x_j = 0 \qquad (10.1)$$

$$\partial\rho U_i/\partial t + \partial\rho U_i U_j/\partial x_j = -\partial P/\partial x_i, \qquad (10.2)$$

where repeated indices (in this case j) implies summation over $j = 1, 2, 3$. The second equation can be rewritten as

$$\rho\partial U_i/\partial t + U_i[\partial\rho/\partial t + \partial\rho U_j/\partial x_j] + U_j\partial U_i/\partial x_j = -\partial P/\partial x_i$$

and, combined with Eq. 10.1, as

$$DU_i/Dt = -(1/\rho)\partial P/\partial x_i, \qquad (10.3)$$

where $D/Dt \equiv \partial/\partial t + U_j\partial/\partial x_j$ (sum over $j = 1, 2, 3$).

The mass equation can be expressed in similar manner,

$$D\rho/Dt = -\rho\partial U_j/\partial x_j. \qquad (10.4)$$

Next, we introduce the acoustic perturbations δ, p, and u_i of the variables and put $\rho = \rho_0 + \delta$, $P = P_0 + p$, and $U_i = U_{0i} + u_i$ with the subscript '0' signifying the unperturbed state. Now, there will products of the (first order) acoustic variables and the mean flow as well as products of acoustic variables. The first type of terms express the coupling or interaction between the sound field and the mean flow. The second type account for the coupling of the sound field with itself, so to speak, and expresses the nonlinearity of the field; in the linear theory of sound, these second order terms are neglected.

As discussed in Chapter 3, the relation between the density and pressure perturbation is $\delta = p/c^2$, where c is the sound speed. Then, if the mean flow is uniform and time independent, the linearized versions of the mass and momentum equations are

$$(1/c^2)(Dp/Dt = \rho_0 \partial u_i/\partial x_i$$
$$Du_i/Dt = -(1/\rho_0)\partial p/\partial x_i. \tag{10.5}$$

(Note that on the right-hand side in the first equation we have replaced j by i which can be done since summing over the the index is involved.) The velocity u_i can now be eliminated between these equations by taking the time derivative of the first and the space derivative $\partial/\partial x_i$ (sum over i) of the second. This results in the wave equation

$$(1/c^2)D^2 p/Dt^2 = \nabla^2 p, \tag{10.6}$$

i.e.,

$$\frac{1}{c^2}(\partial/\partial t + \mathbf{U} \cdot \nabla)^2 p = \nabla^2 p = \partial^2 p/\partial x^2 + \partial^2 p/\partial y^2 + \partial^2 p/\partial z^2. \tag{10.7}$$

For a plane harmonic wave, the sound pressure is $p(x, y, z, t) = \Re\{A \exp(i\mathbf{k} \cdot \mathbf{r}) \exp(-i\omega t)\}$, where $\mathbf{k}$ is the propagation vector with the components k_x, k_y, k_z and a magnitude k (to be determined in terms of ω). From Eq. 10.7 we get the equation for the complex amplitude $p(x, y, z, \omega) = A \exp(i\mathbf{k} \cdot \mathbf{r})$

$$(1/c)^2(\omega - \mathbf{U} \cdot \mathbf{k})^2 p = (k_x^2 + k_y^2 + k_z^2)p = k^2 p. \tag{10.8}$$

It should be noted that the effect of flow is due solely to the flow velocity component U_k in the direction of propagation. It follows from Eq. 10.8 that

$$(1/c)(\omega - kU_k) = \pm k \tag{10.9}$$

and that

$$k = \frac{\pm\omega/c}{1 \pm M_k}, \tag{10.10}$$

where $M_k = U_k/c$ is the Mach number of the flow in the direction of propagation. For subsonic flow, only the plus sign is relevant and we shall consider only this case. The *phase velocity* ω/k in the direction of propagation is then

$$c_p = \omega/k = c + U_k, \tag{10.11}$$

i.e., the sum of the local sound speed c and the component of the flow velocity in the direction of propagation.

The *group velocity* is the vector sum of the local sound velocity and the flow velocity

$$\mathbf{c_g} = c\hat{\mathbf{k}} + \mathbf{U}, \tag{10.12}$$

where $\hat{\mathbf{k}} = \mathbf{k}/k$ is the unit vector in the direction of propagation (normal to the wave font). It is specified by the angles ϕ and θ such that

$$k_x = k \cos\theta, \quad k_y = k \sin\theta \cos\phi, \quad k_z = k \sin\theta \sin\theta. \tag{10.13}$$

The group velocity defines the direction and speed of propagation of acoustic energy as will be explained in Section 10.2.

10.1.1   Sound Propagation in a Duct

The discussion of sound transmission in a duct will now be extended to include the effect of a mean flow. The duct is assumed to be rectangular with its axis along x. Since the flow is in the x-direction, we have $\boldsymbol{k} \cdot \boldsymbol{U} = k_x U$ and Eq. 10.8 reduces to

$$(1/c)^2 (\omega - k_x U)^2 = k^2 = k_x^2 + k_y^2 + k_z^2. \tag{10.14}$$

Quantities k_y and k_z are now determined by the boundary conditions of zero normal velocity amplitude at the duct walls. With these walls located at $y = 0$, $y = d_1$ and $z = 0$, $z = d_2$, these conditions yield $k_y = m\pi/d_1$ and $k_z = n\pi/d_2$, as before. Then, with

$$k_{m,n} = \sqrt{k_y^2 + k_z^2} = \sqrt{(m\pi/d_1)^2 + (n\pi/d_2)^2} \tag{10.15}$$

we get from Eq. 10.14

$$k_x = \frac{\omega/c}{1 - M^2}\left(\pm\sqrt{1 - (k_{m,n}/k_0)^2(1 - M^2)} - M\right), \tag{10.16}$$

where $k_0 = \omega/c$ and $k_{m,n}$ is given in Eq. 10.15. In the absence of flow the cut-off frequency of the (m, n) mode is

$$\omega_{m,n} = ck_{m,n} \tag{10.17}$$

as discussed earlier.

We now introduce

$$\omega'_{m,n} = \omega_{m,n}\sqrt{1 - M^2} \tag{10.18}$$

and rewrite Eq. 10.16

$$k_x = \frac{\omega/c}{1 - M^2}\left(\pm\sqrt{1 - (\omega'_{m,n}/\omega)^2} - M\right). \tag{10.19}$$

If $\omega < \omega'_{m,n}$ the square root in this expression becomes imaginary and $\omega'_{m,n}$ takes the role of a cut-off frequency of the (m, n) mode. The physical explanation for the Mach number dependence in the dispersion relation will be given shortly.

The phase and group velocities are now obtained from Eq. 10.19 and we find

$$c_{px} = \omega/k_x = \frac{c(1 - M^2)}{\pm\sqrt{\cdots} - M}$$

$$c_{gx} = d\omega/dk_x = \frac{c(1 - M^2)\sqrt{\cdots}}{\pm 1 - M\sqrt{\cdots}}, \tag{10.20}$$

where $\sqrt{\cdots} = \sqrt{1 - (\omega'_{m,n}/\omega)^2}$.

The *group velocity* is zero at the cut-off frequency $\omega_{m,n}$ which is smaller than in the absence of flow by a factor $\sqrt{1 - M^2}$. The phase velocity is then negative, and considering the plus sign in the equation above, it goes to $\mp\infty$ as the frequency increases to $\omega_{m,n}$. The group velocity increases monotonically with frequency toward the asymptotic value $(1+M)c$ and the phase velocity decreases toward the same value, as expected. The corresponding asymptotic value for these velocities for propagation in the opposite direction (corresponding to the minus sign in Eq. 10.20) is $-(1-M)c$.

In order to gain some physical insight into these relations, we specialize to the case when $n = 0$ so that the pressure field is uniform across the duct in the z-direction. We can then interpret the mode as resulting from a plane wave traveling in the xy-plane at an angle ϕ with the x-axis. Reflection of this wave from the duct wall results in a wave traveling in the $-\phi$ direction and the superposition of these waves results in the higher mode in the duct as discussed in Section 6.2.2.

The phase velocity in *the direction of propagation* of this contituent plane wave is given by Eq. 10.11 which in this case reduces to

$$c_p = c + U \cos\phi. \tag{10.21}$$

The intersection point of the wave front of this wave and the boundary moves in the x-direction with the trace velocity $c_{px} = c_p / \cos\phi$. This trace velocity is the same as the phase velocity c_{px} in the x-direction of the higher mode and it follows from Eq. 10.21 that

$$c_{px} = \frac{c}{\cos\phi} + U. \tag{10.22}$$

Similarly, the component of the group velocity in the x-direction is, from Eq. 10.12,

$$c_{gx} = c \cos\phi + U. \tag{10.23}$$

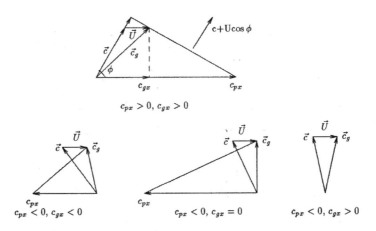

Figure 10.1: Geometrical interpretation of the expressions for the phase and group velocity of a higher order mode in a duct with flow in Eq. 10.20 and illustration of four specific cases including one (the last) in which the phase velocity and the x-component of the group velocity are in opposite direction.

The general geometrical relations between the various velocities involved are shown in Fig. 10.1, where several specific cases demonstrate the condition at cut-off and situations in which the phase and group velocity will have the same or opposite directions.

In the first, the component $U \cos \phi$ in the direction of sound propagation is positive. The velocity of the intersection point of the corresponding wave front with the x-axis is the phase velocity c_{px} of the corresponding higher order wave mode in the duct. The group velocity c_{gx} of this mode is the x-component of $c_g = c + U$ and it is in the same direction as the phase velocity c_{px}.

In the second example in Fig. 10.1, the projection of the mean velocity on the direction of propagation is negative; the intersection point of the corresponding wave front (the line perpendicular to c) with the x-axis again determines the phase velocity c_{px} of the wave mode. Like the group velocity c_{gx}, it is directed in the negative x-direction.

The third example is special in as much as the group velocity c_g is perpendicular to the duct axis and the group velocity c_{gx} of the corresponding higher mode is zero. This represents the cut-off condition of the mode.

In the fourth case, the modal group velocity c_{gx} is positive and the phase velocity c_{px} is negative. The intersection of the wave front with the x-axis in this case falls outside the range of the figure. Actually, if the the wave front is normal to the duct axis, the phase velocity becomes infinite; the group velocity is still positive.

Returning to Eqs. 10.10 and 10.13 we note that with $\theta = \pi/2$

$$k_x = (\omega/c)\frac{\cos \phi}{1 + M \cos \phi}. \qquad (10.24)$$

Thus, expressing $\cos \phi$ in terms of k_x we find that c_{px} and c_{gx} in Eqs. 10.22 and 10.23 are consistent with those in Eq. 10.20.

As already indicated, the cut-off frequency of the $(m, 0)$ mode in the presence of flow is $\omega'_{m,0}$ and it is important to realize that at cut-off it is not the phase velocity which becomes zero but rather the *group velocity*. The corresponding group velocity vector is then perpendicular to the duct axis, as shown in Fig. 10.1, and since the group velocity is the vector sum of the sound velocity and the flow velocity it follows that the phase velocity must point in the upstream directions so that c_{px} will be negative, in fact $c_{px} = -(1 - M^2)/M$. The corresponding direction of the propagation of the wave front is given by $\cos \phi = -M$.

10.2 Conservation of Acoustic Energy; Energy Density and Intensity

In our discussions of waves so far, the field variables such as pressure, density, and velocity have been considered to be small perturbations so that products of these quantities could be omitted in the equations of motion. The equations thus obtained are called 'linearized,' and the solutions are linear or 'first order.' Having obtained a linear solution there is nothing to prevent us from multiplying two first order quantities to produce a second order quantity. Realizing that quantities of this order have been omitted in obtaining a solution in the first place, one should be very cautious in assigning physical significance to such a product.

There is an important exception, however. The product of the first order pressure and velocity in a sound field, $p\boldsymbol{u}$, has the dimension of power per unit area (intensity) and the quantities ρu^2 and $p^2/\rho c^2$ have the dimension of energy density. The reason why these second order quantities (products of first order solutions) are not discarded is that they obey a conservation law. This follows from the linearized fluid equations for conservation of mass and momentum (see Chapters. 3 and 5)

$$\kappa \partial p/\partial t + \operatorname{div} \boldsymbol{u} = 0$$
$$\rho \partial \boldsymbol{u}/\partial t + \operatorname{grad} p = 0, \tag{10.25}$$

where $\kappa = 1/\rho c^2$ is the compressibility. We multiply the first of these equations by p and the second by $\boldsymbol{u}$ (scalar multiplication) and then add the equations to obtain

$$\partial[\kappa p^2/2 + \rho u^2/2]/\partial t + \operatorname{div}(p\boldsymbol{u}) = 0, \tag{10.26}$$

where we have used $p\operatorname{div}(\boldsymbol{u}) + \boldsymbol{u} \cdot \operatorname{grad} p = \operatorname{grad}(p\boldsymbol{u})$.

The quantities $w = \kappa p^2/2 + \rho u^2/2$ and $\boldsymbol{I} = p\boldsymbol{u}$ have the physical dimensions of energy density and energy flux (intensity). Eq. 10.26 states that the time rate of change of w is balanced by the inflow of the quantity $\boldsymbol{I}$ per unit volume and has the same form as the equations for conservation of mass and momentum Eq. 10.25. The existence of this additional conservation law makes the quantities w and $\boldsymbol{I}$ useful and they have been given the names *acoustic energy density* and *acoustic intensity*, respectively. In terms of these quantities, Eq. 10.26 becomes

$$\partial w/\partial t + \operatorname{div} \boldsymbol{I} = 0 \quad \text{where} \tag{10.27}$$
$$\boldsymbol{I} = p\boldsymbol{u}$$
$$w = \rho u^2/2 + \kappa p^2/2. \tag{10.28}$$

In steady state motion, the time average of $\partial w/\partial t$ is zero which means that $\operatorname{div} \boldsymbol{I} = 0$. Recall that the integral of $\operatorname{div} \boldsymbol{I}$ over a closed volume can be expressed as the surface integral of the outward normal component of $\boldsymbol{I}$ over the surface. With this quantity defined as the *acoustic power*, the conservation law tells us that the total acoustic power out of a volume must be balanced by the rate of change of the acoustic energy within the volume (integrate Eq. 10.27 over the volume).[1] As a simple example, consider a spherically symmetrical outgoing wave; the area is proportional to r^2 and in a stationary field ($\partial/\partial t = 0$), it follows that $I(r_1)r_1^2 = I(r_2)r_2^2$, the acoustic intensity must vary as $1/r^2$ and, hence, the pressure, as $1/r$.

Other second order quantities, formed as products of two first order quantities, do not enjoy the same happy fate, however. For example, the product $\delta\boldsymbol{u}$ of the first order perturbations in density δ and velocity $\boldsymbol{u}$ has the dimension as mass flux but is not the mass flux in a sound field, as will be discussed in Section 10.4.

There is one important restriction on the validity of the conservation law (Eq. 10.27); the fluid involved should have no mean motion. If a fluid does have an average velocity and if the equations of motion for sound is in reference to a stationary coordinate

---

[1] Recall also that the physical meaning of $\operatorname{div} \boldsymbol{I}$ can be thought of as the 'yield' of $\boldsymbol{I}$ per unit volume.

system, the conservation law (Eq. 10.27) is no longer valid in the form given. In that case, other expressions for acoustic intensity and energy density can be found, however, which do obey a conservation law as will be shown next.

10.2.1 Effect of Mean Flow

In a moving fluid, the acoustic energy density and flux (intensity) have to be reexamined.

We start with the expressions for the *total enthalpy* per unit mass of a fluid, $B = \boldsymbol{U} \cdot \boldsymbol{U}/2 + H$, where $H = \int dP/\rho$ is the ordinary enthalpy per unit mass and $\boldsymbol{U}$ the total velocity. The total mass flux is $\boldsymbol{J} = \rho \boldsymbol{U}$. The acoustic perturbations of ρ and $\boldsymbol{U}$ are denoted δ and $\boldsymbol{u}$ so that $\boldsymbol{U} = \boldsymbol{U}_0 + \boldsymbol{u}$ and $\rho = \rho_0 + \delta$, where the subscripts 0 indicates unperturbed values (0th order quantities). The corresponding first order perturbations $\boldsymbol{j}$ and b in the mass flux and the total enthalpy are then

$$\boldsymbol{j} = \delta \boldsymbol{U}_0 + \rho_0 \boldsymbol{u}$$
$$b = \boldsymbol{u} \cdot \boldsymbol{U}_0 + p/\rho_0. \tag{10.29}$$

The linearized equations for mass and momentum conservation can then be written

$$\partial \delta/\partial t + \operatorname{div} \boldsymbol{j} = 0$$
$$\partial \boldsymbol{u}/\partial t + \operatorname{grad} b = 0. \tag{10.30}$$

The term $\partial \delta/\partial t$ can also be written $(1/c^2)\kappa \, \partial p/\partial t$. The second follows from linearization of the momentum equation in combination with the mass equation (see Problem 2).

Multiplying the first of the equations in Eq. 10.30 by b and the second by $\boldsymbol{j}$ (scalar multiplication), we get

$$\boldsymbol{u} \cdot \boldsymbol{U}_0 \, \partial \delta/\partial t + \partial \kappa (p^2/2)/\partial t + b \operatorname{div} \boldsymbol{j} = 0$$
$$\delta(\boldsymbol{U}_0 \cdot \partial \boldsymbol{u}/\partial t) + \partial(\rho_0 u^2/2)/\partial t + \boldsymbol{j} \cdot \operatorname{grad} b = 0, \tag{10.31}$$

where we have used the notation u^2 for the scalar product $\boldsymbol{u} \cdot \boldsymbol{u}$. The sum of these equations is

$$\frac{\partial}{\partial t}[\kappa p^2/2 + \rho_0 u^2/2 + \delta \boldsymbol{u} \cdot \boldsymbol{U}_0] + \operatorname{div}(b \, \boldsymbol{j}) = 0, \tag{10.32}$$

which has the form of an energy conservation law for the acoustic perturbations with the energy density and acoustic energy flux (intensity) given by

$$\boxed{\begin{array}{c} \text{Energy density and intensity in a moving fluid} \\ w = \kappa p^2/2 + \rho_0 u^2/2 + \delta(\boldsymbol{u} \cdot \boldsymbol{U}_0) \\ \boldsymbol{I} = b\boldsymbol{j} = (\boldsymbol{u} \cdot \boldsymbol{U}_0 + p/\rho_0)\boldsymbol{j} \end{array}} \tag{10.33}$$

[$\boldsymbol{U}_0$: *Mean velocity.* $\delta = p/c^2$: *Perturbation in density.* p: *Sound pressure.* ρ_0: *Mean density.* $\boldsymbol{u}_0$: *Perturbation in velocity.* b: *Perturbation in total enthalpy.* $\boldsymbol{j}$: *Mass flux.*]

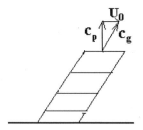

Figure 10.2: Illustration of the difference between phase and group velocity in a high-frequency sound beam projected from a source on the ground surface in the vertical direction in the presence of a horizontal mean flow.

In other words, the acoustic intensity is obtained from the standard expression $p\boldsymbol{u}$ by replacing the sound pressure p by the perturbation in b in total enthalpy and $\boldsymbol{u}$ by the perturbation $\mathbf{j}$ of the mass flux.

Written out in detail, the intensity is

$$\boldsymbol{I} = (p/\rho_0 + \boldsymbol{u}\cdot\boldsymbol{U}_0)(\rho\,\boldsymbol{u} + \delta\,\boldsymbol{U}_0). \tag{10.34}$$

For a traveling wave, with $p = \rho c\,|\boldsymbol{u}|$ and $\boldsymbol{u} = |\boldsymbol{u}|\,\hat{\boldsymbol{k}}$, where $\hat{\boldsymbol{k}}$ is the unit propagation vector, indicating the direction of propagation with respect to the moving fluid, we get

$$\boldsymbol{I} = w(\boldsymbol{c} + \boldsymbol{U}_0), \tag{10.35}$$

where $\boldsymbol{c} = c\,\hat{\boldsymbol{k}}$. The quantity $\boldsymbol{c} + \boldsymbol{U}_0$ is the group velocity, i.e., the vector sum of the sound velocity $c\hat{\boldsymbol{k}}$ and the flow velocity $\boldsymbol{U}_0$. The phase velocity, on the other hand, is the sum of the sound speed c and the projection of the flow velocity on the direction of propagation, $c_p = c + \hat{\boldsymbol{k}} \cdot \boldsymbol{U}_0$. In the simple illustration in Fig. 10.2 we have a high-frequency beam of sound projected up into the atmosphere from the ground surface. In the absence of a wind, the beam will go straight up. In the presence of a horizontal wind velocity $\boldsymbol{U}_0$, the beam will be convected by the wind sideways and takes on the appearance shown in the figure. The group velocity is the vector sum of the flow velocity and the sound (vector) velocity and the energy of the sound is carried by the group velocity and in order to 'catch' the sound beam at an elevated location one has to move sideways, as indicated. The surfaces of constant phase (the horizontal lines in the figure), however, move in the vertical direction, unaffected by the wind velocity since in this case it has no component normal to the surface of constant phase.

Sound Propagation in a Duct with Flow

We now apply the results in Eqs. 10.33 and 10.34 to a plane wave propagating in a duct. In this case $p = \rho cu$, i.e., $\rho_0 u^2/2 = \kappa p^2/2$ ($\kappa = 1/\rho_0 c^2$), and we obtain, with

$\delta = p/c^2$,

$$w = \kappa p^2 (1 \pm M_0)$$
$$I = (p^2/\rho_0 c)(1 \pm M_0)^2, \qquad (10.36)$$

where $M_0 = U_0/c$ is the mean flow Mach number and the plus and minus signs apply to sound propagation with and against the flow, respectively. To get the same intensities in the two directions, it follows that the ratio of the sound pressure amplitudes for the waves in the upstream and downstream directions will be $p_-/p_+ = (1 + M_0)/(1 - M_0)$.

Regarding Measurement of Intensity

In a fluid at rest, the velocity amplitude, being proportional to the gradient of the sound pressure and hence the fluid velocity, can be measured with an intensity probe mentioned briefly in Section 3.2.3. In principle, it consists of two microphones which are separated a small distance (small compared to a wavelength). The *difference* between the sound pressures at the two locations can then be determined and the sound pressure at the midpoint between the microphones is taken to be the average of the *sum* of the two sound pressures. The product of these two quantities yields the intensity and with the aid of a two-channel FFT analyzer, the frequency dependence of the intensity in a sound field can readily be measured. In a moving fluid, as we have seen, the expression for the intensity is not as simple, and this technique is not valid. For small Mach numbers, however, the error involved in measuring the acoustic intensity in this way is small.

10.2.2 Radiation into a Duct with Flow

In some applications, sound is injected into a duct from a source mounted into a side wall of the duct, as shown schematically in Fig. 10.3. The duct carries a mean flow, and due to it, the radiation in the upstream and downstream directions will be different. This problem is analyzed in some detail in Section A.4.4 and in the light of this analysis some experimental results, shown in Fig. 10.3, are discussed. The analysis shows that the radiated sound field depends on what kind of boundary condition is assumed at the loudspeaker source regarded as a piston. One option is to assume that the mean flow is stream lined above the speaker and that the acoustic perturbation of the flow produced by the source results in a displacement of the stream lines. In that case the proper boundary condition to use is continuity of the particle displacement normal to the duct wall. However, if the flow is not laminar, it is more appropriate to use continuity of normal particle velocity. These two boundary conditions are the same when there is no mean flow but in the presence of flow, they are not.[2]

The result obtained from continuity of displacement results in a ratio of the upstream and downstream pressures given by (see Eq. A.101)

$$p_-/p_+ = (1 + M)^2/(1 - M)^2. \qquad (10.37)$$

[2]In the presence of a mean flow and with a displacement ξ, the velocity in the y-direction is $u_y = (\partial/\partial t + U \partial/\partial x)\xi$, where $\xi(x)$ is the displacement of the piston.

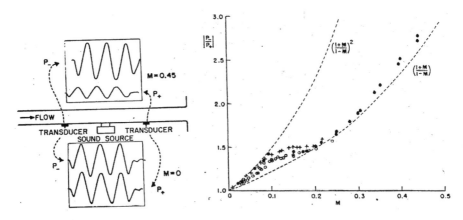

Figure 10.3: Left: Sound source mounted on the side of a duct, emitting harmonic pulse trains with transducers recording the the upstream and downstream signals. Right: Measured ratio of the upstream and downstream pressures.

Continuity of velocity, however, yields the result

$$p_-/p_+ = (1 + M)/(1 - M). \tag{10.38}$$

The experimental results in Fig. 10.3 indicate a transition from a $(1 + M)^2/ (1 - M)^2$-dependence to a $(1 + M)/(1 - M)$-dependence as the mean velocity is increased, presumably related to a transition from laminar to turbulent flow at the source. For further discussion we refer to Appendix A, Section A.4.4.

10.2.3 Problems

1. **Group velocity for higher mode in duct with flow**

   Prove the expression for the group velocity in Eq. 10.20. (Hint: One way of doing it is to differentiate both sides of Eq. 10.19 with respect to k_x, realizing that $\omega = \omega(k_x)$.)

2. **Linearized momentum equation**

   Starting from the momentum equation $\partial \rho U/\partial t + \partial \rho U^2/\partial = -\mathrm{grad}\, P$ in combination with the mass equation $\partial \rho/\partial t + \mathrm{div}\, J = 0$, check the linearized versions of these equations in Eq. 10.30.

3. **Energy equation for sound**

   Carry out in detail the algebraic steps that led to the conservation law for acoustics energy in Eq. 10.32.

4. **Sound radiation into a duct with flow**

   With reference to the discussion in Section 10.2.2 which of the two boundary conditions mentioned there leads to equal radiated power in the upstream and down stream directions?

10.3 Flow-Induced Acoustic Energy Loss

10.3.1 Orifice and Pipe Flow

The steady flow resistance of porous materials generally is a nonlinear function of the flow velocity.[3]

This is due to flow separation and turbulence, and the effect is particularly pronounced in a perforated plate. Due to this nonlinear relation between pressure drop and the steady flow velocity, a superimposed oscillatory flow, as produced by a sound wave, will make the corresponding oscillatory pressure drop proportional to the mean flow velocity. In other words, the mean flow, in effect, provides the perforated plate or an acoustic cavity resonator with an acoustic resistance which in most cases far exceeds the viscous contribution.

The steady flow through an orifice plate or a duct with sudden changes in cross section is discussed in most texts on fluid flow. Because of flow separation and turbulence, the problem is difficult to analyze from first principles in all its details and empirical coefficients usually have to be introduced to express the relation between pressure and flow velocity.

For isentropic flow through an orifice plate in a uniform duct, the pressure falls to a minimum at the location of maximum velocity in the orifice but then recovers to its upstream value sufficiently far from the orifice on the downstream side. However, due to flow separation and turbulence, this recovery is not complete and the pressure loss will be proportional to $U_0^2/2$, where U_0 is the velocity in the orifice. The constant of proportionality depends on the open area fraction of the orifice plate; when it is sufficiently small, the constant is approximately 1. For the purpose of the present discussion, the pressure loss is expressed simply as

$$\Delta P \approx \frac{\rho U_0^2}{2}(1 - s), \qquad (10.39)$$

where s is the open area fraction, the ratio of the orifice area, and the duct area. The factor $(1 - s)$ is a semi-empirical correction which makes pressure drop zero when the orifice area is the same as the duct area.

We now regard a superimposed sound wave as a quasi-static modulation of the pressure loss resulting in a variation $\Delta U_0 = u$ in the velocity, where u is the velocity in the sound. The corresponding variation p is the pressure loss.

$$p \approx \rho U_0 u (1 - s). \qquad (10.40)$$

The semi-empirical factor $(1 - s)$ makes the pressure loss zero if the orifice area is the same as the pipe area.

The acoustic resistance of the orifice is then $r_0 \approx p/u$ and the normalized value is

$$\theta_0 = r_0/\rho c = M_0(1 - s), \qquad (10.41)$$

---

[3] For details, see, for example, Uno Ingard, *Noise Reduction Analysis; Absorbers and Duct Attenuators*, in preparation.

where $M_0 = U_0/c$ is the Mach number of the mean flow in the orifice. If we define the acoustic resistance based on the velocity variation $\Delta U = s\,\Delta U_0$ in the duct rather than in the orifice the corresponding resistance $r/\delta U$ becomes $1/s$ times the value in Eq. 10.41, i.e.,

$$\theta \approx r/\rho c = \frac{1-s}{s}M_0. \tag{10.42}$$

It is illuminating to approach this problem from another point of view, considering the increase in the average energy loss caused by a superimposed harmonic acoustic perturbation of the velocity U_0. Thus, we start from the kinetic energy flux in the (jet) flow from the orifice, $W_0 = \rho U_0^3/2$. With a superimposed acoustic velocity perturbation $u(t)$, the corresponding flux is $W = \rho[U_0 + u(t)]^3/2$. Expanding the bracket, we get $U_0^3 + 3U_0^2 u(t) + 3U_0 u(t)^2 + u(t)^3$. With the time average of the acoustic velocity perturbation $u(t)$ being zero, $\langle u(t) \rangle = 0$, the increase in the time average flux caused by the perturbation is $W - W_0 = (3/2)U_0\langle u(t)^2 \rangle$.

In terms of the acoustically induced increase in the flux, the corresponding acoustic resistance is obtained from $r\langle u(t)^2 \rangle = W - W_0$. This yields an acoustic resistance $(3/2)\rho U_0$, i.e., 50 percent higher than before. This apparent paradox is resolved when we realize that the acoustic perturbation increases the static pressure drop required to maintain the average flow U_0. In the case of harmonic time dependence this static pressure increase multiplied by U_0 will account for the 50 percent increase in $W - W_0$. The remaining contribution is the energy drawn from the sound wave and this leads to the same value for the acoustic resistance as before. It is left for one of the problems to go through the details and show that a harmonic variation of the flow velocity through the orifice does not produce a harmonic variation in the pressure drop and that an increase in its time average value is obtained.

What we have said about the flow-induced acoustic losses in an orifice applies equally well to a flow through a pipe due to the losses at the discharge from the pipe. An illuminating demonstration of this effect is illustrated in Fig. 10.4. A pipe open at

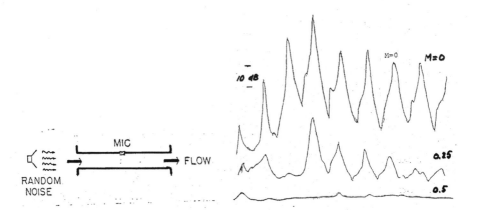

Figure 10.4: Open-ended pipe with flow excited by a random noise field produced by a source outside the pipe. The sound pressure spectra are recorded by a microphone mounted at the center of the pipe as shown (on the right) at different flow Mach numbers.

both ends is connected to a plenum chamber which in turn is connected to a pump which pulls air through the pipe. The flow discharges into the plenum chamber. A loudspeaker is placed in the free field in front of the other end of the pipe and exposes the pipe to random noise. A microphone is mounted in the side wall of the pipe at its center and the spectrum of the sound in the pipe is displayed by means of a narrow band frequency analyzer. When there is no flow present, the acoustic resonances of the pipe which have their pressure maxima at the center (i.e., the odd modes for which the pipe length is an odd number of half wave lengths) are well-defined as pronounced resonance spikes in the spectrum. As the flow speed increases, the amplitude of the resonances is reduced and width increases. For Mach numbers in the pipe larger than approximately 0.5, the resonances have been essentially eliminated. The results thus obtained can be shown to be consistent with a theoretical analysis of the problem. Thus, a pipe with a Mach number in excess of 0.5 would not function as an organ pipe.

10.3.2 Flow-Induced Damping of a Mass-Spring Oscillator

Flow-induced damping can be obtained not only in an acoustic cavity resonator but also in a mechanical mass-spring oscillator, as illustrated schematically in Fig. 10.5.

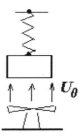

Figure 10.5: Flow damping of mass-spring oscillator.

A sphere or some other appropriate mass element, attached to a spring which is supported in its upper end, oscillates in harmonic motion with the vertical displacement $\xi = |\xi| \cos(\omega t)$ in the downwards direction. The spring constant is K and there is also a dashpot friction force $r\dot{\xi}$. A fan placed under the oscillator supplies a vertical flow of air over the mass element with velocity U_0. The velocity relative to the sphere in the upwards direction is $U_0 + \dot{\xi}$. This velocity is assumed high enough so that the drag force on the sphere is $F = C\rho(U + \dot{\xi})^2 A/2$, where C is the drag coefficient and A the cross-sectional area of the sphere. Normally, $u << U_0$, so that

$$F \approx C\rho U_0 \dot{\xi} A \qquad (10.43)$$

is the driving force acting on the oscillator. It acts in the upwards, negative ξ-direction so that the equation of motion becomes

$$M\ddot{\xi} + K + R\dot{\xi} = -F. \qquad (10.44)$$

From the expression for F in Eq. 10.43 it follows that the force represents a (viscous) damping force with the equivalent friction constant

$$R = C\rho U_0 A. \tag{10.45}$$

The damping effect provided by the flow in this case can readily be demonstrated in a simple table-top experiment. The flow damping can be enhanced if a stiff piece of cardboard is attached to the bottom of the mass element to increase its effective area. Actually, a critically damped oscillator can readily be achieved in this manner.

10.3.3 Problems

1. **Flow-induced orifice resistance**

   Following the outline in the text below Eq. 10.42, discuss the flow-induced acoustic resistance of an orifice based on energy consideration in which the incident sound modulates the energy flux through the orifice. In this context, determine the change in the mean pressure across the orifice as a result of the acoustic modulation of the flow.

2. **Flow damping of an oscillator**

   A sphere of mass M is attached to a spring to form an oscillator as shown in Fig. 10.5. A fan provides a vertical flow velocity U_0 over the sphere. With the drag coefficient for the sphere $C \approx 1$, what velocity will yield critical damping due to the flow-induced resistance. Discuss the feasibility of demonstrating this effect in terms of the size and mass density of the sphere.

10.4 The Mass Flux Paradox

Consider a plane sound wave traveling in the positive x-direction. The first order solutions to the wave equation for the pressure and velocity fields are $p = |p| \cos(kx - \omega t)$ and $u = (|p|/\rho c) \cos(kx - \omega t)$. Since the sound speed is given by $c^2 = dP/d\rho$ (see Chapter 3), the first order density fluctuation is $\delta = |p|/c^2 \cos(kx - \omega t)$. Thus, the total density is $\rho = \rho_0 + \delta$, where ρ_0 is the unperturbed density.

The mass flux in the sound wave is then

$$j = \rho u = \rho_0 u + \delta u = \rho_0 |u| \cos(kx - \omega t) + |p|^2/\rho c^3 \cos^2(kx - \omega t). \tag{10.46}$$

The time average of the first term is zero but not of the second and, according to this equation, there should be a time average mass flux associated with a plane wave given by

$$\langle j \rangle = |p|^2/(2\rho c^3) = \rho |u|^2/2c. \tag{10.47}$$

This clearly makes no sense physically since conservation of the average mass would require the plane piston that generates the wave to be a source of mass as well, which is not the case (the time average displacement of the piston is zero). Thus, we have arrived at a paradoxical result.

10.4.1   Resolution of the Paradox

The fallacy of the argument leading to the result of a time average mass flux in the sound wave (Eq. 10.47) is the omission of second order quantities other than the product of two first order quantities. In balancing an equation like the mass conservation equation $\partial \rho / \partial t + \partial \rho u / \partial x = 0$ correctly to second order, we have to consider each field variable as a sum of terms of different orders of the perturbation. Thus, the first term, the zeroth order term, is the ambient, unperturbed value. The next, the first order term, is the solution to the linearized equations; for the density, for example, the the first order term, which we earlier denoted δ, is now $\rho^{(1)}$. A second order term, $\rho^{(2)}$, is of the order of the product of two first order terms and so on for higher order terms. We are generally not able to solve the nonlinear equations but make the assumption that the solution can be expanded as a series of terms of increasing order. Thus, for the density we put

$$\rho = \rho^{(0)} + \rho^{(1)} + \rho^{(2)} + \cdots \tag{10.48}$$

with a similar expression for the velocity u.

In terms of these quantities the mass flux, correct to second order, will be

$$\begin{aligned} j = j^{(0)} + j^{(1)} + j^{(2)} &= [\rho^{(0)} + \rho^{(1)} + \rho^{(2)}][[u^{(0)} + u^{(1)} + u^{(2)}] \\ &= [\rho^{(0)}u^{(1)} + \rho^{(1)}u^{(0)}] + [\rho^{(0)}u^{(2)} + \rho^{(1)}u^{(1)}]. \end{aligned} \tag{10.49}$$

In our case there is no mean velocity so that $u^{(0)} = 0$ and the time average of $u(1)$ is zero.

In order for the mass conservation law

$$\partial \rho / \partial t + \operatorname{div} j = 0, \tag{10.50}$$

to be satisfied at all times, it is necessary that each order in the expansion of the left-hand side must be zero. For a steady state motion, the time average of the first term is zero and it follows that in order for the time average of the mass flux to be zero, correct to second order, it follows that

$$\langle j \rangle = \langle \rho^{(0)}u^{(2)} + \rho^{(1)}u^{(1)} \rangle = 0. \tag{10.51}$$

This means that the time average of the Eulerian velocity must be negative

$$\langle u \rangle = \langle u^{(2)} \rangle = -(|u|/c)|u|/2, \tag{10.52}$$

where we have used Eq. 10.47.

It remains to understand physically how the average (Eulerian) velocity in the traveling harmonic wave can be negative. There are several ways to do that. First, we refer to Fig. 10.6 where, on the left, the trajectories of five adjacent fluid particles are shown in the transmission of a harmonic wave. The time scale is normalized with respect to the period T. We express that the particle displacement in a traveling wave

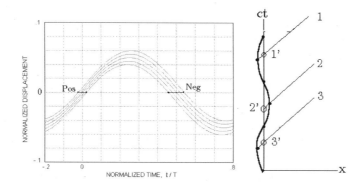

Figure 10.6: Left: The trajectories of fluid elements which at $t = 0$ are at the normalized positions $a_n = -.02, -0.01, 0, 0.01, 0.02$ and moving under the influence of a traveling wave with the displacement $\xi_n = a_n + |\xi| \sin(\omega t - ka_n)$, where $k = 2\pi/\lambda$ and $\omega = 2\pi/T$. A fixed observer at 0 sees a negative time average of the velocity.
Right: Trajectory in a ct-x-diagram of the particle with equilibrium position $x = 0$ in a traveling wave, using wave lines to show regions of positive and negative velocity.

is $\xi(t) = A \sin(\omega t - kx)$ $(k = 2\pi/\lambda)$ so that the trajectory of the particle which is at $x = 0$ at $t = 0$ will be $\xi(0) = A \sin(\omega t)$. It is the middle curve of those shown in the figure, where the displacement is normalized with respect to the wavelength λ. The four other curves refer to the particles which at $t = 0$ are at the normalized positions $a_1 = -0.02$, $a_2 = -0.01$, $a_4 = 0.01$, and $a_5 = 0.02$. With this notation, the middle curve corresponds to $a_3 = 0$. The displacement of the particles then will be $\xi_n(t) = a_n + |\xi| \sin(\omega t - ka_n)$.

The five curves are closer together in the region where the slope is positive, i.e., where the velocity is positive and further apart where the velocity is negative. Consequently, a fixed observer at $x = 0$ will observe a positive velocity in the region marked 'Pos' and a negative velocity in the region 'Neg.' Since the latter clearly is longer than the former and the magnitudes of the velocities in the two regions are the same, the time average will be negative.

Another way of looking at this problem is shown on the right in Fig. 10.6. It gives us an opportunity to review the idea of a wave line, discussed in Chapter 3. The trajectory of the particle with the equilibrium position at $x = 0$ is shown in an ct-x diagram, where c is the sound speed. We recall that along a wave line the state of the motion remains constant. In this diagram the wave lines are straight lines inclined 45 degrees. Thus, as the wave propagates the displacement will have a negative maximum along the wave lines marked 1 and 3 and a positive maximum along the wave line marked 2. The lines (not shown) between 1 and 2 correspond to a negative velocity and the lines (not shown) between 2 and 3 to a positive velocity. The intersections 1', 2', 3' of these lines with the ct-axis marks the regions of positive and negative velocity recorded by a fixed observer at $x = 0$. It is readily seen that the negative region is larger than the positive so that the time average of the velocity will be negative.

10.5 Mean Pressure in a Standing Wave

10.5.1 Fountain Effect and Mode Visualization

In the linear approximation, the time average of a field variable with harmonic time dependence is zero. It has been known for a long time, however, that in a sufficiently intense sound field there are effects that indicate a time average different from zero. The formation of dust patterns in a standing wave in a tube and the associated vortices are typical examples (Kundt's tube). A related phenomenon is demonstrated in Fig. 10.7 where a one-dimensional standing wave in a tube is shown to deform the surface of the liquid that partially fills the tube.

In addition to the static deformation of the liquid surface, that is of primary interest here, we find in the figure that 'fountains' are present at the locations of the acoustic pressure nodes (velocity anti-nodes). The fountains are fun to watch and measuring the distance between them (half a wavelength) at a given frequency, the speed of sound can be determined. The instability, involving the 'rupture' of the liquid surface, occurs above a threshold value of the sound pressure. The mechanism leading to the rupture is complicated and appears to involve a surface wave caused by the second order acoustically induced streaming of the air above the liquid surface in Section 10.6. Instead, we turn to an extension of this phenomenon to a three-dimensional

Figure 10.7: The mean pressure in a standing wave causes deformation of the water surface and creates fountains in the pressure nodes.

mode in a cavity. In our experiment, we used a rectangular box made from 0.5 cm thick Plexiglas walls of dimensions $12.2 \times 6.4 \times 9$ cm. As in Fig. 10.7, a horn type loudspeaker, driven by a 60-W amplifier, was used to excite acoustic modes in the cavity through a hole in one of the walls. The floor of the cavity was covered with a liquid layer, approximately 1 cm deep.

When a mode of the cavity is excited at a sufficiently high level, a static deformation of the liquid surface is visible by the unaided eye. It can be better viewed by projection onto a screen, however, by means of an overhead projector arranged so that the liquid surface is located in the object plane of the projector. Similarly, the deformation pattern of the liquid can be contact printed on photographic paper by exposing it to light that has passed vertically through the liquid. The deformed liquid surface acts like a 'lens' and causes refractions of the light that mimic the surface deformation and the pressure distribution in the cavity. An example of such a contact print is given in Fig. 10.8 of the (3,2,0)-mode in the cavity; it has three pressure nodes in one direction and two in the other.

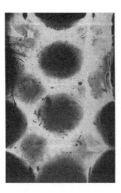

Figure 10.8: Contact print of the deformation of a liquid surface caused by the mean pressure distribution of an acoustic mode in a rectangular cavity.

Another way of visualization is to photograph a rectangular grid graph paper placed under the cavity and in the focal plane of a camera placed above the cavity. The rectangular grid then is distorted to reflect the deformation of the surface. Overhead projection can be used also in this case, of course.

In order to use the images obtained of the surface quantitatively, we start by analyzing the one-dimensional case, corresponding to Fig. 10.7, and compute the static pressure distribution $\langle p \rangle$ in the sound field. The momentum equation is $\partial \rho u / \partial t + \partial \rho u^2 / \partial x = -\partial p / \partial x$ (see Eq. 5.3) (Eulerian) time average of this equation yields

$$\langle p \rangle + \langle \rho u^2 \rangle = \text{constant}, \tag{10.53}$$

since the (long) time average of $\partial \rho u / \partial t$ is zero. Thus, in order to obtain the spatial variation of $\langle p \rangle$ correct to second order, we use the first order solution for the velocity field which is $u = (|p|/\rho c) \sin(kx) \sin(\omega t)$ with the corresponding pressure field $p = |p| \cos(kx) \cos(\omega t)$, where x is the distance from the solid termination of the tube and $k = \omega/c = 2\pi/\lambda$. It follows then from Eq. 10.53 that

$$\langle p \rangle = P + (p_1^2/4\rho c^2) \cos(2kx), \tag{10.54}$$

where P is the spatial average pressure.

In order to relate the observed deformation of the liquid to the sound field in the tube, we shall assume that the deformation is caused solely by the pressure $\langle p \rangle$ at the liquid surface. Thus, the shear stresses on the surface caused by acoustically induced vortex motion will be neglected. Then, if the vertical displacement of the surface is η, the surface tension σ, and the density of the liquid ρ_ℓ, the boundary condition at the surface is $-\rho g \rho_\ell + \sigma \nabla^2 \eta = \langle p \rangle$. In the one-dimensional case, with $\langle p \rangle$ given in Eq. 10.54, this boundary condition yields the surface deformation

$$\eta = \eta_0 \cos(2kx) \quad \text{where}$$

$$\eta_0 = -\frac{p_1^2/4\rho c^2}{\rho_\ell g + (2k)^2 \sigma}. \tag{10.55}$$

This result has been found in good agreement with measurements.[4] Thus, the omission of the effect of shear stresses seems to be justified in this case of a standing wave.

Following the procedure given above, the analysis can be extended to the interaction of higher modes in a rectangular cavity or cylindrical cavity.[5]

10.5.2 Acoustic Levitation

The spatial variation of the mean pressure in a standing wave makes it possible to keep light objects, small droplets or styrofoam balls, for example, suspended and locked in position in a pressure node. This can be readily demonstrated in a vertical tube driven at one end with a loudspeaker, similar to the arrangement in Fig. 10.7. By varying the frequency and hence the location of the nodes, the object can be moved. Such acoustic 'levitation' is possible also in the standing wave in any other cavity.

The levitated objects usually are found to spin, an effect which appears to be caused by a local non-uniformity of the sound field and related to acoustically driven vorticity (streaming) discussed below in connection with Fig. 10.14.

Acoustic levitation has found practical use in the growing of protein crystals and cells under controlled conditions, avoiding chemical and thermal contamination resulting from contact and external objects.[6]

10.5.3 Other Demonstrations

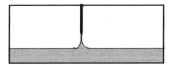

Figure 10.9: Mean pressure distribution of a sound wave in the vicinity of a constricting partition in a tube causing the liquid to be pulled up to the partition.

The deformation of the surface of a liquid can be made even more pronounced by introducing a constriction in a tube, as shown in Fig. 10.9.

In this figure, a partial partition in a rectangular tube creates a constriction in area so that the velocity amplitude is increased in this region. As a result, the mean pressure in the sound field is decreased and the liquid is pulled up to the edge of the partition. As the constriction is thus blocked by the liquid, the sound path is blocked and the mean pressure and the liquid surface return to normal. The cycle then repeats and we get an oscillatory blockage of the constriction caused by the sound wave.

---

[4]J. A. Ross Jr, MS Thesis, M. I. T., August 1969.

[5]Uno Ingard and Daniel C. Galehouse: *Second-Order Pressure Distribution in an Acoustic Normal Mode in a Rectangular Cavity.* (American Journal of Physics, Volume 39/7, 811–813, July 1971).

[6]NASA Tech Briefs, Vol. 23, Nr 3, March 1999, p 78.

In experiments of this kind, the use of a rather viscous liquid prevents 'rupture' of the liquid surface and fountain formation. The height of the liquid in a mean pressure minimum can then be adjusted by varying the sound pressure level.

If the partial partition is placed near the closed end of the tube so that an acoustic (Helmholtz) resonator is formed, the effect can be even further enhanced at the resonance frequency of the resonator. Furthermore, if the corresponding resonator orifice is asymmetrical such that the flow coefficient is different in the two flow directions, a difference in surface level of the liquid can be created inside and outside the resonator.

The nonlinearity of higher modes in a circular tube can also be demonstrated. Of these, the 'sloshing' mode has the lowest frequency; its sound pressure has one nodal diametrical plane across which the velocity amplitude has its maximum at the center, going to zero at the walls. The pressure dependence on the azimuthal angle ϕ is given by $\cos(\phi)$ (the location of the nodal plane, corresponding to $\phi = 0$, could be related to an asymmetry of the sound source). As a result of the nonlinear perturbation (decrease) in the mean pressure, a "sheet" of the liquid is "pulled up" in the nodal plane with the maximum height at the center; theoretically the height goes to zero at the walls of the tube, as indicated schematically in the top left sketch in Fig. 10.10. In reality, there is a threshold acoustic amplitude required to produce a sheet. It is located away from the wall at a location where the acoustic particle velocity is at the critical value for sheet formation.

At a sufficiently high sound pressure level, the ridge of the sheet becomes corrugated, and close inspection reveals rapid steady circulations (streaming) on the surface of the liquid in this region. Thus, nonlinear effects can create strange phenomena, as indicated further in the bottom right photo, where a free floating (levitated) separated portion of the ridge can be seen.

The acoustic mode with *two* nodal diametrical planes in the circular tube, corresponding to the angular dependence $\cos(2\phi)$ of the sound pressure, has the maximum

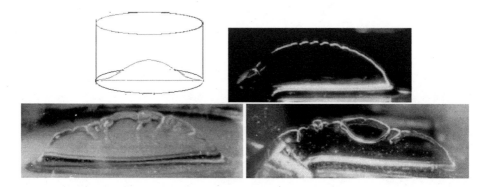

Figure 10.10: Demonstration of the mean pressure distribution and related liquid sheet formation in the first circumferential mode in a cylindrical cavity. Top right: Almost perfect sheet corresponding to the idealized version in top left. Bottom left: Formation of "channels" in the ridge of the sheet. Bottom right: Free-floating channel section.

Figure 10.11: An acoustically induced liquid sheet in a cylindrical cavity by an acoustic mode with two nodal diametrical planes.

(tangential) velocity amplitude across the planes at the wall (the velocity amplitude at the center is zero in this mode). Theoretically, then, due to the nonlinear acoustically induced decrease in the mean pressure, there should be four symmatrically located sheets created with the maximum heights at the walls. In our experiments, only one of these sheets could be captured on film, however, as shown in Fig. 10.11. The other sheets were unstable and decayed. A likely reason is that a sheet, once formed, perturbs the sound field and affects both its amplitude and symmetry.

Figure 10.12: Liquid sheet formation in a circumferential acoustic mode in a cylindrical cavity.

Returning to Fig. 10.7 and the standing wave in a tube creating a water fountain at every nodal plane in the sound field, we replaced the water with a highly viscous liquid. Instead of a fountain, we then got an elevated stable liquid sheet in every nodal plane of the standing sound pressure wave in the tube, as anticipated from the result in Fig. 10.10. What is noteworthy is that the irregularity of the ridge of such a sheet is now more clearly seen (Fig. 10.12), as a quasi-periodic "corrugation" of the top of the sheet.

10.5.4   Acoustic Radiation Pressure

According to Eq. 10.53, the Eulerian time average of the one-dimensional momentum equation leads to $\langle p \rangle + \langle \rho u^2 \rangle =$ constant, where the angle brackets indicate time average. For a plane wave incident on a perfect reflector with $u = 0$, the pressure on the sonified side of the reflector is $p + \rho \langle u^2 \rangle$. Then, if the mean pressure in the wave is the same as the pressure on the other side of the reflection (both equal to the ambient pressure), the rate of momentum transfer from the incident sound wave will be $\langle \rho u^2 \rangle$ which is then the acoustic radiation pressure.

10.5.5   Acoustic 'Propulsion'

An amusing, and perhaps puzzling demonstration, is illustrated schematically in Fig. 10.13 showing two cavity resonators at the end of a bar which is supported at its center by a thin wire with a small torsion constant. This assembly can be driven in rotational motion by sound tuned to the resonance frequency of the resonators. Resonators, ideally suited for this lecture demonstration, are the (brightly colored)

Figure 10.13: Two identical resonators (viewed in a horizontal plane) hung from a thin wire connected to the center of the cross bar, as shown. A sufficiently strong sound wave at the resonance frequency of the resonators will make the resonators move, as indicated, so that the resonator assembly will rotate about its vertical axis of suspension.

thin glass bulbs used for Christmas tree ornaments. With the metal insert (used for support) in the neck removed, such a bulb becomes a small Helmholtz resonator. The two resonators are selected to have identical resonance frequencies. A loudspeaker is placed under this assembly and driven at the resonance frequency of the resonators. At a sufficiently high sound pressure level, the rod with its resonators is set in rotational motion.

One might wonder at first sight, how this can be possible. What about conservation of angular momentum?

At least part of the answer is related to the asymmetry of the orifice in each of the resonators. Although in steady state, the time average mass flow out of each resonator is zero; the time average of the momentum flow is not. Flow separation and emission of vortex rings (as discussed in the next section) should play an important role in causing a corresponding reaction force on each resonator and hence a torque on the assembly.

10.6   Vorticity and Flow Separation in a Sound Field

It is an experimental fact that steady vortex motion or 'streaming' can be produced by a sound field. To understand qualitatively how an oscillatory motion can produce a time independent vortex flow, let us first establish the role of viscosity in such a motion. In the absence of viscosity, the pressure gradient and the acceleration in a sound field are in the same direction. In the presence of viscosity, however, this is generally not the case since the viscous force depends on the spatial variations of the components of the velocity. Furthermore, the sound wave produces an oscillatory pressure gradient and a corresponding density gradient so that the density of a volume element is temporarily larger at one end than at the other; half a period later, the roles are reversed.

With these observations, let us consider a simple mechanical analog of this behavior. It consists merely of a dumbbell with one end heavier than the other. If the dumbbell is accelerated in the direction of the handle, i.e., in the direction of the line between the two end weights, nothing unusual happens; it corresponds to the response of an inviscid gas with the acceleration being in the same direction as the pressure gradient. With viscosity present, there is no longer such an alignment of the acceleration and the pressure gradient, however, and there will be an inertial torque on the dumbbell and an angular displacement, the heavier side lagging behind during one half of the cycle. During the other half, one might think that there would be an angular displacement in the other direction so that after one whole period the net displacement would be zero. This is not so, however. The reason is that the direction of the pressure gradient also changes so that the heavy and light ends of the dumbbell will be reversed. Therefore, the angular displacement during the two half cycles will be in the *same* direction, resulting in a steady rotation caused by the oscillatory acceleration. Qualitatively, this is the origin of the acoustically induced vorticity in the fluid. The inertial torque, being proportional to the product of acceleration and a pressure gradient, is of second order in the field variables. Actually, it can be shown that the diffusion equation for vorticity has a source term which is of second order in the acoustic field variables.

This phenomenon can be significant particularly in regions where the velocity gradients are strong as is the case close to sharp corners in a boundary, for example. In Fig. 10.14 are shown examples of acoustically induced vorticity around an orifice in a plate at the end of a circular tube in which sound is generated by a loudspeaker at the other end.

The linear theory of a sound field in the absence of viscosity and other loss mechanisms, predicts that the velocity amplitude goes to infinite at a sharp corner and it comes as no surprise that nonlinear effects make their presence known in this case. The details of the effect, however, are difficult to foresee, even qualitatively, and the mathematical analysis of the problem is difficult and has not been carried out, as far as I know.

With reference to Fig. 10.14, the form of the induced circulations (vorticity) were made visible by means of smoke generated inside the tube and was found to vary with the level of the incident sound. The images 1 and 2 were obtained using a light sheet for illumination so the pattern refers to a central slice of the circulation. The pattern depends on the level of the sound. It starts out with a small inner vortex with outflow at the center of the orifice (a bubble like formation) and it is still present in frame 1 in the figure. As the sound level is increased the outer vortex becomes prominent with a direction opposite that of the inner and this is what is shown in frame 1. As the level is increased further, the inner vortex pattern grows until flow separation occurs, and frame 2 shows the state when this is about to happen. At a still higher level, a 'jet' is formed, as shown in frame 3. For a different perspective which shows the remarkable extent of the jet, see frame 4.

These photographs were obtained with steady illumination and what looks like a jet

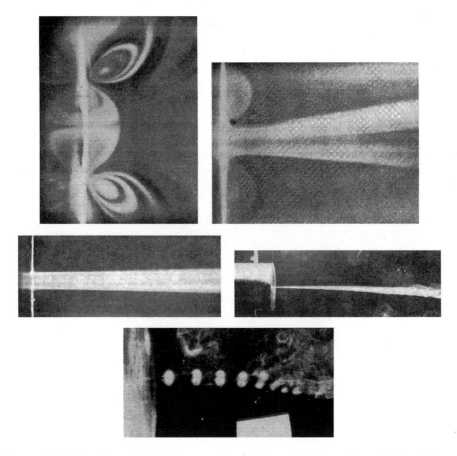

Figure 10.14: 1: Acoustically driven steady vorticity around an orifice. 2: Transition to separated flow. 3,4: Acoustically driven vortex rings, steady illumination. 5: Acoustically driven vortex rings, stroboscopic illumination. Frequency: 234 Hz. Sound pressure level: 120 dB.

actually consists of a succession of vortex rings. These are made visible by stroboscopic illumination, as shown in frame 5. The frequency of the sound is 234 Hz and vortex rings are produced at this rate on both sides of the orifice plate.

10.6.1   Nonlinear Orifice Resistance

The energy required to drive the circulations and the vortex rings is drawn from the sound wave and an orifice plate can then be described acoustically in terms of an equivalent acoustic resistance which depends on the incident sound pressure level.

It is shown in Section 10.3 that the acoustic modulation of a steady separated flow through an orifice or a duct produces an energy transfer from sound to vorticity and that the equivalent normalized acoustic resistance is proportional to the mean flow velocity. For a perforated plate with an open area fraction s, the normalized flow-induced resistance is given by Eq. 10.41, i.e., $\theta \approx (1-s)M_0$, where M_0 is the Mach number of the steady flow in the orifice.

By analogy, we expect the nonlinear acoustic resistance of an orifice due to flow separation and vortex ring generation to be $C(1-s)(|u|/c)$, where $(|u|/c)$ is the Mach number of the oscillatory flow in the orifice and C a constant of the order of unity. Due to the nonlinearity, a harmonic driving pressure will not produce a harmonic velocity. Experiments have shown, however, that if a large amplitude harmonic pressure drives the flow through an orifice, the main contribution to the velocity will be harmonic. For a sufficiently large pressure amplitude, the ratio of the amplitudes of the pressure drop across the orifice and the fundamental component of the velocity corresponds to a nonlinear contribution to the orifice resistance given by

$$\zeta_{nl} \approx C(1-s)|u|/c. \tag{10.56}$$

For a symmetrical orifice and with $|u|$ being the rms value of the velocity, it is a good approximation to put $C \approx 1$. To this nonlinear resistance should be added the linear contribution caused by viscosity, which normally is negligible. Ordinarily it is not the velocity amplitude in the orifice that is known but rather the sound pressure amplitude of the incoming wave and the velocity amplitude has to expressed in terms of it.

We apply these observations to the orifices in a perforated plate with an open area fraction s and an absorber consisting of the plate backed by an air layer. With the velocity through the orifice being u, the average velocity over the phase of the perforated plate is σu, and the average resistance over the plate is then

$$\zeta_{av} = \zeta_{nl}/\sigma. \tag{10.57}$$

At resonance, this constitutes the input impedance of the absorber. With reference to Chapter 4, Eq. 4.50, the pressure reflection coefficient is

$$R_p = (\zeta_{av} - 1)/(\zeta_{av} + 1). \tag{10.58}$$

Then, if the incident pressure at the absorber is p_i, the total pressure amplitude will be $p_t = (1+R)p_i$, or

$$p_t = [(2\zeta_{av}/(\zeta_{av} + 1)]p_i. \tag{10.59}$$

The average velocity over the surface of the absorber at resonance is then $u_{av} = p_t/\rho c\zeta_{av}$ and the corresponding velocity in the orifice is $u = u_{av}/s$. Thus, from Eq. 10.59 then follows

$$su/c = [2/(\zeta_{av} + 1)](p_i/\rho c^2). \tag{10.60}$$

With $\zeta_{av} = [(1-s)/s]|u|/c = (1/s_1)|u|/c$ (see Eqs. 10.56 and 10.57), where $s_1 = s/(1-s)$, we can then solve Eq. 10.60 for $|u|/c$ to obtain, at resonance,

$$|u|/c = (s_1/2)[\sqrt{1 + (8/s_1 s)\delta} - 1], \tag{10.61}$$

where $\delta = p_i/\rho c^2$. If $\delta \ll 1$, this expression reduces to

$$|u|/c \approx (1/s)(2/\gamma)(p_i/P), \tag{10.62}$$

where we have used $1/\rho c^2 = 1/\gamma P$, where P is that static pressure and γ, the specific heat ratio.

The normal incidence absorption coefficient (see Eq. 4.52) at resonance is

$$\alpha = 4\theta_{av}/(1+\theta_{av})^2. \tag{10.63}$$

Total absorption, i.e., $\alpha = 1$, is obtained for $\theta_{av} \approx (1/s^2)(2/\gamma)(p_i/P) = 1$, i.e.,

$$s \approx \sqrt{(2/\gamma)(p_i/P)}. \tag{10.64}$$

As an example, let the incident sound pressure level be 140 dB, which corresponds to an rms pressure amplitude of 2000 dyne/cm^2. Then, with $\gamma \approx 1.4$ and $P \approx 10^6$ dyne/cm^2, we obtain $s \approx 0.053$. In obtaining this result, we used the approximate expression for u/c as given in Eq. 10.62. It is left for one of the problems to improve on this result by use of the complete expression (10.61) and also to extend the study to include the frequency dependence of the absorption.

10.6.2 Problems

1. **Nonlinear absorption characteristics of a perforated plate resonator**

   In obtaining the result in Eq. 10.64 for the open area of a perforated plate absorber for 100 percent resonance absorption, we used the approximate expression Eq. 10.62.
   (a) Improve on this result by using the complete expression (10.61).
   (b) The normalized reactance of the air layer behind the plate is $i\cot(kL)$, where $k = \omega/c$ and L, the thickness of the air layer. Include this reactance in the equation (10.60). Use your favorite software to solve this equation numerically for $|u|/c$ and make a plot of the absorption versus the normalized frequency ω/ω_0, where ω_0 is the first resonance frequency corresponding to $kL = \pi/2$. Hint: Since the expression for u/c in Eq. 10.60 is no longer real, one must account for both the real and imaginary parts on the right-hand side of the equation. As a numeric example, consider the perforated plate with the open area of 5.3 percent and an incident sound wave of 140 dB. Neglect the reactance of the plate.

2. Again consider a perforated plate, as in Problem 1, but this time it is placed in free field. Calculate the normal incidence nonlinear reflection and transmission coefficients and the corresponding coefficient of absorption coefficient (based on the absorption within the plate, i.e., not including the transmitted sound).

10.7   Acoustically Driven Mean Flow of Heat

In the bulk of a gas, away from boundaries, it is a good approximation to consider the change of state of the gas corresponding to a periodic change in pressure as reversible, or more specifically, as isentropic. Although the temperature in the gas varies periodically with the pressure, there is no time average work done on the gas and no net heat production. The reason is that the pressure is in phase with the compression and therefore 90 degrees of phase with the rate of compression so that the time average of their product is zero.

In the vicinity of a boundary, however, there is a thin thermal boundary layer in which the change of state of the gas from isothermal at the boundary to isentropic outside the boundary. In this region, the temperature gradient is much greater than in free field by a factor of the order of the ratio of the wavelength and the thickness of the thermal boundary layer, a factor which depends on frequency but typically is of the order of $1.5 \times 10^5/\sqrt{f}$, where f is in Hz. Thus, at 100 Hz, the factor is $\approx 15{,}000$. Under such conditions, the role of heat conduction can no longer be ignored. In fact, a compression no longer leads to a pressure which is in phase with the compression; heat conduction causes a phase lag in the pressure response. The rate of compression and the pressure no longer are 90 degrees out of phase and there is a time average generation of heat by the sound wave. The corresponding losses at a boundary was determined in Chapter 4.

If there is a static temperature gradient *along* the boundary and a component of the oscillatory fluid velocity in the sound field along the boundary, this effect of heat conduction in delaying the pressure response with respect to the compression causes a net flow of heat along the boundary.

This mechanism has been implemented in acoustic refrigeration devices. It should be kept in mind that it is a second order effect and large amplitude oscillations are required, which can be obtained by means of acoustic resonators. The process is affected by another nonlinear effect, the acoustically induced vorticity or 'streaming,' described in Section 10.6, resulting in convection of heat.

10.8 Formation of a Periodic Shock Wave ('Saw-Tooth' Wave)

Neglecting the effect of surface tension, a surface wave on a body of water is known to have a phase velocity $v_p = \sqrt{gh}$, where g is the acceleration of gravity and h the average depth. Without going through the equations of motion, we note that this is a reasonable expression; it is at least dimensionally correct. Because of the depth dependence, we expect the local wave speed in the crest of the wave to be somewhat larger than in a trough so that a crest tends to catch up with the trough ahead, thus distorting an initially sinusoidal wave. After a sufficiently large distance of travel, the crests will overtake the troughs and 'breakers' will develop. This effect is enhanced when the mean height decreases as waves approach the shore; the waves further out run faster and overtake the slower waves in front. Furthermore, conservation of wave energy requires the wave amplitude to increase as the wave speed decreases.

There is an analogous amplitude dependence of the sound speed in a gas and a corresponding nonlinear distortion of a traveling sound wave. An experimental demonstration of the amplitude dependence of the wave speed is shown in Fig. 10.15. It involves a pressure pulse in a shock tube recorded at a fixed position from the source for two different peak amplitudes of the pulse. The larger amplitude pulse is seen to arrive first, indicating a higher wave speed.

In a sound wave the regions of compression and rarefaction correspond to the

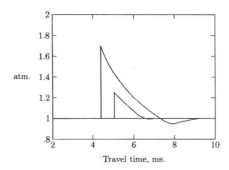

Figure 10.15: Demonstration of the amplitude dependence of the wave speed of two pulses of different amplitudes, ≈ 0.7 and ≈ 0.25 atm. The travel times from the source are ≈ 4.4 ms and ≈ 5 ms, respectively.

crests and troughs in the water wave. In a harmonic sound wave, the reason for the wave speed difference in a crest and a trough is two-fold. First, the particle velocity u in the crest is in the same direction as the local sound speed and in the trough it is in the opposite direction. Second, the temperature in the crest is higher than in the trough. With $c \propto \sqrt{T}$, these effects contribute to make the local wave speed in the crest higher than in the trough. Thus, if the local sound speeds in the crest and in the trough are $c_0 + \Delta c$ and $c_0 - \Delta c$ due to the difference in temperature, the local phase velocities are $c_1 = c_0 + u + \Delta c$ and $c_2 = c_0 - u - \Delta c$, where c_0 is the unperturbed sound speed. To determine Δ_c, we have to obtain the temperatures in the crest and the trough.

The temperature variation can be expressed in terms of the density and pressure fluctuations in the wave from the equation of state, $P = r\rho T$, from which follows

$$\Delta T/T_0 = \Delta p/P_0 - \Delta\rho/\rho_0 = (\gamma - 1)\Delta\rho/\rho_0 = (\gamma - 1)p/(\rho_0 c_0^2) = (\gamma - 1)u/c_0$$

where we have put $\Delta P = p$ and used $u = p/\rho_0 c_0$. We have assumed an isentropic (adiabatic) change of state, with $P \propto \rho^\gamma$, and γ (≈ 1.4 for air) is the specific heat ratio.

Since $c \propto \sqrt{T}$ it follows that $\Delta_c/c_0 = (1/2)\delta T/T_0$ so that $\Delta c = (\gamma - 1)(u/2)$. Thus, $c_1 = c_0 + \Delta c + u$ and $c_2 = c_0 - \Delta c - u$ we get

$$c_1 = c_0 + (\gamma + 1)(u/2), \qquad c_2 = c_0 - (\gamma + 1)(u/2). \tag{10.65}$$

The crest will overtake the trough when it has traveled a distance $\lambda/2$ further than the trough. The corresponding time of travel then follows from $(c_1 - c_2)t \approx \lambda/2$, i.e., $t = (\lambda/2)/[(\gamma + 1)u]$. The corresponding distance traveled is obtained as $c_0 t$. With $|u| = |p|/\rho c$ and $c_0^2 = \gamma P_0/\rho_0$, the shock formation distance x_s can be written

$$x_s \approx \lambda \frac{\gamma}{2(\gamma + 1)} \frac{P_0}{|p|}, \tag{10.66}$$

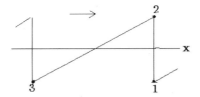

Figure 10.16: Stable shape of a large amplitude sound wave, 'saw-tooth' wave.

where $|p|$ is the magnitude of the sound pressure. Thus, if we start out with a harmonic sound pressure with a level of 120 dB, we have $|p|/P_0 \approx 3 \times 10^{-4}$ and Eq. 10.66 predicts $x_s/\lambda \approx 333$. At a frequency of 1000 Hz, this means that $x_s \approx 333$ ft. We then expect an abrupt change in the wave from the crest to the trough, i.e., a shock wave. Actually, if we go beyond x_s, we would predict a 'breaking' wave, just as for a water wave. In reality, however, losses ease the gradients in the wave and the convective nonlinear effects enhance them; a balance is struck between these competing processes which leads to the formation of a stable, saw-tooth like wave, as indicated in Fig. 10.16. As a result of the wave steepening, the frequency spectrum of the sound changes with the distance of propagation. The wave is harmonic at the beginning ($x = 0$) but develops harmonics as it proceeds. As energy is transferred from the fundamental to higher frequencies, the fundamental will decay and the harmonics will initially grow with the distance to a maximum and then decay as losses become dominant. The distance at which the maximum is reached depends on frequency; the higher the frequency, the shorter the distance.

The attenuation of a wave is related to the gradients of the field variables and the rate of change of these variables. A gradient in pressure and hence temperature induces heat flow and relaxation effects (of molecular modes of motion) and a gradient in velocity produces viscous forces. In the idealized case of a saw-tooth wave in Fig. 10.16, where there is a discontinuous change in pressure in a wave front, the gradient is infinite. Then, although the viscous stress and the heat flow goes to infinity, the related dissipation is finite because the volume of integration goes to zero as the gradient goes to infinity.

Rather than to try to calculate the attenuation of a saw-tooth wave from this point of view, the following possibility should be considered. It relies on the often surprising power of thermodynamics and it goes like this. As a shock passes a fixed location (Fig. 10.16), the state of a fluid element in front of the shock is forced to change from a pressure P_1 to a pressure P_2. The change of state is constrained by the laws of conservation of mass and momentum of the fluid as it goes through the shock and if these constraints are accounted for, it is found that the process of 'lifting' the element from P_1 to P_2 is associated with an increase in entropy. For small values of $\epsilon = (P_2 - P_1)/P_1$, the entropy change can be shown to be

$$S_2 - S_1 \approx r \, \frac{\gamma + 1}{12\gamma^2} \, \epsilon^3 + \cdots . \tag{10.67}$$

Notice that there is no first or second order contribution in ϵ. The corresponding heat production is $dQ = T(S_2 - S_1)$.

Then, by neglecting the losses that occur during the change of state from 2 to 3 in the figure and putting the distance between 1 and 3 equal to one wavelength, the heat production per unit length is

$$\frac{dQ}{dx} = \frac{1}{\lambda} P_0 \frac{\gamma + 1}{12\gamma^2} \epsilon^3, \tag{10.68}$$

where $P_0 = r\rho_0 T_0$, and r is the gas constant per unit mass. By equating this with the spatial rate of change of the wave energy, the decay constant can be obtained.

To compute the wave energy, we note that in a traveling plane wave the kinetic energy density $\rho u^2/2$ is the same as the potential energy density $\kappa p^2/2$, where $\kappa = 1/\rho c^2$ is the compressibility. Thus, the total wave energy density is κp^2. Integration over one wavelength (i.e., from point 3 to point 1 in Fig. 10.16) the wave energy per wavelength is obtained and the corresponding wave energy density is then

$$E = \kappa \frac{(P_2 - P_1)^2}{12} = \frac{\epsilon^2 P_1^2}{12\gamma P_0} \approx \frac{\epsilon^2 P_0}{12\gamma}, \tag{10.69}$$

where we have used $\kappa = 1/\gamma P_0$ and $P_1 \approx P_0$ in the last expression.

Combining Eqs. 10.68 and 10.69, we obtain from $dE/dx = -\partial Q/\partial x$

$$\frac{d\epsilon}{dx} = -\frac{1}{2\lambda} \frac{\gamma + 1}{\gamma} \epsilon^2 \tag{10.70}$$

with the solution

$$\frac{1}{\epsilon(x)} - \frac{1}{\epsilon(0)} = \frac{\gamma + 1}{2\gamma} \frac{x}{\lambda}. \tag{10.71}$$

The attenuation is not exponential because of its dependence of the wave amplitude. Rather, the inverse pressure ratio increases linearly with distance.

10.8.1 Problems

1. **Attenuation of a saw-tooth wave**

   Follow the outline in the text and prove the relations in Eqs. 10.68 and 10.71.

2. **Regular versus shock wave attenuation**

   Consider a plane sound wave with a frequency of 1000 Hz. At what sound pressure level will the attenuation dp/dx of a saw-tooth wave be the same as the attenuation of a harmonic wave due to absorption in the atmosphere (visco-thermal and molecular) at a temperature of 20°C.

3. **Frequency spectrum of a saw-tooth wave**

   Determine the Fourier spectrum of a saw-tooth wave. Make a plot of the sound pressure level of the first ten harmonic components.

10.9 Nonlinear Reflection from a Flexible Porous Layer

When the acoustic amplitude becomes sufficiently high (approximately one percent of the atmospheric pressure), we can expect nonlinear effects to be significant also in the interaction of sound with a boundary.

As an illustration, we describe briefly some results from experiments on the reflection of large amplitude acoustic pulses or shock waves from a flexible porous material. The amplitudes involved are of the order of 1 atm (about 194 dB re 20 microPascal).

The primary motivation for carrying out this study was to simulate the waves generated in a closed loop pulsed laser in which the gas was energized by an electron beam, pulsed at a rate of 125 Hz. The wave produced by the pulsed electron beam had a detrimental effect on the performance of the laser since the acoustic reverberation in the loop made the density in the lasing cavity sufficiently nonuniform to prevent lasing from occurring at the pulse rate of the electron beam. Thus, the problem of attenuating the wave by a substantial amount was essential for the proper functioning of the laser and as a basis for designing an appropriate attenuator, a study of the interaction of shock waves with various (porous) boundaries was called for.

10.9.1 Apparatus

The shock wave was generated in a shock tube shown schematically in Fig. 10.17. Made of steel with a 3 mm wall thickness, the tube was 2 m long and supplied with appropriate flanges and ports for attaching the driver section, transducers, test section, and tube extensions, one 92 cm and the other 213 cm long. One of the extensions was provided with holes over parts of its length to accommodate transducers.

The gas in the tube was air at atmospheric pressure. The driver section was terminated with a properly chosen membrane. We experimented with a variety of membrane materials, particularly Mylar films of different thicknesses, to obtain a

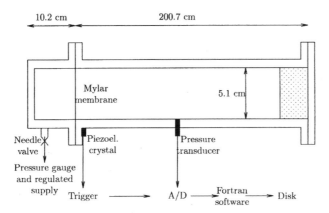

Figure 10.17: Experimental arrangement. As shown, the shock tube is terminated by a porous layer, but other terminations, perforated plates, lined ducts, etc., were used.

peak pressure of the shock wave in the range from 0.2 to 2 atm corresponding to peak pressure levels of $\approx$ 180 to $\approx$ 200 dB. (The dB levels given here are based on a reference pressure 0.0002 dyne/cm^2, although this commonly used reference refers to the rms value of a harmonic wave.) For example, mylar films with thicknesses of 0.013 and 0.025 mm ruptured at driver pressures of 2.45 and 3.7 atm, and another plastic film of thickness 0.0065 mm ruptured at 1.5 atm. The rupture pressure could be made quite repeatable from one membrane to the next with a proper experimental procedure.

The shock tube could be terminated by various elements, such as a rigid plate, various orifice plates, or tube extensions containing porous baffles or lined duct elements. The rigid termination was a 1.5 cm thick steel plate, and the orifice plates were cut from 2.5 mm aluminum stock.

The transducer (piezoelectric) was flush mounted with the interior wall in the shock tube with a resonance frequency of 500 kHz and an excellent transient response (no ringing); it was designed for the study of shock waves. The diameter was 5.5 mm, which determined the 'resolution.' With a shock wave speed of approximately 480 m/sec, the travel time over the transducer was 12 microseconds, which sets an upper limit of 86 kHz on the meaningful sampling rate of the signal from the transducer.

10.9.2   Experimental Data

The studies included a series of measurements dealing with the interaction of a shock wave with a flexible porous material. The peak pressures of the waves ranged from 0.33 to 1.4 atm, as measured at a distance of 1 m from the source.

The pulse reflected from a rigid wall is labeled A on the left in Fig. 10.18. The

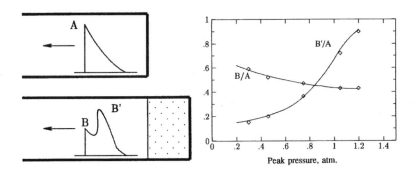

Figure 10.18: Left: Shock wave reflections from a flexible porous layer. The first reflection, amplitude B, is from the surface of the material and the second, amplitude B,' is interpreted as coming from the rigid backing. The amplitude A refers to reflection from a rigid backing without a porous layer.

Right: The pressure reflection coefficients B'/A and B/A versus incident pulse incident peak pressure, measured approximately 1 m from the termination. Material: Solimide, layer thickness=8", flow resistance=0.61 ρc per cm, and mass density=0.030 g/cm^3.

incident pulse has the same form but is traveling in the opposite direction, of course. The pulse reflected from a flexible porous layer actually consists of two components, labeled B and B′ in the figure. They are interpreted as reflections from the surface of the material and from the rigid backing, respectively. The amplitude ratios B/A and B'/A are defined as the corresponding pressure reflection coefficients. These coefficients, obtained at different values of the incident pressure peak amplitude, are shown on the right in the figure. As the amplitude increases, the reflection from the rigid wall becomes dominant.

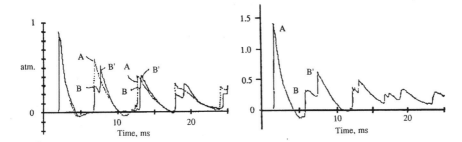

Figure 10.19: Left: Shock wave reflection from a flexible porous layer (4 inch thick layer, flow resistance, $0.61\rho c$ per cm, density, 0.030 g/cm^3). For comparison is shown the reflection from a rigid plate termination (dashed curve). Incident peak pressure, 0.9 atm ($\approx$193 dB). Right: Same as above for an 8 inch thick layer and pulse peak pressure, 1.4 atm ($\approx$197 dB).

Layer thicknesses of up to 8 inches were used and in Fig. 10.19 are shown the recorded pressure traces from the transducer as a pulse travels back and forth between the rigid wall at the driver end and the porous layer at the end of the tube. The trace on the left refers to a layer thickness of 4 inch and on the right, 8 inch. For comparison, in the left graph, is shown also the pulse (dashed line) reflected from a rigid wall termination. The incident peak amplitudes in the two cases are 0.9 and 1.4 atm at the location of the transducer 1.0 m from the membrane in the shock tube, as indicated. Accounting for the nonlinear attenuation of the pressure pulse, the 1.4 atm pulse corresponds to a pressure of 1.05 atm at the surface of the absorber.

The general character of the reflected waves are the same; the essential difference involves the time delay between B and B′ which is larger for the thicker layer, as expected. However, this time delay is much longer than would be expected from the roundtrip time in the porous layer based on the speed of sound in free field. An explanation is that the material is compressed and thus dragged along by the wave so that the effective mass density of the layer is much larger than that for air so that a reduction in the wave speed occurs.

The wave B′ has traveled back and forth through the porous layer, and from the difference in amplitudes between A and B′, we can estimate the attenuation of the wave in the layer. However, to make such comparisons accurately, we have to account for the difference in the nonlinear attenuation of the waves A and B along the path between the termination and the transducer.

The reflected waves referred to above continue toward the source where they are

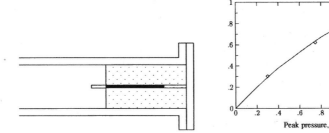

Figure 10.20: Left: Method of measuring compression. A thin rod is inserted in the material and the compression is measured by the marking made on the rod by a dye in the surface of the material.
Right: Measured relative compression of a flexible porous layer as caused by an incident shock wave. The compression is expressed as a fraction of the initial layer thickness and plotted versus the peak pressure of the incident pulse.
Material: Solimide, layer thickness=8″, flow resistance=0.61 ρc per cm, and mass density=0.03 g/cm^3.

reflected from the rigid wall and appear as the next set of pulses in Fig. 10.18. The wave B′, with a larger amplitude and consequently a higher wave speed than B, has now almost caught up with B. Similarly, A appears a little earlier than B′ because of the amplitude and wave speed difference.

To measure the possible compression of the porous material, the following experiment was done. A thin wooden rod was inserted into the material as shown in Fig. 10.20 and the surface of the material was stained with ink. A compression of the material would lead to a staining of the rod, as indicated, and the corresponding compression was determined by removing the rod and measuring the length of the unstained portion of the rod.[7] The results indeed proved that a compression took place, and its magnitude, expressed as a fraction of the initial thickness of the material, is shown as a function of the peak pressure of the incident wave in Fig. 10.20.

Because of the large compression, the flexible material is not likely to withstand repeated exposure to shock waves for an extended period of time. In the closed loop laser application mentioned in the introduction, the pulses were designed to occur at a high rate and under such conditions it is advisable to use a rigid absorption material, for example, porous metal or ceramic, to avoid acoustically induced fatigue failure within the porous layer.

---

[7]A low-budget experiment calls for simple means of measurements.

Chapter 11

Examples

1. **Forced harmonic motion of a particle subject to viscous drag**

   An electrically charged water droplet of mass m is acted on by an oscillating electric field so that the driving force on the droplet is $F = |F| \cos(\omega t)$. The viscous drag force on the particle caused by the air is αu, where u is the velocity.

   (a) What is the differential equation for the velocity $u(t)$ of the particle?

   (b) What is the corresponding equation for the complex velocity amplitude $u(\omega)$?

   (c) Solve the equation for the complex velocity amplitude and determine the amplitude (magnitude) and phase angle of the resulting velocity as functions of frequency.

   (d) What are the amplitude and phase angle of the complex displacement amplitude?

   (e) Do the above without the use of complex amplitudes.

   SOLUTION

   (a) $m\dot{u}(t) + \alpha u = |F| \cos(\omega t)$

   (b) $-i\omega m u(\omega) + \alpha u(\omega) = |F|$

   (c) $u(\omega) = F/(\alpha - i\omega m) = [|F|/\sqrt{\alpha^2 + (\omega m)^2}] e^{i\phi}$,

   where $\tan\phi = \omega m/\alpha$. (The phase angle of the denominator is $\arctan(-\omega m/\alpha)$ so that the phase angle of the inverse is $\phi = \arctan(\omega m/\alpha)$, making the phase angle of $u(\omega)$ equal to $\arctan(\omega m/\alpha)$.

   At low frequencies, $\omega m \ll \alpha$, the viscous force dominates and the velocity is approximately in phase with the driving force, $\phi \approx 0$. At high frequencies, the inertia dominates, and the velocity lags behind the force by a phase angle $\phi \approx \pi/2$ (a positive angle means a lag).

   (d) and (e) are left for you to do.

2. **Spectra**

   (a) The third octave band level of the measured velocity of an oscillator is found to be 100 dB in the frequency band centered at 100 Hz with respect to some reference value u_r. What is the corresponding spectrum density level if the spectrum density is constant over the band?

   (b) If the reference rms value of the velocity is $u_r = 10^{-4}$ cm/sec, what is the actual value of the spectrum density $E(f)$? Indicate its unit.

   (Often in presentation of data in dB, the reference value is not always defined. This is acceptable as long as comparisons of data are concerned but is not acceptable, of course, when the actual rms value of the quantity involved is desired. Even if a standard reference exists, it is not always used. Sometimes equipment manuals are confusing in this respect.)

SOLUTION

(a) According to Eq. 2.102, the bandwidth of the 1/3 octave band centered at 100 Hz is $\Delta f \approx 23$ Hz. The contribution to the mean square value from a bandwidth of one cycle is $E(f) = u^2/\Delta f$ and the corresponding spectrum density level is $10 \log[E(f)/u_r^2] = 10 \log(u^2/u_r^2) - 10 \log \Delta f = 100 - 10 \log(23) \approx 86.4$ dB.

(b) With $10 \log(E(f)/u_r^2) = L = 86.4$, we get

$$E(f) = u_r^2 \, 10^{8.68} = 10^{-8} \times 10^{8.68} = 4.70 \quad \sec(\text{cm/sec})^2.$$

3. Frequency dependent spectrum density

Suppose the spectrum density of the acceleration spectrum is known to increase as the square of the frequency in a certain region of the spectrum. What is the difference in the octave band acceleration levels in two adjacent octave bands in this region?

SOLUTION

The spectrum density is expressed as $E(f) = Af^2$, where A is a constant. Let the lower and upper frequencies of an octave be f_1 and $2f_1$.

The mean square value in the band is then

$$\int_{f_1}^{2f_1} E(f)df = A \int_{f_1}^{2f_1} f^2 \, df = A\, \frac{7}{3}\, f_1^3. \tag{11.1}$$

The first band starts at f_1 and the second, at $2f_1$. Therefore, the level difference is $10 \log(2f_1/f_1)^3 = 10 \log(8) \approx 9$ dB.

4. Springs in series and in parallel

Two spring with the same relaxed lengths have the spring constants $K_1 = K$ and $K_2 = 1.5K$, where $K = 10^5$ N/m. With these springs and a mass $M = 10$ kg, construct four oscillators, each with the upper end(s) of the spring(s) held fixed and for each determine the static deflection and the frequency of oscillation.

SOLUTION

The gravitational force Mg produces a static extension of a spring

$$\xi_{st} = Mg/K = g/\omega_0^2,$$

where $\omega_0 = \sqrt{K/M}$ is the (angular) frequency of the oscillator, K and M being the spring constant and the mass.

Incidentally, we note that the frequency can be expressed in terms of the static deflection,

$$\omega_0 = \sqrt{g/\xi_{st}}.$$

The mass combined with each of the individual springs yields two oscillators with spring constants

$$K_1 = K$$

and

$$K_2 = 1.5\, K.$$

With the springs in parallel, the resulting spring constant will be

$$K_3 = K_1 + K_2 = 2.5\, K,$$

and with the springs in series, it will be

$$K_4 = K_1 K_2 / (K_1 + K_2) = (3/5)\,K.$$

The frequencies of the corresponding oscillators are then found to be 100, 122, 158, and 77.5 Hz.

The static deflection of the first oscillator is

$$\xi_{st,1} = 1 \times 9.81/10^5 = 9.81 \times 10^{-5}\,m,$$

etc.

5. **Motion in the gravitational field inside a sphere**

Imagine a straight tunnel through the center of the Earth, which is here regarded as a sphere with a uniform density ρ. A particle is dropped into the tunnel from the surface at $t = 0$. (Neglect friction)

(a) Show that the subsequent motion will be harmonic.

(b) When will the particle reach the other end of the tunnel?

(c) Show that the motion will be harmonic even if the tunnel does not go through the center of the Earth and that the period will be the same as before.

SOLUTION

(a) At the surface of the Earth (radius R), the gravitational force on a mass m is $mg = G(M/R^2)m$, where M is the mass of the Earth and G the gravitational constant. At a distance ξ from the center, the mass inside the sphere of radius ξ is $M(\xi) = M(\xi/R)^3$, and the force on m will be

$$F(\xi) = GmM(\xi)/\xi^2 = (GMm/R^3)\xi.$$

In other words, the restoring force will be proportional to the displacement ξ from the center of the Earth and, therefore, the free motion of m will be harmonic. The angular frequency of oscillation becomes

$$\omega = (GM/R^3)^{1/2}.$$

(b) The particle will reach the other side after the time $T/2$, where T is the period of oscillation

$$T = 2\pi/\omega = 2\pi(R^3/GM)^{1/2}.$$

Using the data $R = 6.4 \times 10^6$ m, $M = 6 \times 10^{24}$ kg, $G = 6.67 \times 10^{-11}\,Nm^2/(kg)^2$ we find $T/2 \approx 2500$ sec.

(c) With the tunnel not going through the center of the Earth, let the displacement of m from the center of the tunnel be ξ and the distance to the center of the Earth at that position be $R(\xi)$. The force on m toward the center of the Earth is then $F = (GMm/R^3)R(\xi)$, by analogy with the discussion in (a). The component of this force along the tunnel, however, will be $(\xi/R(\xi))F$, which is proportional to ξ with the same constant of proportionality as before. Again, the motion will be harmonic, and the period is the same as before.

6. **Thomson model of the atom. Plasma oscillations**

J. J. Thomson (British physicist, 1856-1940) imagined the atom as a swarm of electrons contained within a uniform spherical distribution of positive charge equal in magnitude to the total charge of the electrons. In this atomic model, consider the case of one

electron within the sphere under the influence of the electric field of the uniform positive
charge distribution (Hydrogen atom).
(a) Show that the motion is harmonic.
(b) If the sphere diameter chosen is one Ångström ($= 10^{-10}$ m), what then would be
the frequency of oscillation of the electron?

(J. J. Thomson's son, G. P. Thomson, was one of the pioneers in establishing the **wave**
nature of the electron. His father's work dealt with the **particle** nature of the electron.)

SOLUTION

Let the total electric charge of the sphere be Q and the radius R. Since the charge
distribution is uniform, the charge within a sphere of radius ξ will be

$$Q(\xi) = Q(\xi/R)^3.$$

Then, according to Coulomb's law, the force on a negative charge of magnitude q, a
distance ξ from the center, will be

$$F(\xi) = (1/(4\pi\epsilon_0)(Q(\xi)q/\xi^2)$$

directed toward the center. (ϵ_0: permittivity constant $= 8.85 \times 10^{-12}$ F/m).

Combining the equations, we note that the 'restoring' force F on the charge q is proportional to the displacement ξ from the origin, and, therefore, the motion of q will be
harmonic with the frequency of oscillation

$$\omega_0 = [1/(4\pi\epsilon_0)](Qq/mR^3)^{1/2},$$

where m is the mass of the particle (electron) of charge q.
(b) Using the value $R = 10^{-10}$ m and the electric charge and mass of an electron,
$Q = q = e = 1.6 \times 10^{-19}$ Coul, $m = 9.1 \times 10^{-31}$ kg, and $1/(\epsilon_0 4\pi) \approx 9 \times 10^9$ m/Farad,
we obtain $f_0 = \omega_0/(2\pi) \approx 10^{15}$ 1/sec, which is of the same order of magnitude as the
frequency of radiation from Hydrogen.

Plasma oscillations. Although the Thomson model of the atom was able to explain
radiation from an atom (due to the oscillating electron), the radiation contained only a
single frequency and not a spectrum of frequencies, as observed. The model, however,
explains the oscillations in ionized gases, plasmas, known as *plasma oscillations*.

In a plasma (fully ionized gas) in thermal equilibrium the speed of the electrons is much
greater than the speed of the much heavier ions. Therefore, it is a good approximation
to consider the ions to act as a stationary background of positive charge in which the
electrons are moving. Thus, the plasma is like the Thomson model of the atom, and
Thomson type electron oscillations are to be expected. Indeed, such (plasma) oscillations
occur and are known to have a frequency given by

$$\omega_p = \sqrt{N_0 e^2/\epsilon_0 m},$$

where N_0 is the number of electrons per m^3, e the electron charge, and m its mass.

7. Thermal vibrations

The Young's modulus of a solid is defined by the relation $\sigma = Y\Delta L/L$, where σ is the
stress, i.e., force per unit area of a uniform rod, and $\Delta L/L$ the strain, i.e., the relative
change in the length of the rod produced by the stress. For steel, $Y \approx 2 \times 10^{11}$ N/m^2.

(a) What is the corresponding 'spring' constant of a bar of length 1 m and a cross-sectional area of 1 cm^2?

(b) Imagine the bar composed of parallel atomic 'strings' consisting of mass element (atoms) connected by springs. The atoms are assumed to be arranged in a cubical lattice with an interatomic distance $d = 10^{-8}$ cm. What then is the spring constant of one of the strings?

(c) The density of steel is $\rho \approx 7.8$ g/cm^3.

(b) From the result estimate the frequency of thermal atomic oscillations.

SOLUTION

(a) Let the area of the rod be A. The force F corresponding to the stress σ is then $F = A\sigma$. The extension ΔL produced by the force is then given by $\sigma = F/A = Y(\Delta L/L)$. By definition of the spring constant $K = F/\Delta L$ we obtain

$$K = YA/L.$$

(b) The number of atomic 'strings' in an area A is $n = A/d^2$ and the spring constant per string is

$$K_1 = K/n = Y(d^2/L).$$

(c) The number of interatomic 'springs' which make up one atomic string is L/d. These springs are all in 'series,' and the spring constant of each of these springs is then

$$k = (L/d)K_1 = Yd.$$

There are $1/d^3$ atoms per unit volume and the relation between the mass m of an atom and the mass density ρ is $m = d^3\rho$.

The frequency of oscillation of an atom is $\omega_0 = \sqrt{k/m}$, and introducing the expressions for k and m obtained above, we get

$$\omega_0 = \sqrt{Y/d^2\rho}.$$

Using the numerical values given above, we find $\omega \approx 5 \times 10^{13}$ sec^{-1}, in good agreement with the upper range of the thermal spectrum of oscillation.

In reality, there is a broad range of frequencies of oscillation corresponding to the various modes of oscillation of the particle system constituting the solid. The frequency obtained here represents the upper limit of this frequency range. For further details about thermal oscillations, we refer to the treatment of the specific heat of solids in standard physics texts.

8. **Initial value problem and energy considerations**

Consider a horizontal mass-spring oscillator (mass M, spring constant K) on a frictionless table. The end of the spring is attached to a rigid wall. At $t = 0$ a bullet of mass m is shot at a speed v into the block in the direction of the spring. The collision takes place in a very short time, so that the spring can be considered to be relaxed during the collision.

(a) Determine the subsequent time dependence of the displacement of M.

(b) What is the total mechanical energy of oscillation?

(c) What is the energy lost in the collision between the bullet and the block?

(d) When should a second bullet be shot into the block in order to increase (decrease) the subsequent amplitude of oscillation as much as possible? In each case, determine

ACOUSTICS

the new expression for the time dependence of the motion (both amplitude and phase) using the same origin of time t as before.

(e) With $M = 1$ kg, $m = 5$ g, $v = 300$ m/sec, and $K = 400$ N/m, determine the numerical answers in (a) and (b).

SOLUTION

(a) The momentum of the bullet p=mv represents the total momentum of the system both before and after the collision. Thus, the velocity V_1 of the block after the collision is given by $(M + m)V_1 = mv$ so that $V_1 = mv/(M + m)$.

The initial conditions for displacement and velocity ($t = 0$) in the subsequent oscillation of the mass-spring oscillator are $\xi_1(0) = 0$ and $v_1(0) = V_1$. The corresponding velocity and displacement of the oscillator are then

$$v_1(t) = V_1 \cos(\omega t)$$
$$\xi_1(t) = (V_1/\omega) \sin(\omega t).$$

(b) Since the initial displacement is zero, the initial total energy is equal to the initial (maximum) kinetic energy of the oscillator

$$\frac{1}{2}(M + m)V_1^2 = \frac{1}{2}(mv^2/2)[m/(M + m)].$$

Neglecting friction, this total energy will be conserved during the oscillations.

(c) The mechanical energy lost in the (inelastic) collision between the bullet and the block is $mv^2/2 - (M + m)V_1^2/2 = (mv^2/2)[M/(M + m)]$.

(d) At the moment of the second collision, at time t, the momentum of the block is $p_1(t) = mv_1(t)$ and the total momentum after the collision is $p + p_1(t)$, where, as before, p=mv is the momentum of the bullet. The largest velocity amplitude after the collision is obtained when $p_1(t) = +p$, i.e., when the block has maximum velocity in the same direction as the bullet. The velocity of the oscillator after the collision then will be $V_2 = 2p/(M + 2m)$, with two bullets embedded in the block.

The expressions for the subsequent velocity and displacement will be the same as in Eqs. 11.1 and 11.2 with V_1 replaced by V_2.

If $p_1(t) = -p$ at the second shot, the oscillator will be stopped.

(e) The numerical solutions are
for (a): $v_1(t) \approx 1.5 \sin(\omega_1 t)\, m/sec$,
for (b): 1.12 joule and
for (d): $v_2(t) \approx 3.0 \sin(\omega_2 t)$ m/sec, where

$$\omega_1 = \sqrt{K/(M + m)} \approx \sqrt{K/M}\,(1 - m/M) \approx 20\,sec$$
$$\omega_2 = \sqrt{K/(M + 2m)} \approx \sqrt{K/M}\,(1 - 2m/M)$$

with $m/M = 0.005$ and $\sqrt{K/M} = 20$ sec

9. **Complex frequencies**

Solve the frequency equation for the free oscillation of a mass-spring oscillator with $M = 1$ kg and $K = 100$ N/m for the following values of γ:
(a) 0.5, (b) 1, (c) 2.
In each case indicate the location of the complex frequencies in a complex frequency plane and determine the decay rate and period of oscillation.

SOLUTION

The frequency equation is

$$\omega^2 + i2\gamma\omega - \omega_0^2 = 0 \tag{11.2}$$

with the solution

$$\omega = -i\gamma \pm \sqrt{\omega_0^2 - \gamma^2}, \tag{11.3}$$

where $\gamma = R/2M$. Numerically, with $\omega_0 = \sqrt{100/1} = 10\,\text{sec}^{-1}$, we obtain $\gamma = 5,\ 10,\ 20$ for $\gamma = 0.5,\ 1,\ 2\omega_0$. The corresponding values for the complex frequencies, obtained from Eq. 11.56, are then

$$\omega_1 = -i5 \pm 10\sqrt{3/4},\ -i10 \pm 0,\ -i20 \pm i10\sqrt{3}. \tag{11.4}$$

In the case of oscillatory decaying motion, the actual physical frequency has the value

$$\omega_0' = \sqrt{\omega_0^2 - \gamma^2} \tag{11.5}$$

with a corresponding period

$$T' = \frac{2\pi}{\omega'}. \tag{11.6}$$

In the present case the motion will be oscillatory only for $\gamma = 0.5$, and the period is $2\pi/(10\sqrt{3/4}) = 0.73\,\text{sec}^{-1}$. The decay constant is $\gamma = 5\,\text{sec}^{-1}$, corresponding to a 'life-time' or decay time of the oscillation $\tau = 1/\gamma = 1.4\,\text{sec}$.

With $\gamma = \omega_0$, the oscillator is critically damped, the decay constant being $\gamma = \omega_0$, and for $\gamma = 2\omega_0$, the oscillator is overdamped with the two decay constants $\gamma_1 = 20 + 10\sqrt{3}$ and $\gamma_2 = 20 - 10\sqrt{3}$.

Comments

The fact that there are generally two solutions for the complex frequency, as expressed by the plus and minus signs in front of the square root, is merely a reminder that there are two independent solutions for the displacements and that the general solution is a linear combination of the two. This leads to an expression for the general displacement which contains two constants, which have to be adjusted to meet the initial conditions, as explained in the text.

10. **Frequency response. Maximum displacement amplitude**

Consider the frequency dependence of the displacement amplitude of a mass-spring oscillator, driven by a harmonic force $|F|\cos(\omega t)$.
(a) Determine the maximum value of the displacement amplitude and the corresponding value of the normalized frequency $\Omega = \omega/\omega_0$, where $\omega_0 = \sqrt{K/M}$.
(b) For what value of the damping factor $D = R/M\omega_0$ will a maximum occur at zero frequency?
(c) Discuss the conditions for maximum velocity amplitude.

SOLUTION

(a) Let $|\xi'| = |F|/K$, where $|F|$ is the amplitude of the driving force and K the spring constant. As shown in Chapter 2, the frequency dependence of the displacement amplitude $|\xi|$ is then given by

$$|\xi|/|\xi'| = \frac{1}{\sqrt{(1 - \Omega^2)^2 + D\Omega^2}}. \tag{11.7}$$

The maximum (minimum) value of the displacement is obtained when the quantity in the denominator is a minimum (maximum), i.e., when $dG/d\Omega = 0$, where $G = (1 - \Omega^2)^2 + (D\Omega)^2$.

Carrying out the differentiation, we obtain

$$dG/d\Omega = -4\Omega(1 - \Omega^2)\Omega + 2D^2\Omega = 0.$$

The solution to this equation is $\Omega_1 = 0$ or the value of Ω that satisfies the equation $\Omega^2 + D^2 - 2 = 0$, which is $\Omega_2^2 = 1 - (D^2/2)$. If $D^2 < 2$, the value $\Omega = \Omega_1$ corresponds to a minimum and $\Omega = \Omega_2$ to a maximum.

Insertion of $\Omega = \Omega_2$ into Eq. 11.7 yields the maximum value of the displacement amplitude

$$|\xi|_m/|\xi'| = \frac{1}{D\sqrt{1 - (D/2)^2}}$$

and the minimum value, for $\Omega = 0$, is simply ξ'.

(b) If $D^2 > 2$, the frequency Ω^2 becomes imaginary, and the only real solution is $\Omega = 0$. This corresponds to and overdamped oscillator, with the maximum amplitude ξ' at $\Omega = 0$.

(c) The maximum of the displacement amplitude occurs at a frequency somewhat below $\Omega = 1$ (by an amount that depends on the damping), whereas the maximum of the velocity amplitude always occurs at $\Omega = 1$, regardless of the damping. If the damping is small, the difference between the two frequencies generally is negligible and vanishes if there is no damping. That the frequencies *are* different for a damped oscillator, however, has lead to some confusion as to the meaning of 'resonance' and resonance frequency. We define the resonance frequency to correspond to $\Omega = 1$, for which the *velocity* amplitude is a maximum, and we have to keep in mind that the maximum of the displacement amplitude in a damped oscillator occurs at a lower frequency.

In an electrical circuit this ambiguity rarely is encountered, since we are generally interested only in the current amplitude, which is analogous to the velocity. The electrical quantity, which is analogous to the displacement, is the electric charge, and the amplitude of oscillation of it generally is of little interest.

11. **Forced harmonic motion. Centrifugal fan**

A centrifugal fan with its motor and concrete slab foundation (total mass of system is 1000 kg) is mounted on springs at the corners of the slab. The static compression of the springs is 1 cm. Due to an imbalance in the fan, the system is oscillating in the vertical direction. The rotational speed of the fan is 1200 rpm (revolutions per minute) and the damping factor $D = R/M\omega_0$ is 0.15. The amplitude of the vertical force component due to the imbalance is .001 of the total weight of the system.

(a) What is the steady state displacement amplitude of the system?

(b) What will be the amplitude if the number of springs under the concrete slab is doubled (the springs are in parallel)?

(c) What is the purpose of the springs in the first place?

SOLUTION

(a) The static displacement of the system is $\xi' = Mg/K = g/\omega_0^2$, and we can express the angular resonance frequency in terms of it,

$$\omega_0 = \sqrt{g/\xi'}. \tag{11.8}$$

Introducing the known value for the static displacement, 1 cm, we obtain $\omega_0 = 31.3\,sec^{-1}$. The forcing frequency is $1200/60 = 20$ Hz and the corresponding angular frequency is $\omega = 2\pi 20 = 125.7$ 1/sec. Thus, the normalized forcing frequency is $\Omega = \omega/\omega_0 = 4.0$. With $|\xi'| = F/K$, the displacement amplitude is given by

$$|\xi|/|\xi'| = \frac{1}{\sqrt{(1 - \Omega^2)^2 + (D\Omega)^2}}. \tag{11.9}$$

With the force amplitude $|F| = .001 \times Mg$, we get $|F|/K = .001 \times \xi'$, and, with $\Omega = 4$, the displacement amplitude becomes

$$|\xi| \approx .001 \times \xi'/16 \quad \text{cm}.$$

(b) Doubling of the number of springs (assumed to be in parallel) increases the equivalent spring constant by a factor of 2 and the resonance frequency by a factor of $\sqrt{2}$. The normalized frequency is then reduced by the same factor, and the amplitude of oscillation becomes

$$|\xi| \approx .001 \cdot \xi'/7 \quad \text{cm},$$

where $\xi' = 1$ cm.

At zero frequency, the amplitude is controlled by the spring stiffness, which plays the dominant role at frequencies below resonance. As the frequency increases, the influence of the inertia of the system in effect reduces the spring stiffness, (the inertial force and the spring force are 180 degrees out of phase) and the amplitude increases. At resonance, the spring force and the inertial mass force cancel each other and the amplitude is controlled by friction alone.

At frequencies above the resonance frequency, the oscillator is mass controlled; the inertia of the system is the major factor controlling the amplitude and the amplitude decreases with increasing frequency.

(c) The reason for using springs under a piece of equipment is to reduce the amplitude of the oscillatory force transmitted to the floor supporting the equipment. Neglecting friction, this force amplitude is the product of the displacement amplitude and the spring constant. A reduction of the force amplitude is obtained only if the forcing frequency is larger than the resonance frequency of the system by a factor $\sqrt{2}$. In the present case, the introduction of additional springs reduces the normalized frequency from 4 to $4/\sqrt{2} \approx 2.8$ with a corresponding increase in the vibration amplitude and the amplitude of the force transmitted to the floor.

In addition to the vertical mode of oscillation considered here, there are other possible modes, corresponding to the remaining degrees of freedom of the system. In this case there are two horizontal translation modes of motion and three rotational modes, which correspond to oscillatory motions about two horizontal axes and the vertical axis. These must be considered also in a thorough analysis of the response of the system to oscillatory forces.

12. **Forced motion of an oscillator by non-harmonic force**

A damped oscillator (mass M, resistance R, spring constant K) is driven by a force $f(t) = A \exp(-\gamma t)$ for $t > 0$ and $f(t) = 0$ for $t < 0$, where $\gamma = R/2M$.
(a) Calculate the resulting displacement as a function of time.
(b) Determine the maximum displacement and the time when it occurs.
(c) Discuss separately the case when the oscillator is critically damped.
(d) Do the same for the overdamped oscillator.

SOLUTION

The general expression for the response of the oscillator to an arbitrary driving force $F(t)$ in Eq. 2.53 involves an integration from $-\infty$ to t to account for all impulses in the past. In this case, there is no force until $t = 0$ so that integration starts at $t = 0$, i.e.,

$$\xi(t) = \int_0^t F(t')h(t, t')dt',$$

where

$$h(t, t') = (1/\omega_0'M)e^{-\gamma(t-t')}\sin[\omega_0'(t - t')]$$

is the impulse-response function and $\omega_0' = \sqrt{\omega_0^2 - \gamma^2}$.

It is important to realize that t' is the variable of integration and t is to be regarded as a parameter, representing the upper limit of integration and the time of observation.

In this case, with $F(t') = Ae^{-\gamma t'}$, the integral in reduces to

$$\xi(t) = (A/\omega'M)e^{-\gamma t}\int_0^t \sin[\omega_0'(t - t')]dt'$$

or

$$\xi(t) = (A/\omega_0'^2 M)\,e^{-\gamma t}\,[1 - \cos(\omega_0't)],$$

where $1 - \cos(\omega_0't)$ can be expressed as $2\sin^2(\omega_0't_1/2)$.

Note that $\xi(0) = \xi(\infty) = 0$, as it should.

(b) The maximum value of the displacement is $(A/\omega_0^2 M)\exp(-\gamma t_1)$ at the time $t_1 = (2/\omega_0')\arctan(\omega_0'/\gamma)$.

(c) Critical damping corresponds to $\omega_0' = 0$, and the expression for the displacement then reduces to $\xi = (A/2M)\,t^2\,e^{-\gamma t}$.

(d) For the overdamped oscillator, with $\gamma > \omega_0$ and $\omega' = i\sqrt{\gamma^2 - \omega_0^2} = i\epsilon$, we get

$$\xi = (2A/\epsilon^2 M)\exp(-\gamma t)\sinh^2(\epsilon t/2).$$

13. **Beats between steady state and transient motion of an oscillator**

   Refer to the discussion of an oscillator driven by a harmonic force which starts at $t = 0$ and is zero for $t < 0$. For a given difference in frequency between the steady state and transient motions, discuss the influence of the damping factor $D = R/(\omega_0 M)$ on the amplitude variation in the beats resulting from the interference of these motions.

SOLUTION

The frequencies ω and ω_0' in Eq. 2.56 are generally different and if they are sufficiently close, low-frequency beats can be observed, as long as the transient motion has an amplitude which is not negligible compared to the steady state amplitude. The amplitude of the transient is determined largely by the factor $\exp(-\gamma T_b)$, where T_b is the beat period. Consequently, in order for the beats to be clearly seen, the product γT_b cannot be too large. This means, that even if the damping is rather small, it will be difficult to see beats with very long periods (i.e., when the driving frequency is very close to the frequency of free oscillation).

The maximum amplitude occurs when the two motions are in phase and the minimum when they are 180 degrees out of phase. The ratio of the two is

$$r = \frac{1 + e^{-\gamma T_b}}{1 - e^{-\gamma T_b}}.$$

Thus, in order for this ratio to be larger than a prescribed value r_1, the following condition must be obtained

$$\gamma T_b < \log \frac{1 + r_1}{1 - r_1}.$$

Introducing the expression for $T_b = 2\pi / |\omega_0|\Omega - 1||$, given above, and $\gamma = \omega_0 D/2$, the corresponding condition on the damping parameter $D = R/(\omega_0 M)$ is

$$D < \frac{1}{\pi |\Omega - 1|} \log \frac{1 + r_1}{1 - r_1}.$$

In the present case, we have $r_1 = 0.8$, so that $D < 2.2/(\pi |\Omega - 1|)$.

14. **Nonlinear oscillator**

In most mechanical systems, a small displacement of a particle from its equilibrium position will result in a restoring force proportional to the displacement. There are exceptions, however, and a simple example is a particle of mass M which is attached to the center of a horizontal spring which is clamped at both ends. The spring, with a spring constant K, has a length $2L$ and is initially slack. The body is set in motion perpendicular to the spring (neglect friction).
(a) What is the potential energy of the particle $V(\xi)$, where ξ is the displacement?
(b) What is the restoring force for small displacements, $\xi \ll L$? Will the motion be harmonic?
(c) Qualitatively, how does the period depend on the amplitude of oscillation?

SOLUTION

(a) A transverse displacement ξ increases the length of the spring to $2L' = 2\sqrt{L^2 + \xi^2}$ so that the elongation becomes $\Delta = 2L' - 2L$ and the potential energy

$$V(\xi) = (1/2)K\Delta^2,$$

where K is the spring constant.
For small displacements, $\xi \ll L$, we get $\Delta \approx L(\xi/L)^2$ and the potential energy

$$V(\xi) \approx (1/2)KL^2(\xi/L)^4.$$

(b) The restoring force on the particle can be obtained simply as $F = dV/d\xi$, which, for small displacements, becomes

$$F(\xi) \approx (2K/L^2)\xi^3.$$

(c) For a harmonic oscillator the potential energy is proportional to the second power of ξ and the restoring force to the first power. This is not the case for this oscillator, even at small amplitudes; the transverse oscillatory motion will not be harmonic.

One way of looking at the problem is to say that the equivalent spring constant increases with amplitude. Consequently, we expect the period of oscillation to decrease with increasing amplitude.

(d) For a pendulum the potential energy is $MgL(1 - \cos(\theta))$, where θ is the angle of deflection. The restoring torque is $MgL \sin(\theta)$. It has a weaker than linear dependence of θ. Consequently, the equivalent spring constant will decrease, and the period will increase with increasing amplitude.

15. **Spring pendulum**

A body, hung at the end of a vertical spring, stretches the spring statically to twice its initial length. This system can be set into oscillation either as a simple pendulum or as a mass-spring oscillator (in the pendulum mode, assume the length of the spring to be constant). Determine the ratio of the periods of these motions.

SOLUTION

Let the original length be L. The extension of the spring by the gravitational force Mg is then

$$L = Mg/K = g/\omega_0{}^2,$$

where $\omega_0 = \sqrt{K/M}$.

For the pendulum with a length 2L, the angular frequency of oscillation is

$$\omega_p = \sqrt{g/2L} = \omega_0/\sqrt{2}.$$

If the length of the spring is not constant, there will be a coupling between the axial and pendulum modes of oscillation. Both centrifugal and Coriolis forces have to be accounted for.

Problems of this kind, involving several degrees of freedom, are best solved in a systematic and unified manner by using the Lagrangian formulation of mechanics.

16. **Wave kinematics; harmonic wave**

The end of a string is driven at $x = 0$ with the transverse displacement $\eta(0, t) = \eta_0 \cos(\omega t)$ with the frequency 10 Hz and an amplitude 0.2 m. The wave speed is 10 m/sec. Determine

(a) the displacement as a function of time at $x = 1$ m.
(b) the shape of the string at $t = 0.5$ sec. What is the wavelength?
(c) the velocity and accelerations waves at $x = 1$ m.
(d) the phase difference between the harmonic oscillations at $x = 0$ and $x = 0.2$ m.

SOLUTION

The angular velocity is $\omega = 2\pi f = 2\pi 10 = 20\pi$.

(a) $\eta(1, t) = 0.2 \cos(20\pi t - 2\pi) = 0.2 \cos(20\pi t)$.

(b) $\eta(x, 0.5) = 0.2 \cos(20\pi 0.5 - 20\pi x/10) = 0.2 \cos(10\pi - 2\pi x) = 0.2 \cos(2\pi x)$
The wavelength is $\lambda = vT = v/f = 10/10 = 1$ m

(c) The velocity and acceleration are

$$u(x, t) = \partial \eta(x, t)/\partial t = -\eta_0 \omega \sin[\omega(t - x/v)] = -4 \sin(2\pi t - \pi/5)$$
$$a(x, t) = \partial^2 \eta(x, t)/\partial t^2 = -\omega^2 \eta(x, t) = 80\pi^2 \cos(2\pi t - \pi/5),$$

where the units are m/sec and m/sec^2, respectively.

(d) The phase angle at x is $\phi(x) = \omega x/v$. Thus, $\phi(0.2) - \phi(0) = 2\pi 0.2/10 = 0.4\pi$.

17. **Wave pulse on a string**

The transverse wave speed on a stretched string is 10 m/sec. The transverse displacement at x=0 is $\eta(0, t) = 0.1 \cdot (t^2 - t^3)$ m for $0 < t < 1.0$ sec and zero at all other times (time t is measured in seconds).

(a) Plot the transverse displacement as a function of t at $x = 0$.
(b) Plot the transverse displacement as a function of x at $t = 1.0$ sec.
(c) What is the mathematical expression for the displacement as a function of time at $x = 10$ m? What are the displacements at this point at $t = 1$, 1.5 and 3 sec?

(d) What is the transverse velocity of the string at $x = 10$ m and $t = 1.5$ sec?

(e) What is the slope of the string at $x = 10$ m and $t = 1.5$ sec?

SOLUTION

(a) The initial 'pulse' is shown on the left in the figure. The maximum displacement occurs at $t = 2/3$ sec and has the value $\eta_{max} = 0.4/27$ m.

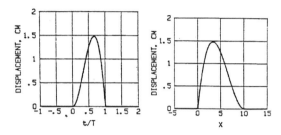

Figure 11.1: The transverse displacement of a string. Left: versus t at $x = 0$. Right: versus x at $t = 1.0$ sec.

(b) A wave traveling in the positive x-direction is such that if the displacement at $x = 0$ is $\eta(0, t)$, the displacement will be the same at the position x at the time $t - (x/v)$, where v is the wave speed and x/v the travel time of the wave from $x = 0$ to x. Expressed mathematically

$$\eta(x, t) = \eta(0, t - x/v).$$

In this particular case we get

$$\eta(x, t) = 0.1\big((t - x/v)^2 - (t - x/v)^3\big),$$

where $t = 1$ sec. The particular wave pulse considered here is such that the displacement is different from zero only if the argument fulfills the condition $0 < t - x/v < 1$. With $t = 1$, this means that $0 < x < v \cdot 1$. This formula merely expresses the obvious fact that at time t the front of the pulse has reached the location $x = vt$ and the tail the position $x = v(t - 1) = 0$, since $t = 1$. With $v = 10$, the front of the pulse will be located between $x = 0$ and $x = 10$ m, as shown on the right in the figure.

The shapes of the t- and x-dependence of the displacement are essentially mirror images of one another, as discussed further below.

(c) At $x = 1$, the time dependence of the displacement is $\eta(10, t) = 0.1[(t - 1)^2 - (t - 1)^3]$, since $x/v = 10/10 = 1$ sec. At $t = 1$ we get $\eta(10, 1) = 0$. This merely means that the front of the pulse has just reached $x = 10$ at $t = 1$. At $t = 1.5$, the argument in the wave function is 0.5 and the displacement is $\eta(10, 1.5) = .0125$ m. At $t = 3$ the argument is 2, i.e., outside the range $0 - 1$. This means that at $t = 3$ the pulse has passed $x = 10$, and the displacement is zero.

(d) The transverse velocity of the string at a fixed location x is $\partial\eta(x, t)/\partial t$. With $\eta(x, t)$ given above, we obtain

$$u(x, t) = \partial\eta(x, t)/\partial t = 0.1[2(t - x/v) - 3(t - x/v)^2].$$

Inserting $x = 10$ and $t = 1.5$, we get $t - x/v = 1.5 - 1 = 0.5$.

$$u(10, 1.5) = 0.1(2 \cdot 0.5 - 3 \cdot 0.25) = 0.025 \text{ m}$$

(e) The slope of the string at a fixed time t is $\partial\eta(x, t)/\partial x$. Since the argument of the wave function for a single wave traveling in the positive x-direction is $t - x/v$, the partial derivatives with respect to t and x are related,

$$\partial\eta/\partial x = -(1/v)\partial\eta/\partial t.$$

Therefore, using the result in (d), we obtain $\partial\eta/\partial x = -(1/v)0.025 = -0.0025$ at $x = 10$ and $t = 1.5$.

18. **Sum of traveling waves**

Consider two harmonic waves $\xi_1 = A\cos(\omega t - kx)$ and $\xi_2 = 2A\cos(\omega t - kx - \pi/4)$. Prove that the sum is a traveling wave $B\cos(\omega t - kx - \phi)$ and determine B and ϕ.

SOLUTION

We denote $\omega t - kx$ by Ψ. One way of solving the problem is to express ξ_1 and ξ_2 as $\xi_1 = A\cos(\omega t - kx) = \cos(\Psi)$ and $\xi_2 = 2A\cos\Psi\cos(\pi/4) + 2A\sin\Psi\sin(\pi/4) = \sqrt{2}A\cos\Psi + \sqrt{2}A\sin\Psi$. The sum is then

$$\xi_1 + \xi_2 = A(1 + \sqrt{2})\cos\Psi + \sqrt{2}A\sin\Psi). \qquad (11.10)$$

To put this into the desired form $B\cos(\Psi - \phi)$, we rewrite it as

$$B\cos\Psi\cos\phi + B\sin\Psi\sin\phi. \qquad (11.11)$$

Comparison with Eq. 11.55 yields

$$B\cos\phi = (1 + \sqrt{2})A$$
$$B\sin\phi = \sqrt{2}A.$$

From these relations follows

$$B = A\sqrt{(1 + \sqrt{2})^2 + 2}$$

and

$$\tan\phi = \sqrt{2}/(1 + \sqrt{2}), \qquad (11.12)$$

which proves that the sum is a traveling wave with the amplitude B and the phase angle ϕ.

19. **Sum of traveling waves once again**

Reconsider Example 18, now with the use of complex amplitudes. These amplitudes are $\xi_1(x, \omega) = A\exp(ikx)$ and $\xi_2(x, \omega) = 2A\exp(ikx)\exp(i\pi/4)$. The sum is a wave with the amplitude

$$\xi(x, \omega) = A[1 + 2e^{i\pi/4}]e^{ikx} = A[1 + \sqrt{2} + i\sqrt{2}]e^{ikx}, \qquad (11.13)$$

which can be written $\xi(\omega) = B\exp(i\phi)e^{ikx}$ where

$$B = A\sqrt{(1 + \sqrt{2})^2 + 2} \qquad (11.14)$$

and

$$\tan\phi = 2/(1 + \sqrt{2}) \qquad (11.15)$$

which is the same as before, but obtained in a much more straight-forward manner. The real wave function is $\xi(x, t) = B\cos(\omega t - kx - \phi)$.

20. **Wave diagram; moving source and Doppler effect**

A source of sound emits short pulses at a rate of f per second as it moves through the air in the x-direction. The duration τ of each pulse is short compared to the time between pulses $T = 1/f$.

Indicate in a wave diagram the wave lines representing the waves emitted in the positive and negative x-directions. From the diagrams, determine geometrically the time between two successive pulses as recorded by observers on the x-axis ahead of and behind the source. Determine also the corresponding number of pulses per second and express the result in terms of f, the velocity U of the source, and the sound speed c when

(a) $U < c$. Subsonic motion of the source.

(b) $U > c$. Supersonic motion of the source.

(c) Is it possible that wave interference can occur in (a) or (b) between waves emitted in the two directions? If so, indicate in the (t, x)-plane the regions where such interference takes place.

SOLUTION

The wave lines corresponding to waves traveling in the positive and negative x-directions have the trajectories $x = \pm ct$ or $t = \pm x/c$ in the (t, x)-plane. The trajectory of the source is $t = x/U$.

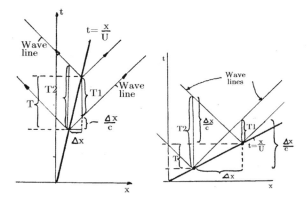

Figure 11.2: Wave lines for sound emitted from a moving source. Left: Subsonic speed. Right: Supersonic speed.

(a) If the source speed is subsonic, the source trajectory will have a slope larger than that of a wave line. The waves lines emitted in the forward and backwards direction from the source are indicated in the figure. The time interval between two successive pulses, as measured in the source frame of reference, is denoted T. The corresponding time intervals recorded by an observer ahead of and behind the source are denoted T_1 and T_2, as shown. Each of these times is obtained as the distance between the intercepts of a vertical line through x by two successive wave lines. The corresponding frequencies are $f_1 = 1/T_1$ and $f_2 = 1/T_2$.

Using the notation indicated in the figure, we obtain $T = \Delta x/U$ and

$$T_1 = T - \Delta x/c = T(1 - M)$$
$$f_1 = f/(1 - M),$$

where $M = U/c$ is the Mach number of the source.

Similarly, for an observer behind the source, we obtain

$$T_2 = T + \Delta x/c = (1 + M)T$$
$$f_2 = f/(1 + M).$$

(b) In the case of supersonic motion, the trajectory of the source is always below any of the wave lines emitted. It follows from the figure that

$$T_1 = \Delta x/c - T = T(M - 1)$$
$$f_1 = f/(M - 1),$$

where M is now greater than 1.

In a similar manner we obtain for an observer behind the source

$$T_2 = T + \Delta x/c = T(M + 1)$$
$$f_2 = f/(M + 1).$$

The shift in frequency due to the relative motion of the source and the receiver is known as the *Doppler effect*.

(c) Interference will occur where wave lines intersect each other. This can occur only when the source speed is supersonic. In that case the region above the source trajectory is one of interference between waves traveling in the positive and negative directions. An intersection between two wave lines is shown in the diagram on the right in the figure.

21. **Sound speed and molecular thermal motion**

The thermal kinetic energy of a molecule for each translational degree of freedom is $m\langle v_x^2 \rangle/2 = k_B T/2$, where k_B is the Boltzmann constant, T, the absolute temperature, and m the mass of a molecule. (The angle brackets signify average value.) What is the relation between the sound speed and the rms value of the thermal speed $\sqrt{\langle v_x^2 \rangle}$?

SOLUTION

With reference to Eq. 3.21, the speed of sound in a gas can be written

$$c = \sqrt{\gamma P/\rho},$$

where $\gamma = c_p/c_v$ (≈ 1.4 for air) is the ratio between the specific heats at constant pressure and constant volume, P the static pressure, and ρ the density.

According to the equation of state for a gas, we have

$$P = k_B n T = nm\langle v_x^2 \rangle = \rho\langle v_x^2 \rangle,$$

where n is the number of molecules per unit volume and $\rho = nm$. Using this expression for P, we get

$$c = \sqrt{\gamma \langle v_x^2 \rangle},$$

which shows that the sound speed is proportional to the root mean square value of the thermal molecular speed, larger by a factor of $\sqrt{\gamma} \approx 1.18$.

22. **Power carried by superimposed waves**

Consider two pressure waves on a transmission line, $p_1 = A\cos(\omega t - kx - \alpha)$ and $p_2 = B\cos(\omega t - kx - \beta)$. When alone, the average intensity transmitted by the first wave is I_1 and by the second, I_2.

(a) What is the average intensity when both waves are present simultaneously? Express the dependence of the intensity on the phase difference $\alpha - \beta$.

(b) If the phase difference is varied, what are the resulting maximum and minimum values of the resulting intensity?

(c) If N waves of equal amplitude are present, what then is the maximum value of the total intensity in terms of the power of a single wave?

SOLUTION

(a) To simplify writing somewhat, we let $\phi = \omega t - kx$. The total pressure is $p = p_1 + p_2$. The total intensity can be expressed in terms of the mean square value of the total pressure

$$I = (1/Z)(1/T)\int_0^T p^2(t)\,dt,$$

where Z is the wave impedance of the transmission line and T the period of oscillation, $T = 2\pi/\omega$. The intensity of the first wave alone is $I_1 = (A^2/2Z)$ and of the second $I_2 = B^2/2Z$.

We have $p^2 = (p_1 + p_2)^2 = p_1^2 + p_2^2 + 2p_1 p_2$. The first term contributes I_1 to the integral and the second I_2. The integral of the cross product $p_1 p_2$ becomes

$$I_{12} = (1/Z)(1/T)(AB)\int_0^T \cos(\phi - \alpha)\cos(\phi - \beta)\,dt$$
$$= (1/Z)(AB/2T)\int_0^T \big(\cos(\alpha - \beta) + \cos(2\phi - \alpha - \beta)\big)dt$$
$$= (1/2)(AB/Z)\cos(\alpha - \beta).$$

The integral of the term containing 2ϕ covers two full periods and is zero. Thus, the total intensity becomes

$$I = I_1 + I_2 + 2I_{12},$$

where $I_{12} = (1/2)(1/Z)AB\cos(\alpha - \beta)$.

(b) If the phase difference is zero, we get $I_{12} = AB/2Z$, and if $A = B$ it follows that $I_{12} = I_1 = I_2$ so that the total intensity becomes $I = 4I_1$. If the phase difference is $\pi/2$, we get $I = 0$.

(c) Similarly, if N waves of equal amplitude are superimposed, the power will be N^2 times the power of a single wave if the waves are all in phase.

Laser light. In ordinary light, the phases of the light waves which are emitted from the individual atoms in spontaneous emission are random, and this means that the time average of all cross products in the squared electric field will vanish. The total wave power then will be the sum of the powers in the individual waves. Thus, with N waves, the total intensity will be $I_{incoherent} = N I_{single}$.

In laser light, on the other hand, the waves are brought in phase through stimulated emission, and the total power is $I_{coherent} = N^2 I_1$, i.e., an increase by a factor N. When we deal with atomic radiators, N is a very large number indeed.

23. **Wave pulse on a spring**

A spring of length 2 m and mass 0.5 kg has the spring constant $K = 50$ N/m. The end $(x = 0)$ of a long spring of the same material is driven by a longitudinal displacement

$\xi(0, t) = 0.02[(t/T) - (t/T)^2]$ during the time between $t = 0$ and $t = T = 0.02$ sec. The displacement is zero at all other times.
(a) What is the wave speed?
(b) What is the velocity of the spring at $x = 0$?
(c) During what time interval is the wave pulse passing $x = 10$ m?
(d) What region of the spring is occupied by the wave pulse at $t = 2$ sec?

SOLUTION

(a) The spring constant per unit length (1 m) is $K = 100$ N/m and the corresponding compliance per unit length is $\kappa = 1/K$. The mass per unit length is $\mu = 0.25$ kg. First we determine the longitudinal wave speed which can be expressed as

$$v = \sqrt{1/\mu\kappa}. \tag{11.16}$$

The numerical value is $v = \sqrt{100/0.25} = 20$ m/sec.
The wave impedance is

$$Z = \mu v = \sqrt{\mu\kappa} \tag{11.17}$$

with the numerical value $Z = 20$ kg/sec.
(b) The velocity of the spring is

$$u(0, t) = \partial\xi(0, t)/\partial t = 0.02(1 - 2t/T)/T \quad \text{for} \quad 0 < t/T < 1. \tag{11.18}$$

Since in this case we are dealing only with a single wave traveling in the positive x-direction, the driving force $F(0, t)$ can be expressed simply as the product of the velocity of the driven endpoint of the spring and the wave impedance.

$$F(0, t) = Zu(0, t) = 20(1 - 2t/T) \tag{11.19}$$

where the unit is Newton.
The velocity and the driving force start out positive, with the values 1 m/sec and 20 N, go through zero at $t = T/2 = .01$ sec, and then become negative.
(c) The front of the pulse reaches $x = 10$ m at the time $x/v = 0.5$ sec. Since the duration is 0.02 sec the time interval is between 0.5 sec and 0.52 sec.
(d) The length of the pulse is $vT = 20 \cdot 0.02 = 0.4$ m. At $t = 2$ sec, the front of the pulse is at $x = vt = 20 \cdot 2 = 40$ m. The region occupied by the pulse at $t = 2$ sec, therefore, is given by $39.6 < x < 40$.

24. **Radiation damping**

A mass M is attached to the beginning of a long coil spring. It is given an impulse at $t = 0$ so that the initial velocity becomes $u(0)$. The spring is sufficiently long so that wave reflection from the other end of the spring can be ignored. Determine the subsequent motion of M. How far does it move?

SOLUTION

As M moves, a wave is produced on the spring which carries energy from M. The corresponding reaction force on M is $-Zu$, where the wave impedance is $Z = \sqrt{\mu/C}$, μ is the mass and C the compliance, both per unit length of the spring. The equation of motion of M is then $M\dot{u} = -Zu$ with the solution

$$u(t) = u(0) \exp[-(Z/M)t]. \tag{11.20}$$

The velocity decays exponentially.

The total displacement of M in this motion will be

$$\int_0^\infty u\,dt = u(0)M/Z. \tag{11.21}$$

In other words, the mass moves a finite distance.

25. **Longitudinal and torsional waves**

The Young's modulus of steel is $Y = 1.7 \times 10^{11}$ and the shear modulus $G = 7.6 \times 10^{10}\,\text{N/m}^2$.

(a) What is the ratio of the longitudinal and torsional wave speed on a steel shaft?

(b) What is the stress amplitude in a harmonic longitudinal wave in the shaft when the velocity amplitude is 0.1% of the longitudinal wave speed?

(c) Suppose the shaft diameter is $D = 2.5\,\text{cm}$, What is the torque amplitude required to produce a torsion wave with an angular displacement amplitude of 0.001 radians in a harmonic traveling wave? The frequency of oscillation is 100 Hz.

SOLUTION

(a) With reference to the discussion in the text, the longitudinal and torsional wave speeds are $\sqrt{Y/\rho}$ and $\sqrt{G/\rho}$, where Y is Young's modulus and G the shear modulus. Therefore, the ratio of the speeds is simply $v/v_t = \sqrt{Y/G}$. The numerical value of this ratio is 1.5, independent of the diameter of the shaft.

(b) For a single traveling wave, the stress is simply the product of the wave impedance $Z = \rho v$ and the velocity $u = \partial\xi/\partial t$. With $u_0 = .001\,v$, we obtain the stress amplitude $\sigma_0 = 0.001\,\rho v_\ell^2 = 0.001\,Y$, where the numerical value of Y is given above.

(c) Again, with reference to the text, a comparison of the longitudinal and torsional waves shows that the mass density ρ in the longitudinal wave corresponds to the moment of inertia $J = \rho I$ in the torsional wave. Similarly, the velocity u corresponds to the angular velocity $\dot\theta$ of the angular displacement θ, and the longitudinal force function F corresponds to the torque τ. Consequently, the wave impedance in the longitudinal wave $Z = F/u = \rho v$ corresponds to $Z_t = \tau/\dot\theta = \rho I v_t = J v_t$.

In the present case, the moment of inertia per unit length is

$$J = \rho \int_0^a r^2 2\pi r\,dr = \mu a^2/2, \tag{11.22}$$

where $a = D/2$ and $\mu = \rho\pi a^2$ is the mass per unit length of the shaft.

In a harmonic angular displacement wave $\theta(x, t) = \theta_0 \cos(\omega t - kx)$, the angular velocity amplitude $\frac{\partial\theta}{\partial t}$ is $\omega\theta_0$, where ω is the angular frequency. Thus, with the torsional wave impedance being $Z_t = v_t J$, the torque amplitude becomes, with $A = \pi a^2$,

$$\tau_0 = Z_t(\omega\theta_0) = v_t \rho A(a^2/2)\omega\theta_0.$$

With $\rho = 7.8 \times 10^3\,\text{kg/m}^3$ and the value of G given above, we get $v_t = 3121\,\text{m/sec}$. Introducing the numerical values for the remaining variables, we obtain $\tau_0 = 8.62\,\text{Nm}$.

26. **Longitudinal and transverse waves on a spring**

The relaxed length of a coil spring is ℓ, the mass M, and the spring constant K. The spring is stretched to a length L and kept at this length.

a). What is the ratio of the transverse and longitudinal wave speed on the spring? Can the transverse wave speed be larger than the longitudinal?

b). What is the time of travel of the longitudinal wave from one end of the spring to the other, and how does it depend on the length L?

SOLUTION

(a) The tension of the stretched spring is $S = K(L - \ell)$ and the mass per unit length $\mu = M/L$. The compliance per unit length of the stretched spring is $\kappa = 1/KL$.

The transverse and longitudinal wave speeds are

$$v_t = \sqrt{S/\mu} = \sqrt{K(L-\ell)L/M}$$
$$v_\ell = \sqrt{1/\kappa\mu} = \sqrt{L^2 K/M}.$$

The ratio of these speed is

$$v_t/v_\ell = \sqrt{1 - \ell/L}, \tag{11.23}$$

which is always less than one. In other words, regardless of the tension in the spring, the transverse wave speed is always less than the longitudinal speed.

(b) The time of travel is $\Delta t = L/v_\ell = 1/\omega_0$, which is independent of the length L.

27. **Reflection of sound**

A tube is filled with air and Helium and we assume that there is a well-defined boundary between the two gases provided by a thin (sound transparent) membrane perpendicular to the tube axis. A harmonic sound wave is incident on this boundary from the air.

(a) Determine the pressure reflection and transmission coefficients at the boundary.

(b) If the sound pressure amplitude of the incident wave is A, what then is the sound pressure amplitude at the interface between the two gases?

(c) What are the minimum and maximum values of the magnitude of the sound pressure amplitude in the air?

(d) Repeat (a) and (b) with the roles of air and Helium interchanged.

Densities: Air: 1.293 kg/m^3. He: 0.170 kg/m^3. Sound speeds: Air: 340 m/sec. He: 998 m/sec.

SOLUTION

(a) The wave impedances in the two gases are $Z_A = 1.23 \cdot 340 = 418.2 \, \text{kg/m}^2\text{sec}$ and $Z_H = 0.170 \cdot 998 = 169.7$. The pressure reflection coefficient is

$$R_p = \frac{Z_H - Z_A}{Z_H + Z_A} = -0.42.$$

The negative sign indicates that the reflected pressure wave is 180 degrees out of phase with the incident pressure at the boundary.

The pressure transmission coefficient is

$$T_p = \frac{2Z_H}{Z_H + Z_A} = 0.58. \tag{11.24}$$

(b) The total pressure at the boundary is the sum of the incident and the reflected pressures, $A + R_p A = A(1 - 0.42) = 0.58 \, A$, and this must also equal the pressure of the transmitted wave which is $T_p A = 0.58 \, A$. We note, incidentally, that $T_p = 1 + R_p$. On the helium side, we have assumed no reflection so that we have only a single wave traveling in the positive x-direction with an amplitude (magnitude) 0.42A.

(c) On the air side of the boundary, on the other hand, the total pressure is the sum of the incident and reflected pressures. Using complex amplitudes, the complex sound pressure amplitude can be written

$$p_i(\omega) = A[e^{ikx} + R_p e^{-ikx}]. \tag{11.25}$$

With $\exp(\pm ikx) = \cos(kx) \pm i \sin(kx)$, the squared magnitude is

$$|p_i(\omega)|^2/A^2 == (1 + R_p)^2 \cos^2(kx) + (1 - R_p)^2 \sin^2(kx) = 1 + R_p^2 + 2R_p \cos(2kx), \tag{11.26}$$

where we have used $\cos^2(kx) - \sin^2(kx) = \cos(2kx)$. Since R_p is negative, the minimum magnitude is obtained at $x = 0$, i.e., at the interface, where $|p(0)|^2/A^2 = (1 + R_p)^2$, $|p(0)| = (1 + R_p)A = 0.58A$. The maximum is obtained where $\cos(2kx) = -1$ at which point $|p| = (1 - R)A = 1.42A$.

It is recommended that you carry out the solution also without the use of complex amplitudes.

28. **Sound transmission from air into water**

What fraction of the incident power of a sound wave in air is transmitted into water at normal incidence?

SOLUTION

We denote the wave impedances by $Z_a = \rho_a c_a$ and $Z_w = \rho_w c_w$. With reference to Eq. 4.5 the power transmission coefficient can be written $\tau = 4x/(1 + x)^2$, where $x = Z_w/Z_a$.

The data for air and water are: $\rho_a = 1.29 \, \text{kg/m}^3$, $\rho_w = 1000$, $c_a = 331 \, \text{m/sec}$, and $v_w = 1480 \, \text{m/sec}$.

It follows that $x = 3466$ and $\tau = 1.15 \times 10^{-3}$. The corresponding transmission loss in decibels is $= 10 \log(1/\tau) = 30.6 \, \text{dB}$

About one-tenth of one percent of the incident power is transmitted into the water. The same is true for transmission into the air from the water.

29. **Reflection of longitudinal and transverse waves**

Two long coil springs, A and B, of equal length and mass but with different spring constants K_A and K_B are connected and stretched. It is found that the length of A is doubled and the length of B tripled. A wave is incident on the junction from A. What are the force reflection and transmission coefficients if the wave is a), transverse, and b), longitudinal?

SOLUTION

Let the initial length of each spring be L and the tension in the springs S. It follows that

$$S = K_A(2L - L) = K_B(3L - L),$$

i.e., $K_A = 2K_B$.

(a) In the stretched state of the springs, the values of the mass per unit length are such that $\mu_A = (3/2)\mu_B$. The corresponding wave impedances for transverse waves are $Z_A = \mu_A v_A = \sqrt{S\mu_A}$ and $Z_B = \sqrt{S\mu_B}$.

The force reflection coefficient is

$$R_F = \frac{Z_B - Z_A}{Z_A + Z_B} = \frac{\sqrt{\mu_B} - \sqrt{\mu_A}}{\sqrt{\mu_B} + \sqrt{\mu_A}} = -0.10$$

and the transmission coefficient is $T_F = 1 + R_F = 0.9$.

(b) The compliance per unit length of a spring is $\kappa = 1/(KL)$. Thus, the ratio of the κ-values for the two springs is

$$\kappa_B/\kappa_A = K_A L_A/(K_B L_B) = 4/3.$$

The wave impedance for longitudinal waves is $Z = \mu v = \sqrt{\kappa/\mu}$ and we have, with $\mu_B/\mu_A = 2/3$,

$$Z_B/Z_A = \sqrt{\kappa_B \mu_B/(\kappa_A \mu_B)} = 8/9.$$

The reflection coefficient has the same form as above, and we get $R_F = -1/17$. The transmission coefficient becomes $R_T = 1 + R_F = 16/17$.

30. **Reflection of a wave pulse on a string**

Two strings, with the masses per unit length $\mu_1 = 0.1$ kg/m and $\mu_2 = 0.4$, are connected and kept under a tension $\tau = 10$ N. At $t = 0$ a transverse displacement of the end of the light string is started with a constant velocity $u(0)$ immediately followed by a displacement back to the origin with a velocity $-u(0)/2$. The duration of this 'triangular' displacement pulse is 0.3 sec.

(a) Sketch the time dependence of the transverse displacement at $x = 0$.

(b) What is the minimum length of the light string in order that the entire wave pulse be carried by the string without interference from the reflected pulse?

(c) Sketch the x-dependence of the pulse in b) at $t = 0.3$ sec.

(d) Sketch the reflected and transmitted pulses after the process of reflection is completed.

SOLUTION

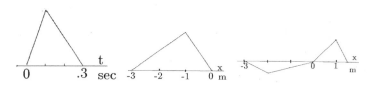

Figure 11.3: Left: (a). Middle: (c). Right: (d).

(a) The time dependence $\eta(0, t)$ of the displacement at $x = 0$ is shown on the left in the figure. The peak displacement is $\eta_{max} = 0.1 \cdot u(0)$.

(b) The spatial extent of the pulse on the first string will be $0.3 \cdot v_1$, where $v_1 = \sqrt{\tau/\mu_1} = 10$ m/sec is the wave speed. Thus, the length of the first string must be at least 3 m long in order to accommodate the pulse without interference from the reflection from the junction of the two strings.

The wave speed on the second string is $v_2 = 5$ m/sec.

(c) The x-dependence of the displacement of the string at $t = 0.3$ sec is shown in the figure. If the length of the string is chosen to be 3 m, the entire pulse is confined to the first string and the front of the pulse has just reached the junction of the two strings.

(d) The wave impedance for a transverse wave on a string is $Z = \mu v = \sqrt{\tau \mu}$ and the values for the two strings are $Z_1 = 1\,\text{kg/sec}$ and $Z_2 = 2$. The reflection coefficient for displacement is

$$R_d = \frac{Z_1 - Z_2}{Z_1 + Z_2} = -1/3$$

and the transmission coefficient is $T = 1 + R_p = 2/3$.

The process of reflection is completed at $t = 0.6\,\text{sec}$ and the shapes of the strings at that time is shown in the figure. The peak value of the transmitted pulse will be $(2/3)(0.3 \cdot u(0))$ and the length of the pulse will be $0.3 \cdot v_2 = 1.5\,\text{m}$.

31. **Wave reflection on bars**

    One end of a Copper bar is joined to the end of an Aluminum bar. The cross sections of the two bars are the same. A wave pulse with a total energy of 10 joule in the Copper bar is incident on the junction.
    (a) How much energy is transmitted into the Aluminum bar?
    (b) If instead the wave is incident on the junction from the Aluminum bar, what energy will be transmitted into the Copper bar?
    (c) Which of the two cases will yield the larger velocity amplitude at the junction?

    ### SOLUTION

    The wave impedance for a longitudinal wave in a solid bar is $Z = \rho v = \sqrt{Y\rho}$, where Y is Young's modulus. We have $\rho_C = 8900\,\text{kg/m}^3$, $Y_C = 1.26 \times 10^{11}\,\text{N/m}^2$. and $\rho_A = 270\,\text{kg/m}^2$, $Y_A = 0.72 \times 10^{11}$.
    (a) The reflection coefficient for the velocity wave is

    $$R_{uCA} = \frac{Z_C - Z_A}{Z_C + Z_A}.$$

    With $Z_C/Z_A = 7.6$, we obtain $R_{uCA} = 0.77$. The corresponding transmission coefficient for velocity is

    $$T_{uCA} = 1 + R_{uCA} = \frac{2Z_C}{Z_C + Z_A} = 1.77.$$

    The power in the longitudinal wave (not time averaged) is $P = Zu^2$. With the incident power being $P_i = Z_C u_i^2$, the transmitted power will be

    $$P_t = Z_A(T_{uCA}^2 u_i^2) = (Z_A/Z_C)T_{uCA}^2 \, P_i = \frac{4Z_A Z_C}{(Z_A + Z_C)^2} \, P_i = 0.41 \, P_i. \qquad (11.27)$$

    (b) If we reverse direction, the velocity transmission coefficient will be $T_{uAC} = 2Z_A/(Z_A + Z_C)$, but the power transfer will be the same as before.
    (c) For a given transmitted power, the velocity in the transmitted wave will be $u = \sqrt{P/Z}$, where Z is the wave impedance of the bar carrying the transmitted wave. This velocity must equal the velocity of the junction. Since $Z_A < Z_C$, the largest velocity is obtained when the wave is incident from the Copper bar.

32. **Power transmission**

    In a particular case of wave reflection at the junction of two transmission lines, it is desired that 20% of the incident wave energy be transmitted across the junction. If the wave impedance of the transmission line of the incident wave is Z_1, how should the impedance of the second line be chosen? Is there more than one possible choice?

SOLUTION

Let the impedance of the second transmission line be Z_2. The power transmission coefficient is (see Problem 32)

$$T_w = \frac{4Z_1 Z_2}{(Z_1 + Z_2)^2} = \frac{4x}{(1+x)^2},$$

where $x = Z_2/Z_1$. It follows that

$$x = Z_2/Z_1 = \frac{2}{T_w} - 1 \pm \frac{2}{T_w}\sqrt{1 - T_w}. \qquad (11.28)$$

Thus, there are two possible values x_+ and x_- of the impedance ratio for which the same power transfer is obtained. The product of these values is 1. For example, if $T_w = 0.2$ we get $x_\pm = 9 \pm 10\sqrt{0.8}$ or $x_+ = 17.9$ and $x_- = 0.056$.

33. **Spring with a lossy termination**

A coil spring of mass M is driven at one end in transverse motion and the other end is attached to a ring, also with a mass M, which can slide on a horizontal bar, normal to the direction of the spring. We assume that the friction force on the ring from the bar is Ru, where u is the velocity of the ring. The spring has a spring constant K and is stretched to a length which is much longer than the relaxed length.
(a) Show that the wave impedance of the spring is approximately $M\omega_0$, where $\omega_0 = \sqrt{K/M}$.
(b). Show that the absorption coefficient of the termination can be expressed as $\alpha = 4D/((1 + D)^2 + (\omega/\omega_0)^2)$, where $D = R/\omega_0 M$.

SOLUTION

(a) If the stretched spring (length L) is much longer than the relaxed spring, the extension can be considered to be L and the tension $S = KL$. The mass per unit length is $\mu = M/L$ and the wave speed $v = \sqrt{S/\mu} = \sqrt{KL^2/M}$. The wave impedance is

$$Z_0 = \mu v = (M/L)\sqrt{KL^2/M} = M\omega_0,$$

where $\omega_0 = \sqrt{K/M}$.
(b) With reference to the text, the absorption coefficient at a termination can be expressed as

$$\alpha = \frac{4\theta}{(\theta + 1)^2 + (\chi)^2},$$

where θ and χ are the normalized termination resistance and reactance, respectively. The normalization is made with respect to the wave impedance Z_0 of the transmission line. In our case, the mechanical termination (ring) impedance is $Z = R - i\omega M$, the resistance and reactance being R and $X = -i\omega M$. With the wave impedance being $Z_0 = \omega_0 M$ the corresponding normalized values are $\theta = R/\omega_0 M = D$ and $\chi = -\omega/\omega_0$. It follows then from Eq. 11.2 that the absorption coefficient becomes

$$\alpha = \frac{4D}{(1 + D)^2 + (\omega/\omega_0)^2}.$$

In order to make the absorption coefficient in this example equal to 100%, the mass of the ring must be made negligibly small and the resistance R must match the wave impedance of transverse waves on the spring so that $\theta = 1$.

It is possible, though, to achieve 100% absorption at a particular frequency even with a massive ring, if a spring, fixed at one end, is attached to the ring to form an oscillator. Then, if the spring constant K_r is chosen such that the resonance frequency $\sqrt{K_r/M}$ of this oscillator is equal to the incident frequency, we get 100% absorption at that frequency.

34. Q-value of an acoustic tube resonator

Accounting for visco-thermal losses at the boundary, show that the Q-value of an acoustic tube resonator can be expressed approximately by Eq. 4.38.

SOLUTION

For a simple harmonic oscillator (spring constant K, mass M, and resistance constant R) driven by a harmonic force with frequency independent amplitude, the frequency dependence (response) of the velocity amplitude $u(\omega)$ is characterized by the familiar resonance at the frequency $\omega_r = \sqrt{\omega_0^2 - \gamma^2}$, where $\omega_0 = \sqrt{K/M}$ and $\gamma = R/2M$. For small damping, $\omega_r \approx \omega_0$ and the sharpness of the resonance curve is often expressed in terms of the Q-value, $Q = \omega_0 M/R$. It can be interpreted as the ratio of the resonance frequency and the total width of the response curve at the 'half-power point,' defined by $|u(\omega)/u(\omega_0)|^2 = 1/2$. $Q = \omega_0 M/R = \omega_0 M|u|^2/R|u|^2$ can be interpreted also as ω_0 times the ratio of the time average of the energy of oscillation (being twice the kinetic energy average) and the dissipation rate or, apart from a factor of 2π, as the ratio of the energy of oscillation and the dissipation in one period. This relation is valid also for an acoustic cavity resonator in the vicinity of a resonance.

The cavity under consideration is a straight tube of length L, area A, perimeter S, open at one end, and terminated by a rigid wall at the other (at $x = L$). With the pressure amplitude at the wall being $p(L)$, the amplitude at a distance x from the wall is $p(x) = p(L)\cos(kx)$, where $k = \omega/c$. Then, if the 'driving pressure' at the open end ($x = 0$) is p_0, we get $p(x) = p_0 \cos[k(L - x)]/\cos(kL)$. Similarly, the velocity amplitude distribution is $|u(x)| = |u_0| \sin[k(L - x)]/\cos(kL)$, where $|u_0| = |p_0|/\rho c$. Integrating the kinetic and potential energy densities $\rho|u|^2/2$ and $|p|^2/2\rho c^2$ over the volume of the tube gives the total energy

$$E = (AL/2)|p_0|^2/\rho c^2. \tag{11.29}$$

Using these expressions for $|u|$ and $|p|$ and integrating over the tube walls including the thermal losses at the end wall, we get for the total loss rate

$$\begin{aligned} W &= (SL/2)[|u_0|^2(kd_v\rho c/2) + (|p_0|^2/\rho c)(\gamma - 1)kd_h/2] \\ &\quad + A(|p_0|^2/\rho c)(\gamma - 1)kd_h/2 \\ &= \omega(SL/2)(|p_0|^2/\rho c^2)(d_{vh}/2)[1 + (A/SL)(\gamma - 1)d_h/d_{vh}] \\ &\approx \omega SL/2)(|p_0|^2/\rho c^2)(d_{vh}/2), \end{aligned}$$

where

$$d_{vh} = d_v + (\gamma - 1)d_h = d_v[1 + (\gamma - 1)/\sqrt{P_r}] \approx 1.46 d_v \tag{11.30}$$

is the 'visco-thermal' boundary layer thickness and $P_r = \mu C_p/K \approx 0.77$ (air), the Prandtl number. The term which contains the area A expresses the heat conduction loss at the rigid wall termination which is usually small compared to the rest.[1] Without

---

[1] There is no tangential velocity at the end wall and no viscous losses.

this contribution and with $\omega = \omega_0$, the Q-value of the resonator becomes

$$Q = \omega_0 E / W \approx \frac{2A}{S d_{vh}}, \qquad (11.31)$$

which can be interpreted as twice the ratio of the volume of the tube and the volume occupied by the visco-thermal boundary layer.

As an example, consider a circular tube of radius a. With $A = \pi a^2$ and $S = 2\pi a$ we get

$$Q = a / d_{vh} \approx 3.11 a \sqrt{f}, \qquad (11.32)$$

i.e., simply the ratio of the radius and the boundary layer thickness. The approximate numerical expression was obtained by using $d_{vh} \approx 1.46 d_v$, and $d_v \approx 0.22/\sqrt{f}$. Thus, a 100 Hz quarter wavelength circular tube resonator with a one-inch diameter will have a Q-value of ≈ 39.5.

For the channel between two parallel plates, separation d and width w, the area dw and $S = 2(w + d) \approx 2w$. Therefore,

$$Q \approx d / d_{vh} \quad \text{(parallel plates)}. \qquad (11.33)$$

35. **Acoustic impedance of the air column in a tube closed at the end**

As an example of a boundary value problem we consider the important case of an air column in a straight, uniform tube of length L driven in harmonic motion at one end by a piston. The velocity amplitude of the piston is U independent of frequency. The other end of the tube is closed by a rigid wall. As an example of the use of Eqs. 5.14 and 5.15, determine

(a) the impedance of the column.

(b) the sound pressure and velocity fields in the tube as a function of position.

(c) What is the difference in the frequency dependence of the pressure amplitude in the tube if the piston provides a constant force (rather than) velocity amplitude?

SOLUTION

(a) We place $x = 0$ at the rigid termination so that the piston is located at $x = -L$ (just to make the algebra a bit simpler). The velocity (see Eq. 5.15) must be zero at $x = 0$ which requires $A = B$. Thus, the pressure field will be

$$p(x, \omega) = 2A \cos(kx) \qquad (11.34)$$

and the velocity

$$u(x, \omega) = 2i (A/\rho c) \sin(kx). \qquad (11.35)$$

The impedance is

$$z(-L, \omega) = p(-L, \omega)/u(-L, \omega) = i\rho c \cot(kL). \qquad (11.36)$$

Since $i = \exp(i\pi/2)$, the impedance has the phase angle $\pi/2$ and since a positive angle means a phase lag, the pressure $p(-L)$ will lag behind the velocity $u(-L)$ by this angle. Thus, if the time dependence of the piston velocity is $U \cos(\omega t)$ the time dependence of the pressure will be $\cos(\omega t - \pi/2) = \sin(\omega t)$.

(b) If the piston velocity is $U \cos(\omega t)$, its complex amplitude is $u(-L, \omega) = U$. It follows from Eq. 11.35 that $A = \rho c u(-L, \omega)/(-2i \sin(kL)) = \rho c U /[-2i \sin(kL)]$ and from Eqs. 11.34 and 11.35 that

$$p(x, \omega) = i\rho c U \frac{\cos(kx)}{\sin(kL)} \qquad (11.37)$$

and

$$u(x, \omega) = -U \frac{\sin(kx)}{\sin(kL)}. \tag{11.38}$$

As a check, we note that for $x = -L$, $u(x, \omega) = U$ and for $x = 0$ (at the wall), $u = 0$.
(c) If U is independent of frequency, the frequency dependence of the pressure amplitude is contained in the factor $1/\sin(kL)$ (Eq. 11.37). Thus, theoretically, in this loss-free case, the pressure will be infinite when $\sin(kL) = 0$, i.e., for $kL = n\pi$ or $L = n\lambda/2$. The corresponding (resonance) frequency is $f = nc/2L$.
The driving force per unit area of the piston is $P = p(-L, \omega) = zu(-L, \omega)$, where $z = i\rho c \cot(kL)$. Thus, $\rho cu(-L, \omega) = -iP/\cot(kL)$ and it follows then from Eq. 11.37 that

$$p(x, \omega) = P \cos(kx)/\cos(kL). \tag{11.39}$$

In this case of frequency independent driving force (rather than velocity), the pressure at the wall ($x = 0$) goes to infinity for $kL = (2n-1)\pi/2$ ($n = 1, 2, \ldots$) or $L = (2n-1)\lambda/4$, i.e., whenever the tube length is an odd number of quarter wavelengths. This is understandable since the pressure distribution is such that it goes from a maximum at $x = 0$ to zero at the piston. But if the pressure at the piston is to be finite, the pressure at the wall must be infinite (leaving the product of 0 and ∞ a finite number P). These two cases correspond to source impedances of ∞ and 0, respectively (constant velocity and constant pressure sources), and illustrate the importance of the internal impedance of the source in problems of this kind.

36. **Sheet absorber, the hard way. Without complex variables**

As problems become a bit more complicated, the analysis without the use of complex amplitude becomes increasingly more cumbersome as shown in this example. This is about as far as we can go in problem complexity without making the algebra repulsive. Thus, carry out the calculation of the absorption coefficient for the sheet absorber in Section 4.2.6 and in Fig. 4.5, now without the use of complex amplitudes. Limit the analysis to sound at normal incidence. For oblique incidence, see Section 4.2.6.

Two configurations are shown in Fig. 4.5, one with and the other without a honeycomb structure in the air layer. The honeycomb has a cell size assumed much smaller than a wavelength and it forces the fluid velocity in the air layer to be normal to the wall, regardless of the angle of incidence of the sound. The two types of absorbers are called *locally* and *nonlocally* reacting absorbers, as indicated.

As we shall see, either configuration can be considered to be a form of acoustic resonator but unlike the resonator absorber in the previous example, it has multiple resonances. One cell in the partitioned air backing can be regarded as a tube of length L terminated by a rigid wall. We start the analysis by determining the relation between the velocity and pressure at the beginning of this tube. The wall is then chosen to be at $x = 0$ and the sheet at $x = -L$. The velocity field in the tube is expressed as the sum of two traveling waves, $u(x, t) = A \cos(\omega t - \phi_1 - kx) + B \cos(\omega t - \phi_2 + kx)$ where $k = \omega/c = 2\pi/\lambda$. The velocity must be zero at the rigid wall at, $x = 0$, which means $A \cos(\omega t - \phi_1) + B \cos(\omega t - \phi_2) = 0$ at all times. Expressing this quantity as a sum of $\sin(\omega t)$ and $\cos(\omega t)$ terms and requiring that the coefficient for each must be zero, we find $\phi_1 = \phi_2$ and $B = -A$. We can always choose the origin of time such that $\phi_1 = 0$ in which case $u(x, t) = A[\cos(\omega t - kx) - \cos(\omega t + kx)] = 2A \sin(\omega t) \sin(kx)$.

Recalling that $p = \pm\rho c u$, where the plus and minus signs refer to a wave traveling in the positive and negative direction, respectively, the corresponding pressure field is

$$p(x, t) = \rho cA[\cos(\omega t - kx) + \cos(\omega t + kx)] = 2A\rho c \cos(kx) \cos(\omega t).$$

Thus, the velocity and pressure at the beginning of the tube, at $x = -L$, are

$$u_t(t) = -2A \sin(kL) \sin(\omega t)$$
$$p_t(t) = 2A\rho c \cos(kL) \cos(\omega t). \tag{11.40}$$

The incident and reflected pressures at the absorber are denoted p_i and p_r. The pressure in front of the screen is then $p_i + p_r$ and the pressure behind it is p_t, as given in Eq. 11.40. The screen is assumed to be rigid (immobile) and the velocity through it is the same as u_t, the velocity at the entrance to the cavity behind it. This velocity must also equal the velocity in the sound field in front of the absorber, $p_i/\rho c - p_r/\rho c = u_t$. Since the difference in pressure at the two sides of the sheet is $p_i + p_r - p_t$, it follows from the definition of the flow resistance that

$$p_i + p_r = p_t + r u_t$$
$$p_i - p_r = \rho c u_t. \tag{11.41}$$

Adding and subtracting these equations, respectively, yields $2p_i = p_t + (r + \rho c)u_t$ and $2p_r = p_t + (r - \rho c)u_t$. Then use Eq. 11.40 for p_t and u_t and the trigonometric identity $C \cos(\omega t - \phi) = C \cos(\omega t) \cos \phi + C \sin(\omega t) \sin \phi$ to obtain

$$2p_i = p_t + (r + \rho c)u_t = 2A\rho c \sqrt{(1+\theta)^2 \sin^2(kL) + \cos^2(kL)} \, \cos(\omega t - \phi_p)$$
$$2p_r = 2A\rho c \sqrt{(1-\theta)^2 \sin^2(kL) + \cos^2(kL)} \, \cos(\omega t - \phi_u). \tag{11.42}$$

Of main interest is the ratio of the squared magnitudes of the reflected and incident pressures, i.e., the reflection coefficient R_I for intensity and the corresponding absorption coefficient $\alpha = 1 - R_I^2$,

$$R_I = |p_r|^2/|p_i|^2 = [(1-\phi)^2 + \cot^2(kL)]/[(1+\phi)^2 + \cot^2(kL)]$$
$$\alpha = \frac{4\phi}{(1+\phi)^2 + \cot^2(kL)}. \tag{11.43}$$

The result is valid for sound at normal incidence on the absorber. For oblique incidence we refer to Section 4.2.6, where complex variables are employed.

37. **Field from a pulsating sphere (the hard way, i.e., no complex amplitudes)**

Having established in the introduction of Section 5.1.2 that the pressure from a pulsating sphere is inversely proportional to r in the far field, complete the analysis of the field for all values of r, analogous to what was done in Section 5.1.2 but this time without the use of complex variables.

SOLUTION

The starting point is the pressure for harmonic time dependence of the source and large r

$$p(r, t) = |A|(a/r) \cos(\omega t - kr' - \phi), \tag{11.44}$$

where $r' = r - a$ and $k = \omega/c$. The term kr' is the phase lag due to the time of wave travel from the surface of the sphere at $r = a$ to the field point at r. This travel time is $(r - a)/c$, where c is the sound speed.

The amplitude $|A|$ and the phase angle ϕ are yet to be determined. Although we have justified this form of the wave only for large r, it will be demonstrated shortly, and shown rigorously in the complex amplitude description in Section 5.1.2, that this pressure field is valid for all values of r. For the moment, accepting Eq. 11.44 for the sound pressure field, the radial velocity is obtained from the momentum equation $\rho \partial u / \partial t = -\partial p / \partial r$

$$u(r, t) = \frac{|A|}{\rho c} \frac{a}{r} \left[\cos(\omega t - kr' - \phi) + \frac{1}{kr} \sin(\omega t - kr' - \phi) \right]. \tag{11.45}$$

We specify the radial velocity of the surface of the sphere to be $|u| \cos(\omega t)$ and by matching this velocity to Eq. 11.45, $|A|$ and ϕ can be determined. Thus $|u| \cos(\omega t) = (|A|/\rho c)[\cos(\omega t - \phi) + (1/ka) \sin(\omega t - \phi)]$. With $\cos(\omega t - \phi) = \cos(\omega t) \cos \phi + \sin)\omega t) \sin \phi$ and $\sin(\omega t - \phi) = \sin(\omega t) \cos(\phi) - \cos(\omega t) \sin \phi$, the right-hand side can be rewritten as a sum of $\cos(\omega t)$ and $\sin)\omega t)$ terms. The sum of the $\cos(\omega t)$-terms must equal $|u| \cos(\omega t)$ and the sum of the $\sin(\omega t)$-terms must be zero, i.e.,

$$|u| = (A/\rho c)[\cos \phi - (1/ka) \sin \phi]$$
$$0 = \sin \phi + (1/ka) \cos \phi. \tag{11.46}$$

It follows from these equation that

$$\tan \phi = -1/ka$$
$$|A| = \rho c |u| ka / \sqrt{1 + (ka)^2}. \tag{11.47}$$

With these values for $|A|$ and ϕ used in Eq. 11.44, the pressure at the surface of the sphere can be written $p(a, t) = |A| \cos(\omega t - \phi) = |A| \cos \phi \cos(\omega t) + |A| \sin \phi \sin(\omega t)$, or, as

$$p(a, t) = \rho c |u| \left[\frac{(ka)^2}{1 + (ka)^2} \cos(\omega t) - \frac{ka}{1 + (ka)^2} \sin(\omega t) \right]. \tag{11.48}$$

If $ka << 1$, we then get $p \approx -\rho c |u| (ka) \sin(\omega t) = -\rho a \omega |u| \sin(\omega t)$ which is the same as obtained for the incompressible fluid in Eq. 5.23. With $-\sin(\omega t) = \cos(\omega t + \pi/2)$, the velocity $|u| \cos(\omega t)$ at the surface of the sphere is seen to lag behind the pressure by an angle of $\pi/2$, characteristic of a mass reactive load.

The pressure component in phase with the velocity, the first term in Eq. 11.48, is the power producing component. The factors in front of $\cos(\omega t)$ and $\sin(\omega t)$ are usually denoted

$$\theta_r = \frac{(ka)^2}{1 + (ka)^2} \quad \text{and} \quad \chi_r = -\frac{ka}{1 + (ka)^2}. \tag{11.49}$$

Quantity θ_r is the normalized *radiation resistance* and χ the normalized *radiation reactance* of the pulsating sphere. We have chosen to include the negative sign in the expression for χ_r to indicate, by convention, that it represents a mass (rather than stiffness) reactive load, thus giving rise to a phase lag in the velocity. This is clarified further in Section 5.1.2. The resistance and reactance are plotted as functions of ka in Fig. 5.1. The radiated power Π is the time average of $(4\pi a^2) p(a, t) u(a, t)$; only the $\cos(\omega t)$ term in Eq. 11.48 will contribute, and we obtain

$$\Pi = (4\pi a^2) \rho c \theta_r \langle u^2(a, t) \rangle = (1/2) \rho c |u|^2 (4\pi a^2) \frac{(ka)^2}{1 + (ka)^2}, \tag{11.50}$$

where the angle brackets signify time average.

ACOUSTICS

38. Pressure and velocity fields from an acoustic point source

A pulsating sphere of radius a and a surface velocity $|u|\cos(\omega)$ is small enough to be treated as an acoustic point source. Determine
(a) the complex amplitude of the acoustic source strength?
(b) the complex amplitudes of the pressure field?
(c) the real sound pressure $p(r, t)$.

SOLUTION

(a) The complex amplitude of the surface velocity is $U(\omega) = |u|$, since the phase angle of the velocity is zero. The complex amplitude of the acoustic source strength is then $q = \rho(-i\omega|u|)4\pi a^2$.

(b) The complex amplitude of the sound pressure is

$$p(r\omega) = \frac{q}{4\pi r}e^{ikr} = \frac{\rho(\omega|u|)4\pi a^2}{4\pi r}e^{ikr-i\pi/2}, \qquad (11.51)$$

where we have used $-i = \exp(-i\pi/2)$.

(c) The phase angle $-\pi/2$ indicates that the pressure runs ahead of the velocity (remember that lag corresponds to a positive phase angle). Therefore,

$$p(r, t) = \frac{|q|}{4\pi r}\cos(\omega t - kr + \pi/2) = -\frac{|q|}{4\pi r}\sin(\omega t - kr), \qquad (11.52)$$

where $|q| = \rho\omega|u|4\pi a^2$.

39. Intensity and power from a point force

(a) Calculate the intensity in the far field of a harmonic point force at the origin, $f_x = |f|\cos(\omega t)$. The force is directed along the x-axis.
(b) What is the corresponding radiated power?

SOLUTION

(a) The intensity in the far field is

$$I(r) = \frac{p^2}{\rho c} = \frac{\omega^2 f^2}{(4\pi rc)^2}\cos^2\phi, \qquad (11.53)$$

where p and f are rms values.

(b) The radiated power is obtained from the integral is

$$\Pi = \int_0^\pi I\, 2\pi r^2 \sin\phi\, d\phi = (1/6\pi)(1/\rho c)(\omega^2/c^2)f^2. \qquad (11.54)$$

40. Sound radiation by oscillating sphere; induced mass

This discussion is a slightly different way of approaching the problem analyzed in Section 5.3.2. A rigid sphere of radius a oscillates back and forth in harmonic motion, angular frequency ω, along the x-axis with the velocity amplitude U. For $\omega a/c << 1$, where c is the sound speed in the surrounding fluid, determine the pressure amplitude distribution over the surface of the sphere and calculate the corresponding reaction force on the fluid. Show that it is the same as the force required to oscillate a fluid mass equal to half of the fluid mass displaced by the sphere volume (i.e., induced mass equal to $(1/2)[\rho 4\pi a^3/3]$.

SOLUTION

The complex amplitude of the sound field pressure from a point force was found in Eq. 5.42 to be

$$p(r) = (A/r)(1 + \frac{i}{kr})e^{ikr}\cos(\theta), \tag{11.55}$$

where A/r is the far field and θ is the angle measured from the direction of the force (chosen here to be the x-axis).

When the sphere oscillates back and forth, there will be a push on the fluid and an outflow on one side of the sphere and a corresponding pull and inflow occurs on the other side. Thus, if the sphere is small, its effect on the surrounding fluid is expected to be the same as from a point force. Thus, we use Eq. 11.55 for the pressure field, where the constant A is now to be determined from the known velocity amplitude of the sphere.

Thus, the next step is to derive the expression for the velocity amplitude. We are going to match this expression for the velocity with the known velocity of the sphere in order to determine the constant A. The radius of the sphere is a and with $ka << 1$, the dominant term of interest in the expression for the pressure corresponds to the second term within the parenthesis in the equation,

$$p(r) \approx \frac{iA}{kr^2}\cos(\theta) \qquad kr << 1 \tag{11.56}$$

and the corresponding radial velocity amplitude is

$$u(r) = \frac{1}{i\omega\rho}\frac{\partial p}{\partial r} \approx -\frac{2A}{\omega\rho kr^3}\cos(\theta) \qquad kr << 1, \tag{11.57}$$

where we have used $\exp(ika) \approx 1$.

The radial velocity component of the velocity of the surface of the sphere is the component of $\mathbf{U}$ along the normal to the sphere, and this component is $u(a) = U\cos(\theta)$. Equating this expression to $u(r)$ in Eq. 11.61 at $r = a$, we obtain

$$A = -(\omega\rho a^3 k/2)U. \tag{11.58}$$

Thus, the pressure field in Eq. 11.56 becomes

$$p(r) \approx -i\omega\frac{\rho a U}{2}\cos(\theta) \qquad kr << 1. \tag{11.59}$$

The force component along the x-axis caused by the pressure on a surface element $2\pi a \sin(\theta)a\,d\theta$ (a ring of radius $a\sin(\theta)$ and width $ad\theta$) will be $dF = -p(a)\cos^2(\theta)2a^2\pi\sin(\theta)\,d\theta$, and the corresponding total force *on the fluid*

$$F = -i\omega\rho a(U/2)\int_0^\pi 2\pi a^2\cos^2(\theta)\sin(\theta)\,d\theta = [-i\omega\rho\,U](V/2), \tag{11.60}$$

where $V = 4\pi a^3/3$.

In other words, the force is the same as that required to oscillate a mass $M = V\rho/2$, and this is the 'induced' mass of the fluid surrounding the fluid. It can be thought of as the mass being pushed back and forth between the pressure and suction sides of the sphere.

41. **Sound field from a finite random line source**

Reconsider the line source problem in Section 5.4.2 with a random uniform source distribution along a line between $y = -L$ and $y = L$ treated in the text. This time, however, determine the mean square pressure
(a) along the y-axis outside the source, i.e., for $y > L$ and $y < -L$.
(b) along a line perpendicular to the y-axis and starting at $y = L$.
(c) If $L = 1$ m and the power level of the line source is 100 dB re $W_r = 10^{-12}$ w, what then is the sound pressure level at $y = 2L$ in (a)?

SOLUTION

(a)
$$\langle p(x)^2 \rangle = \frac{\langle Q^2 \rangle}{16\pi^2} \int_{-L}^{L} \frac{dy'}{(y-y')^2} = \frac{2L\langle Q^2 \rangle}{16\pi^2(y^2 - L^2)}$$

If $y \gg L$ the result is the same as for a point source with the source strength equal to the total source strength of the line.
As $y \to L$, $\langle p(y)^2 \rangle \approx \langle Q^2 \rangle/(y-L)$.

(b)
The perpendicular distance is denoted h.

$$\langle p(h)^2 \rangle = \frac{\langle q^2 \rangle}{16\pi^2} \int_{-L}^{L} \frac{dx'}{h^2 + (L-y')^2} = \frac{\langle q^2 \rangle}{16\pi^2 h} \arctan(2L/h)$$

If $h \gg 2L$, $\langle p(h)^2 \rangle \approx 2L\langle Q^2 \rangle/(16\pi^2 h^2)$, i.e., the same as for a point source with the total source strength of the line.
If $h \ll 2L$ $\langle p(h)^2 \rangle = \langle Q^2 \rangle/(8\pi h)$.

(c)
The total power is denoted by W and the power level PWL.

$$PWL = 10 \log(W/W_r)$$

The total power from the source can be expressed as

$$W = \frac{2LQ^2}{4\pi\rho c},$$

which means that
$$2LQ^2 = 4\pi\rho c\, W.$$

It follows from Eq. 11.1 that the mean square pressure at $y = 2L$ is

$$\langle p(2L)^2 \rangle = \frac{2L\langle q^2 \rangle}{16\pi^2 (4L^2 - L^2)} = \frac{\rho c W}{4\pi\, 3L^2}.$$

With the reference power level written as $W_r = (p_r^2/\rho c)A_r, (A_r = 1 \text{ m}^2)$ as explained above, the sound pressure level becomes

$$SPL = 10 \log(p(2L)^2/p_0^2) = PWL - 10 \log(4\pi\, 3L^2/A_r) = 84.2 \text{ dB}.$$

42. **Angular resolution of antenna array**

A radio astronomical interferometer consists of an array of 32 antennas, seven meters apart, placed along a straight line. What is the angular resolution of the array for the 21 cm line of radiation? (The angular resolution is defined as the half-width of the main lobe of the array, i.e., the angular separation of the maximum and the adjacent minimum in the directivity pattern of the array.)

SOLUTION

The intensity distribution from an array of N line sources is

$$\frac{I(\theta)}{I(0)} = \left(\frac{\sin(N\delta)}{N\sin(\delta)}\right)^2$$
$$\delta = (kd/2)\sin(\theta),$$

where $k = 2\pi/\lambda$ and d is the distance between two adjacent antennas. The first zero in the patter is obtained when $N\delta = \pi$, i.e., $(Nkd/2)\sin(\theta_1) = \pi$ or,

$$\sin(\theta_1) \approx \theta_1 = \frac{\lambda}{Nd}. \tag{11.61}$$

In our case, with $\lambda = 21$ cm, $N = 32$, and $d = 700$ cm, we get $\theta_1 = 9.38 \times 10^{-4}$ rad $= 0.034$ deg. This, by definition, is the resolution of the antenna.

Comments

It is interesting in this context to compare the intensity distribution for an array of N sources with that of a continuous source distribution. For a given total length $b = Nd$ of the antenna we obtain from Eq. 11.2 $\delta = (kb/2N)\sin(\theta)$ or $N\delta = \beta = (kb/2)\sin(\theta)$. As $N \rightarrow \infty$, $N\sin(\delta) \rightarrow N\delta = \beta$ and the expression for the intensity distribution reduces to

$$\frac{S(\theta)}{S(0)} \rightarrow \left(\frac{\sin(\beta)}{\beta}\right)^2,$$

which is the result for a continuous source of length b.

43. **Interference pattern over ground with bands of noise**

In atmospheric acoustics involving the effect of ground reflections, the sound generally is not a pure tone but a band of random noise. Thus, consider a point source a distance h above a totally reflecting plane boundary emitting an octave band of random noise. Determine the rms pressure in this band as a function of distance from the source at the same height as the source.

SOLUTION

The solution is carried out for an arbitrary time dependence of the pressure without the use of complex amplitudes. (For harmonic time dependence, see Problem 44.) Only a totally reflecting boundary is considered with the source S and its image S', as shown in Fig. 9.3. Then if the direct sound field at the receiver is $p_1(t)$, the reflected pressure will be the same except for a reduction in amplitude by the factor r_1/r_2 and a time delay $\tau = (r_2 - r_1)/c$, where c is the sound speed. Thus, the reflected pressure becomes $p_2(t) = (r_1/r_2)p_1(t - \tau)$.

The total mean square pressure of the total sound field is then

$$\langle [p_1(t) + (r_1/r_2)p_1(t - \tau)]^2 \rangle = p_1^2 + (r_1/r_2)^2 p_1^2 + 2(r_1/r_2)\Psi(\tau), \tag{11.62}$$

where $\psi(\tau) = \langle p_1(t)p_1(t - \tau)\rangle$ is the correlation function (see Chapter 2). The angle brackets signify time average over a time T long compared to the characteristic fluctuation time of the signal. It is assumed that the signal is stationary so that the time average does not depend on when it was taken. Therefore, the time average of $p_1(t)^2$ and $p_1^2(t - \tau)$ are both equal to p_1^2, where p_1 is the rms value.

We have already shown in Chapter 2, Eq. 2.85, that if the (one-sided) spectrum density of a signal is E_1, the correlation function can be expressed as

$$\Psi(\tau) = \int_0^\infty E(f)\cos(2\pi f \tau)df. \tag{11.63}$$

In this case the spectrum density is zero at all frequencies except in the octave with the center frequency f_c where it is a constant E_0. If the upper and lower frequency limits of the octave are f_2 and f_1, we have $f_2 = 2f_1$ and $f_c = \sqrt{f_2 f_2}$ from which follows that $f_1 = f_c/\sqrt{2}$ and $f_2 = f_c\sqrt{2}$. Thus, carrying out the integration in Eq. 11.63, we get

$$\Psi(\tau) = \frac{E_0}{2\pi\tau}[\sin(2\pi f_2\tau) - \sin(2\pi f_1\tau)] = \frac{2E_0}{2\pi\tau}\sin[\pi(f_2 - f_1)\tau]\cos[\pi(f_1 + f_2)\tau]. \tag{11.64}$$

With $p_s^2 = E_0(f_2 - f_1)$, we obtain from Eq. 5.37,

$$p^2 = p_d^2\left[1 + (r_1/r_2)^2 + 2(r_1/r_2)\frac{sin[\pi(f_2 - f_1)\tau}{\pi(f_2 - f_1)\tau}\cos[\pi(f_1 + f_2)\tau]\right], \tag{11.65}$$

where p_d is the direct field $p_d^2 = A/r_1^2 = A/x^2$. In the absence of a boundary this is the only contribution to the field.

In carrying out the numerical computations, we note that $r_2 = \sqrt{x^2 + (2h)^2}$ and $r_1 = x$. Thus, $r_2 - r_1 = x(\sqrt{1 + (2h/x)^2} - 1)$ and the argument $\pi(f_2 - f_c)\tau = \pi(x/\lambda_c)(\sqrt{2} - 1/\sqrt{2})(\sqrt{1 + 4(h/x)^2} - 1)$. Thus, if we plot the sound pressure level versus x/h, we have to specify the parameter h/λ_c. Using x/h as a variable the direct field is $p_d \propto 1/(x/h)^2$. In Fig. 9.4 is plotted $10\log(p^2/p_d^2)$ versus $h/2$ for the parameter value $h/\lambda_c = 4$.

As the bandwidth goes to zero, with $f_1 = f_2$, we obtain the result for a pure tone with a frequency f_c

$$p^2 = p_d^2[1 + (r_1/r_2)^2 + 2(r_1/r_2)\cos(2\pi f_c\tau)]. \tag{11.66}$$

For comparison, this is also included in the figure.

The minima in the interference pattern of the sound pressure level are clearly seen for the pure tone, corresponding to a path difference of an integer number of half wavelengths. For the octave band, they are less pronounced.

The path difference $r_2 - r_1 = x[\sqrt{1 + (2h/x)^2} - 1] \approx 2h^2/x$, where the approximation applies to large value of x, i.e., $h/x \ll 1$. The last minimum then occurs if this quantity is $\lambda/2$, i.e., $x \approx 4h^2/\lambda$. In this example, with $h = 4\lambda_c$, this means $x/h \approx 16\lambda_c$, which is consistent with the result in the figure.

44. **Reflecting boundary with a finite impedance**

Extend the analysis in Problem 43 to include a boundary with a finite rather than infinite impedance. In particular, let the impedance be purely real with the normalized resistance $\theta = 2$. The height of the source, as before, is $h = 4\lambda_c$, where λ_c is the wavelength at the center frequency of the octave band.

SOLUTION

We now use complex amplitude description of the fields. Thus, the direct pressure wave from the source is expressed by the complex amplitude $p_1 = (A/r_1) \exp(ikr_1)$ and the reflected field by $p_2 = R(A/r_2) \exp(ikr_2)$ for harmonic time dependence at the angular frequency ω, where $k = \omega/c$. As before, the distance from the source to the receiver is r_1 and from the source to the image source is r_2 (see Fig. 9.3). The total pressure is then

$$p = A\frac{e^{ikr_1}}{r_1}[1 + (r_1/r_2)Re^{ik(r_2-r_1)}]. \tag{11.67}$$

For noise with a spectrum density $E(f)$ and a filter function $F(f)$, the mean square pressure will be

$$p^2 = |p_d|^2 \int_0^\infty E(f)F(f)|1 + R(r_1/r_2)e^{ik(r_2-r_1)}|^2 \, df, \tag{11.68}$$

where $p_d = A\exp(ikr_1)/r_1$ is the direct pressure field, and A a constant signifying the source strength.

For a frequency band between f_1 and f_2, the filter function in this case is $F(f) = 0$ for f outside the band and $F(f) = w_0$ within the band.

If the normalized impedance of the boundary is ζ, the pressure reflection coefficient is

$$R = \frac{\zeta \cos \phi - 1}{\zeta \cos \phi + 1}, \tag{11.69}$$

where ϕ is the angle of incidence given by $\tan \phi = 2h/x$.

In Fig. 9.4 is shown the computed pressure distribution over a boundary with a purely resistive impedance of $2\rho c$, for both a pure tone and an octave of random noise. It is significantly different from that for a totally reflecting boundary which is due to the variation of the reflection coefficient with the angle of incidence of the sound. As explained in connection with Eq. 9.19, the pressure distribution in the far zone, beyond the interference zone, is quite different from that over the totally reflecting boundary as the sound pressure decreases as $1/x^2$ rather than $1/x$ which means a slope of the SPL curve versus distance of $40 \log 2 \approx 12$ dB per doubling of distance rather than the 6 dB for the hard boundary. In the interference zone, the maxima decrease with distance as ≈ 6 dB per doubling of distance and the change to the 12 dB slope in the far zone is quite apparent.

The distance to the last minimum has been reduced in comparison with that for the hard boundary. The reason is that for an impedance $\zeta = 2$, the incident sound will be totally absorbed by the boundary at an angle of incidence of 60 degrees and the reflected wave that causes an interference at the last minimum is weakened considerably as is the destructive interference.

45. **Complex compressibility; plane wave in a lined duct**

Derive the expression (6.40) for the average complex compressibility of the air in a lined duct to explain that the attenuation of the fundamental mode goes to zero with increasing frequency.

SOLUTION

We start from the mass conservation equation $\partial\rho/\partial t + \rho_0 \text{div}\, \boldsymbol{u} = 0$. By introducing the compressibility of air, $(1/p)\rho/\rho = 1/\rho c^2$, this equation can be expressed as

$$\kappa \partial t/\partial + \text{div}\, \boldsymbol{u} = 0. \tag{11.70}$$

The air channel in the duct has a width D_1 and it is lined on one side where the normalized admittance is η. Integrate Eq. 11.70 over a volume element $D_1 dx$ of length dx and unit height. We consider harmonic time dependence with angular frequency ω so that $\partial/\partial t \to -i\omega$. Then, if the average pressure in this volume is denoted p_a, the first term becomes $(-i\omega\kappa p_a)D_1 dx$. The volume integral of the second term is converted into a surface integral of the normal velocity u over all the surfaces of the volume element. The contribution from the surfaces normal to the x-axis is $D_1[u(x+dx) - u(x)]$ and the contribution from the lined surface is $(\eta/\rho c)p_a\, dx$, where we have expressed the normal velocity into the liner as $(\eta/\rho c)p_a$, approximated the pressure at the surface by the average pressure. Eq. 11.70 then reduces to

$$(-i\omega\kappa D_1 + \eta/\rho c)p_a + \partial u/\partial x = 0. \tag{11.71}$$

This has the form of the one-dimensional equation for mass conservation with a complex compressibility $\tilde{\kappa}$

$$(-i\omega\tilde{\kappa})p_a + \partial u/\partial x = 0, \tag{11.72}$$

where

$$\tilde{\kappa} = \kappa(1 + i\eta/\rho c\kappa D_1) = \kappa(1 + i\eta/kD_1), \tag{11.73}$$

where $k = \omega/c$.

46. **Sound radiation from a moving corrugated board**

A corrugated board moves with a velocity U in the y-direction, as shown in Fig. 5.5. The amplitude of the corrugation is $|\xi|$.
Determine the sound pressure field generated by the board if
(a) $U > c$.
(b) $U < c$.

SOLUTION

The coordinate along the board and in the direction of motion is y and normal thereto it is x. (a) The complex amplitude of the fluid displacement in the x-direction produced by the board is

$$\xi(\omega, y) = |xi|\exp(iK_y y),$$

where $K_y = 2\pi/\Lambda$ and Λ are the 'wave length' of the corrugation. The frequency of the field produced by the board is $f = U/\Lambda$, where the corresponding (first order) velocity amplitude in the x-direction at $x = 0$ is then

$$u(\omega, 0, y) = -i\omega\xi\exp(iK_y y). \tag{11.74}$$

Let the radiated plane pressure wave be of the form

$$p(\omega, x, y) = |p|\exp(ik_x x)\exp(ik_y y).$$

From the wave equation it follows

$$k_x^2 + k_y^2 = k^2 \equiv (\omega/c)^2.$$

The corresponding normal particle velocity in the x-direction is obtained from the momentum equation $-i\omega\rho u_x = -\partial p/\partial x$ and the velocity at $x = 0$ becomes

$$u(\omega, 0, y) = (|p|/\rho c)(k_x/k)\exp(ik_y y). \tag{11.75}$$

Equating this velocity with that in Eq. 11.74, we get

$$|p| = (\rho c \, \omega |\xi|/|k_x/k| \tag{11.76}$$

and $k_y = K_y$. This means, $k \sin \phi = K_y = \omega/U$ or

$$\sin \phi = c/U.$$

From Eq. 11.46 it follows $k_x = k\sqrt{1 - sin^2\phi} = k \cos \phi$ so that $|p| = \rho c |u_0|/ \cos \phi = \rho c \, \omega |\xi|/ \cos \phi$ and

$$p(\omega, x, y) = |p|e^{ikx \cos \phi} e^{iky \sin \phi}.$$

(b) Formally, the solution given above is valid also for $U < c$. The important difference is that $k_x = k\sqrt{1 - \sin^2 \phi} = k\sqrt{1 - (c/U)^2}$ now becomes imaginary

$$k_x = ik\sqrt{(c/U)^2 - 1}$$

and the pressure field becomes

$$p(\omega, x, y) = |p|e^{-kx\sqrt{(c/U)^2-1}} e^{iky(c/U)}.$$

In other words, it decays exponentially with distance x from the board.

47. **Dispersion relation for a higher mode in a fan duct**

Derive the modal cut-off value of the blade Mach number in Eq. 8.12 for a higher order mode in a fan duct with flow.

SOLUTION

The 'unwrapped' annular duct becomes the duct between two parallel walls separated by the distance $y = d$. The axial flow is in the x-direction and the swirling flow is in the z-direction. An acoustic mode in the duct has the y-dependence expressed by a standing wave function $\cos k_y y$, with $k_y = n\pi/d$ and the z-dependence by a traveling wave function $\exp(ik_z z)$ with $k_z = m2\pi/2\pi r_a = m/r_a$, corresponding to periodic boundary conditions, the period being the average circumference $2\pi r_a$. If we neglect reflections, the x-dependence is $\exp(ik_x x)$, where k_x is determined by k_y, k_z, and ω from the wave equation (dispersion relation). If reflections are present, a wave traveling in the negative x-direction must be included with an amplitude which is determined by boundary conditions.

In the presence of flow with components in both the axial and transverse directions with the Mach numbers M_x and M_z, the wave equation for the complex pressure amplitude $p(\omega)$ at the angular frequency ω and in the absence of sources takes the form (see Chapter 10, for example, Eq. 10.8)

$$\nabla^2 p + \left(\frac{\omega}{c}\right)^2 (1 - M_x K_x - M_z K_z)^2 p = 0, \tag{11.77}$$

where $K_x = k_x/(\omega/c)$ and $K_z = k_z/(\omega/c)$. Inserting $\nabla^2 p = -(K_x^2 - K_y^2 - K_z^2)(\omega/c)^2 p$, we note that the wave equation imposes a condition on K_x,

$$-K_x^2 - K_y^2 - K_z^2 + (1 - M_x K_x - M_z K_z)^2 = 0 \tag{11.78}$$

with the solutions

$$K_{x\pm} = -\frac{M_x(1 - M_z K_z)}{1 - M_x^2} \pm \frac{1}{1 - M_x^2}\sqrt{(1 - M_z K_z)^2 - (K_y^2 + K_z^2)(1 - M_x^2)}. \tag{11.79}$$

In this case, M_z exists only on the downstream side of the fan; on the upstream side, $M_z = 0$. The plus and minus signs in the second term correspond to a wave in the positive and negative x-direction, respectively.

It should be noted that the propagation constants has been normalized with respect to ω/c. The frequency of the mth mode in the force function in Eq. A.54 is $\omega = mB\Omega$, $k_z = mB/r_a$, and $k_y = n\pi/d$. Thus, the normalized value of k_z is $K_z = (mB/r_a)/(mB\Omega/c) = 1/M$, independent of m, where $M = r_a\Omega/c$ is the Mach number of a blade. Similarly, $K_y = n\pi r_a/mdBM$.

It follows from the dispersion relation A.55 that the pressure will decay exponentially if the square root argument in Eq. A.55 is negative ('cut-off' condition). The critical condition for cut-off of a particular mode is determined by putting the argument equal to zero. This leads to the following condition for the critical Mach number of the blade (in the present approximation the tip speed is the same as the average speed of the blade)

$$M_c = M_z \pm \sqrt{[1 + (n\pi a/mBd)^2](1 - M_x^2)}. \tag{11.80}$$

The plus sign is used if the blades and the swirling flow move in the same direction, otherwise the minus sign applies. It should be recalled that a is the mean radius of the fan and d is the width of the annulus. The integer n is the number of pressure nodes of the wave function in the span-wise direction and m the harmonic order of the blade passage frequency. For the lowest order, $n = 0$, the critical Mach number M_c will be independent of the harmonic m of the blade passage frequency. If, in addition, $M_x = 0$, a mode will propagate if the relative Mach of the blades relative to the swirling flow exceeds unity, as expected. This result is modified by the axial flow speed which reduces the critical Mach number. As an example, with $M_z = M_x = 0.5$, we get $M_c = 1.37$ for downstream radiation. For the field radiated in the upstream direction, however, where $M_z = 0$, the critical fan Mach number is only $M_c = 0.87$.

The x-dependence of the complex sound pressure amplitude is $p \propto e^{ik_x x}$, where $k_x = (\omega/c)K_x$ and $\omega = B\Omega$ is the (angular) blade passage frequency. If K_x, as obtained from Eq. A.55, has an imaginary part, $i\alpha$, the pressure decays exponentially with x,

$$|p| \propto e^{-\alpha(\omega/c)x} = e^{-\alpha mBM(x/r_a)}, \tag{11.81}$$

where r_a is the mean radius of the fan. In other words, the sound pressure level decreases $\approx 8.7\alpha mBM(x/r_a)$ dB in a distance x (we have put $20\log(e) \approx 8.7$).

48. Tyndall's paradox

In our explanation of Tyndall's paradox in Chapter 9 sound traveling over the ocean downwind was found to have a shorter range than sound traveling over the ocean against the wind; we proposed that this result was a result of 'molecular' absorption in the air. Determine whether this explanation makes sense quantitatively as it was found that the corresponding values of the 'audibility range' of a single frequency sound could vary from 12 to 4 miles.

SOLUTION

The distance dependence of the sound pressure from the source is given by $(1/r)\exp(-\alpha r)$. Thus, if $r_1 = 12$ and $r_2 = 4$ miles, we must have $(1/r_1)\exp(-\alpha_1 r_1) = (1/r_2)\exp(-\alpha_2 r_2)$, or

$$(\alpha_2 r_2 - \alpha_1 r_1)_{dB} = 20\log(r_1/r_2) \approx 9.5. \tag{11.82}$$

Consider, for example, a frequency of 500 Hz. Then, at a relative humidity of 20 percent and a temperature of $50°F$, the attenuation $(\alpha_2 r_2)_{dB}$ in 4 miles (downwind) is ≈ 18 dB and at a humidity of 80 percent and the same temperature, the attenuation $(\alpha_1 r_1)_{dB}$ in 12 miles (upwind) is about 9 dB. In other words, we see that with a small adjustment in temperature and/or humidity, Eq. 11.82 can be satisfied. This shows that this explanation makes quantitative sense.

49. **Refraction in an atmosphere with a constant temperature gradient**

In Section 9.6.2, the travel distance of an acoustic ray from a high altitude source to the ground in a vertically stratified atmosphere was expressed by numerical integration.
(a) Obtain an analytical expression for the travel distance in a windless atmosphere in which the temperature decreases linearly with height from T_0 at the ground to $T_1 = T(H)$ at the height H.
(b) Do the same for the travel time.

SOLUTION

(a) If $\Delta T = T_0 - T_1$, we have

$$T(z) = T_0 - (z/H)\Delta T \tag{11.83}$$

and $c^2/c_1^2 = T/T_1$. With reference to Fig. 9.15 and Eq. 9.26, we have

$$\cos^2 \psi = \cos^2 \psi_1 [1 + (\Delta T/T_1)(1 - z/H)]. \tag{11.84}$$

The critical angles Ψ_c for shadow formation are then given by

$$\cos \Psi_c = \pm\sqrt{T_1/T_0}. \tag{11.85}$$

For sound rays that reach the observer on the ground, the distance of wave travel between source and observed can be written

$$r = \int_0^H dz/\sin\psi = \int_0^H dz/\sqrt{\sin^2\psi_1 - (\Delta T/T_1)(1 - z/H)\cos^2\psi_1}$$
$$= (2T_1 H/\Delta T)[1 - \sqrt{1 - (\Delta T/T_1)\cos^2\psi_1}/(\sin\Psi_1 \cot^2\Psi_1). \tag{11.86}$$

(b) The travel time is

$$t_r = \int_0^H \frac{dx}{c(z)\sin\psi} = \frac{H}{c_1}\int_0^1 \frac{dy}{(c/c_1)\sin\psi}, \tag{11.87}$$

where $y = z/H$.

Again, assuming a linear decrease of temperature with height, we have

$$c/c_1 = \sqrt{1 + \epsilon(1 - y)}, \tag{11.88}$$

where $\epsilon = \Delta T/T_1$ and it follows that

$$t_r = \frac{H}{c_1}\int_0^1 \frac{dy}{\sqrt{[1 + \epsilon(1 - y)][1 - \cos^2\psi_1(1 + \epsilon(1 - y))]}}. \tag{11.89}$$

This can be rewritten in the form of a standard elementary integral with the result

$$t_r = \frac{H}{c_1}\frac{1}{\epsilon\psi_1}[\arcsin(2\epsilon\cos^2\psi_1 + \cos(2\psi_1)) - \arcsin(\cos(2\psi_1))] \equiv \frac{H}{c_1}F_1(\psi_1). \tag{11.90}$$

It the zero point $t = 0$ on the time scale is chosen to be when the source is seen straight overhead, the location of the source is at time t is Vt. The source has then traveled a distance Vt_r. We can then express the horizontal distance traveled by the source as $Vt_r - Vt$ which must equal the horizontal distance traveled by the sound. Thus, we can express the time of arrival of the sound from the relation

$$Vt_r - Vt = \int_0^H \frac{dz \cos \psi}{\sin \psi} \equiv H F_2(\psi_1).$$

(11.91)

The corresponding normalized time is $\tau = t/(H/c_1)$. Then, from Eq. 11.90 it follows

$$\tau = F_1(\psi_1) - (1/M) F_2(\psi_1),$$

(11.92)

where $M = V/c_1$.

Appendix A

Supplementary Notes

A.1 Fourier Series and Spectra

A.1.1 Fourier Transform. Spectrum of Finite Harmonic Wave Train

Consider an oscillation of finite length, a 'chirp,' such that $F(t) = A\cos(\omega_0 t)$ between $t = -t_0$ and $t = t_0$ and zero elsewhere. According to Eq. 2.67, the Fourier amplitude of this function is

$$F(v) = A \int_{-t_0}^{t_0} \cos(\omega_0 t) e^{i\omega t} dt = (A/2) \int_{-t_0}^{t_0} [e^{i(\omega+\omega_0)t} + e^{i(\omega-\omega_0)t}] dt$$

$$= (At_0/2)[\tfrac{\sin(X_+)}{X_+} + \tfrac{\sin(X_-)}{X_-}], \tag{A.1}$$

where $\omega = 2\pi v$ and $X_+ = (\omega+\omega_0)t_0$ and $X_- = (\omega-\omega_0)t_0$, which, for computational purposes, we express as $X_\pm = (\Omega \pm 1)2\pi t_0/T_0$, where $T_0 = 2\pi/\omega_0$ is the period and $\Omega = \omega/\omega_0$. With reference to Section 2.6.3, the energy spectrum density is $E(v) = 2|F(v)|^2$ and it should be recalled that only positive frequencies are involved in this expression. In Fig. A.1 are shown the energy spectra for signal lengths from 0.5 to 8 periods. The 0.5 period signal covers the central maximum of the signal and, unlike the other cases, has a time average different from zero. This is the reason why the corresponding spectrum is quite different from the others; it has a maximum at zero frequency. Such a pulse is typical for explosive events which are rich in low frequencies. As the signal covers an integer number of periods, so that the time average is zero, the width of the spectrum decreases with an increasing number of periods, illustrating the uncertainty relation.

A.1.2 Fourier Transform and Energy Spectrum

With reference to Eq. 2.78, the proof of the relation

$$\int_0^\infty I^2(t)dt = \int_0^\infty E(v)dv, \tag{A.2}$$

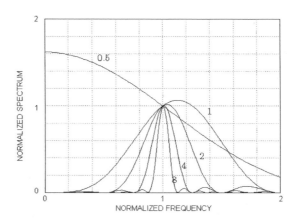

Figure A.1: The spectrum of the finite harmonic wave train in Eq. A1, between $-t_0$ and t_0, for values of t_0/T equal to 0.5, 1, 2, 4, and 8, corresponding to signal lengths from half a period to eight periods. The normalized frequency is $\Omega = \omega/\omega_0$ and the normalized spectrum is $E(\nu)/E(\nu_0)$.

where $E(\nu) = 2|I(\nu)|^2$, can be carried out as follows. The function $I(t)$, which we think of here as a current, is expressed in terms of its Fourier transform

$$I(t) = \int_{-\infty}^{\infty} I(\nu)\,d\nu. \tag{A.3}$$

The squared function can then be expressed as

$$I^2(t) = \int_{-\infty}^{\infty} I(\nu)e^{-i2\pi\nu t}\,d\nu \int_{-\infty}^{\infty} I(\nu')e^{-i2\pi\nu'}\,d\nu' = \int\int e^{-i2\pi(\nu+\nu')t}\,d\nu d\nu'. \tag{A.4}$$

Integrating over time from minus to plus infinity then leads to

$$\int_{-\infty}^{\infty} e^{-i2\pi(\nu+\nu')t}\,dt = \delta(\nu + \nu'), \tag{A.5}$$

which follows from Eq. 2.71 in the text.

From the property of the delta function $\int \delta(t - t')F(t')\,dt' = F(t)$ it follows that the integration over ν' in Eq. A.4 yields a contribution only if $\nu' = -\nu$. This means that the factor $I(\nu)I(\nu') = I(\nu)I(-\nu)$. It was shown in connection with Eq. 2.69, $I(-\nu) = I^*(\nu)$. Then, with $I(\nu)I^*(\nu) = |I(\nu)|^2$, Eqs. A.4 and A.5 show that

$$\int_{-\infty}^{\infty} I^2(t)\,dt = \int_{-\infty}^{\infty} |I(nu)|^2 = 2\int_{0}^{\infty} |I(\nu)|^2\,d\nu \equiv \int_{0}^{\infty} E(\nu)\,d\nu. \tag{A.6}$$

In the next to the last step in this equation we used the property that $I(\nu)I^*(\nu)$ does not change if ν is changed to $-\nu$.

A.1.3   Measurement of Intensity by Means of a Probe

The basic idea for the intensity probe is outlined in Section 3.2.3; the formal mathe-matical basis is similar to that in Sectione A.1.2.

The sound pressure $p(x, t)$ is expressed in terms of its Fourier amplitude $p(\nu)$, i.e.,

$$p(x, t) = \int p(x, \nu)e^{-i2\pi \nu t}d\nu. \tag{A.7}$$

In an analogous manner the particle velocity $u(x, t)$ in the x-direction is expressed in terms of its Fourier amplitude $u(x, \nu)$. Then, from the momentum equation $\rho du/dt = -\partial p/\partial x$ it follows that $u(x, \nu) = (1/i\omega\rho)\partial p/\partial x$.

The intensity in the x-direction is

$$I(t) = p(x, t)u(x, t) = \int p(x, \nu)e^{-i2\pi \nu t}d\nu \int (1/i\omega\rho)\partial p(x, \nu')/\partial x e^{-i2\pi \nu' t}d\nu'$$
$$= (1/i\omega\rho)\int\int e^{-i2\pi(\nu+\nu')t}d\nu d\nu'. \tag{A.8}$$

Integrating $I(t)$ over all times produces $\delta(\nu + \nu')$ in the integral on the right-hand side and integration over ν' yields a contribution only if $\nu' = -\nu$ and we obtain

$$\int I(t)dt = (1/i\omega\rho)\int p(x, \nu)[\partial p(x, -\nu)/\partial x]\,d\nu. \tag{A.9}$$

The microphones are located at $x - d/2$ and $x + d/2$ at which points the pressures are p_1 and p_2. We put $p(x) = (p_1 + p_2)/2$ and express the gradient as $\partial p(x)/\partial x = (p_2 - p_1)/d$. With $p(-\nu) = p^*(\nu)$, the integrand in Eq. A.9 becomes $(p_1 + p_2)(p_2^* - p_1^*)$. Neglecting the term $|p_2|^2 - |p_1|^2$ and realizing that $p_2 p_1^*$ is the complex conjugate of $p_1 p_2^*$, the remaining $p_1 p_2^* - p_2 p_1^*$ is twice the imaginary part of $p_1 p_2^*$. Thus, we obtain

$$\int I(t)dt = (1/i\omega\rho d)\int 2\Im\{p_1(\nu)p_2^*(\nu)\}\,d\nu \equiv \int I(\nu)d\nu, \tag{A.10}$$

where the intensity spectrum is

$$I(\nu) = (2/\omega\rho d)\Im\{p_1(\nu)p_2^*(\nu)\}. \tag{A.11}$$

With the signals from the two microphones analyzed with a two-channel analyzer, the quantity $p_1(\nu)p_2^*(\nu)$, the cross spectrum density, is obtained directly from the two-channel FFT analyzer.

A.2   Radiation from a Circular Piston in an Infinite Wall

The piston has a radius a and the velocity amplitude U is uniform over the piston. The normal velocity amplitude over the surrounding rigid wall is zero. This boundary condition can be satisfied by the piston pair in Fig. 5.2 radiating into free field (no baffle). By symmetry, this field will have zero normal velocity over the infinite vertical plane containing the pistons. Thus, the fields in the hemispheres to the right of the pistons will be the same in the two cases.

The piston pair can then can be described in terms of a uniform monopole dis-tribution with a flow strength $2\rho U$ per unit area of the piston and a corresponding acoustic source strength $-i\omega 2\rho U = 2Q$, where U is the velocity amplitude, ω, the angular frequency, and $Q = -i\omega\rho U$ the acoustsic source strength per unit area.

A.2.1 The Far Field. Radiated Power

Then, using the known sound pressure field from a monopole, the sound field from the piston is

$$p = 2Q \int_0^a \frac{e^{ikh}}{4\pi h} dS', \tag{A.12}$$

where $Q = -i\omega\rho U$, $h = |\boldsymbol{r} - \boldsymbol{r}'|$, and $\boldsymbol{r}$ and $\boldsymbol{r}'$ are the position vectors of the field point and the source point, respectively. With θ and ϕ being the polar and azimuthal angles of the field point, the far field approximation for h is $h \approx r - r' \cos\phi \sin\theta$ ($r \gg r'$) and for p,

$$p \approx \frac{2Qe^{ikr}}{2\pi r} \int_0^{2\pi} \int_0^a e^{-ikr' \cos\phi \sin\theta} r'dr'd\phi \quad \text{(far field)}. \tag{A.13}$$

The integral over ϕ is known to be the 0th order Bessel function

$$J_0(z) = (1/2\pi) \int_0^{2\pi} e^{-iz \cos\phi} d\phi. \tag{A.14}$$

Thus,

$$p \approx \frac{2Qe^{ikr}}{r} \int_0^a J_0(kr' \sin\theta) r'dr'. \tag{A.15}$$

The Bessel function of 1st order is given by

$$z J_1(z) = \int_0^z z J_0(z) \, dz. \tag{A.16}$$

The sound pressure amplitude in the far field can then be written

$$p = 2Q \frac{e^{ikr}}{r} a^2 \frac{J_1(ka \sin\theta)}{ka \sin\theta}. \tag{A.17}$$

From the power expansion

$$J_1(z) = z/2[1 - (z/2)^2/(1! \cdot 2) + (z/2)^4/(2! \cdot 2 \cdot 3) - \ldots] \tag{A.18}$$

it follows that with $ka \ll 1$, $p \approx (2Q) \exp(ikr)/(4\pi r)$, which is the field from a monopole with the acoustic source strength $2Q$.

The angular distribution of the intensity is expressed by

$$I(\theta)/I(0) = \left[\frac{2J_1(ka \sin\theta)}{ka \sin\theta} \right]^2. \tag{A.19}$$

It is left as a problem to calculate the total radiated power into the hemisphere. If r is the radiation resistance per unit area of the piston, this power can be expressed also as $(1/2)r(\pi a^2)U^2$ and a part of the problem is to show that the corresponding normalized radiation resistance is

$$\theta_r = 1 - \frac{J_1(2ka)}{ka} \approx (ka)^2/2. \tag{A.20}$$

Thus, at low frequencies, the radiated power $\rho c\theta_r \pi a^2 U^2/2$ is proportional to the square of both the frequency and the area of the piston.

A.2.2 Near Field and the Radiation Impedance

Although the energy consideration above enabled us to determine the radiation resistance, it yielded no information about the reactive part of the radiation impedance.
To determine it, we have to use the near field and proceed as follows.

Figure A.2: Concerning the calculation of the radiation impedance of a circular piston source
in an infinite baffle.

With reference to Fig. A.2, we calculate the sound pressure at a location dS on
the piston by adding the field contributions from all the other elements dS' and then
integrate over the piston to determine the total (reaction) force on the piston. Thus,

$$2Q \int_S \int_{S'} \frac{e^{ikh}}{4\pi h} \, dS \, dS', \qquad (A.21)$$

where $h = \sqrt{r^2 + r'^2 - 2rr' \cos(\phi - \phi')}$.

The integration can be carried out in a different way which makes it possible to
express the result in terms of known functions as follows. Consider the circle of
radius r in the figure. The integrated pressure over this circle produced by the piston
element dS can be expressed in terms of h and the angle α by varying h from zero
out to the rim of the circle (where $h = 2r \cos \alpha$) as α goes from $-\pi/2$ to $\pi/2$. The
elementary area in this integration is then $h\alpha \, dh$ and the force one the circle of radius r

$$2Q \int_{-\pi/2}^{\pi/2} \int_0^{2r \cos \alpha} \frac{e^{ikh}}{4\pi h} h \, d\alpha \, dh = (2Q/4\pi) \int_0^{\pi/2} (2/ik)[\exp(ik2r \cos \alpha) - 1].$$
$$(A.22)$$

We now make use of the reciprocity theorem that says that the force produced by
the piston element dS on the circular area (radius r) is the same as the force pdS
produced on the element dS by the radiation from the circular area. This reciprocity
is a consequence of the symmetry of Green's function, examples of which are given
in Eq. 3.82 and A.59. Thus, having obtained an expression for the force on dS as a
function of r, the total pressure on the piston is then obtained by integrating over r
and multiplying by a factor of 2 to account for equality of the force on the circular
area of radius r and the force on dS. Thus,

$$\text{force} = 2 \cdot (Q/\pi) \int_0^a 2\pi r \, dr \int_0^{\pi/2} (1/ik)[\exp(ik2r \cos \alpha) - 1]. \qquad (A.23)$$

Note that

$$\int_0^{\pi/2} \cos(z\cos\alpha)\,d\alpha = \tfrac{\pi}{2}J_0(z) = \tfrac{\pi}{2}(1 - z^2/2^2 + z^4/2^2\cdot 4^2\ldots) \qquad (A.24)$$

$$\int_0^{\pi/2} \sin(z\cos\alpha)\,d\alpha = \tfrac{\pi}{2}S_0(z) = \tfrac{\pi}{2}(z - z^3/1^2 3^2 + z^5/1^2 3^2 5^2 - \ldots),$$

where J_0 and S_0 are the Bessel and Struve functions of 0th order. Thus,

$$\int_0^{\pi/2} (e^{ik2r\cos\alpha} - 1)\,d\alpha = -\frac{\pi}{2}[1 - J_0(2kr) - i\,S_0(2kr)]. \qquad (A.25)$$

Furthermore,

$$\int_0^a [1 - J_0(2kr) - i\,S_0(2kr)]r\,dr = \frac{a^2}{2}[1 - \frac{2J_1(2ka)}{2ka} - i\frac{2S_1(2ka)}{2ka}], \qquad (A.26)$$

where

$$\int_0^z J_0(z)z\,dz = zJ_1(z)$$
$$\int_0^z S_0(z)z\,dz = zS_1(z)$$
$$J_1(z) = (z/2)[1 - \frac{(x/2)^2}{1!\cdot 2} + \frac{(z/2)^4}{2!\cdot 2\cdot 3 - \ldots}]$$
$$S_1(z) = (2/\pi)[\frac{z^2}{1^2\cdot 3} - \frac{z^4}{1^2\cdot 3^2\cdot 5} + \ldots]. \qquad (A.27)$$

Then, with $Q = -i\omega\rho U = -ik\rho cU$, the total sound pressure on the piston in Eq. A.24 becomes

$$\pi a^2 U\rho c\left[1 - \frac{2J_1(2ka)}{2ka} - i\frac{2S_1(2ka)}{2ka}\right] \qquad (A.28)$$

from which follows the normalized specific radiation impedance

$$\zeta_r = p/(\rho c\,U) \equiv \theta_r + i\kappa = \left[1 - \frac{2J_1(2ka)}{2ka} - i\frac{2S_1(2ka)}{2ka}\right]. \qquad (A.29)$$

The resistance term is the same as found in Eq. A.20.

At low frequencies, $ka \ll 1$, we have $J_1(2ka) \approx ka - (ka)^3/2$ and $S_1(2ka) \approx *2/\pi)(2ka)^2/3$ with corresponding values

$$\theta_r \approx (ka^2/2)$$
$$\chi_r \approx 8ka/(3\pi), \quad (ka \ll 1). \qquad (A.30)$$

This reactance corresponds to a mass end correction of $(8/3\pi)a \approx 0.85a$.

A.3   Radiation from Pistons into Ducts

A.3.1   Rectangular Piston in a Rectangular Duct

A rectangular piston source is located in the end wall at $x = 0$ of a rectangular duct, as shown in Fig. A.3. The wall is acoustically hard so that the only axial velocity at $x = 0$ is contributed by the piston and it is assumed to be uniform with the amplitude U.

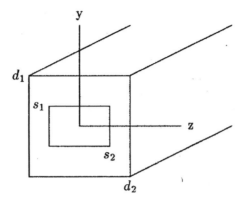

Figure A.3: Rectangular piston source radiating into a rectangular duct.

The coordinate axes are chosen as shown with $y = 0$ and $z = 0$ in the center of the duct between the walls at $y = \pm a_1 = \pm d_1/2$ and $z = \pm a_2 = \pm d_2/2$. The piston boundaries are $y \pm s_1$ and $z = \pm s_2$. Since the source distribution is symmetrical with respect to $y, z = 0$, the wave field will be symmetrical also. Accounting for this symmetry, the appropriate solution to the wave equation $\nabla^2 p + (\omega/c)^2 p = 0$ for the complex sound pressure amplitude $p(x, y, \omega)$ is of the form

$$p(x, y, \omega) = \sum_{m,n=0}^{\infty} F(y, z)e^{ik_x x}, \tag{A.31}$$

where $F(y, z)$ corresponds to standing waves,

$$F(y, z) = \cos(k_y y)\cos(k_z z) \tag{A.32}$$

and the x-depedence to a traveling wave. The solution is the sum of modes with amplitudes $P_{m,n}$ which at this point are unknown. It should be noted that we have used only $\cos(k_y y)\cos(k_z z)$ functions (and not $\sin()$- functions) terms) in $F(y, z)$ to make $F(y, z)$ an even function in y, z since the source function is even, as mentioned earlier. If this symmetry had not been anticipated, the coefficients for the modes with $\sin()$ functions would have been found to be zero in the formal calculations below.

The velocity amplitude normal to the duct walls, being proportional to $\partial p/\partial y$ and $\partial p/\partial z$, must be zero. Thus, $\sin(k_y a_1) = 0$ and $\sin(k_z a_2) = 0$, i.e.,

$$k_y = m\pi/a_1, \qquad k_z = n\pi/a_2, \tag{A.33}$$

where m, n are integers. We have assumed that the duct is infinitely long so that we need not be concerned with a reflected wave traveling in the negative x-direction.

Of particular interest is the propagation constant k_x, and it follows from the wave equation that $k_x^2 + k_y^2 + k_z^2 = (\omega/c)^2$ and hence,

$$k_x = \sqrt{(\omega/c)^2 - k_y^2 - k_z^2} = (\omega/c)\sqrt{1 - (\omega_{m,n}/\omega)^2}, \tag{A.34}$$

where $\omega_{m,n} = c\sqrt{k_y^2 + k_z^2}$ with k_y, k_z given in Eq. A.33.

The mode $(1,0)$ has the cut-off frequency $f_{1,0} = \omega_{1,0}/2\pi = 1/2a_1 = 1/d_1$. (Because of the symmetry of the wave field, as discussed above, the wave function with a cut-off wavelength equal to $2d_1$ and other asymmetrical functions are not excited).

The piston source is located in an acoustically hard wall at the beginning of the duct at $x = 0$ and is described in terms of the axial velocity amplitude u_x which is assumed to be constant U over the piston and zero over the rest of the wall.

From the equation of motion $-i\omega\rho u_x = -\partial p/\partial x$ and the expression for p in Eq. A.31 we obtain for the axial velocity field

$$u_x(x, y, z, \omega) = \sum_{mn=0}^{\infty} (k_x/k)(P_{m,n}/\rho c)F(y, z)e^{ik_x x}, \tag{A.35}$$

where $F(y, z)$ is given in Eq. A.32. At $x = 0$, this velocity must equal the velocity distribution of the source, and to be able to utilize this equality for determination of $P_{m,n}$, we expand the source velocity distribution in a Fourier series,

$$u(0, y, \omega) = \sum_{mn=0}^{\infty} U_{m,n}F(y, z). \tag{A.36}$$

For $m \neq n$, the normal mode functions $F(y, z) = \cos(k_y y)\cos(k_z z)$ are orthogonal, and by multiplying both sides by $F(y, z)$ and integrating over the area of the duct, we obtain, for $m, n > 0$,

$$U_{m,n} = \int_{-s_1}^{s_1}\int_{-s_2}^{s_2} U F(y, z)\,dydz = U\frac{s_1 s_2}{a_1 a_2}\beta_{m,n} S(k_y s_1)S(k_z s_2), \tag{A.37}$$

where $S(\xi) = \sin(\xi)/\xi$ and $\beta_{0,0} = 1$, $\beta_{m,0} = \beta_{0,n} = 2$, $\beta_{m,n} = 4$.

By equating the velocities in Eqs. A.36 and A.37 at $x = 0$ we obtain

$$P_{m,n} = (\rho cU/K_x)U_{m,n} \quad \text{where}$$
$$K_x = k_x/k = \sqrt{1 - (k_{m,n}/k)^2} = \sqrt{1 - (\omega/\omega_{m,n})^2}. \tag{A.38}$$

The radiation impedance of the piston is

$$z_r = p_a/U, \tag{A.39}$$

where p_a is the average pressure over the piston and is readily obtained by integrating p in Eq. A.31 over the piston. We then obtain

$$z_r = \rho c \sum_{m,n} U_{m,n}\beta_{m,n}S(k_y s_1)S(k_y s_2)/K_x, \tag{A.40}$$

where $\beta_{m,n}$ and $S(\xi$ are given in Eq. A.37.

The contribution from the plane wave is the $0, 0$-term, which is simply the wave impedance of the plane wave multiplied by the ratio of the piston area and the duct area, as expected and consistent with Eq. 6.37. The nature of the contribution from the (m, n) mode, characterized by $k_{m,n}$ and the corresponding cut-off frequency $\omega_{m,n}$, depends on whether ω is less than or greater than the cut-off value $\omega_{m,n}$. If $\omega > \omega_{m,n}$, K_x is real, and the wave propagates and carries energy and contributes to the radiation resistance.

On the other hand, if $\omega < \omega_{m,n}$, we have $K_x = i\sqrt{(\omega_{m,n}/\omega)^2 - 1}$, and the contribution to the impedance in Eq. A.37 will be a proportional to -i, which corresponds to a mass reactance.

Let us express explicitly this mass reactance in the important case when $\omega << \omega_n$ so that $K_x \approx \omega_{m,n}/\omega = k_{m,n}/k$,

$$\zeta_r \approx \rho c (s_1 s_2/a_1 a_2) - i\omega\rho\delta$$
$$\delta = \sum_{m,n} \frac{1}{k_{m,n}} \beta_{m,n} U_{m,n} S(k_y s_1) S(k_z s_2). \tag{A.41}$$

The mass end correction δ of the piston is contributed by all modes m, n except the plane wave $0, 0$. Thus, the mass load is contributed by the evanescent modes and can be thought of as the mass of a layer of air on the piston with a thickness δ, consistent with the observation in Eq. 6.37.

A.3.2 Circular Piston in a Circular Tube

Another example involves a circular concentric piston in a circular tube which can be used to simulate an ordinary loudspeaker. The radii of the piston and the duct are r_0 and R, respectively.

With reference to Section 6.2.1, the general expression for the sound pressure field is now

$$p(r, x, \omega) = \sum_{0}^{\infty} P_n J_0(k_r r) e^{ik_x x}, \tag{A.42}$$

where $J_0(k_r r)$ is the Bessel function of zeroth order, corresponding to $\cos(k_y y)$ for the rectangular duct. Both these functions equal unity when the argument is zero.

The related radial velocity field is a similar sum containing the first order Bessel function $J_1(k_r r)$, which[1] corresponds to $\sin(k_y y)$ in the rectangular case, and since we consider rigid walls, the velocity amplitude in the radial direction is zero and k_r is determined by

$$J_1(k_r R) = 0. \tag{A.43}$$

The first root is known to be $k_r R = 3.83$; it corresponds to $k_y a = \pi = 3.14$ in the rectangular case; the second root is $k_r R = 6.93$, which corresponds to $k_y a = 2\pi = 6.28$.

The axial velocity field is

$$u_x = \sum_{0}^{\infty} (P_n/\rho c) K_x J_0(k_r r) e^{ik_x x} \tag{A.44}$$

---

[1] $d J_0(z)/dz = -J_1(z)$

and the coefficients P_n are determined by equating this velocity with the known velocity in the plane of the piston at $x = 0$. To do that, we expand this velocity in a Bessel function series, i.e., in terms of the characteristic modes in the duct,

$$u(0) = \sum U_n J_0(k_r r), \tag{A.45}$$

where the coefficients are

$$U_n = (1/N) \int_0^{r_0} u(r) J_0(k_r r) r \, dr, \tag{A.46}$$

where r_0 is the piston radius and $N = \int_0^R J_0^2(k_r r) r \, dr$.

Since $u(r) = 0$ for $r > r_0$, the integration over u extends only over the piston. With $u(r) = U$ being a constant for $r < r_0$, the integral over $u(r)$ becomes $[2J_1(k_r r_0)/k_r r_0]r_0^2/2$. For small values of the argument we have $J_1(x) \approx x/2$. Similarly, the integral in the denominator is $N = J_0^2(k_r R)R^2/2$ so that

$$U_n = [2J_1(k_r r_0)/k_r r_0]/J_0^2(k_r R)R^2. \tag{A.47}$$

By equating the velocities in Eqs. A.45 and A.46 at $x = 0$, we get for the amplitudes of the nth pressure mode

$$P_n = \rho c U A_n / K_x. \tag{A.48}$$

The average pressure over the piston is

$$p_a = \sum (1/\pi r_0^2) \int_0^{r_0} 2\pi P_n J_0(k_r r) r \, dr = \sum P_n 2[J_1(k_r r_0)/k_r r_0]. \tag{A.49}$$

The radiation impedance of the piston is then

$$z_r = p_a/U = \rho c \sum_0^\infty \frac{r_0^2}{R^2} \frac{1}{K_x} \left(\frac{2J_1(k_r r_0)}{k_r r_0} \right)^2 \frac{1}{J_0^2(k_r R)}. \tag{A.50}$$

The plane wave contribution, corresponding to $n = 0$, is $(r_0/R)^2 \rho c$, as expected.

We consider now the low frequencies much lower than the first cut-off frequency. In that case $K_x = ik_r/k$ and the impedance becomes

$$z_r = \rho c(r_0^2/R^2) - i\rho ck\Delta, \tag{A.51}$$

where the mass end correction δ is

$$\delta = r_0 \sum_1^\infty \frac{r_0}{R} \left[\frac{2J_1(k_r r_0)}{k_r r_0} \right]^2 \frac{1}{k_r R}. \tag{A.52}$$

As $r_0 \to R$, $J_1(k_r r_0) \to 0$, and we note that $\delta \to 0$.

In Fig. 6.5 we have plotted $\Delta/\sqrt{A_p}$, where A_p is the piston area, as a function of r_0/R together with similar results for radiation of a circular and square piston into a square duct. The normalized radiation impedance can be expressed as

$$\zeta_r = A_p/A - ik\Delta, \tag{A.53}$$

where A is the duct area and A_p the piston area.

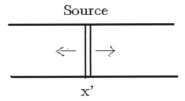

Figure A.4: Concerning the one-dimensional Green's function; a 'pillbox' source with two pistons moving in opposite directions with the same amplitude.

A.4 One-Dimensional Green's Functions

A.4.1 Free Field or Infinite Duct and no Mean Flow

As an introduction to our discussion of Green's function, we consider the sound field produced by a source which can be regarded as a thin 'pillbox' with the side walls consisting of plane pistons moving 180 degrees out of phase in harmonic motion, as indicated in Fig. A.4. Thus, the pistons move outward at the same time, thus simulating a source of oscillatory flow out of the box. A similar oscillatory flow can be created by time dependent mass or heat injection. The box is centered at $x = x'$. If the velocity amplitude of the piston on the right is u it will be $-u$ for the piston on the left. The amplitude of the corresponding oscillatory mass flow rate out of the box per unit area is then $2\rho u$. If this rate were to be created by a source of mass injection at a rate Q_f per unit volume distributed uniformly over the box volume it follows that[2]

$$\int Q_f \, dx = 2\rho u. \tag{A.54}$$

If sound is to be generated, Q_f must be time dependent and we define the *acoustic source strength* per unit volume as

$$Q = \frac{dQ_f}{dt}. \tag{A.55}$$

For harmonic time dependence, the acoustic source distribution which is equivalent to the pillbox source must be such that

$$\int Q \, dx = -i\omega 2\rho u. \tag{A.56}$$

We now let the pillbox source be so thin that the equivalent acoustic source function becomes a delta function, i.e., $Q = Q_0 \delta(x - x')$, where x' is the coordinate of the center of the box. From Eq. A.56 then follows

$$Q_0 = -i\omega 2\rho u(x'). \tag{A.57}$$

---

[2]For a heat source, Q_f has to be replaced by $(\gamma - 1)H/c^2$, where H is the rate of energy absorption per unit volume in the source region, γ the specific heat ratio, and c the sound speed.

In calculating the sound pressure field produced by this source, we consider first the case when the source is in free field, or, equivalently, in an infinitely long duct with acoustically hard walls. The sound pressure field then consists of two waves traveling in opposite directions. Each of these waves has the amplitude $\rho c u(x') = i q_0 c/2\omega = i q_0/2k_0$, i.e.,

$$p(x, \omega) = \frac{i Q_0}{2k_0} e^{ik_0|x-x'|}, \tag{A.58}$$

where $k_0 = \omega/c$ and $|x - x'| = x - x'$ for $x > x'$ and $|x - x'| = -(x - x')$ for $x < x'$.

If $Q_0 = 1$, the total source strength is unity (in the particular units used) and the acoustic source function is the delta function $\delta(x - x')$.

For this unit source function, the pressure field is called the one-dimensional Green's function

$$G(x, x', \omega) = \frac{i}{2k_0} e^{ik|x-x'|}. \tag{A.59}$$

Explicitly, this has the dimension of length.

Since we are dealing with a linear system we can use Green's function to express the sound pressure field produced by an arbitrary source distribution $Q(x')$ per unit volume by linear superposition

$$p(x, \omega) = \int q(x')G(x, x', \omega)\, dx'. \tag{A.60}$$

Since Green's function is always used in conjuction with a source function in a final calculation of a field, the problem of physical dimensions resolves itself (the source function has the same dimension $[p/L^2]$, Green's function and has the dimension $[1/k] = [L]$, and dx' the dimension $[L]$ so that the integral will have the dimension of pressure).

In the discussion above, we constructed Green's function from well-known elementary properties of waves. In a formal study, Green's function generally is introduced via the wave equation with source terms included. Thus, to account for a volume source distribution of mass flow injection (creation) into a fluid at a rate of Q_f per unit volume, the conservation of mass equation takes the form

$$\frac{\partial \rho}{\partial t} + \rho \operatorname{div} \vec{u} = Q_f. \tag{A.61}$$

By taking the divergence of the linearized momentum equation

$$\rho \frac{\partial \vec{u}}{\partial t} = -\operatorname{grad} p \tag{A.62}$$

and the time derivative of Eq. A.61 and eliminating $\vec{u}$ between them, we obtain the equation for the complex sound pressure amplitude $p(x, \omega)$

$$\nabla^2 p + (\omega/c)^2 p = -Q(x, y, z, \omega), \tag{A.63}$$

where $Q = pda\, Q_f t$, as given by Eq. A.55, is the acoustic source function. Then, by putting the source function equal to a delta function, we define Green's function as the solution to Eq. A.63.

In our one-dimensional problem, we note that although Green's function is continuous at $x = x'$, the derivative dG/dx is not. On the right-hand side of x' we get $dG/dx = -1/2$ and on the left-hand side $dG/dx = 1/2$ so that the discontinuity in dG/dx is -1. This property can be obtained also from the wave equation

$$\frac{d^2G}{dx^2} + (\omega/c)^2 G = -\delta(x - x') \tag{A.64}$$

by integrating both sides over x over the source region (pillbox) to yield, with $G' = dG/dx$,

$$G'_+ - G'_- = -1. \tag{A.65}$$

Although our construction of Green's function above may be helpful from a physical standpoint it is not the simplest way to obtain it. By using the Fourier transform (expansion in terms of plane wave components)

$$G(x) = \frac{1}{2\pi} \int_{-\infty}^{\infty} G(k)e^{ikx} \, dk \tag{A.66}$$

and $\delta(x - x') = (1/2\pi) \int \exp(ik(x - x') \, dk$, insertion into Eq. A.64 yields

$$G(k) = \frac{e^{-kx'}}{k^2 - (\omega/c)^2} \tag{A.67}$$

and from Eq. A.66

$$G(x) = \frac{1}{2\pi} \int_{-\infty}^{\infty} \frac{e^{ik(x-x')}}{(k - \omega/c)(k + \omega/c)} \, dk. \tag{A.68}$$

The poles in the integrand are at ω/c and $-\omega/c$ and accounting for the physical requirement of a slight damping so that ω/c has an imaginary part $i\epsilon$, the first pole lies a little above and the second a little below the real k-axis in the complex k-plane. Then, for $x - x' > 0$ we can evaluate the integral by closing the integration path along a semi-circle in the upper half of the k-plane (where k has a positive imaginary part) to obtain, from the residue theorem,

$$G(x, x', \omega) = \frac{i}{2(\omega/c)} e^{i(\omega/c)(x-x')}. \tag{A.69}$$

For $x - x' < 0$ the path is closed in the lower half-plane and the expression for G is the same with $\exp[i(\omega/c)(x - x')]$ replaced by $\exp[-i(\omega/c)(x - x')]$. This is the same result as obtained earlier.

A.4.2 Finite Duct

The Green's function that we have discussed so far referred to an infinitely long duct so that only outgoing waves were involved. In a duct of finite length, reflections from the ends will usually occur and this will be accounted for in the following discussion.

The source is the same as before, i.e., the delta function pillbox at $x = x'$ and the pressure field produced by this source will be Green's function for the duct. The ends of the duct are placed at $x = 0$ and $x = L$ and the boundary conditions at these ends will be specified in terms of the pressure reflection coefficients $R(0)$ and $R(L)$ (Fig. A.5).

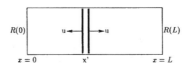

Figure A.5: Concerning the one-dimensional Green's function of a tube with walls at $x = 0$ and $x = L$ with the pressure reflection coefficients $R(0)$ and $R(L)$.

We now construct Green's function as follows. To the right and to the left of the source, in the regions $x > x'$ and $x < x'$, respectively, the sound pressure is of the form

$$G(x, x', \omega) = \begin{cases} A \exp(ik_0 x) + B \exp(-ik_0 x), & \text{if } x > x' \\ C \exp(-ik_0 x) + D \exp(ik_0 x), & \text{if } x < x', \end{cases} \qquad (A.70)$$

where $k_0 = \omega/c$.

The boundary conditions at the ends $x = 0$ and $x = L$ are expressed in terms of the pressure reflection coefficients $R(0)$ and $R(L)$ so that $B \exp(-ik_0 L) = R(L)A \exp(ik_0 L)$ and $D = R(0)C$.

At the source, it follows from the momentum equation by integrating it over the source region that $G(x, x')$ is continuous and from the wave equation that $\partial G_+/\partial x - \partial G_-/\partial x = -1$, as before.

Applying these conditions to the wave field in Eq. A.70, we can express the constants A and C (with $B = AR(L) \exp(2ikL)$ and $D = R(0)C$, as above) in terms of the reflection coefficients, and hence obtain Green's function. We find

$$G(x, x', \omega) = \frac{i}{2k_0 D} e^{ik_0(x-x')}[1 + R(L)e^{i2k_0(L-x)}][1 + R(0)e^{i2k_0 x'}] \qquad (x > x'),$$
$$(A.71)$$

where $k_0 = \omega/c$ and $D = 1 - R(0)R(L) \exp(i2k_0 L)$. To obtain the function for $x < x'$, merely interchanging x and x'.

As a check we put $R(0) = R(L) = 0$ and we recover the free field Green's function. With $R(0) = R(L) = 1$, we get Green's function for a duct closed at both ends by acoustically hard walls

$$G(x, x', \omega) = -\frac{1}{k_0} \frac{\cos(k_0 x)\cos(k_0 x')}{\sin(k_0 L)}. \qquad (A.72)$$

A.4.3 Effects of Mean Flow

The one-dimensional equation for mass conservation is

$$\frac{\partial \rho}{\partial t} + \frac{\partial \rho U}{\partial x} = Q_f, \tag{A.73}$$

where $U = U_0 + u$, U_0 is the time independent flow, u is the acoustic perturbation in the fluid velocity, and Q_f is the source flow strength per unit volume. In the first term, it is often convenient to express the density perturbation in terms of the sound pressure so that $\partial \rho / \partial t \equiv (1/c^2) \partial p / \partial t$.

Does an injection of mass result in an injection of momentum? It depends. If the injected mass has the same mean velocity as the flow, there will be a time dependent source of momentum $Q_f U_0$ per unit volume (we assume here that Q_f has zero time average). The momentum equation then takes the form

$$\frac{\partial \rho U}{\partial t} + \frac{\partial \rho U^2}{\partial x} = -\frac{\partial p}{\partial x} + Q_f U_0. \tag{A.74}$$

However, with use of Eq. A.73, the left side becomes $\rho(\partial U/\partial t + U \partial U/\partial x) + U Q$ so that in the linear approximation the term $U_0 Q$ cancels the source term in the equation. Thus, the linearized momentum equation becomes

$$\rho(\frac{\partial}{\partial t} + U_0 \frac{\partial}{\partial x})u = -\frac{\partial p}{\partial x}. \tag{A.75}$$

Combining this equation with Eq. A.73 we obtain

$$\frac{\partial^2 p}{\partial x^2} - \frac{1}{c^2}(\frac{\partial}{\partial t} + U_0 \frac{\partial}{\partial x})^2 p = -(\frac{\partial}{\partial t} + U_0 \frac{\partial}{\partial x})Q_f \tag{A.76}$$

and the corresponding equation for the complex pressure amplitude $p(x, \omega)$ in the case of harmonic time dependence is

$$(1 - M^2)\frac{d^2 p}{dx^2} + i(2M\omega/c)\frac{dp}{dx} + (\omega/c)^2 p = -(-i\omega Q_f + U_0 \frac{dQ_f}{dx}). \tag{A.77}$$

If the mass is injected with zero mean flow, there is no source $Q_f U_0$ on the right-hand side in Eq. A.74 to cancel the term $Q_f U_0$ on the left-hand side and the linearized momentum equation becomes

$$\rho(\frac{\partial}{\partial t} + U_0 \frac{\partial}{\partial x})u = -\frac{\partial p}{\partial x} - Q U_0. \tag{A.78}$$

Using Eq. A.73 to eliminate u, we obtain

$$\frac{\partial^2 p}{\partial x^2} - \frac{1}{c^2}(\frac{\partial}{\partial t} + U_0 \frac{\partial}{\partial x})^2 p = -(\frac{\partial}{\partial t} + U_0 \frac{\partial}{\partial x})Q - U_0 \frac{\partial Q}{\partial x}, \tag{A.79}$$

i.e., the same as Eq. A.76 except for an additional term $-U_0 \partial Q_f / \partial x$ on the right-hand side. In either case, the mean flow gives rise to a source term involving the spatial derivatative of Q which corresponds to a dipole contribution to the wave field.

Free Field or Infinite Duct

If the right-hand side of Eq. A.77 is replaced by $-\delta(x - x')$, the solution becomes the Green's function (by definition) and it can be obtained in much the same way as for no mean flow. Thus, introducing the Fourier transform of $G(x, \omega)$ and $\delta(x - x')$ as before, the expression for the Fourier amplitude $G(k)$ now becomes

$$G(k) = \frac{e^{-ikx'}}{(1-M^2)k^2 + 2M\omega k/c - (\omega/c)^2} = \frac{e^{-ikx'}}{(1-M^2)(k-k_1)(k-k_2)}$$

$$k_1 = \frac{\omega/c}{1+M}, \qquad k_2 = -\frac{\omega/c}{1-M}. \tag{A.80}$$

The poles are now located at $k_1 \equiv k_+ = (\omega/c)/(1 + M)$ and $k_2 \equiv -k_- = -(\omega/c)(1 - M)$ and by analogy with the derivation of Eq. A.69, we obtain for the Green's function

$$G(x, x', \omega) = \begin{cases} \frac{i}{2(\omega/c)} e^{ik_+(x-x')} & if \text{ x-x'}>0, \\ \frac{i}{2(\omega/c)} e^{-ik_-(x-x')} & if \text{ x-x'}<0. \end{cases} \tag{A.81}$$

Having obtained Green's function, we return to Eq. A.79 to determine the sound field produced as a result of mass injection at a rate Q_f per unit volume without injection of momentum. The corresponding acoustic source strength is $Q = \partial Q_f/\partial t$ which in the case of harmonic time dependence is $Q = -i\omega Q_f$. We shall consider the case when $Q = Q_0\delta(x - x')$. The right-hand side of Eq. A.79 then becomes

$$rhs = -Q_0\delta(x - x') - 2U\frac{Q_0}{-i\omega}\frac{\partial\delta(x - x')}{\partial x}. \tag{A.82}$$

Apart from the factor Q_0, the first of these source terms produces the Green's function in Eq. A.81 and the second yields the solution

$$\frac{Q_0 2U}{-i\omega}\frac{\partial G(x, x'\omega)}{\partial x} = -k_+ Q_0 2U G/\omega = -\frac{2M}{1+M}Q_0G. \tag{A.83}$$

Adding the two solutions Q_0G and $(-2M/(1 + M)Q_0G$ corresponding to the two source terms in Eq. A.82, we obtain

$$p(x, x', \omega) = \begin{cases} \frac{iQ_0}{2(\omega/c)}\frac{1-M}{1+M} e^{ik_+(x-x')} & \text{for } x - x' > 0 \\ \frac{iQ_0}{2(\omega/c)}\frac{1+M}{1-M} e^{-ik_-(x-x')} & \text{for } x - x' < 0 \end{cases}. \tag{A.84}$$

It is interesting to note that the amplitude in the upstream direction resulting from a mass injection source of this kind will be larger than in the downstream direction by a factor $(1 + M)^2/(1 - M)^2$. As we shall see later, this is consistent with some experimental observations.

Finite Duct

Again, we seek a solution to Eq. A.77 when the right-hand side is $-\delta(x - x')$. Now, however, we wish to satisfy the boundary conditions at $x = 0$ and $x = L$ for the pressure reflection coefficient. We refer to the similar analysis carried out above in the case of no flow and, as in that case, the boundary conditions at the ends of the duct at $x = 0$ and $x = L$ are expressed in terms of the pressure reflection coefficients $R(0)$ and $R(L)$. The Green's function will be continuous at $x = x'$ (see the expression for the free field Green's function in Eq. 2.3.9) but for the derivative we find $G'_+ - G'_- = -1/(1 - M^2)$.

We build up Green's function from plane waves as follows

$$G(x, x', \omega) = \begin{cases} Ae^{ik_+x} + Be^{-ik_-x} & \text{for } x > x' \\ Ce^{-ik_-x} + De^{ik_+x} & \text{for } x < x' \end{cases}, \qquad (A.85)$$

where, as before, $k_\pm = (\omega/c)/(1 \pm M)$. From the end conditions we have $B \exp(-ik_-L) = R(L)A \exp(ik_+L)$ and $D = R(0)C$ and from the remaining two conditions expressing continuity of G at $x = x'$ and a jump of its derivative by $-1/(1 - M^2)$, we can determine A, B, C, D.

Insertion into Eq. A.85 then yields

$$G(x, x', \omega) = \tfrac{i}{2(\omega/c)N} e^{ik_+(x-x')}[1 + R(L)e^{i(k_+ + k_-)(L-x)}][1 + R(0)e^{i(k_+ + k_-)x'}]$$
$$N = 1 - R(0)R(L)e^{i(k_+ + k_-)L} \qquad (x > x'). \qquad (A.86)$$

To obtain Green's function for $x < x'$, interchange x and x'.

The resonance frequencies of the duct are determined by setting the denominator N equal to zero. For example, if the ends of the duct open into free field and if we neglect losses, the reflection coefficients are close to -1. The resonance frequencies then correspond to $(k_+ + k_-)L = n2\pi$, and with $k_+ + k_- = 2(\omega/c)/(1 - M^2)$ we obtain $\omega_n/2\pi = (nc/2L)(1 - M^2)$ and the corresponding wavelength $\lambda_n = (2L/n)/(1 - M^2)$.

Apart from a factor Q_0, Green's function thus obtained is the contribution p_1 to the solution to the wave equation A.79 which corresponds to the first source term in Eq. A.82. The second source term produces the field $p_2 = (2U_0Q_0/ - i\omega)dG/dx$. The derivative of the Green's function dG/dx in turn consists of two parts. The first part ik_+G, when added to p_1, gives

$$\begin{array}{ll} \frac{1-M}{1+M}q_0G(x, x', \omega) & (x > x') \\ \frac{1+M}{1-M}q_0G(x, x', \omega) & (x < x'), \end{array} \qquad (A.87)$$

where G is given in Eq. A.86. The second part of dG/dx results in the p_2-contributions

$$Q_0G \frac{R(L)}{R(L)+\exp[-i(k_+ + k_-)(L-x)]} \frac{4M}{1-M^2} \quad (x > x')$$
$$-q_0G \frac{R(0)}{R(0)+\exp[-i(k_+ + k_-)x]} \frac{4M}{1-M^2} \quad (x < x'). \qquad (A.88)$$

A.4.4 Radiation from a Piston in the Side Wall of a Duct

In some applications, sound is injected into a duct from a source mounted in a side wall of the duct and we shall now consider this problem in the case of a rectangular

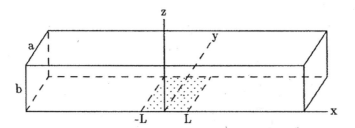

Figure A.6: Sound radiation from a piston in the side wall of a duct carrying a mean flow. The displacement in the z-direction of the source is $\zeta(x, y, t)$.

duct. As before, we let the x-axis run along the length of the duct and locate the side walls in the planes $y = 0$, $y = a$ and $z = 0$, $z = b$. The piston source is located in the xy-plane and extends from $x = -L$ to $x = L$, as indicated in Fig. A.6. The source has harmonic time dependence and a complex displacement amplitude $\zeta(x, y, t)$. The corresponding velocity amplitude it $-i\omega\zeta$. The duct carries a flow with a mean velocity U.

Extending the wave equation (A.76) for the sound pressure field to three dimensions, we get

$$(1 - M^2)\frac{\partial^2 p}{\partial x^2} + \frac{\partial^2 p}{\partial y^2} + \frac{\partial^2 p}{\partial z^2} - (2M/c)\frac{\partial^2 p}{\partial x \partial t} - 1/c^2)\frac{\partial^2 p}{\partial t^2} = 0, \qquad \text{(A.89)}$$

where c is the sound speed and $M = U/c$ is the Mach number of the flow in the duct. The coordinates x, y, z refer to a stationary (laboratory) frame of reference with respect to which the unperturbed fluid moves with a speed U in the positive x-direction.

The duct walls are assumed to be acoustically hard everywhere except in the source region. Consequently, the normal components of the velocity field and the corresponding pressure gradients are zero at the walls except at the source where the displacement is a prescribed function of x and y.

The perturbation of the fluid flow in the duct by the displacement at the source is assumed to result in a laminar flow with displaced stream lines, and the corresponding velocity perturbation is

$$u_z(x, y, t) = (\partial/\partial t + U\partial/\partial z)\zeta(x, y, t). \qquad \text{(A.90)}$$

We shall pay attention only to the plane wave mode in the duct. The corresponding sound pressure is the average of $p(x, y, z, t)$ over the area of the duct, i.e.,

$$p_0(x, t) = (1/ab) \int \int p(x, y, z, t)\, dx\, dy. \qquad \text{(A.91)}$$

Our first objective is to obtain an equation for p_0 and we return to Eq. A.89 and integrate each term over the area.

The integral of $\partial^2 p/\partial y^2$ over y is the difference between the values of $\partial p/\partial y$ at $y = 0$ and $y = a$. Both these values are zero, however, since the walls are acoustically hard at $y = 0$ and $y = a$. Similarly, the integral of $\partial^2 p/\partial z^2$ over z is the difference between $\partial p/\partial z$a evaluated at $z = b$ and $z = 0$. Again, this is zero except in the source region. Here we obtain $\partial p/\partial z$ from the momentum equation for the fluid

$$\rho(\frac{\partial}{\partial t} + U\frac{\partial}{\partial x})u_z = -\frac{\partial p}{\partial z}, \tag{A.92}$$

where u_z is given by Eq. A.90 and

$$(1/ab)\int\int \frac{\partial^2 p}{\partial z^2}\,dy\,dz = (\rho/ab)(\frac{\partial}{\partial t} + U\frac{\partial}{\partial x})\int u_z(x, y, 0, t)\,dy. \tag{A.93}$$

From now on, to simplify notation, we drop the subscript on the plane wave pressure component in Eq. A.91 and denote it simply $p(x, t)$. Thus, carrying out the integration of Eq. A.89 over the area of the duct and considering Eq. A.93, we obtain the following equation for the plane wave pressure field

$$(1 - M^2)\frac{\partial^2 p}{\partial x^2} - 2(M/c)\frac{\partial^2 p}{\partial x\partial t} - (1/c^2)\frac{\partial^2 p}{\partial t^2} = s(x, t), \tag{A.94}$$

where the source term is

$$s(x, t) = -(\rho/ab)(\frac{\partial}{\partial t} + U\frac{\partial}{\partial x})\int u_z(x, y, 0, t)\,dy. \tag{A.95}$$

Comparing this result with Eq. A.91, we note that the piston source is equivalent to a volume source distribution

$$Q(x, t) = (\rho/ab)\int u_z(x, y, 0, t)\,dy \tag{A.96}$$

in an acoustically hard duct.

Example

As a special case, we consider a piston with a uniform harmonic displacement amplitude $\zeta_0(x, \omega)$. According to Eq. A.90, the corresponding velocity amplitude in the fluid outside the piston is $u_z(x, \omega) = (-i\omega + U\partial/\partial x)\zeta(x, \omega)$ and the source function in Eq. A.95 becomes

$$s(x, \omega) = (\rho/b)(-i\omega + U\frac{\partial}{\partial x})^2\zeta(x, \omega). \tag{A.97}$$

In terms of the Green's function in Eq. A.81 the pressure becomes

$$p(x, \omega) = (-i\omega + U\frac{\partial}{\partial x})^2\int_{\ell}^{\ell}\zeta(x', \omega)G(x, x', \omega)dx'. \tag{A.98}$$

Considering first the sound field downstream of the piston, so that $x > \ell$, we have $G(x, x', \omega) = [i/2(\omega/c)] \exp(ik_+(x - x'))$. Furthermore, with $\zeta(x', \omega)$ being a constant $\zeta_0(\omega)$ between $-\ell$ and ℓ and zero elsewhere,

$$\int G(x, x', \omega)\zeta(x', \omega)\, dx' = e^{ik_+x}2\ell\frac{\sin(k_+\ell)}{k_+\ell}. \tag{A.99}$$

Then, with $k_+ = (\omega/c)/(1+M)$ and $(-i\omega+U\partial/\partial x) = (-i\omega+iUk_+) = -i\omega/(1+M)$ it follows that

$$p(x, \omega) = \rho c u_0 \frac{A_p}{2A}\frac{1}{(1+M)^2}\frac{\sin(k_+\ell)}{k_+\ell}, \tag{A.100}$$

where $u_0 = -i\omega\zeta_0$ is the velocity amplitude of the piston.

The sound pressure field in the upstream direction is obtained by replacing M by $-M$ and, accordingly, k_+ by $k_- = (\omega/c)/(1 - M))$. It is noteworthy, that if the extension of the source is much smaller than a wavelength so that $\ell/\lambda << 1$, i.e., $\sin(k_+\ell)/k_+\ell \approx 1$, the ratio between the pressure amplitudes of the waves in the upstream and downstream directions is

$$\frac{p_-}{p_+} = \frac{(1 + M)^2}{(1 - M)^2}. \tag{A.101}$$

This is the same ratio as was obtained for the point source in a duct with flow, which follows from Eq. A.84.

We obtained the result in Eq. A.100 by using the operator $(-i\omega + U\partial/\partial x)^2$ on the solution obtained with a surce function containing only ζ_0. A lengthier but direct approach would have been to carry out the differentiation explicitly on the source function in Eq. A.97. In that case we express $\zeta_0(x, \omega)$ as in terms of unit step functions U_s, i.e., $\zeta_0(x, \omega) = \zeta_0[U_s(x + \ell) - U_s(x - \ell)]$. The derivative of this function will be the sum of delta functions at $x = -\ell$ and $x = \ell$ and the second derivative will be the sum of the derivatives of these functions. After integration of the product of this source with Green's function we obtain the same result as in Eq. A.100.

With reference to Section 10.2.1 on the acoustic energy flow in a moving fluid, we note that the intensity of a plane wave in a duct carrying a flow of Mach number M is $I_\pm = (p_+^2/\rho c)(1 \pm M)^2$, where the plus and minus sign refers to downstream and upstream propagation, respectively. Consequently, the result in Eq. A.101 shows that the radiatiom from the piston source produces the same acoustic intensity in both directions. It should be emphasized that in the presence of flow, this means that the sound pressure in the upstream direction is larger than in the downstream direction, as indicated in Eq. A.101.

It should be kept in mind that in our analysis the normal velocity perturbation of the fluid at the piston was not the same as the velocity amplitude of the piston because of the displacement of the stream lines in the steady flow, thus producing a component of the mean flow normal to the piston. If the flow is turbulent, however, there will be a random distribution of these components over the piston which tend to average to zero. Under such conditions, in the low-frequency approximation, the average normal component of the velocty in the fluid is the same as that of the piston. If we use this boundary condition in our analysis we find that the ratio between the upstream and downstream pressure amplitudes is $(1 + M)/(1 - M)$. As illustrated in Fig. 10.3, there is experimental evidence that this ratio is obtained at sufficiently high flow velocities whereas at low flow velocities the ratio is close to $(1 + M)^2/(1 - M)^2$.

A.5   Sound from an Axial Fan in Free Field

A.5.1   Point Force Simulation of Axial Fan in Free Field

We consider B identical point dipoles distributed uniformly on a circle of radius a' and moving with a constant angular velocity Ω, as indicated schematically in Fig. A.7. The radial positions a' of the sources may be different and not uniformly spaced. (An improved version of this model is given in Section A.5.2 where swirling dipole line forces are used in which the span-wise load distribution on the blades are accounted for.)

Each force is a result of the interaction of a blade with the surrounding fluid and is determined by the fluid velocity relative to the blade. The flow incident on the fan is now assumed to be uniform and axial so that there is no time dependence of the interaction force in the frame of reference moving with the fan. The coordinates in

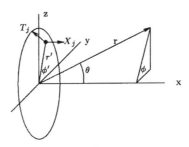

Figure A.7: Coordinates used in the analysis of sound radiation from point dipoles moving along a circle. Each source is specified acoustically in terms of an axial and a tangential force amplitude, X_j and T_j.

the frame of reference attached to and moving with the propeller are denoted r' and ϕ' and the plane of the propeller is placed in the yz-plane at $x = 0$.

The sound field from a given volume force distribution can be calculated as shown in Eq. 5.39 in terms of the force per unit volume and an integral over the force distribution. To follow this approach in this problem, we have to express the point forces in terms of a force density distribution. To do that, we enclose the fan in a pillbox type control volume and express the point forces in terms of delta functions. If the box is thin, we can approximate the x-dependence of the delta function by $\delta(x)$, assuming that the propeller plane is $x = 0$. With $\int \delta(x)dx = 1$, the volume integral over the source volume is transformed into a surface integral over a force distribution per unit area which we denote f with axial and tangential (circumferential) components f_x and f_t.

Our first task is to determine this force distribution for point forces in the source plane. We label the sources by j, where j runs from 1 to B, the number of blades. The axial and tangential forces associated with the jth source are X_j and T_j at the radial and angular positions a' and ϕ'_j.

Each of the B point forces is described by a delta function, $\delta(r' - r'_j)(1/r')\delta(\phi' - \phi'_j)$, as illustrated here for the axial component of the force per unit area of the pillbox

$$f_x(r', \phi') = \sum_{j=1}^{B} X_j \delta(r' - a'_j)(1/r')\delta(\phi' - \phi'_j). \qquad (A.102)$$

(The factor $1/r'$ goes with $\delta(\phi' - \phi_j)$ to assure that the integration over the angle variable, which involves the differential $r'd\phi'$, will be unity.) We note that integration over the source area yields $\sum_j X_j$ which is the total thrust of the propeller.

The factor $\delta(\phi' - \phi'_j)$ is periodic in ϕ' with the period 2π and using the Fourier expansion of this function, we obtain

$$f_x(r', \phi') = \frac{1}{2\pi} \left[\sum_{n=-\infty}^{\infty} \sum_{j=1}^{B} X_j e^{-in\phi'_j} \delta(r' - a'_j) \right] (1/r')e^{in\phi'}. \qquad (A.103)$$

The force function for the tangential (ϕ'-) component of the force is obtained in an analogous manner with X_j replaced by the tangential force T_j.

We consider here only the normal case of identical and uniformly spaced blades, so that $X_j = X$ is constant, $a'_j = a'$, and $\phi'_j = j2\pi/B$. The amplitude within the brackets will be different from zero (and equal to XB) only if n is an integer multiple of B, say $n = sB$ (if $n \neq sB$, the complex numbers in the sum lie on the corners of a closed polygon and add up to zero; if $s = nB$, they all point in the same direction). If follows that

$$f_x(r', \phi') = \frac{BX}{2\pi} \sum_{s=-\infty}^{\infty} e^{isB\phi'}(1/r')\delta(r' - a'). \qquad (A.104)$$

The distribution of tangential force is obtained by replacing X by the tangential force component T.

So far, the analysis has referred to the frame of reference rotating with the propeller and if the flow is uniform, the blade-flow interaction will be time independent in this frame. In the laboratory frame of reference, however, there will be a time dependent perturbation of the fluid as the blades pass by. With the angular velocity of the fan denoted Ω, this time dependence is obtained from by making use of the transformation

$$\phi' = \phi - \Omega t \qquad (A.105)$$

from the moving (primed) coordinates to the stationary (laboratory) frame (unprimed). The force as described in the laboratory frame is then

$$f_x(r, \phi, t) = \frac{BX}{2\pi} \sum_{s=-\infty}^{\infty} e^{isB\phi} e^{-isB\Omega t} (1/r')\delta(r' - a'). \qquad (A.106)$$

In other words, the frequencies of the Fourier components of the force function are multiples of the blade passage frequency $B\Omega/2\pi$. This is no longer true if the blades are not identical and/or their spacing nonuniform. Then the fundamental angular frequency in the spectrum will be the angular velocity of rotation Ω (shaft frequency).

Sound Field from a Single Point Source

Next we recall the expression for the complex amplitude of the sound pressure generated at location $\boldsymbol{r}$ by a simple harmonic source (monopole) with angular frequency ω and located at $\boldsymbol{r}'$

$$g_0(\omega, \boldsymbol{r}, \boldsymbol{r}') = \frac{e^{ik|\boldsymbol{r} - \boldsymbol{r}'|}}{4\pi|\boldsymbol{r} - \boldsymbol{r}'|}, \qquad (A.107)$$

where $k = \omega/c$, c being the sound speed. The distance from the source point to the field point is

$$|\boldsymbol{r} - \boldsymbol{r}'| = \sqrt{r^2 + r'^2 - 2\boldsymbol{r} \cdot \boldsymbol{r}'} \approx r - \boldsymbol{r} \cdot \boldsymbol{r}'/r, \qquad (A.108)$$

where the approximation is valid in the far field $r \gg r'$.

The x, y, z-components of $\boldsymbol{r}'$ are $x' = 0$, $y_1 = r' \cos\phi'$, and $z_1 = r' \sin\phi'$ and the components of $\boldsymbol{r}$ are $x = r \cos\theta$, $y = r \sin\theta \cos\phi$, and $z = r \sin\theta \sin\phi$ (see Fig. A.7). Expressing the scalar product in terms of these components, we readily see that $\boldsymbol{r} \cdot \boldsymbol{r}' = rr' \sin\theta \cos(\phi - \phi')$ so that

$$|\boldsymbol{r} - \boldsymbol{r}'| \approx r - r' \sin\theta \cos(\phi - \phi'). \qquad (1.2.8)$$

In the far field, $r \gg a$, the complex amplitude of the sound pressure can then be expressed as

$$g_0 \approx \frac{e^{ikr}}{4\pi r} e^{-ikr' \sin\theta \cos(\phi - \phi')}. \qquad (A.109)$$

This function is periodic in the variable ϕ and it is convenient to work with its Fourier series. The coefficients in this series are known to be Bessel functions, as expressed by the formula

$$e^{iz \cos(\phi - \phi')} = \sum_{-\infty}^{\infty} i^m J_m(z) e^{im(\phi - \phi')}, \qquad (A.110)$$

where J_m is the Bessel function of order m.

Eq. A.109 can then be written

$$g_0 = \frac{e^{ikr}}{4\pi r} \sum_{-\infty}^{\infty} i^m J_m(kr' \sin\theta) e^{im(\phi - \phi')}. \qquad (A.111)$$

A harmonic point force can be thought of as a dipole (two monopoles of opposite sign close together) and the sound field from it is obtained by differentiating the field from the monopole with respect to the source coordinates. We note, however, that the derivative with respect to the source coordinates is the negative of the derivative with respect to the field coordinates. Thus, we can express the field, the Green's function, from a harmonic point force of unit strength, acting on the fluid in the x-direction as

$$g_{1x} = -\frac{\partial g_0}{\partial x} = -\frac{\partial g_0}{\partial r}\cos\phi = -ik\cos\phi \frac{e^{ikr}}{4\pi r}\sum_{-\infty}^{\infty} i^m J_m(kr'\sin\theta)\, e^{im(\phi-\phi')}.$$

(A.112)

Similarly, a unit force in the ϕ'-direction produces the field

$$g_{1\phi} = -\frac{1}{r'}\frac{\partial g_0}{\partial\phi} = -\frac{i}{r'}\frac{e^{ikr}}{4\pi r}\sum_{-\infty}^{\infty} mi^m J_m(kr'\sin\theta)e^{im(\phi-\phi')},$$

(A.113)

where, as before, $k = \omega/c$.

The sound field from a force distribution is determined by the spatial variation in the force (a space independent force cannot produce any deformation of the material involved and hence no sound). For the axial component the characteristic length involved in this variation is the wavelength, as expressed by the factor $k = (2\pi/\lambda)\cos\theta$. For the tangential component, the characteristic length is a' and the factor $k\cos\theta$ in the g_{1x} is replaced by m/a'.

The Sound Field from the Force Distribution

The sound pressure field from the axial force distribution in Eq. A.104 is now obtained by multiplying by Green's function (A.112) and carrying out the integration over the source coordinates (see also Eq. 5.39). In a similar manner the sound field produced by the tangential force distributions is obtained using Green's function (A.113). The corresponding sound fields will be referred to as the X-field and the T-field, respectively. Their sum is the total complex sound pressure amplitude of the nth harmonic and is found to be

$$p_n(r,t) = FnB\frac{M\cos\theta + (aT)/(a'X)}{2(r/a)}J_{nB}(nB\eta M\sin\theta)\,\sin[nB(\phi-\Omega t+kr+\pi/2)],$$

(A.114)

where F is the total thrust per unit area and $\eta = a'/a$. Function J_{nB} is the Bessel function of order (nB) and M is the tip Mach number of the fan, $M = a\Omega/c$.

A.5.2 Fan Simulation by Swirling Line Forces

To improve the acoustic simulation of a fan, we replace the point dipole forces in the previous section by dipole line sources which are assumed identical and uniformly spaced. As before, X will be the axial force on one of the blades and, with $\eta = r'/a$, the force distribution in the span-wise (radial direction) of the blade is described by a distribution function $\beta_x(\eta)$ such that $(X/a)\beta_x(\eta)$ is the force per unit length, where

a is the span of the blade (in this case the radius of the fan). In an analogous manner, we introduce a force distribution function $\beta_t(\eta)$ for the tangential component of the force so that the force per unit length is $(T/a)\beta_t(\eta)$. It follows that

$$\int_0^1 \beta_x(\eta)\,d\eta = \int_0^1 \beta_t(\eta)\,d\eta = 1. \tag{A.115}$$

The expression for the axial force per unit area over the propeller plane is then obtained by replacing $\delta(r'-a')$ in Eq. A.102 by $(X/a)\beta_x(\eta)$. Following the procedure in Section A.5.1, we arrive at the analogue of Eq. A.106

$$f_x(r',\phi,t) = \frac{BX}{2\pi a}\sum_{-\infty}^{\infty}\beta_x(\eta)(1/r')e^{isB\phi}e^{-isB\Omega t}. \tag{A.116}$$

The tangential force distribution is obtained in a similar manner. The next step involves carrying out the integration in Eq. 5.39 in complete analogy with the previous section on the point force simulation. The sound pressure at the far field at r, θ, ϕ is then found to be

$$p = \sum_{n=1}^{\infty} p_n$$
$$p_n(r,\theta,\phi,t) = P_n(r,\theta)\sin[nk_1 r + nB(\phi + \pi/2) - nB\Omega t]$$
$$P_n(r,\theta) = F\frac{nB}{2(r/a)}[I_x M\cos\theta + I_t(T/X)]$$
$$I_x = \int_0^1 \beta_x(\eta)J_{nB}(nB\eta M\sin\theta)\,d\eta$$
$$I_t = \int_0^1 (1/\eta)\beta_t(\eta)J_{nB}(nB\eta M\sin\theta)\,d\eta \tag{A.117}$$

as already given and discussed in the text.

A.5.3   Nonuniform Flow

The analysis so far has dealt with the idealized situation in which the flow into the fan is uniform. As a result, the interaction force on a blade will be independent of its angular position and thus will be independent of time in the frame of reference of the fan. One important consequence of this idealization is that the sound pressure produced by the fan is always zero on the axis regardless of the fan parameters. Although, in reality, there is generally an observed minimum of the sound pressure amplitude on the axis, it is far from zero. One important reason for this is the effect of nonuniform flow into the fan on sound generation; another is its effect on sound propagation. In this section, the problem of sound generation will be analyzed.

The flow is assumed nonuniform but stationary. The interaction force X on a blade is approximately proportional to the squared flow speed relative to the blade. In uniform flow, the force is constant, independent of the angular position ϕ, but in nonuniform flow, this is no longer so. In general, the force will depend on both r and ϕ, but since the flow is stationary, the dependence on ϕ will be periodic with the period 2π. In the propeller frame of refence, the force on a blade will be time dependent, which is the significant effect of the flow nonuniformiy; it not only affects sound generation but it also causes vibration of the fan.

The force on a blade due to the nonuniform part of the flow will be expressed as a fraction of the force caused by the uniform part. This fraction γ generally depends on both r and ϕ so that the force due to the nonuniformity will be expressed formally by replacing the force distribution function $\beta(\eta)$ by $\gamma(\eta, \phi)\beta(\eta)$. Thus, denoting this force f', we get

$$f'(r', \phi, \eta, t) = \frac{BX}{2\pi a} \sum_{-\infty}^{\infty} \gamma(\eta, \phi)\beta(\eta)(1/r')e^{isB\phi}e^{-isB\Omega t}. \tag{A.118}$$

Since γ is periodic in ϕ, it can be Fourier expanded as

$$\gamma(\eta, \phi) = \sum_{-\infty}^{\infty} \gamma_q(\eta)e^{iq\phi}, \tag{A.119}$$

where $\gamma_0 = 0$, since the circumferential average of the nonuniformity is zero. The force f' then takes the form

$$f' = \frac{BX_0}{2\pi a} \sum_{q,s=-\infty}^{\infty} \gamma_q(\eta)\beta(\eta)\, e^{i(sB+q)\phi}\, e^{-isB\Omega t}\,(1/r'). \tag{A.120}$$

Next, we multiply by Green's function (A.112) and integrate over the source coordinates (i.e., replace ϕ in the expression for f_x by ϕ'). Integration over ϕ' then yields zero unless $m = (sB + q)$, in which case the integral is 2π. The resulting sound field becomes

$$p' = -i\frac{BX\cos\theta}{a}\frac{1}{4\pi r} \sum_{s,q=-\infty}^{\infty} sk_1\, i^{sB+q-1}\, e^{i(sB+q)(\phi+\pi/2)}\, I_{x,sq}\, e^{isk_1 r}\, e^{-isB\Omega t}, \tag{A.121}$$

where $k_1 = B\Omega/c = BM/a$ and

$$I_{x,sq} = \int_0^1 \gamma_q\beta(\eta)\, J_{sB+q}(sBM\eta\sin\theta)\, d\eta. \tag{A.122}$$

Similarly, sound field from the tangential force distribution is obtained by using Green's function (A.113). The distribution functions corresponding to β and γ_q, are denoted β_t and γ_{tq}. The sound pressure field becomes

$$p'_t = -i\frac{BT_0}{a}\frac{1}{4\pi r} \sum_{s,q=-\infty}^{\infty} (sB + q)\, e^{(sB+q)(\phi+\pi/2)}\, I_{t,sq}e^{isk_1 r}\, e^{-isB\Omega t}, \tag{A.123}$$

where

$$I_{t,sq} = \int_0^1 \gamma_{tq}\beta_t(\eta)\,(1/\eta)\, J_{sB+q}(sBM\xi\sin\theta)\, d\eta. \tag{A.124}$$

In the numerical examples given in the text, the span-wise force distribution functions for the axial and tengential forces $\beta(\eta)$ and $\beta_t(\eta)$ are assumed to be the same as are γ_q and γ_{tq}.

For numerical calculations, it is convenient to rewrite the expressions for p' and p'_t pairing terms with positive and negative values of s and q and we do it in the following manner. First, we add the terms with s, q positive and s, q negative, i.e., $s = n, q = \ell$ and $s = -n, q = -\ell$ and then do the same for $s = n, q = -\ell$ and $s = -n, q = \ell$, where n and ℓ are positive integers. The sum of the exponentials in each of these groups of terms then becomes of the form $i2 \sin(\ldots)$, which, when multiplied by the factor $-i$ (the first factor in Eqs. A.121 and A.124) becomes the real function $2 \sin(\ldots)$.

The resulting total sound pressure p field can then be written

$$p = p_- + p_+$$
$$p_\pm = (\Delta P) \sum_{n=1}^{\infty} \sum_{\ell=0}^{\infty} P_{n\ell} \sin(nk_1 r + (nB \pm \ell)(\phi + \pi/2) - nB\Omega t)$$
$$P_{n\ell} = [X_0 M \cos\theta I_x(n, \pm\ell) + T_0(1 \pm \ell/(nB))I_t(n, \pm\ell)]/(2r/a)$$
$$I_x(n, \pm\ell) = \int_0^1 \gamma_{x\ell}(\eta)\beta_x(\eta)J(nB \pm \ell)(nBM\xi \sin\theta)\,d\xi$$
$$I_t(n, \pm\ell) = \int_0^1 \gamma_{t\ell}\beta_t(\eta)(1/\xi)J_{nB\pm\ell}(nBM\xi \sin\theta)\,d\xi. \tag{A.125}$$

In general, the spin velociy of a mode in the angular direction is the ratio of the factors $nB\Omega$ and $(nB \pm \ell)$ in the t- and ϕ-terms in the argument of the sin-function, i.e.,

$$\Omega_\pm = \frac{nB\Omega}{nB \pm \ell} = \frac{\Omega}{1 \pm (\ell/nB)}. \tag{A.126}$$

Thus, $|\Omega_+| < \Omega$ and $|\Omega_-| > \Omega$; the pressure pattern p_+ spins slower than the propeller and p_- faster. If $\ell > nB$, the p_--field spins in the negative direction, i.e., opposite to the direction of the fan (compared with the example in Fig. 8.3, where we had $B = 2$ and $\ell = 3$). The radiation efficiency of the latter is greater and this is expressed in terms of the lowering of the order of the Bessel function. In fact, for $\ell = nB$, the order is reduced to zero. In this case ℓ blades interact in phase with the ℓth Fourier component of the ϕ-dependence of the flow so that, in effect, the axial components of the force combine to make the fan act like a piston oscillating in the axial direction.

The radiation from the tangential force, however, will be zero in this case. The reason is that when all the tangential forces act in phase, the propeller is acoustically equivalent to a piston that oscillates in the angular direction. In that case there will be no periodic compression of the surrounding fluid and no sound radiation.

Of the two contributions, p_- and p_+, to the total sound pressure in Eq. A.125, p_- is by far the most important. With its higher angular velocity, the radiation efficiency is higher than for the slow moving p_+ particularly when the number of angular periods of the flow inhomogeneity is equal to the number of blades of the propeller or a multiple thereof (i.e., $\ell = nB$). Formally, this means a reduction of the order of the Bessel function from nB to 0 and a corresponding increase in the amplitude. If nB is large, this normally means an increase in the sound pressure by several orders of magnitude.

The role of inhomogeneous flow is demonstrated in Figs. 8.14 and 8.15 in Chapter 8 where angular sound pressure distributions are shown and discussed for fans with 2 and 16 blades.

Appendix B

Complex Amplitudes

Some of the derivations so far without the use of complex numbers have been algebraically quite cumbersome since, repeatedly, it was necessary to break up a harmonic function $\cos(\omega t - \phi)$ in terms of a sum of $\cos(\omega t)$- and $\sin(\omega t)$ terms and to use elementary trigonometric identities to arrive at a final answer. In problems that go much beyond the simple harmonic oscillator, this can become a considerable burden.

Euler's formula makes it possible to express a harmonic function of time in terms of an exponential function and to define a complex amplitude which contains the characteristics of the harmonic function under consideration.

In a similar manner, time derivatives of the harmonic function can also be expressed in terms of the complex amplitude and differential equations of motion which describe the problems that will be converted into algebraic equations for the complex variable. Once the solution for the complex variable has been obtained, the time dependent functions expressing the real solution can readily be retrieved. The use of complex amplitudes thus avoids many of the algebraic difficulties mentioned above and has been illustrated in examples in the text.

B.1   Brief Review of Complex Numbers

B.1.1   Real Numbers

First, a reminder about the real number system and the role of the basic operations of addition and multiplication in the process of building up the set of real numbers from the set of positive integers which runs from 1 to infinity.

Consider, for example, the relation $A + B = C$. For any two positive integers A and B, the number C can always be found amongst the set of positive integers to satisfy this equation. On the other hand, for any two positive integers B and C, say 5 and 3, we cannot find a corresponding A in this set. Then, to make the relation valid, we use it to define *negative numbers*, thus extending the number system. In this particular case with $B = 5$ and $C = 3$, the number $A = -2$ is defined as the number, which, when added to 5 yields 3 as a result. The number 0 is an defined in analogous manner. Furthermore, with $5 + (-2) = 3$ it follows that $5 = 3 - (-2)$ so that $-(-2) = 2$. The numbers are visualized geometrically on the number line, the positive numbers being marked off to the right of 0 and the negative numbers to the left.

In an analogous manner, rational and irrational numbers are introduced to make the operation $AB = C$ valid for arbitrary integers. For example, with $B = 3$ and $C = 2$, the *fraction* $A = 2/3$ is defined as the number, which, when multiplied by 3 yields 2 as a result. If $A = B$ and $C = 3$, so that $A^2 = C$, the *irrational number* $A = \sqrt{3}$ (or $A = -\sqrt{3}$) is defined as the number, which, when multiplied by itself yields 3. In this manner the set of *real numbers* can be generated, comprising positive and negative integers, fractions, and irrational numbers. Note that an irrational number, like $\sqrt{3}$, cannot be expressed in terms of a rational number by means of the simple operations of addition, subtraction, and multiplication.

All these numbers form the set of *real* numbers which can be represented geometrically as points on a real number line.

B.1.2 Imaginary Numbers

If C is negative, there is no real number A such that $A^2 = C$. Thus, there is need for an extension of the number system to 'save' $AB = C$ and a number i, called the *imaginary unit number*, is introduced such that $i^2 = -1$. Similarly, with y a real number, the imaginary number $A = iy$ has the property that $A^2 = -y$. The imaginary number iy is represented geometrically as a point on a new number line, *the imaginary axis*, perpendicular to the real number line and the point is a distance y from the origin.

We also require the associative laws of algebra of real numbers $(A + B) + C = A + (B + C)$ and $(AB)C = A(BC)$ to be valid also for imaginary numbers and ABC is defined either as $(AB)C$ and $A(BC)$. Thus, with $i^2 = -1$, we have $i^3 = (i^2)i = -i$ and $i^4 = 1$. From these relations and with $b = i$ it follows, as an example, that $z = b^4 + b = 1 + i$, which contains both a real and an imaginary part.

B.1.3 Complex Numbers

A number $z = x + iy$ with a real part x and an imaginary part y defines a *complex number* z. It is represented geometrically by a point in the *complex plane*, which is spanned by the real and imaginary axes, as shown in Fig. B.1. It can be described also in terms of the *magnitude* or *amplitude* $|z|$ and the phase angle ϕ given by

$$|z| = \sqrt{x^2 + y^2}, \quad \tan\phi = y/x, \quad z = x + iy = |z|\cos\phi + i|z|\sin\phi. \quad (B.1)$$

With the number system extended to include complex numbers, an algebraic equation of nth order always has n roots (Theorem of Algebra). For example, the second order equation $z^2 + 2z + 5 = 0$ has the two roots $z = -1 \pm i2$, and the fourth order equation $z^4 = -1$ has the four roots $(\pm 1 \pm i)/\sqrt{2}$, located symmetrically in the complex plane.

The *sum* of two complex numbers

$$z = z' + z'' = x' + x'' + i(y' + y'') = x + iy \qquad (B.2)$$

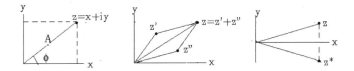

Figure B.1: Complex numbers. Left: Representation of a complex number in the complex plane. Middle: Addition of complex numbers z' and z''. Right: The complex conjugate $z* = x - iy$ of $z = x + iy$.

and the difference is defined in an analogous manner. The geometrical interpretation of addition is shown in Fig. B.1.

We require the distributive law of multiplication $A(B+C) = AB + AC$ to be valid also for complex numbers. Thus, the *product* $w = z_1 z_2$ of two complex numbers z_1 and z_2 is obtained by using the distributive law for the components, accounting for $i^2 = -1$,

$$w = z_1 z_2 = (x_1 + iy_1)(x_2 + iy_2) = x_1 x_2 - y_1 y_2 + i(-x_1 y_2 + x_2 y_1). \tag{B.3}$$

If $z = x + iy$, the number $z* = x - iy$ is called the *complex conjugate* of z. It follows that $zz* = x^2 + y^2 = |z|^2$, where $|z|$ is the amplitude (magnitude) of z. The geometrical interpretation of $z*$ in the complex plane is simply the mirror image of z with respect to the x-axis as shown on the right in Fig. B.1.

The *ratio* of two complex numbers z_1 and z_2, denoted z_1/z_2, is defined by the requirement that the product of this number and z_2 equals z_1.

B.1.4 Euler's Identity

By repeated use of multiplication of complex numbers, we can determine the complex number z^n, where n is a positive or negative integer. The meaning of a function $f(z)$ can then be defined in terms of the power series expansion of the function. For the exponential function the power series is

$$\exp(z) = 1 + z + z^2/2! + z^3/3! + z^4/4! + \dots. \tag{B.4}$$

When the argument is purely imaginary, $z = i\phi$, the even terms in the series will be real and the odd, imaginary. The series has the remarkable property thatthe even terms forms the series expansion for $\cos\phi$ and the odd, the expansion for $\sin\phi$. Thus

$$e^{i\phi} = \cos\phi + i\sin\phi \tag{B.5}$$

This relation is known as *Euler's formula..* It follows form Eq. B.1 that a complex number can be written

$$z = |z|e^{i\phi} \tag{B.6}$$

Furthermore, if $\Re\{\dots\}$ stands for 'the real part of' $\{\dots\}$, we have

$$\cos\phi = \Re\{e^{i\phi}\} = Re\{e^{-i\phi}\} \tag{B.7}$$

The exponential function $\exp(i\phi)$ is periodic in ϕ with the period 2π so that $\exp(i\phi) = \exp[i(\phi + q2\pi)]$. This latter extended form of the exponential must be used in determining all the roots to the equation $z^n = w$, where $w = |w|\exp[i(\phi + q2\pi)]$ is a complex number and n an integer. The n roots are

$$z = |w|^{1/n}\, e^{i(\phi + q2\pi)/n}, \quad q = 0, 1, 2, \ldots (n-1) \tag{B.8}$$

As a simple example consider the equation $z^4 = -1$. We rewrite -1 as the complex number $\exp[i(\pi + q2\pi)]$. The four solutions are then $z = \exp[i(\pi/4 + q\pi/2]$ with $q = 0, 1, 2$, and 3. For further discussions, we refer to section B2.

B.1.5 The Complex Amplitude of a Harmonic Function

The usefulness of complex numbers in the description and analysis of oscillations and waves is linked to Euler's formula, which makes it possible to express a harmonic function as the real part of a complex exponential function.

We take the harmonic function to be the displacement of a mass-spring oscillator, for example, $\xi(t) = |\xi|\cos[\omega t - \phi]$, where $|\xi|$ and ϕ generally depend of ω. This function can be expressed as the real part of $|\xi|\exp[i(\omega t - \phi)]$ or of $|\xi|\exp[-i(\omega t - \phi)]$. As will be shown shortly, one can present good arguments for choosing one or the other of these options. We shall use the latter, so that

$$\xi(t) = |\xi|\cos(\omega t - \phi) = \Re\{|\xi|e^{-i(ot-\phi)}\} \equiv \Re\{\xi(\omega)\, e^{-i\omega t}\}$$
$$\xi(\omega) = |\xi|e^{i\phi} \tag{B.9}$$

The quantity $\xi(\omega)$ is the *complex displacement amplitude*. It is a complex number which uniquely defines the harmonic motion (at a given frequency) since it contains both the amplitude $|\xi|$ and the phase angle ϕ.

From Eq. B.9 follows that the velocity $u(t) = \Re\{(-i\omega)\xi(\omega)\exp(-i\omega t)\}$ which means that the complex velocity amplitude $u(\omega) = (-i\omega)\xi(\omega)$. Each differentiation with respect to t brings down a factor $-i\omega$ so that the complex amplitude of the acceleration becomes $-\omega^2\xi(\omega)$.

A harmonic wave traveling in the positive x-direction is given by $p(x, t) = |p|\cos(\omega t - kx - \phi)$ and the corresponding complex amplitude is $p(x, \omega) = p(0, \omega)\exp(ikx)$, where $p(0, \omega) = |p|\exp(i\phi)$ is the complex amplitude at $x = 0$. A wave traveling in the negative direction will have the spatial dependence expressed by $\exp(-ikx)$.

B.1.6 Discussion. Sign Convention

It is a great advantage to use complex amplitudes in dealing with most problems of sound and vibration and it is used almost exclusively in modern treatments of the subject. Familiarity with it is often a prerequisite in handling modern instrumentation and understanding related instruction manuals. Furthermore, the complex amplitude approach in acoustics serves as a valuable preparation for dealing with conceptually more difficult problems in other areas of physics and engineering.

A harmonic displacement $\xi(t)$ with the amplitude A, angular frequency ω and the phase angle ϕ is of the form

$$\xi(t) = A\cos(\omega t - \phi) = A\cos(\phi - \omega t) \qquad (B.10)$$

This can be expressed as the real part of the complex number $A\exp[i(\omega t - \phi)]$ or $A\exp[i(\phi - \omega t)]$ and if harmonic time dependence is implied (so that it can be suppressed), the oscillation is uniquely determined either by the complex number $A\exp(-i\phi)$ or $A\exp(i\phi$ depending on whether time factor $\exp(i\omega t)$ or $\exp[-i\omega t)$ is used. In the case of a harmonic wave travelling in the positive x-direction the latter choice of the time factor, i.e. $\exp(-i\omega t)$, yields the complex wave amplitude $A\exp(ikx)$ which clearly has the advantage since the sign of the argument of the exponential correctly indicates the directlion of wave travel. For a wave in the netative didrection the corresponding complex amplitude becomes $A\exp(-ikx)$

Linearity. The equation of motion of the mass-spring oscillator is linear; it contains only first order terms of the displacement and its derivatives and no products of these quantities. As a result, two driving force F_1 and F_2 which, when acting separately, produce the displacements ξ_1 and ξ_2 will yield a total displacement $\xi_1 + \xi_2$. This property of superposition can be taken as a definition or 'test' of linearity.

An important consequence of linearity is that a harmonic driving force with an angular frequency ω produces a harmonic displacement of the same frequency. If, on the other hand, an equation contains a nonlinear term, say ξ^3, this term produces not only a harmonic term with a frequency ω but also one with a frequency 3ω if a solution of the form $|\xi|\cos(\omega t - \phi)$ had been assumed. Thus, a harmonic driving force with a frequency ω would not yield a harmonic displacement of a nonlinear oscillator. In practice, such nonlinearity frequently is present in the friction force and the spring constant.

B.2 Examples

1. *Elementary operations*

   With $z_1 = 1 - i\sqrt{3}$ and $z_2 = \sqrt{3} + i$, calculate

   a). $z_1 + z_2$

   b). $z_1 - z_2$

   c). $z_1 z_2$

   d). z_1/z_2

   In each case, indicate the locations of these quantities in the complex plane.

   SOLUTION

   (a)
   $$z_1 + z_2 = 1 + \sqrt{3} + i(1 - \sqrt{3})$$

   (b)
   $$z_1 - z_2 = 1 - \sqrt{3} - i(1 + \sqrt{3})$$

   (c)
   $$z_1 z_2 = (1 - i\sqrt{3})(\sqrt{3} + i) = 2\sqrt{3} - i2$$

   (d)
   $$z_1/z_2 = (1 - i\sqrt{3})/(\sqrt{3} + i) = 0 - i4$$

2. *Amplitude and phase angle*

Express the following complex numbers in the form $Ae^{i\alpha}$, and determine A and α, where A and ϕ are real.

a). $3 + i4$
b). $(3 + i4)/(4 + i3)$
c). $1 + i$
d). $\sqrt{1 + i}$
e). i

SOLUTION

(a)

$$3 + i4 = 5\,e^{i0.93}$$

$A = \sqrt{3^2 + 4^2} = 5$
$\tan(\alpha) = 4/3 \quad \alpha = \arctan(4/3) = 0.93\,rad$

(b)

$$(3 + i4)/(4 + i3) = e^{-i0.28},$$

$A = 1$ and $\alpha = \arctan(-7/24) = -0.28$

(c)

$$1 + i = \sqrt{2}\,e^{i\pi/4}$$

(d)

$$\sqrt{1 + i} = (\sqrt{2}\,exp(i\pi/4))^{1/2} = 2^{1/4}\,e^{i\pi/8}$$

(e)

$$i = e^{i\pi/2}$$

3. *Euler's identity*

(a) Use Euler's identity and express $\cos()$ in terms of exponentials and show that $2\cos(\alpha)\cos(\beta) = \cos(\alpha + \beta) + \cos(\alpha - \beta)$.
(b) Show that $1 + exp(i\alpha) = 2\,e^{i\alpha/2}\cos(\alpha/2)$.

SOLUTION

(a) With

$$\cos() = (1/2)(e^{i0} + e^{-i0}) \tag{1}$$

we obtain

$$2\cos(\alpha)\cos(\beta) = (1/2)(e^{i\alpha} + e^{-i\alpha})(e^{i\beta} + e^{-i\beta}).$$

Carrying out the multiplication on the right-hand side yields

$(1/2)(e^{i(\alpha+\beta)} + e^{-i(\alpha+\beta)}) + (1/2)(e^{i(\alpha-\beta)} + e^{-i(\alpha+\beta)}).$

The proof is completed by replacing the sum of the exponentials by the corresponding $\cos()$ -terms, according to Eq. B.1.

(b)

$$1 + e^{i\alpha} = e^{i\alpha/2}(e^{-i\alpha/2} + e^{i\alpha/2}) = 2\,e^{i\alpha/2}\cos(\alpha/2)$$

4. *Complex integral*

Determine the amplitude and phase angle of $\int_0^b e^{i\beta}\cos(\beta)\,d\beta$.

SOLUTION

The simplest approach appears to be to put
$e^{i\beta}$ as $\cos(\beta) + i\sin(\beta)$

to give

$$\int_0^b (\cos^2(\beta) + i \sin(\beta)\cos(\beta))\, d\beta = \left(\frac{2b + \sin(2b)}{4}\right) + i\left(\frac{\sin^2(b)}{2}\right).$$

It is somewhat more cumbersome to solve the problem by expressing $\cos(\beta)$ as $(1/2)(e^{i\beta} + e^{i\beta})$ to give

$$\int_0^b e^{i\beta}\cos(\beta)\, d\beta = (1/2)\int_0^b e^{i\beta}(e^{i\beta} + e^{-i\beta})\, d\beta$$
$$= (1/2)\int_0^b (1 + e^{i2\beta})\, d\beta$$
$$= R + iX = Ae^{i\alpha}, \tag{B.11}$$

where

$$R = \tfrac{b}{2} + \tfrac{1}{2}\sin(b)\cos(b)$$
$$X = \tfrac{1}{2}\sin^2(b). \tag{B.12}$$

Amplitude

$$A = \sqrt{R^2 + X^2}.$$

Phase angle

$$\alpha = \arctan(X/R).$$

5. *Complex amplitude of a harmonic motion*
   A harmonic displacement is of the form $\xi(t) = A\cos[\omega(t-t_1)]$, where $A = 10\,cm$, $t_1 = T/10$, and the period $T = 0.1\,sec$. What are the complex amplitudes of displacement, velocity, and acceleration?

SOLUTION
Complex amplitude of displacement,

$$\xi(\omega) = A\,e^{i\omega t_1} = 10\,e^{i\pi/5}\,cm,$$

velocity,

$$v(\omega) = (-i\omega)\cdot\xi(\omega) = 20\pi\,e^{i(\pi/5 - \pi/2)}\,cm/sec$$

and acceleration,

$$a(\omega) = -(\omega)^2\cdot\xi(\omega) = 20\pi^2\,e^{i(\pi/5 - \pi)}\,cm/sec^2.$$

6. *Complex conjugate*
   Consider the complex numbers
   $z_1 = 1 + i3$ and $z_2 = 4e^{i\pi/6}$. Determine
   (a) z_1^2
   (b) $z_1 z_1^*$
   (c) $z_2 z_2^*$
   (d) $z_2 + z_2^*$
   and indicate the locations in the complex plane.

SOLUTION
(a)

$$z_1^2 = (1 + i3)^2 = -8 + i6$$

(b)
$$z_1 z_1^* = (1 + i3)(1 - i3) = 10$$

(c)
$$z_2 z_2^* = 16$$

(d)
$$z_2 + z_2^* = 8 \cos(\pi/6) = 4$$

7. *Complex roots*

Determine the roots of the equations
(a) $z^2 - 4z + 8 = 0$
(b) $z^4 = 1 + i$
and indicate the location of the roots in the complex plane.

SOLUTION

(a)
$$z = 2 \pm \sqrt{4 - 8} = 2 \pm i2$$

(b) $z^4 = 1 + i = 2^{1/2} e^{i(\pi/4 + 2n\pi)}$

$$z = 2^{1/8} e^{i(\pi/16 + n\pi/2)}, \quad n = 0, 1, 2, 3$$

8. *Hyperbolic functions*

Prove that
(a) $\sin(ix) = i \sinh(x)$
(b) $\cos(ix) = \cosh(x)$

SOLUTION

(a)
$$\sin(ix) = \left(\frac{1}{2i}\right)(e^{i(ix)} - e^{-i(ix)}) = i \sinh(x)$$

(b)
$$\cos(ix) = \left(\frac{1}{2}\right)(e^{i(ix)} + e^{-i(ix)}) = \cosh(x),$$

where $i(ix) = -x$.

9. *Complex phase and the meaning of z^w*

It is sometimes convenient to express a complex number $|z| \exp(i\phi)$ as a single exponential $\exp(i\Psi)$, where Ψ is a complex phase. Determine the real and imaginary parts of Ψ. Also explain the meaning of z^w, where both z and w are complex numbers.

Appendix C

References

C.1 Fourier series in chapter 2

(i) http://en.wikipedia.org/wiki/Fourier-series

(ii) *Die Grundlagen der Akustik* Chapter II (32–60), by E. Skudrzyk, Wien, May 1954.

C.2 Loudness. Figure 3.4 in chapter 3

(i) H. Fletcher and W.A. Munson, Loudness, its definition, measurement, and calculation, J.A.S.A. 5, (1933), 82–108

C.3 'Molecular' sound absorption in chapter 9

(i) H.O. Kneser, Ann. der Physik 33. 277 (1938), J. Acoust. Soc. Am., 5, 122 (1933), Ann. Physik 39, 261, (1941)

(ii) Knudsen, J. Acoust. Soc. Am, 18, 90–96 (1946); 3, 126 (1931), 5, 64, 113 (1933), 6, 199 (1935)

(iii) Markham, Beyer, and Lindsay, "Absorption of sound in fluids," Rev. Mod. Physics 23, 353–411 (1951)

(iv) E.G. Richardson, Ultrasonic Physics, Elsevier Publishing Company, (1952)

(v) Knudsen and Kneser, Ann. Physik, 21 (1934)

C.4 Books

A short list starting with the old classic standout:

(i) Lord Rayleigh: *The Theory of Sound.* Macmillan and Co. 1929 (first edition printed 1877).

425

(ii) Allan Pierce, *Acoustics, an introduction to its physical principles and applications.* Acoustical Society of America 1989.

(iii) Robert T. Beyer, *Nonlinear Acoustics*, Naval Ship Systems Command, 1974.

(iv) Alexandder Wood, *Acoustics*, Blackie and Son Limited, 1945.

(v) P.M. Morse: *Vibration and Sound*, McGraw-Hill Book Company, 1936, 1948.[1]

(vi) Philip M. Morse and K. Uno Ingard, *Theoretical Acoustics*, McGraw-Hill, New York, 1968.

---

[1] This was one of several other books which were reading and problem solving assignments as part of the requirement for my Licentiate Degree at Chalmers University, Sweden. It got me interested in acoustics which, in turn, led me to apply to the physics department at M.I.T. where I worked with Philip Morse.

Index

T